Nitrogen fixation by free-living micro-organisms

THE INTERNATIONAL BIOLOGICAL PROGRAMME

The International Biological Programme was established by the International Council of Scientific Unions in 1964 as a counterpart of the International Geophysical Year. The subject of the IBP was defined as 'The Biological Basis of Productivity and Human Welfare', and the reason for its establishment was recognition that the rapidly increasing human population called for a better understanding of the environment as a basis for the rational management of natural resources. This could be achieved only on the basis of scientific knowledge, which in many fields of biology and in many parts of the world was felt to be inadequate. At the same time it was recognised that human activities were creating rapid and comprehensive changes in the environment. Thus, in terms of human welfare, the reason for the IBP lay in its promotion of basic knowledge relevant to the needs of man.

The IBP provided the first occasion on which biologists throughout the world were challenged to work together for a common cause. It involved an integrated and concerted examination of a wide range of problems. The Programme was co-ordinated through a series of seven sections representing the major subject areas of research. Four of these sections were concerned with the study of biological productivity on land, in fresh water, and in the seas, together with the processes of photosynthesis and nitrogen fixation. Three sections were concerned with adaptability of human populations, conservation of ecosystems and the use of biological resources.

After a decade of work, the Programme terminated in June 1974 and this series of volumes brings together, in the form of syntheses, the results of national and international activities.

INTERNATIONAL BIOLOGICAL PROGRAMME 6

Nitrogen fixation by free-living micro-organisms

EDITED BY

W. D. P. Stewart

Professor and Head of Department of Biological Sciences, University of Dundee, UK

CAMBRIDGE UNIVERSITY PRESS

CAMBRIDGE

LONDON · NEW YORK · MELBOURNE

CAMBRIDGE UNIVERSITY PRESS
Cambridge, New York, Melbourne, Madrid, Cape Town,
Singapore, São Paulo, Delhi, Tokyo, Mexico City

Cambridge University Press
The Edinburgh Building, Cambridge CB2 8RU, UK

Published in the United States of America by Cambridge University Press, New York

www.cambridge.org
Information on this title: www.cambridge.org/9780521279321

First published 1975
First paperback edition 2011

A catalogue record for this publication is available from the British Library

Library of Congress Cataloguing in Publication data
Main entry under title:
Nitrogen fixation by free-living micro-organisms
At head of title: International Biological Programme 6.
'Papers presented at an IBP synthesis meeting on nitrogen fixation held in Edinburgh in September 1973, together with contributions submitted by other[s].'
Includes index.
1. Nitrogen-Fixation–Congress. 2. Micro-organisms, Nitrogen-fixing–Congresses. I. Stewart, William Duncan Patterson. II. International Biological Programme.
QR89.7.N57 589'.7'04133 75–2731

ISBN 978-0-521-20708-9 Hardback
ISBN 978-0-521-27932-1 Paperback

Contents

Table des matières

Содержание

Contenido

Parte III. La técnica de reducción del acetileno

Parte IV. Bioquímica de la fijación del nitrógeno

Contributors

Editor

W. D. P. Stewart

Co-editors

F. J. Bergersen, R. H. Burris & E. G. Mulder

Authors

Alexander, Vera A.	Institute of Marine Sciences, University of Alaska, Fairbanks, Alaska 99701, USA
Balandreau, J.	Centre de Pédologie du CNRS, B.P. 5, Vandœuvre-les-Nancy 54, France
Bezdicek, D.	Department of Agronomy and Soils, Washington State University, Pullman, Washington 99163, USA
Bergersen, F. J.	CSIRO Division of Plant Industry, Canberra, Australia
van Berkum, P.	Soil Microbiology Department, Rothamsted Experimental Station, Harpenden, Herts AL5 2JQ, UK
Boonkerd, N.	Division of Agricultural Chemistry, Bankhen, Bangkok-9, Thailand
Burns, R. C.	Central Research Department, Experimental Station, E.I. Du Pont de Nemours & Co., Wilmington, Delaware 19898, USA
Burris, R. H.	Department of Biochemistry, College of Agriculture and Life Sciences, University of Wisconsin, Madison, Wisconsin 53706, USA
Coventry, D. R.	Department of Soil Science and Plant Nutrition, University of Western Australia, Nedlands, Western Australia 6009
Dart, P. J.	Soil Microbiology Department, Rothamsted Experimental Station, Harpenden, Herts, UK
DaSilva, E. J.	Department of Microbiology, St Xavier's College, University of Bombay, Bombay-1, India

Day, J. M. — Soil Microbiology Department, Rothamsted Experimental Station, Harpenden, Herts AL5 2JQ, UK

Dharmawardene, M. W. N. — Department of Botany, Vidyodaya University of Ceylon, Nugegoda, Ceylon

Dilworth, M. J. — Department of Soil Science and Plant Nutrition, University of Western Australia, Nedlands, Western Australia 6009

Dixon, R. O. D. — Department of Botany, University of Edinburgh, Edinburgh EH9 3JH, UK

Dobereiner, Johanna — Instituto de Pesquisa Agropecuária Centro Sul, Rio de Janeiro, Brazil

Dommergues, Y. — Centre de Pédologie du CNRS, B.P. 5, Vandœuvre-les-Nancy 54, France

Eady, R. R. — A.R.C. Unit of Nitrogen Fixation, University of Sussex, Brighton BN1 9QJ, UK

El-Nawawy, A. S. — Section of Microbiology, Institute of Soil and Water Research, Agricultural Research Centre, Orman, Giza, ARE

Evans, H. J. — Department of Botany and Plant Pathology, Oregon State University, Corvallis, Oregon, USA

Fares-Hamad, Ibitissam — Centre de Pédologie du CNRS, B.P. 5, Vandœuvre-les-Nancy 54, France

Gallon, J. R. — Department of Biochemistry, University College of Swansea, Singleton Park, Swansea SA2 8PP, UK

Godfrey, C. A. — Department of Soil Science and Plant Nutrition, University of Western Australia, Nedlands, Western Australia 6009

Granhall, U. — Department of Microbiology, Agricultural College, Uppsala, Sweden

Hamdi, Y. A. — Section of Microbiology, Institute of Soil and Water Research, Agricultural Research Centre, Orman, Giza, ARE

Hardy, R. W. F. — Central Research Department, Experimental Station, E.I. du Pont de Nemours & Co., Wilmington, Delaware 19898, USA

Harris, D. — Soil Microbiology Department, Rothamsted Experimental Station, Harpenden, Herts AL5 2JQ, UK

Haystead, A. — Hill Farming Research Organisation, Bush Estate, Penicuik, Midlothian EH26 0PY, UK

Henriksson, Elisabeth	Institute of Physiological Botany, University of Uppsala, S-751, Uppsala, Sweden
Henriksson, L. E.	Institute of Physiological Botany, University of Uppsala, S-751, Uppsala, Sweden
Holm, Esther	Department of Microbiology and Microbial Ecology, Royal Veterinary and Agricultural University, Copenhagen, Denmark
Jensen, V.	Department of Microbiology and Microbial Ecology, Royal Veterinary and Agricultural University, Copenhagen, Denmark
Jump, A.	Department of Biology, University of South Florida, Tampa, Florida 33620, USA
Knowles, R.	Department of Microbiology, Macdonald Campus of McGill University, Ste Anne de Bellevue 800, Quebec, Canada
Kurz, W. G. W.	Prairie Regional Laboratory, National Research Council, Saskatoon S7N 0W9, Canada
LaRue, T. A.	Prairie Regional Laboratory, National Research Council, Saskatoon, S7N 0W9, Canada
Milberg, R.	Plant Physiology Institute, Plant Nutrition Laboratory, Northeastern Region, Agricultural Research Service, US Department of Agriculture, Beltsville, Maryland 20705, USA
Mishustin, E. N.	Institute of Microbiology, USSR Academy of Sciences, Moscow, USSR
Mulder, E. G.	Laboratory of Microbiology, Agricultural University, Hesselink van Suchtelenweg 4, Wageningen, The Netherlands
O'Toole, P.	Faculty of Agriculture, University College Dublin, Glasnevin, Dublin 9, Eire
Parshall, G. W.	Central Research Department, Experimental Station, E.I. du Pont de Nemours & Co., Wilmington, Delaware 19898, USA
Paul, E. A.	Department of Soil Science, University of Saskatchewan, Saskatoon, Canada
Pearson, H. W.	Department of Botany, University of Liverpool, Liverpool L69 3BX, UK
Phillips, D. A.	Department of Life Sciences, Indiana State University, Terre Haute, Indiana, USA

Phillips, W. D.	Central Research Department, Experimental Station, E.I. du Pont de Nemours & Co., Wilmington, Delaware 19898, USA
Postgate, J. R.	A.R.C. Unit of Nitrogen Fixation, University of Sussex, Brighton BN1 9QJ, UK
Renaut, J.	Faculty of Sciences, Avenue Moulay Cherif, Rabat, Morocco
Rinaudo, G.	Orstom, B.P. 1386, Dakar, Senegal
Ruinen, Jacoba	Laboratory of Microbiology, Agricultural University, Hesselink van Suchtelenweg 4, Wageningen, The Netherlands
Sasson, A.	Faculty of Sciences, Avenue Moulay Cherif, Rabat, Morocco
Silver, W. S.	Department of Biology, University of South Florida, Tampa, Florida 33620, USA
Sloger, C.	Plant Physiology Institute, Plant Nutrition Laboratory, North Eastern Region, Agricultural Research Service, US Department of Agriculture, Beltsville, Maryland 20705, USA
Smith, B. E.	A.R.C. Unit of Nitrogen Fixation, University of Sussex, Brighton BN1 9QJ, UK
Stasny, J. T.	Central Research Department, Experimental Station, E.I. du Pont de Nemours & Co., Wilmington, Delaware 19898, USA
Stewart, W. D. P.	Department of Biological Sciences, University of Dundee, Dundee DD1 4HN, Scotland, UK
Thorneley, R. N. F.	A.R.C. Unit of Nitrogen Fixation, University of Sussex, Brighton BN1 9QJ, UK
Venkataraman, G. S.	Microbiology Division, Indian Agricultural Research Institute, New Delhi 110012, India
Yates, M. G.	ARC Unit of Nitrogen Fixation, University of Sussex, Brighton BN1 9QJ, UK
Yemtsev, V. T.	Institute of Microbiology, USSR Academy of Sciences, Moscow, USSR

Preface

Nitrogen fixation was one of the main themes of the Production Processes section of the International Biological Programme, and this volume and the companion one, *Symbiotic nitrogen fixation in plants*, bring together much of the work which has resulted, directly or indirectly, from the stimulus given to nitrogen fixation by the IBP. These volumes contain a series of papers presented at an IBP synthesis meeting on nitrogen fixation held in Edinburgh in September 1973, together with contributions submitted by other workers who participated in the IBP nitrogen fixation programme.

This volume is concerned with the following themes: nitrogen fixation by free-living bacteria; nitrogen fixation by free-living, blue-green algae, and biochemical aspects of nitrogen fixation, many studies of which have been carried out using free-living micro-organisms. In addition, information on the acetylene-reduction technique is provided because this technique, more than any other single technique, has made possible most of the studies reported on in this volume. The idea of having the various themes was to stimulate research in these particular areas and to bring together available information on these topics. Related information on nitrogen fixation by algae in lakes and on nitrogen fixation in the sea will be presented in later volumes in this series.

As will be seen, the contributors to the volume come from all five continents and the contributions vary in scope, depth, detail, and presentation. It is hoped, nevertheless, that the two volumes will collectively provide a useful focal point for further studies on nitrogen fixation; a process which not only overcomes many of the environmental problems associated with the use of synthetic nitrogen fertiliser but which is also independent of natural gas reserves for substrate and energy.

W. D. P. STEWART

University of Dundee, March 1974

PART I

Nitrogen fixation by free-living bacteria

1. Physiology and ecology of free-living, nitrogen-fixing bacteria

E. G. MULDER

Ever since the isolation of *Clostridium pasteurianum* by Winogradsky in 1893 and the isolation of *Azotobacter chroococcum* and *A. agilis* by Beijerinck in 1901 there has been a great deal of interest in the study of free-living, nitrogen-fixing micro-organisms. The reason for this interest undoubtedly was the belief that these organisms contribute substantially to the nitrogen supply of higher plants. As a result of the stimulated research, many new free-living, nitrogen-fixing micro-organisms have been isolated and studied. However, for several decades little progress was made in the understanding of the process of nitrogen fixation and in the development of methods for checking the nitrogen-fixing ability of the organisms. Biochemical research on the enzymology of nitrogen fixation and the development of the ^{15}N and acetylene-reduction techniques during the last 10–20 years have greatly contributed to a much better understanding of nitrogen fixation and of the conditions under which this process proceeds optimally.

The acetylene-reduction technique enables not only a ready estimation of the nitrogen-fixing capacity of pure or mixed cultures of organisms under laboratory conditions, but it can also be used for measuring nitrogen fixation in natural systems like terrestrial and aquatic environments. Associations between free-living, nitrogen-fixing bacteria and higher plants growing under natural conditions like the rhizosphere and phyllosphere, and various types of symbiosis between bacteria and higher plants (root and leaf nodules) can also easily be tested for nitrogen fixation by using this technique.

Survey of free-living, N_2-fixing micro-organisms

Careful retesting of the nitrogen-fixing ability, by using the acetylene-reduction and ^{15}N assays, has shown that several micro-organisms described in the literature as free-living, nitrogen fixers were unable to fix nitrogen (N_2). This is, amongst others, true of some yeasts and fungi which had earlier been described as N_2 fixers (cf. Postgate, 1971). It thus appears that up till now only a number of prokaryotic organisms (bacteria and blue-green algae) have been shown to possess the ability to fix N_2. As it is not very probable that nitrogen-fixing eukaryotes will ever be found in nature, an interesting question arises as to why N_2 fixation is confined to prokaryotic micro-organisms.

The free-living, N_2-fixing bacteria belong to a limited number of families or genera. They include: (*a*) the aerobic azotobacters; (*b*) the facultatively anaerobic klebsiellas (formerly aerobacters); (*c*) the facultatively anaerobic bacilli of the *Bacillus polymyxa* and *B. macerans* group; (*d*) most of the anaerobic saccharolytic clostridia (including *Cl. pasteurianum*, *Cl. butyricum*, *Cl. butylicum*, *Cl. pectinovorum* and several other *Clostridium* species); (*e*) the anaerobic sulphate-reducing bacteria of the genera *Desulphovibrio* and *Desulphotomaculum*; (*f*) the photosynthetic bacteria.

The Azotobacteraceae

Azotobacters represent the main group of aerobic, free-living, N_2-fixing bacteria. The family includes different types of organisms of which the ability to fix N_2 is the main uniting characteristic. *Bergey's Manual* (1957) lists three genera, *Azotobacter*, *Beijerinckia* and *Derxia*, but DNA analyses and DNA-hybridization tests, performed by De Ley & Park (1966) and De Ley (1968) showed that the genus *Azotobacter* includes three genetically different groups of organisms (Table 1.1) and therefore should be split up in three different genera. The generic name *Azotobacter* should be retained for the chroococcum–beijerinckii–vinelandii–paspali group whilst De Ley & Park (1966) proposed the generic name *Azotomonas* for the insignis–macrocytogenes group and *Azotococcus* for the agilis group.

Bacteria of the Azotobacteraceae, with the exception of those of the genus *Beijerinckia*, are large organisms with cells of $2\text{–}3 \times 3\text{–}6\ \mu$m. *A. macrocytogenes*, *A. insignis* and *A. agilis* have almost coccoid cells. Most species form large amounts of slime, with the exception of the aquatic species *A. insignis* and *A. agilis* which form little or no slime. *A. macrocytogenes* forms tetrads or large clusters of cells, surrounded by slime capsules.

Azotobacters of the chroococcum, vinelandii, beijerinckii and paspali types may form large numbers of cysts during the early stationary phase. Cyst formation is closely related to the accumulation within

Table 1.1. *Some characteristics of the Azotobacteraceae*

Species	% (G+C)	Slime	Habitat
Azotobacter chroococcum	64.8–66.0	Moderate	Soil + phyllosphere
A. vinelandii	65.8–66.5	Moderate	Soil + water
A. beijerinckii	66.2	Moderate	Soil
A. paspali	63.2–64.6	Moderate	Rhizoplane *Paspalum* spp.
A. macrocytogenes*	58.2–58.6	Abundant	Soil
A. insignis*	56.9–57.9	Absent	Water
*A.** agilis*	52.5–53.5	Little	Water
Beijerinckia indica	54.7	Abundant	Soil + phyllosphere
Derxia gummosa	70.4	Abundant	Soil

* *Azotomonas*; ** *Azotococcus* (De Ley & Park, 1969; De Ley, 1968).

the cells of large amounts of poly-β-hydroxybutyric acid (PHB) which are used for a substantial part in cyst formation. A cyst usually contains one living cell, possessing a cytoplasmic membrane and cell wall, surrounded by a heavy coat which protects the cell, particularly from physical influences such as desiccation and to a small extent heat. Calcium is required for the formation of the cyst coat (Stevenson & Socolofsky, 1973). Upon the transfer of cysts to media favourable for growth, germination takes place. When the carbon source of the nutrient medium is readily assimilable, the overall process of germination takes about 8 h, i.e. a germination phase of 4–5 h and an outgrowth phase of 3–4 h (Loperfido & Sadoff, 1973).

Azotobacters utilize many carbon compounds for growth and N_2 fixation. In addition to several sugars, polysaccharides, primary, secondary and polyhydroxyalcohols, and organic acids, aromatic compounds like benzoate, *p*-hydroxybenzoate and shikimate may function as sources of carbon and energy (cf. Mulder & Brotonegoro, 1974).

Enterobacteriaceae

N_2 fixation was found to occur in many strains of *Klebsiella pneumoniae* (formerly *Aerobacter aerogenes*; Pengra & Wilson, 1958), and in some strains of *Enterobacter cloacae* (formerly *Aerobacter cloacae*; Raju, Evans & Seidler, 1972) and of *Escherichia intermedia* (Line & Loutit, 1971). Some authors have succeeded in transferring the cluster of the *nif* genes (which code for nitrogen fixation) from *Klebsiella pneumo-*

niae to a strain of *Escherichia coli* (Dixon & Postgate, 1972). Nitrogen fixation by organisms of the Enterobacteriaceae occurs only under anaerobic conditions or at a low P_{O_2}.

Bacillaceae, genus Bacillus

Many strains of the facultatively anaerobic *B. polymyxa* and *B. macerans* are able to fix N_2 under anaerobic conditions (Grau & Wilson, 1962; Witz, Detroy & Wilson, 1967).

Bacillaceae, genus Clostridium

This genus comprises two groups of strictly anaerobic heterotrophic organisms, the saccharolytic clostridia which use carbohydrates and some organic acids as carbon source, and the proteolytic clostridia which utilize proteins and amino acids. N_2 fixation is presumably confined to the saccharolytic species. Many of these species were found by Rosenblum & Wilson (1949) to fix N_2. One of these organisms is *Cl. pasteurianum*, the first free-living N_2 fixer isolated and studied (Winogradsky, 1893). This organism has been widely used in biochemical studies on N_2 fixation. It closely resembles *Cl. butyricum* from which it differs mainly by its ability to fix larger amounts of N_2.

Sulphate-reducing bacteria

These strictly anaerobic bacteria belong to two genera, *Desulphovibrio* which comprises the well-known *D. desulfuricans*, and *Desulphotomaculum*. The organisms of the latter genus differ in several respects from those of the former; they form spores and differ in DNA-base content, type of cytochromes and flagellation from those of the genus *Desulphovibrio*. Of the five species of *Desulphovibrio* tested, three have been found to fix N_2; in the case of *Desulphotomaculum* two of the three species tested have been shown to fix N_2 (cf. Dalton, 1974).

Photosynthetic bacteria

This group of bacteria includes the Thiorhodaceae, the Chlorobacteriaceae and the Athiorhodaceae. The organisms belonging to these families are able to use light as the source of energy under anaerobic conditions. Organisms of the first two families use hydrogen sulphide or sulphur as the electron donor for carbon dioxide assimilation whereas organisms of the Athiorhodaceae use organic material for that purpose. Nitrogen fixation has been found to occur in representatives from each family (cf. Dalton, 1974).

N_2-fixing representatives of other families than those listed above have sometimes been recorded in the literature. In some cases the taxonomic position of these isolates is not clear. *Mycobacterium flavum*, isolated and tested for N_2 fixation by Fedorov & Kalininskaya (1961) and also studied by Biggins & Postgate (1969), has not convincingly been shown to be a *Mycobacterium* species. The same uncertainty concerns Smyk's N_2-fixing *Arthrobacter* strains isolated from rock formations (Smyk & Ettlinger, 1963; Smyk, 1970). More than one hundred *Arthrobacter* strains isolated in the author's laboratory from different soils and aquatic sources were found to be unable to fix N_2 (Antheunisse, unpublished results). Gogotov & Schlegel (1974) recently found that two corynebacteria, which had the capacity to oxidize hydrogen, were able to fix N_2 when supplied with succinic acid and when kept at a reduced P_{O_2}.

V. Jensen's N_2-fixing *Achromobacter* strain (1958) was shown by Mahl, Wilson, Fife & Ewing (1965) to be a *Klebsiella pneumoniae*.

Although some authors have described N_2-fixing *Pseudomonas* strains (see for instance Voets & Debacker, 1956), no convincing evidence has so far been provided concerning the occurrence of N_2 fixation among members of this family. Voets's *Ps. azotogenis* was shown by De Ley & Park (1966) not to be a *Pseudomonas* strain. The same is true of Coty's *Ps. methanitrificans* which would fix N_2 using methane as carbon and energy source (Coty, 1967). Whittenbury, Phillips & Wilkinson (1970), retesting Coty's 'pure culture', isolated *Methylosinus trichosporium* from that culture which would have been responsible for the observed N_2 fixation in Coty's experiments. However, the claimed N_2 fixation by this strain of *Methylosinus* was derived from an experiment in which the organism was exposed to an atmosphere containing 4.4 % methane and 1.8 % acetylene for 7 or 14 days. Conversion of 25 % of the acetylene, without giving data for ethylene formation during this prolonged period, is no evidence of N_2 fixation as the acetylene may have disappeared by co-oxidation. In addition, for assessing the presence of nitrogenase only short-term exposure times should be used in the acetylene-reduction assay (Hardy, Burns & Holsten, 1973).

Experiments carried out by De Bont in the author's laboratory (De Bont & Mulder, 1974) have shown that a methane-oxidizing organism resembling *Methylosinus sporium*, isolated from soil, was able to fix N_2.

To demonstrate this capacity by way of the acetylene-reduction assay, the organism had to be grown on methanol as this assay cannot be used with methane as the carbon and energy source. The fact that bacteria of this type when growing on methane are able to co-oxidize ethylene may be the cause of the negative acetylene-reduction assay. A different explanation is that the large amounts of acetylene which are used in the acetylene-reduction test inhibit the oxidation of methane. Both effects do not occur when methanol is the energy source. To demonstrate the fixation of N_2 with methane as the energy source, the $^{15}N_2$ technique was used.

A further example of a single N_2-fixing species (*Thiobacillus ferrooxidans*) belonging to a large family of apparently non-nitrogen fixers (Thiobacillaceae) has been recorded by Mackintosh (1971). Nitrogen-fixing *Spirillum* species have been described by Rodina (1956) and Becking (1963). The latter author demonstrated N_2 fixation by his isolates (*Spirillum* or *Vibrio* species) in experiments with $^{15}N_2$.

Physiology of N_2 fixation

Although large differences exist in morphology and physiology of different types of free-living, N_2-fixing micro-organisms, the central reaction according to which nitrogen fixation proceeds and the enzyme system (nitrogenase) responsible for this reaction are similar in all types of N_2 fixers, free-living as well as symbiotic forms (1).

$$N_2 + 6e^- + 6H^+ + n\text{ATP} \rightarrow 2NH_3 + n\text{ADP} + nP_i \qquad (1)$$

The ATP-driven nitrogenase reaction also catalyses the conversion of acetylene to ethylene (2) which is generally used for measuring the nitrogenase activity.

$$C_2H_2 + 2e^- + 2H^+ \rightarrow C_2H_4 \qquad (2)$$

From (1) it is seen that N_2 fixation requires ATP, in addition to six electrons. It is generally accepted that per $2e^-$, at least two molecules of ATP are required. If it is assumed that 38 moles of ATP are derived from the aerobic breakdown of 1 mole of glucose (180 g) and further that $6e^-$ are equivalent to 9 molecules of ATP, then 1 mole of glucose would theoretically allow the fixation of 1.8 moles of N_2, assuming that all of the reducing power and energy could be used in the N_2 fixation reaction. This means that nearly 280 mg N_2 could be fixed per 1 g of sugar consumed provided that all of the available carbohydrate was available for N_2 fixation. This may be the case in symbiotic systems like the root nodules of bacteria of the *Rhizobium* type and leguminous plants where nitrogen fixation proceeds in non-growing bacteroids and oxygen supply is regulated by leghaemoglobin (see for instance Bergersen & Goodchild, 1973*a*, *b*). In free-living N_2 fixers, however, efficiency values are found of 5–20 mg fixed N_2 per gram of glucose consumed. More details on the efficiency of N_2 fixation in free-living and symbiotic systems are given below.

Effect of external factors on N_2 fixation

In addition to substrate supply, N_2 fixation by free-living bacteria may be affected by several other factors, including mineral nutrition, oxygen supply and the presence of combined nitrogen, particularly NH_4^+.

Mineral nutrition

Of the essential elements, directly or indirectly involved in nitrogen fixation, molybdenum is the most specific. It is contained in one of the two components of which nitrogenase consists. In the absence of adequate amounts of molybdenum, no N_2 fixation takes place. This is true of all types of nitrogen fixers, free-living as well as symbiotic systems. Addition of ammonium-nitrogen eliminates the requirement for molybdenum in azotobacters and presumably also in other free-living N_2 fixers (Mulder, 1948). Nitrate cannot be used as nitrogen source in molybdenum-deficient, free-living, N_2-fixing bacteria because nitrate reductase, similar to nitrogenase, is a molybdenum-containing enzyme. However, the molybdenum requirement of azotobacters supplied with NO_3^- is much smaller than that of N_2-fixing cultures. From this result it is improbable that the molybdenum-containing subunit of nitrate reductase is identical with a Mo-containing subunit of nitrogenase as has been suggested by some authors (cf. Evans & Russell, 1971).

The molybdenum requirement of leguminous plants can also be eliminated by providing the plants with NH_4^+; nitrate assimilation also requires molybdenum. However, there are clear indications that in higher plants, at least in some types, molybdenum has an additional function (Mulder, 1954).

In some types of free-living, N_2-fixing bacteria, vanadium can be substituted for molybdenum. This is true of most of the strains of *A. chroococcum* and *A. vinelandii* but not of *A. agilis* and *Beijerinckia* spp. (Becking, 1962). The majority of strains of *Clostridium butyricum* were found to fix N_2 with molybdenum as well as with vanadium (Jensen & Spencer, 1947). There is disagreement about the explanation of the Mo–V interrelationship. Some authors believe that in the vanadium-treated organisms a V-nitrogenase occurs (Burns, Fuchsman & Hardy, 1971). Others are of the opinion that the organisms still contain traces of molybdenum, which are responsible for the nitrogenase activity. The effect of vanadium would be to stabilize the molybdenum-protein binding in nitrogenase (Benemann *et al.*, 1972).

The other metal component of nitrogenase, iron, is less unique in its requirement by nitrogen-fixing bacteria, because it also forms part of other electron carriers, like ferredoxin and the cytochromes. Ferredoxin is involved in electron transport in saccharolytic clostridia and is required for N_2 fixation by bacteria of the *Clostridium pasteurianum* type. In the absence of adequate amounts of iron, the organism synthesizes and uses a flavoprotein (flavodoxin), which functions as electron carrier instead of ferredoxin (Knight & Hardy, 1966).

Potassium, calcium and magnesium are essential elements for free-living, N_2-fixing bacteria. Submitting a culture of *Azotobacter* cells to potassium-deficiency reduces and ultimately stops the growth of the cells. N_2 fixation follows this trend, although it is less severely affected by potassium-deficiency than growth (Brotonegoro, 1974). This indicates that the effect of potassium-deficiency on nitrogenase activity is indirect, for instance by the accumulation of soluble nitrogenous compounds like NH_4^+, or by shortage of reductants and (or) ATP resulting from decreased metabolic activity.

Calcium was shown by Norris & Jensen (1958) to be required by azotobacters of the chroococcum, vinelandii, beijerinckii and insignis types but not by *A. agilis*, *A. macrocytogenes* and *Beijerinckia* spp. No difference in calcium requirement of the former group of species was found between cultures growing in the absence or presence of combined nitrogen. However, Jakobsons, Zell & Wilson (1962) found that lower amounts of calcium were needed by *A. vinelandii* when the organism was growing with combined nitrogen. Brotonegoro (1974) transferred growing *A. chroococcum* cells to a medium without Ca^{2+} and observed that both cell growth and N_2 fixation ceased.

Although Mg^{2+} is required for the N_2 fixation reaction, this element is also involved as a co-factor in several other enzymic reactions not directly concerned in nitrogen fixation.

Oxygen supply

Oxygen is required by aerobic, free-living, N_2-fixing bacteria for energy supply. This pertains not only to general growth reactions, but also to N_2 fixation. This can be easily shown by incubating pre-cultivated *Azotobacter* cultures at a different aeration rate under an atmosphere containing 10% acetylene (Brotonegoro, 1974). At a low P_{O_2} nitrogenase activity, measured by acetylene reduction, was practically negligible, owing to lack of ATP generation. With increased oxygen supply, nitrogenase activity rose many times, but when the P_{O_2} exceeded a certain value, acetylene reduction dropped to a value which was hardly higher than that at the low P_{O_2}. In contrast to nitrogenase activity, the respiration rate of these *Azotobacter* cultures attained its highest value at the highest P_{O_2}.

The adverse effect of increased P_{O_2} on the nitrogenase activity of living *Azotobacter* cells is partly due to the oxidation of reductants required in N_2 fixation (cf. reaction 1) and partly to the sensitivity of nitrogenase (particularly the iron-protein component) to a prolonged exposure of the cells to excess oxygen. Under such conditions nitrogenase may be irreversibly inactivated.

Upon a short exposure to excess oxygen, nitrogenase activity of the cells, although being completely eliminated, may readily be restored after lowering the P_{O_2} to an optimum value (conformational protection; Dalton & Postgate, 1969).

A further mechanism of aerobic N_2 fixers of the *Azotobacter* type to protect the nitrogenase system against excess oxygen is the strongly increased utilization of substrate (reductants) to remove the excess oxygen (respiratory protection). This leads to strongly increased respiration rates (high Q_{O_2} values) with increasing P_{O_2}. As a result of this mechanism, azotobacters may consume considerable amounts of substrate which are not directly used for synthesis of cellular material or for N_2 fixation. This results in strongly decreased efficiency values of N_2 fixation.

The formation of excessive amounts of extracellular slime which favours clump formation, and presumably also the large cell size of most azotobacters, may be seen as characters enhancing the efficiency of nitrogen fixation by impeding oxygen uptake.

The adverse effect of excess oxygen on nitrogen fixation by aerobic bacteria has also been shown in experiments with a methane-oxidizing, N_2-fixing bacterium of the *Methylosinus sporium* type (De Bont &

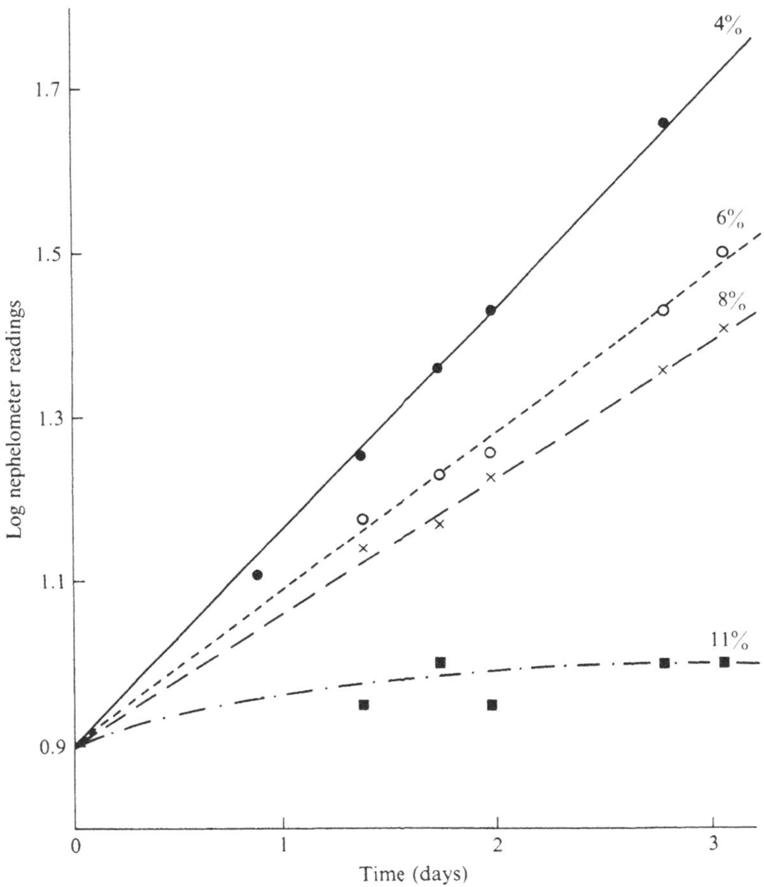

Fig. 1.1. Effect of oxygen (4, 6, 8 and 11 %) on a growing N_2-fixing culture of a methane-oxidizing bacterium.

Mulder, 1974; see also above, p. 7). When incubated in a nitrogen-free medium with methane as the carbon and energy source, N_2 fixation and growth occurred only at moderate P_{O_2} (see Figs. 1.1 and 1.2). Optimum nitrogenase activity in living cells was obtained at 4 % O_2. At 11 % O_2, nitrogen fixation, and as a consequence growth, were completely suppressed. In the presence of nitrate as the nitrogen source, no depressing effect of oxygen on growth was observed. When streaked on agar plates and exposed to methane, the bacterium grew only at the edges of a streak where an accumulation of inoculation material had promoted N_2 fixation by removing excess oxygen.

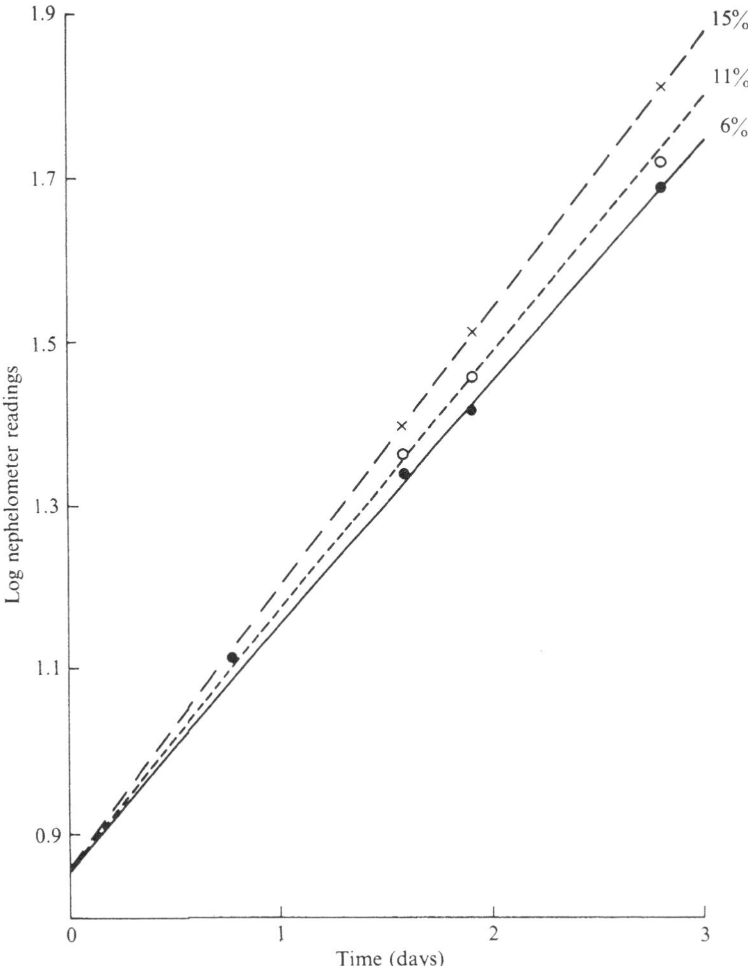

Fig. 1.2. Effect of oxygen (6, 11 and 15 %) on a growing culture of a methane-oxidizing bacterium supplied with nitrate.

The inhibitory effect of oxygen on N_2 fixation by free-living bacteria is clearly shown in some facultatively anaerobic bacteria (*Bacillus polymyxa*, *Klebsiella pneumoniae*) which fix N_2 only under anaerobic conditions and in some instances at low P_{O_2} values.

In symbiotic systems like root nodules of leguminous plants, oxygen supply is apparently optimal for N_2 fixation. The dense packing of the bacteroids surrounded by envelope membranes within plant root cells prevents the entry into the bacteroids of excess O_2. The presence of leghaemoglobin within the envelope membranes and surrounding the bacteroids apparently provides for an appropriate medium to transfer adequate amounts of O_2 to the non-growing bacteroids to enable N_2 fixation at a maximum degree of efficiency. So far no details are available concerning the localization and the mechanism of ATP generation in the root nodules.

Effect of combined nitrogen, particularly NH_4^+

When free-living N_2-fixing bacteria are supplied with NH_4^+, a rapid uptake and assimilation of this compound takes place. This was shown for *Azotobacter vinelandii* by Burris & Wilson (1946) and for *Clostridium pasteurianum* by Daesch & Mortenson (1972). The latter authors found that immediately after the addition of NH_4^+, repression of nitrogenase synthesis took place. As cell growth and cell division went on, nitrogenase decreased by dilution. Depression of the nitrogenase activity (e.g. by feedback inhibition or stimulated degradation of the enzyme) by NH_4^+ did not occur so that the nitrogenase activity of the entire culture remained unchanged for four to five hours. From this result it can be concluded that the anaerobic N_2 fixers of the *Clostridium butyricum* type are able to fix N_2 in the presence of added NH_4^+. A further conclusion is that a temporary over-production of NH_4^+ from N_2 by living *Clostridium* cells will not lead to an immediate depression of N_2 fixation as is the case with aerobic N_2-fixing bacteria of the *Azotobacter* type (see below) but will presumably give rise to excretion of NH_4^+.

Aerobic bacteria of the *Azotobacter* type respond in a different way to added NH_4^+. Immediately after the addition of small amounts of an ammonium salt to an N_2-fixing culture, nitrogenase activity declines and after one or two hours comes to a complete standstill (Fig. 1.3; Brotonegoro, 1974). No effect of NH_4^+ was observed when the nitrogenase activity was estimated in the usual way in cell-free extracts (Brotonegoro, 1974). This result confirmed earlier work by Strandberg & Wilson (1968).

To decide whether the rapid decline of nitrogenase activity in living azotobacters supplied with NH_4^+ was due to a stimulated degradation of the enzyme by NH_4^+, Brotonegoro (1974) added small amounts of an ammonium salt to a living culture of *A. chroococcum* and measured the nitrogenase activities of living cells and of cell-free extracts prepared from the same samples at the same time. The results obtained (Fig. 1.4)

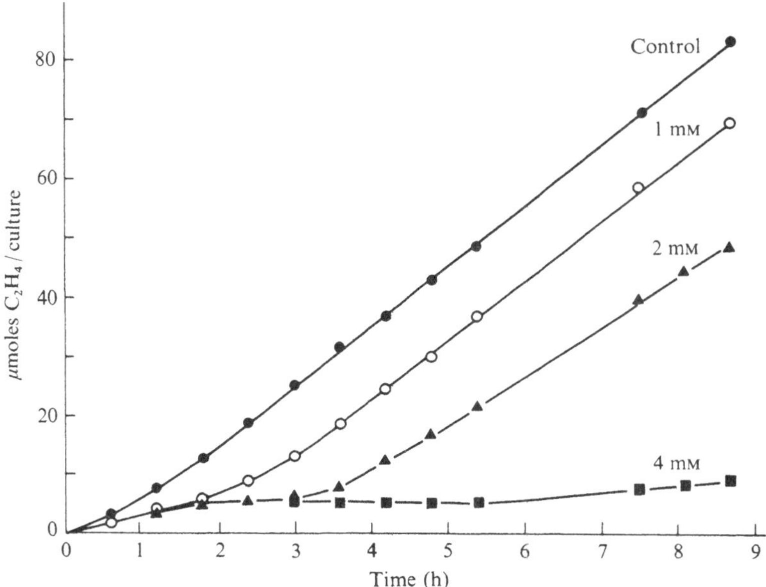

Fig. 1.3. Effect of different amounts of ammonium acetate (final concentrations 1, 2 and 4 mM) on the nitrogenase activity of a growing culture of *A. chroococcum* (Brotonegoro, 1974).

demonstrate that the rapid decline of nitrogenase activity in living azotobacters supplied with NH_4^+ is not due to (*a*) repression of nitrogenase synthesis or (*b*) feedback inhibition. Competition for reductants and/or ATP between nitrogenase activity and assimilation of ammonia were thought to be responsible for the immediate effect of NH_4^+ on nitrogenase activity.

The impairment of nitrogenase in living cells by NH_4^+ explains why non-growing azotobacters are unable to fix N_2. The NH_4^+ derived from N_2 fixation, which in growing cells is readily assimilated, would accumulate in non-growing cells and as a consequence would inactivate the nitrogenase system. When azotobacters are exposed to an atmosphere consisting of air with 10 % C_2H_2, no N_2 fixation and thus no accumulation of NH_4^+ occurs. Under such conditions, the nitrogenase system can be kept at a high level for a prolonged period, in spite of the fact that the organisms do not grow (Figs. 1.5 and 1.6; Brotonegoro, 1974).

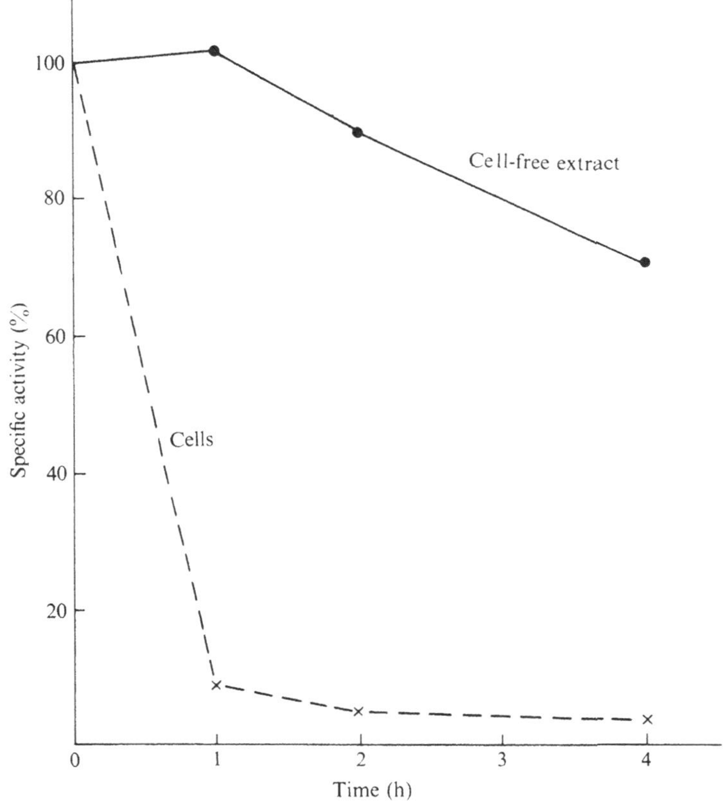

Fig. 1.4. Nitrogenase activity of living cells of *A. chroococcum*, measured at different intervals after the addition of ammonium acetate (final concentration 5 mM) to the culture, as compared to the activity of cell-free extracts prepared from the same culture at the same time (Brotonegoro, 1974).

Comparison of N_2 fixation in free-living and symbiotic systems

Although the nitrogenases of both systems are functioning similarly, pronounced differences are found to occur in several respects. These differences concern (*a*) the growth phase of the nitrogen fixers during which N_2 fixation proceeds; (*b*) the life time of the organisms during which N_2 fixation proceeds; (*c*) the amount of N_2 fixed per gram of cellular material; (*d*) the efficiency of N_2 fixation, calculated as mg N fixed per g of glucose consumed, (*e*) the specific activity of N_2 fixation, calculated as mg N fixed per g of protein per h, and (*f*) the fate of the fixed N_2 (Table 1.2).

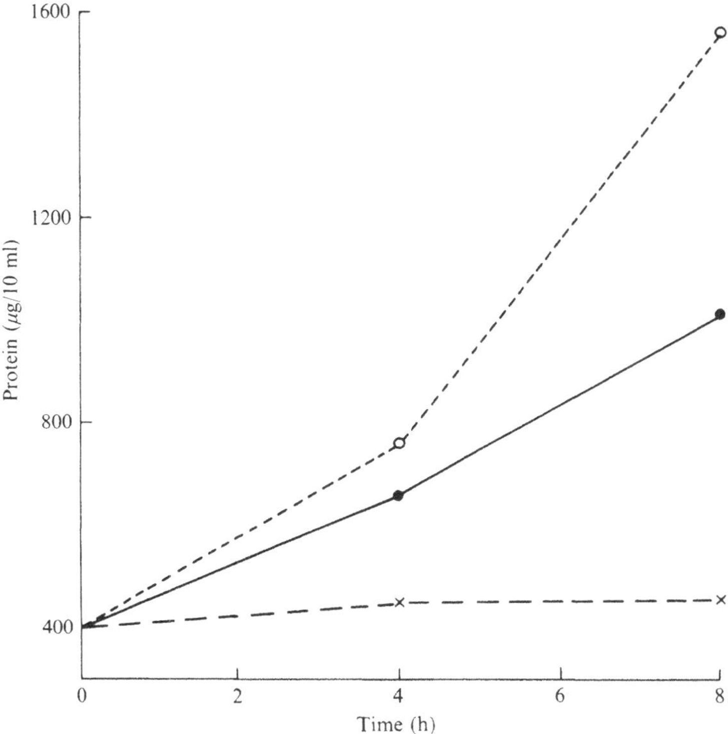

Fig. 1.5. Growth of *A. chroococcum* (expressed as yield of protein), incubated under acetylene, flasks closed with rubber caps (×); air, rubber caps (●); and air, cotton stoppers (○) (Brotonegoro, 1974).

Growth phase

In free-living, N_2-fixing bacteria of the *Azotobacter* type and presumably also in other types, nitrogen fixation occurs only in growing cells in which the fixed N_2 is readily converted into cell protein. In non-growing cells, accumulation of soluble nitrogenous compounds, including NH_4^+, occurs which suppresses N_2 fixation (see above, p. 14).

In contrast to free-living bacteria, N_2 fixation of the *Rhizobium*–legume association takes place mainly in non-growing bacteroids. This was shown in soybean nodules studied by Bergersen & Goodchild (1973*b*). Although N_2 fixation of the nodules (measured by acetylene reduction) started when the number of bacteroids was still increasing, this type of growth (one or two multiplications within one week) is not comparable to that of exponentially growing N_2-fixing cells of free-

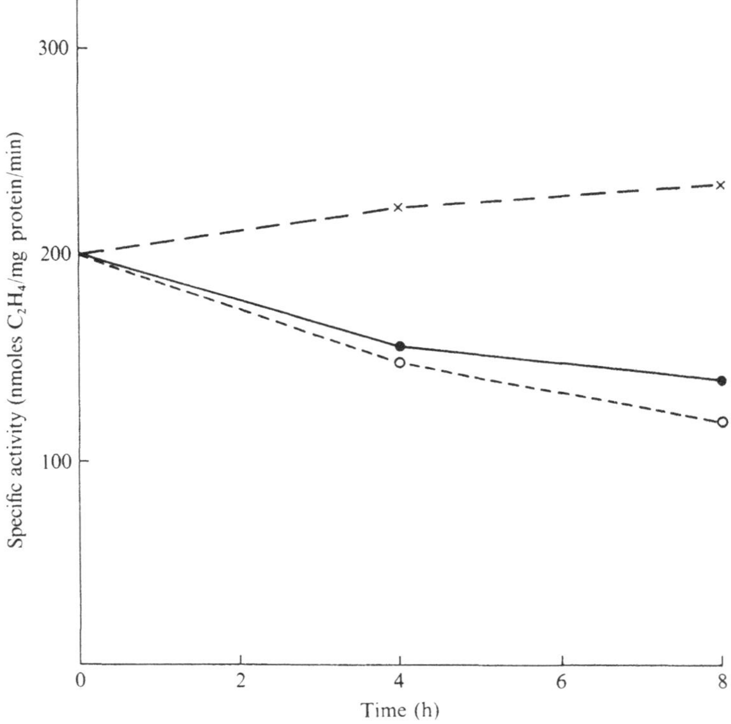

Fig. 1.6. Specific activity of nitrogenase in living, non-growing cells of *A. chroococcum*, incubated under acetylene (×) and of growing cells under air (● and ○, as in Fig. 1.5). For the conditions for growth of these cells see Fig. 1.5 (Brotonegoro, 1974).

Table 1.2. *Comparison of free-living and symbiotic N_2-fixing systems*

	Azoto-bacters	*Rhizobium*-legume system
Growth phase of bacteria during which N_2 is fixed	Exponential	Stationary
Life time of N_2-fixing organisms (h)	2–4	About 1 month
gN fixed/g cell material (entire lifetime)	0.1	1.0–2.5
Efficiency (mg N fixed/g carbon compound consumed)	10–20	250–300
Specific activity (mg N fixed/g bacterial protein/h)	25–50	2–5 (pea) 3 (soybean)
Fate of fixed N_2 (immobilized in bacteria or bacteroids, %)	90–95	5–10

living bacteria. Maximum acetylene-reducing activity of the soybean nodules was maintained from the thirteenth to the forty-third day, i.e. for about four weeks, when no nodule growth occurred.

Lifetime of the organisms during which N_2 fixation proceeds

When it is assumed that N_2 fixation in free-living bacteria occurs only when the cells are in the exponential growth phase, then the lifetime of a N_2-fixing cell is only a few hours (the time between two multiplications, i.e. the generation time). As seen in the preceding paragraph, a bacteroid may retain its N_2-fixing capacity for more than four weeks.

Amount of N_2 fixed per g of cellular material

After one cell division one gram of cell material has formed another gram of cell material, containing approximately 0.1 g of N_2. This is several times less than the 1.0–2.5 g of N_2 fixed per gram of bacteroid during its entire life period.

Efficiency of N_2 fixation

Free-living, nitrogen-fixing bacteria fix relatively small amounts of N_2: viz azotobacters 10–20, klebsiellas approximately 5 and clostridia 5–10 mg N per gram of carbon compound consumed. These values are much lower than those obtained with the symbiotic systems.

For the *Rhizobium*–pea nodules a value of 270 mg of N_2 fixed per gram of carbon consumed was found by Minchin & Pate (1973). Even higher values have been calculated by Gibson (1966) in nodulated *Trifolium subterraneum.*

The large differences in efficiency values between both types of N_2 fixers are mainly due to two phenomena. (1) Nitrogen-fixing, free-living bacteria are growing organisms. This means that a large percentage of the carbon and energy compounds has to be used for the synthesis of cellular material. This is in contrast to the bacteroids of leguminous plants which fix N_2 in the stationary phase (see above, p. 17). (2) Free-living aerobes of the *Azotobacter* type require a large proportion of the available carbon compounds for excluding oxygen from the nitrogenase system in the cell (respiratory protection, see above, p. 11). The higher the P_{O_2} of the atmosphere under which the bacteria are incubated, the larger the amount of carbon compound respired and the lower the amount left for cell synthesis and N_2 fixation.

Specific activity of N_2 fixation

The specific activity of N_2 fixation is much higher for azotobacters than for bacteroids. Assuming that a N_2-fixing *Azotobacter* cell in the exponential growth phase has a generation time of 2–4 h, then during this period 1 g of bacterial protein has increased with the same amount of protein containing 160 mg nitrogen. For the fixation of this amount of nitrogen at an average 1.5 g of bacterial protein was available (1 g at the start, 2 g at the completion of the generation time) so that the calculated specific activity equals 25–50 mg N/g protein/h, a value which approximates to the amount experimentally found by using the acetylene-reduction assay (Brotonegoro, 1974).

For symbiotic systems considerably lower values are found for the specific activity, viz 2–5 mg N/g bacteroid protein/h (Bergersen, 1969; Bergersen & Turner, 1968, with soybean; van den Berg, unpublished results, Wageningen, with pea). Assuming that a nodule maintains its N_2-fixing capacity for approximately one month (Bergersen & Goodchild, 1973*b*), an amount of 1.5–3.5 g of N_2 is fixed per gram of nodule protein during this period. This value is much higher than the 0.16 g of N_2 fixed per gram of cell protein during the short lifetime of an *Azotobacter* cell.

Fate of fixed N_2

Azotobacters and presumably all of the other free-living bacteria use the main part of the fixed nitrogen for cell synthesis. In an experiment with five different strains of azotobacters 7–13 % of the total amount of fixed N_2 was found to be excreted by the cells (Brotonegoro, 1974). This is in contrast to the non-growing symbiotic systems which excrete most of the fixed N_2. Minchin & Pate (1973) found that during the period that nodule initiation had ceased, more than 90 % of fixed N_2 was translocated to the shoots.

Ecology

The occurrence and functioning of free-living, N_2-fixing bacteria in natural environments (soil, water) demand the fulfilment of a number of growth requirements. These requirements include (*a*) the presence of available carbon compounds; (*b*) the presence of adequate amounts of inorganic nutrients like calcium, magnesium, molybdenum, etc.; (*c*) an optimum pH for growth of the organisms; (*d*) a relatively low oxygen supply.

Presence of available carbon compounds

The development of free-living, N_2-fixing bacteria in soil is **favoured** by the presence of considerable amounts of available carbon compounds and very low amounts of combined nitrogen so that the C/N ratios are high. If the environments are well-provided with combined nitrogen, non-nitrogen-fixing micro-organisms will readily develop and will successfully compete with the nitrogen fixers for carbon compounds (Macura & Kunc, 1961).

Examples of environments with large numbers of free-living N_2 fixers are given by Jensen (1965), Abd-el-Malek (1971), Rouquérol (1962), Dobereiner, Day & Dart (1972) and many others.

Jensen (1965) treated nitrogen-poor, water-saturated soil with 1–2 % of oats or wheat straw which had a high C/N ratio and reported the fixation of considerable amounts of N_2 by free-living soil bacteria of the *Azotobacter* type. The numbers of these organisms amounted to 10^6–10^7 per gram of soil. Similar high numbers of azotobacters have been counted in irrigated alkaline clay soils of the Nile Valley, containing large amounts of easily decomposable crop residues (Vančura, Abd-el-Malek & Zayed, 1965; Abd-el-Malek, 1971) and in rice soils of the Camargue in the Rhône delta (Rouquérol, 1962).

The occurrence of azotobacters in numbers amounting to 10^6 per gram of soil indicates that considerable amounts of nitrogen are fixed so that in this case the free-living N_2 fixers contribute substantially to the nitrogen economy of the soil. Normally, much lower numbers of azotobacters are counted (10^3–10^4 per gram of soil when the pH is 7 or higher) and then the contribution of these organisms to soil fertility is insignificant (Brouzes, Mayfield & Knowles, 1971; Knowles, 1975).

Soils amended with large amounts of glucose and incubated under anaerobic conditions contained more than 10^8 *Clostridium* cells per gram which fixed considerable amounts of N_2 (Brouzes *et al.*, 1971). Similar results were obtained by Paul, Myers & Rice (1971) with waterlogged soil supplied with large amounts of straw.

A further example of stimulated N_2 fixation by specific free-living bacteria in soil resulting from an increased supply of particular carbonaceous compounds may be encountered in the vicinity of leaking natural-gas mains. The most important component of this gas, methane, is readily oxidized by micro-organisms of the *Methylosinus* type (Adamse, Hoeks, De Bont & Van Kessel, 1972). As some strains of this type are able to fix N_2 (Fig. 1.1) they will obviously accumulate selectively and will give rise to an increase in the nitrogen content of such environments (Harper, 1939).

Selective enrichment of certain N_2-fixing bacteria has been observed in the rhizosphere and in the phyllosphere of certain plants, particularly under tropical conditions. Beijerinckias have been found to accumulate in the rhizosphere of sugar cane (Dobereiner & Alvahydo, 1959) and in the phyllosphere of several tropical shrubs (Ruinen, 1961; 1974). A particular rhizosphere organism is *Azotobacter paspali* which has been found by Dobereiner and co-workers to occur mainly in the rhizosphere of *Paspalum notatum*, a tropical grass growing in large areas of Brazil (Dobereiner & Campelo, 1971). The relatively large amounts of N_2 fixed in this environment by *A. paspali* suggest that a particular association between *A. paspali* and *Paspalum notatum* may exist (Dobereiner *et al.*, 1972).

Klebsiellas have been found by Evans, Campbell & Hill (1972) to occur in large numbers in the rhizosphere of nodulated leguminous plants. Enterobacters were found in the rhizosphere of maize plants (Raju *et al.*, 1972). Bessems (1973) isolated klebsiellas from the leaf surface and from the liquid inside leaf sheaths of maize and of tropical grasses, particularly guatemala grass. He made a study of the carbohydrate content and of the C/N ratio of the sheath liquid of these plants (Table 1.3). It will be seen that maize plants growing under tropical conditions excreted much greater amounts of carbohydrates than similar plants growing in temperate regions. Removal of the carbohydrate-containing solution from the guatemala grass by washing stimulated the excretion of carbohydrates so that within a few hours considerably higher carbohydrate concentrations were attained than those existing before the washing started (Table 1.4). In spite of these apparently highly favourable conditions for N_2-fixing bacteria, the amount of N_2 fixed (measured by the acetylene-reduction and ^{15}N assays) was low, indicating that an essential factor for N_2 fixation was lacking.

Presence of inorganic nutrients

Although several soils are poor in available inorganic nutrients like calcium, magnesium, molybdenum and some others, very little is known about the effect of adding these nutrients on nitrogen fixation and growth of free-living bacteria in such soils. The adverse effect of a low

Table 1.3. *Carbohydrate content of sheath water of four different leaves of guatemala grass. Recovery after washing* (Bessems, 1973)

Treatment	Carbohydrates (mg/l)			
Before washing	1790	640	465	680
15 min after the first washing	410	142	392	357
15 min after the second washing	385	53	285	340
45 min after the second washing	3070	965	475	1570
105 min after the second washing	4300	2570	640	2070
165 min after the second washing	—	—	2000	6430

Table 1.4. *C/N ratios in sheath water of plants growing under temperate and tropical conditions* (Bessems, 1973)

Plant	Climatic conditions	C/N ratio
Maize	Temperate	26 (3–103)
	Tropical	206 (36–654)
Sorghum bicolor	Tropical	182 (30–640)
Guatemala grass	Tropical	
No N dressing		257 (55–771)
200 kg N/ha		105 (18–277)

soil pH on soil azotobacters may partly be due to calcium deficiency, as acid soils are very low in available Ca^{2+}. Very low amounts of available molybdenum are found in iron-stone-containing soils in Europe and in Australia. Although symbiotic N_2 fixation of several leguminous plants growing on such soils suffers badly from Mo deficiency (Mulder, 1954), no data are available concerning free-living N_2 fixers.

Effect of pH

Most azotobacters require a neutral or alkaline reaction for growth. A survey of 264 Danish soils of widely different pH revealed that practically 100 % of the soils above pH 7.5 contained azotobacters in numbers varying between 10^2 and 10^4 per gram, whereas below pH 6.5 only a small fraction of the soils tested contained a few *Azotobacter* cells (Jensen, 1965). Although beijerinckias, when grown in pure culture, tolerate a wide range of pH values, they are found under natural conditions mainly in acid tropical soils (Becking, 1961). Derxias grow in acid as well as in neutral soils (Dobereiner & Campelo, 1971). For more details on the effect of pH as ecological factor see Mulder & Brotonegoro (1974).

Oxygen supply

Oxygen does not only adversely affect anaerobic and facultatively anaerobic N_2 fixers but to some extent it also interferes with N_2 fixation of the aerobic bacteria. This might indicate that azotobacters would prefer deeper layers of terrestrial and aquatic environments, where the oxygen supply may be expected to be lower. In this connection the experiments of Tschapek & Giambiagi (1954) may be mentioned. These authors found that in nitrogen-free liquid media, *A. chroococcum* contained in tubes, developed a growth zone at a certain distance from the surface where the P_{O_2} apparently was optimum. When the carbon-compound content of the nutrient solution was higher, the zone moved upward, owing to increased oxygen demand.

That a reduced oxygen supply may favourably affect N_2 fixation by rhizosphere organisms was shown by Dobereiner, Day & Dart (1972) in experiments with sugar cane which contained beijerinckias as free-living N_2 fixers in the rhizosphere.

References

Abd-el-Malek, Y. (1971). Free-living nitrogen-fixing bacteria in Egyptian soils and their possible contribution to soil fertility. *Plant and Soil, Special Volume*, 423–42.

Adamse, A. D., Hoeks, J., De Bont, J. A. M. & Van Kessel, J. F. (1972). Microbial activities in soil near natural gas leaks. *Arch. Mikrobiol.*, **83**, 32–51.

Becking, J. H. (1961). Studies on nitrogen-fixing bacteria of the genus *Beijerinckia. Pl. Soil*, **14**, 49–81.

(1962). Species differences in molybdenum and vanadium requirements and combined nitrogen utilization by Azotobacteriaceae. *Pl. Soil*, **16**, 171–201.

(1963). Fixation of molecular nitrogen by an aerobic *Vibrio* or *Spirillum. Antonie van Leeuwenhoek*, **29**, 326.

Beijerinck, M. W. (1901). Über oligonitrophile Mikroben. *Zentbl. Bakt. ParasitKde Abt. II*, **7**, 561–82.

Benemann, J. R., McKenna, C. E., Lie, R. F., Taylor, T. G. & Kamen, M. D. (1972). The vanadium effect in nitrogen fixation by *Azotobacter. Biochim. biophys. Acta*, **264**, 25–38.

Bergersen, F. J. (1969). Nitrogen fixation in legume root nodules: biochemical studies with soybean. *Proc. R. Soc. Lond.* **B**, **172**, 401–16.

Bergersen, F. J. & Goodchild, D. J. (1973*a*). Aeration pathways in soybean root nodules. *Aust. J. biol. Sci.*, **26**, 729–40.

Bergersen, F. J. & Goodchild, D. J. (1973*b*). Cellular location and concentration of leghaemoglobin in soybean root nodules. *Aust. J. Biol. Sci.*, **26**, 741–56.

Bergersen, F. J. & Turner, G. L. (1968). Comparative studies of nitrogen fixation by soybean root nodules, bacteroid suspensions and cell-free extracts. *J. gen. Microbiol.*, **53**, 205–20.

Bergey's Manual of Determinative Bacteriology (1957) (eds. Breed, R. S., Murray, E. G. D. & Smith, N. R.), 7th edn. Williams and Wilkins Co.: Baltimore.

Bessems, E. P. M. (1973). Nitrogen fixation in the phyllosphere of Gramineae. Ph.D. Thesis, Wageningen.

Biggins, D. R. & Postgate, J. R. (1969). Nitrogen fixation by cultures and cell-free extracts of *Mycobacterium flavum* 301. *J. gen. Microbiol.*, **56**, 181–93.

Brotonegoro, S. (1974). Nitrogen fixation and nitrogenase activity of *Azotobacter chroococcum*. Ph.D. Thesis, Wageningen.

Brouzes, R., Mayfield, C. I. & Knowles, R. (1971). Effect of oxygen partial pressure on nitrogen fixation and acetylene reduction in a sandy loam soil amended with glucose. *Plant and Soil, Special Volume*, 481–94.

Burns, R. C., Fuchsman, W. H. & Hardy, R. W. F. (1971). Nitrogenase from vanadium-grown *Azotobacter*: isolation, characteristics, and mechanistic implications. *Biochem. biophys. Res. Commun.*, **42**, 353–8.

Burris, R. H. & Wilson, P. W. (1946). Ammonia as an intermediate in nitrogen fixation by *Azotobacter*. *J. Bacteriol.*, **52**, 505–12.

Coty, V. F. (1967). Atmospheric nitrogen fixation by hydrocarbon-oxidizing bacteria. *Biotechnol. Bioeng.*, **9**, 25–32.

Daesch, G. & Mortenson, L. E. (1972). Effect of ammonia on the synthesis and function of the N_2-fixing enzyme system in *Clostridium pasteurianum*. *J. Bacteriol.*, **110**, 103–9.

Dalton, H. (1974). Fixation of dinitrogen by free-living microorganisms. *Crit. Revs. Microbiol.*, **3**, 183–220.

Dalton, H. & Postgate, J. R. (1969). Effect of oxygen on growth of *Azotobacter chroococcum* in batch and continuous cultures. *J. gen. Microbiol.*, **54**, 463–73.

De Bont, J. A. M. & Mulder, E. G. (1974). Nitrogen fixation and co-oxidation of ethylene by a methane-utilizing bacterium. *J. gen. Microbiol.*, **83**, 113–21.

De Ley, J. (1968). DNA base composition and classification of some more free-living nitrogen-fixing bacteria. *Antonie van Leeuwenhoek*, **34**, 66–70.

De Ley, J. & Park, I. W. (1966). Molecular biological taxonomy of some free-living nitrogen-fixing bacteria. *Antonie van Leeuwenhoek*, **32**, 6–16.

Dixon, R. A. & Postgate, J. R. (1972). Genetic transfer of nitrogen fixation from *Klebsiella pneumoniae* to *Escherichia coli*. *Nature, Lond.*, **237**, 102–3.

Dobereiner, J. & Alvahydo, R. (1959). Sôbre a influência da canade açúcar na ocorrência de "Beijerinckia" no solo. II. Influência das diversas partes do vegetal. *Rev. Bras. Biol.*, **19**, 401–12.

Dobereiner, J. & Campelo, A. B. (1971). Non-symbiotic nitrogen-fixing bacteria in tropical soils. *Plant and Soil, Special Volume*, 457–70.

Dobereiner, J., Day, J. M. & Dart, P. J. (1972). Nitrogenase activity and oxygen sensitivity of the *Paspalum notatum–Azotobacter paspali* association. *J. gen. Microbiol.*, **71**, 103–16.

Evans, H. J., Campbell, N. E. R. & Hill, S. (1972). Asymbiotic nitrogen-fixing bacteria from the surfaces of nodules and roots of legumes. *Can. J. Microbiol.*, **18**, 13–21.

Evans, H. J. & Russell, S. A. (1971). Physiological chemistry of symbiotic nitrogen fixation by legumes. In *The Chemistry and Biochemistry of Nitrogen Fixation* (ed. Postgate, J. R.), pp. 191–244. Plenum Press: London, New York.

Fedorov, M. V. & Kalininskaya, T. A. (1961). A new species of nitrogen-fixing *Mycobacterium* and its physiological properties. *Mikrobiologiya*, **30**, 7–11.

Gibson, A. H. (1966). The carboydrate requirements for symbiotic nitrogen fixation: a "whole plant" growth analysis approach. *Aust. J. biol. Sci.*, **19**, 499–515.

Gogotov, J. N. & Schlegel, H. G. (1974). N_2-fixation by chemoautotrophic hydrogen bacteria. *Arch. Mikrobiol.*, **97**, 359–62.

Grau, F. H. & Wilson, P. W. (1962). Physiology of nitrogen fixation by *Bacillus polymyxa*. *J. Bacteriol.*, **83**, 490–6.

Hardy, R. W. F., Burns, R. C. & Holsten, R. D. (1973). Applications of the acetylene–ethylene assay for measurement of nitrogen fixation. *Soil Biol. Biochem.*, **5**, 47–81.

Harper, H. J. (1939). The effect of natural gas on the growth of microorganisms and the accumulation of nitrogen and organic matter in the soil. *Soil Sci.*, **48**, 461–6.

Jakobsons, A., Zell, E. A. & Wilson, P. W. (1962). A re-investigation of the calcium requirement of *Azotobacter vinelandii* using purified media. *Arch. Mikrobiol.*, **41**, 1–10.

Jensen, H. L. (1965). Nonsymbiotic nitrogen fixation. In *Soil Nitrogen* (eds. W. V. Bartholomew and F. E. Clark), *Am. Soc. Agron. Monogr.*, **10**, pp. 436–80.

Jensen, H. L. & Spencer, D. (1947). The influence of molybdenum and vanadium on nitrogen fixation by *Clostridium butyricum* and related organisms. *Proc. Linnean Soc., N.S.W.*, **72**, 73–86.

Jensen, V. (1958). A new nitrogen fixing bacterium from a Danish water-course. *Arch. Mikrobiol.*, **29**, 348–53.

Knight, E. Jr. & Hardy, R. W. F. (1966). Isolation and characteristics of flavodoxin from nitrogen-fixing *Clostridium pasteurianum*. *J. biol. Chem.*, **241**, 2752–6.

Knowles, R. (1975). The significance of asymbiotic dinitrogen fixation by bacteria. In *Dinitrogen (N_2) Fixation* (ed. Hardy, R. W. F.). Wiley–Interscience: New York (in press).

Line, M. A. & Loutit, M. W. (1971). Non-symbiotic nitrogen-fixing organisms from some New Zealand tussock-grassland soils. *J. gen. Microbiol.*, **66**, 309–18.

Loperfido, B. & Sadoff, H. L. (1973). Germination of *Azotobacter vinelandii* cysts: sequence of macromolecular synthesis and nitrogen fixation. *J. Bacteriol.*, **113**, 841–6.

Mackintosh, M. E. (1971). Nitrogen fixation by *Thiobacillus ferrooxidans* species. *J. gen. Microbiol.*, **66**, i.

Macura, J. & Kunc, F. (1961). Continuous flow method in soil microbiology. II. Observations on glucose metabolism. *Folia Microbiol., Prague*, **6**, 398–407.

Mahl, M. C., Wilson, P. W., Fife, M. A. & Ewing, W. H. (1965). Nitrogen fixation by members of the tribe Klebsielleae. *J. Bacteriol.*, **89**, 1482–7.

Minchin, F. R. & Pate, J. S. (1973). The carbon balance of a legume and the functional economy of its nodules. *J. exp. Bot.*, **24**, 259–71.

Mulder, E. G. (1948). Importance of molybdenum in the nitrogen metabolism of microorganisms and higher plants. *Pl. Soil*, **1**, 94–119.

Mulder, E. G. (1954). Molybdenum in relation to growth of higher plants and micro-organisms. *Pl. Soil*, **5**, 368–415.

Mulder, E. G. & Brotonegoro, S. (1974). Free-living heterotrophic nitrogen-fixing bacteria. In *The Biology of Nitrogen Fixation* (ed. Quispel, A.), pp. 37–85. North-Holland: Amsterdam.

Norris, J. R. & Jensen, H. L. (1958). Calcium requirements of *Azotobacter*. *Arch. Mikrobiol.*, **31**, 198–205.

Paul, E. A., Myers, R. J. K. & Rice, W. A. (1971). Nitrogen fixation in grassland and associated cultivated ecosystems. *Plant and Soil, Special Volume*, 495–507.

Pengra, R. M. & Wilson, P. W. (1958). Physiology of nitrogen fixation by *Aerobacter aerogenes*. *J. Bacteriol.*, **75**, 21–5.

Postgate, J. R. (1971). Relevant aspects of the physiological chemistry of nitrogen fixation. In *Microbes and Biological Productivity* (eds. Hughes, D. E. & Rose, A. H.), pp. 287–307. Cambridge University Press: London.

Raju, P. N., Evans, H. J. & Seidler, R. J. (1972). An asymbiotic nitrogen-fixing bacterium from the root environment of corn. *Proc. natn. Acad. Sci. USA*, **69**, 3474–8.

Rodina, A. G. (1956). Aquatic spirillae fixing molecular nitrogen. *Mikrobiologiya*, **25**, 145–9.

Rosenblum, E. D. & Wilson, P. W. (1949). Fixation of isotopic nitrogen by *Clostridium*. *J. Bacteriol.*, **57**, 413–14.

Rouquérol, T. (1962). Sur le phénomène de fixation de l'azote dans les rizières de Camargue. *Ann. Agron.*, Paris, **13**, 325–46.

Ruinen, J. (1961). The phyllosphere. I. An ecologically neglected milieu. *Pl. Soil*, **15**, 81–109.

Ruinen, J. (1974). Nitrogen fixation in the phyllosphere. In *The Biology of Nitrogen Fixation* (ed. Quispel, A.), pp. 121–67. North-Holland: Amsterdam.

Smyk, B. (1970). Fixation of atmospheric nitrogen by the strains of *Arthrobacter. Zentrbl. Bakt. ParasitKde Abt. II*, **124**, 231–7.

Smyk, B. & Ettlinger, L. (1963). Recherches sur quelques espèces d'*Arthrobacter* fixatrices d'azote isolées des roches karstiques alpines. *Ann. Inst. Pasteur*, **105**, 341–8.

Stevenson, L. H. & Socolofsky, M. D. (1973). Role of poly-β-hydroxybutyric acid in cyst formation by *Azotobacter. Antonie van Leeuwenhoek*, **39**, 341–50.

Strandberg, G. W. & Wilson, P. W. (1968). Formation of the nitrogen-fixing enzyme system in *Azotobacter vinelandii. Can. J. Microbiol.*, **14**, 25–31.

Tschapek, M. & Giambiagi, N. (1954). The formation of Liesegang's rings by Azotobacter due to O_2-inhibition. *Trans. 5th Int. Congr. Soil Sci., Leopoldville*, III, **17**, 97–103.

Vančura, V., Abd-el-Malek, Y. & Zayed, M. N. (1965). *Azotobacter* and *Beijerinckia* in the soils and rhizosphere of plants in Egypt. *Fol. Microbiol., Prague*, **10**, 224–9.

Voets, J. P. & Debacker, J. (1956). *Pseudomonas azotogensis nov.sp.* a new free-living nitrogen-fixing bacterium. *Naturwissenschaften*, **43**, 40–1.

Whittenbury, R., Phillips, K. C. & Wilkinson, J. F. (1970). Enrichment, isolation and some properties of methane-utilizing bacteria. *J. gen. Microbiol.*, **61**, 205–18.

Winogradsky, S. (1893). Sur l'assimilation de l'azote gazeux de l'atmosphère par les microbes. *C. r. Acad. Sci. Paris*, **116**, 1385–88.

Witz, D. F., Detroy, R. W. & Wilson, P. W. (1967). Nitrogen fixation by growing cells and cell-free extracts of the Bacillaceae. *Arch. Mikrobiol.*, **55**, 369–81.

2. Anaerobic nitrogen-fixing bacteria of different soil types

E. N. MISHUSTIN & V. T. YEMTSEV

The composition of microbial coenoses in different types of soil and the ecological variability of soil micro-organisms have been studied for a number of years at the Laboratory of Soil Microbiology of the Institute of Microbiology, USSR Academy of Sciences and at the Microbiology section of the Timiryazev Academy of Agriculture, USSR. These studies showed that the zones of optimum growth are different for different micro-organisms. As a consequence, the groups of dominant micro-organisms differ with soil type. This enables the diagnosis of soil types on the basis of their microbiological composition to be determined. At the same time it is accepted that certain species of micro-organisms, although found in some soils in greater or smaller numbers than in others, are widely distributed over the surface of the Earth, but even then different ecological strains of the same species occur in different ecological situations.

Recently we have begun experiments with a group of anaerobic nitrogen-fixing bacteria whose ecology so far has not been adequately studied. These micro-organisms represented by *Clostridium pasteurianum*, first described by S. N. Winogradsky, are of special interest for our work in view of their widespread distribution. Some of our results have been published elsewhere (Mishustin & Yemtsev, 1973). This paper covers both published and unpublished findings.

Methods

Molecular nitrogen is assimilated by many forms of *Clostridium* and to obtain information on the conditions for growth of a number of species in soil we employed the following mineral medium (MB): KH_2PO_4, 0.5 g; K_2HPO_4, 0.015 g; $MgSO_4.7H_2O$, 0.5 g; NaCl, 0.015 g; $FeSO_4$, 0.01 g; $MnSO_4$, 0.01 g; trace elements; distilled water, 1 l.

Cl. pasteurianum was then detected on the following medium: MB, 1 l; yeast autolysate, 0.2 mg; glucose, 20 g; $CaCO_3$, 10 g; peptone, 5 g. The medium was sterilized at 1.5 atm for 30 min. The pH of the medium was 7.0 and the incubation temperature was 25–26 °C.

Cl. butyricum was detected on a medium of the following composition: MB, 1 l; yeast autolysate, 0.2 mg; potato starch, 20 g; $CaCO_3$, 10 g; peptone, 5 g. The medium was sterilized at 1.5 atm for 30 min. The pH of the medium was 7.5 and the temperature was 30 °C.

Cl. acetobutylicum and *Cl. butylicum* were counted on a medium based on corn meal. Fifty grams of this meal were rapidly heated in 1 l of water at 90 °C for 10 min, then the medium was boiled for 30 min and the pH adjusted to 5.6–5.7. The medium was poured into test tubes and boiled and sterilized at 2 atm for 60 min. For detecting *Cl. acetobutylicum* the temperature of incubation was 37 °C, and for *Cl. butylicum* 30 °C. To reduce the oxidation–reduction potential, 0.05 % of thioglycolic acid was added to all of the media. Neutral red, at a concentration of 0.004 %, was used as an indicator.

The quantitative abundance of the above groups of micro-organisms was measured by serial dilution. The inoculated media were checked daily over a 12–14 day period for foam formation and colour change of the indicator. The presence of typical clostridial cells was checked by microscopic observation.

Pectolytic anaerobes were estimated using a modification of the medium of Kaiser (1961). Tests for protopectinase were carried out using the following medium: potato broth, 500 ml; yeast autolysate, 5 ml; thioglycolic acid, 5 mg; tap water, 500 ml. The test tubes containing this medium were supplied with pieces of carrot (0.3 × 1.5 cm). The medium was sterilized for 10 min at 1 atm. The temperature of incubation was 37 °C. The development of the anaerobes was checked 4–7 days after inoculation by examining for maceration of the carrot pieces and for gas formation.

Tests for bacteria producing pectinesterase and polygalacturonase were carried out using the following medium: potato broth, 1000 ml; pectin, 7 g for pectinesterase and 13 g for polygalacturonase; thioglycolic acid, 1 ml (giving a concentration of 0.004 %); bromothymol blue, 1 ml of 0.5 % (as indicator for pectinesterase and 1 ml of 0.04 % neutral red for polygalacturonase). The pectin was dissolved in 96 % alcohol and the solution was added to the broth which was then heated to 80 °C. After sterilization for 10 min at 1 atm, 0.2 % $CaCl_2$ was added to the medium to promote gel formation. The temperature of incubation was 37 °C. The development of the anaerobes was estimated during days 4–7 by checking for gas formation, colour change of the indicator, and pectin liquefaction.

Results and discussion

From the results obtained the following conclusions can be drawn. Soils of the northern zone (podzols, soddy-podzolic soils, etc.) contain in the upper horizon (*A*) a far greater number of bacteria of the genus *Clostridium* than those from the south. In northern soils the bacteria are mainly represented by *Cl. pasteurianum*, followed by *Cl. butyricum*. The number of pectolytic bacteria in the soils is relatively small. Bacteria possessing protopectinase are more numerous than those with pectinesterase and polygalacturonase. Acetobutylic bacteria (*Cl. acetobutylicum*) are present in northern soils in very small numbers (Table 2.1).

Table 2.1. *Numbers* of bacteria* ($\times 10^3$/*g of soil*) *of the genus* Clostridium *in virgin soils from different latitudes*

Soil type	Butyric acid bacteria: *Cl. pasteurianum*	Butyric acid bacteria: *Cl. butyricum*	Pectolytic bacteria	Acetobutylic bacteria
Soddy-podzolic	7000	700	20	1
Chernozem	100	30	30	1
Chestnut	10	5	—	10
Zierozem	0.5	2.5	5	15

* Average values.

Further to the south, the number of bacteria of the genus *Clostridium* decreases, mainly due to the reduction in *Cl. pasteurianum* and *Cl. butyricum* and partially to decreases in pectolytic bacteria. The number of acetobutylic bacteria, however, increases in southern soils.

Similar results to those from different latitudes were observed in soils of different altitudes. Table 2.2 shows data on the changes in the composition of a number of bacterial species of the genus *Clostridium* in the *A* horizon of mountain soils of Tien Shan (Kazakhstan). The absolute figures are slightly different but the trend is the same as that in the soils of different latitudes.

Figure 2.1 shows the relative numbers of the two most frequently occurring species of the genus *Clostridium* (*Cl. pasteurianum* and *Cl. acetobutylicum*) in the *A* horizon of soils from the mountains and plains of the Caucasus. The above-mentioned regularity is also found here. Bacteria of the genus *Clostridium* are largely distributed in the upper horizons of all of the soils rich in organic matter. They obviously

Table 2.2. *Numbers* of bacteria* ($\times 10^3$/*g of soil*) *of the genus* Clostridium *in soils of Tien Shan of different altitudes*

Soil type	Butyric acid bacteria: *Cl. pasteurianum*	Butyric acid bacteria: *Cl. butyricum*	Pectolytic bacteria	Acetobutylic bacteria
Mountain forest	500	100	50	10
Chernozem	150	25	50	25
Light chestnut	10	—	12	—
Sierozem	0.5	2.5	2	15

* Average values.

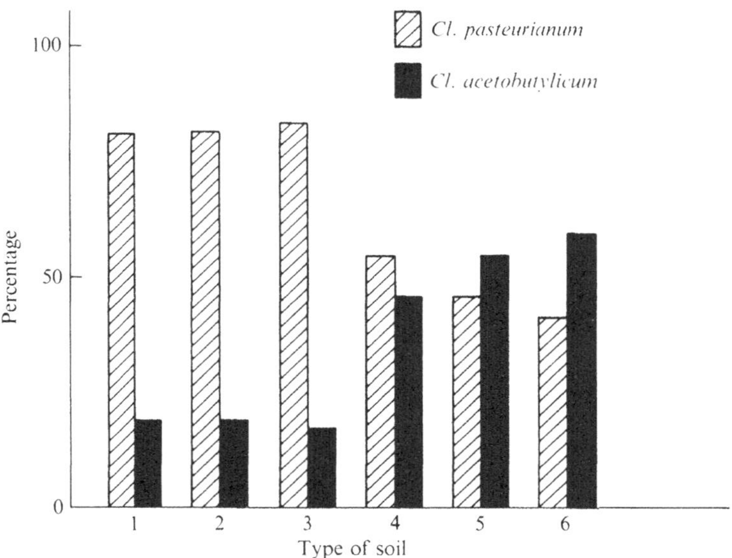

Fig. 2.1. Bacteria of the genus *Clostridium* at different altitudes of the Caucasus (%). 1, Mountain-forest; 2, mountain chernozem; 3, forest cinnamonic; 4, plain chernozem; 5, chestnut; 6, krasnozem.

develop in microzones having no access to oxygen or being well protected against its penetration. Figure 2.2 presents data on the distribution of some species of bacteria of the genus *Clostridium* in the chernozem soil of the Voronezh region.

The numbers of bacteria of different *Clostridium* species in soil fluctuate during the year suggesting that considerable numbers of cells of these micro-organisms are present as vegetative cells in the soil.

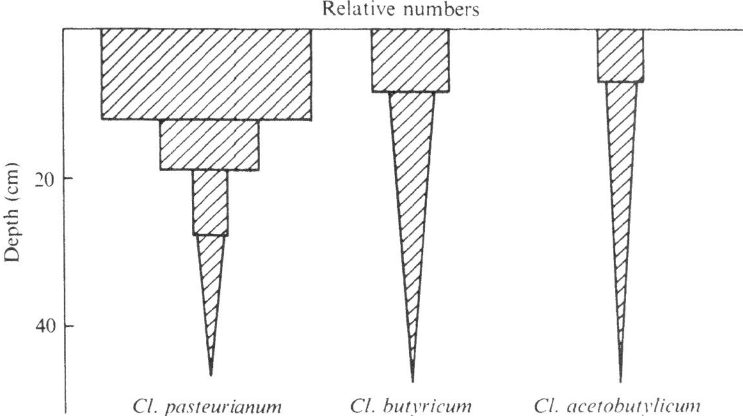

Fig. 2.2. Distribution of bacteria of the genus *Clostridium* in the chernozem soil profile.

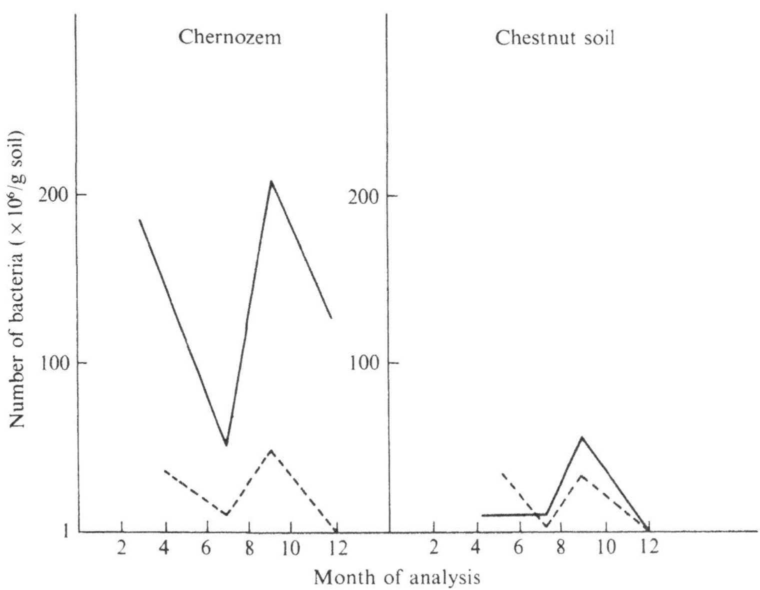

Fig. 2.3. Dynamics of bacteria of the genus *Clostridium* in soil. Solid line, *Cl. pasteurianum*; broken line, *Cl. acetobutylicum*.

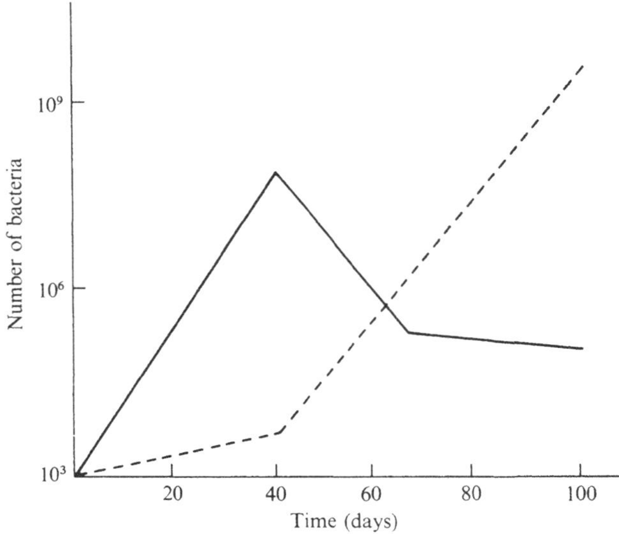

Fig. 2.4. Succession of species of the genus *Clostridium* during the decomposition of organic matter in compost. Solid line, *Cl. pasteurianum*; broken line, *Cl. acetobutylicum*.

Generally *Clostridium* cells are most numerous in spring and autumn when the soil contains much fresh organic matter (see Fig. 2.3).

What is the reason for the abundant proliferation of *Cl. pasteurianum* in northern soils and the predominance of *Cl. acetobutyricum* in southern soils? Model experiments involving composting of plant residues have shown that the butyric acid bacteria (*Cl. pasteurianum*, etc.) are the first to proliferate in the decomposing mass. Later, when the compost is enriched by microbial protein and possibly by other compounds of microbial origin (amino acids, vitamins, etc.), the butyric acid bacteria are replaced by the acetobutylic organisms (Fig. 2.4). The latter are more exacting in their substrate requirements and due to their proteolytic properties they proliferate predominantly in protein-rich media.

The investigations conducted by Lazarev (1949) and by the present authors show that southern soils are richer in micro-organisms, possibly because the favourable climatic conditions result in a rapid transformation of organic compounds. In the north the process is slower so that more favourable conditions for the proliferation of butyric acid bacteria are present. In the southern zone the period during which *Cl. pasteurianum* and physiologically related bacteria predominate in the decomposing mass is short. Thus there are prerequisites for *Cl. acetobutylicum* to predominate.

Table 2.3. *Products of glucose fermentation* (%) *by* Clostridium *species of different origin*

Bacterial species	Origin of culture	Volatile acids		Alcohols	
		Butyric	Acetic	Butyl	Ethyl
Clostridium pasteurianum	Northern soils	80.0–88.0	8.0–9.0	2.5–5.0	2.0–7.0
	Southern soils	58.0–67.0	17.0–20.0	12.0–16.5	5.5–8.0
Clostridium butyricum	Northern soils	*c.* 88	6.5–8.0	4.1–4.7	0.0–1.5
	Southern soils	*c.* 86	6.0–8.0	3.3–3.8	2.5–3.5
Clostridium butylicum	Northern soils	58.0–60.0	22.0–24.0	17.0–20.0	0
	Southern soils	53.0–58.5	18.0–22.0	20.0–29.0	0
Clostridium acetobutylicum	Northern soils	*c.* 56.0	23.0–25.0	13.0–14.5	*c.* 6.0
	Southern soils	46.0–52.0	20.0–22.0	19.0–20.5	8.0–12.0

Our studies show that cultures of the same *Clostridium* species isolated from different soils may give different fermentation products. The data in Table 2.3 present the overall picture of some fermentation products, volatile acids and alcohols which we have found. Acetone and the various gaseous products are not considered here. It is seen that the products of fermentation of southern cultures of *Cl. pasteurianum* contain considerably less butyric acid and more alcohol than those of northern strains. A similar shift is found in the cultures of *Cl. acetobutylicum*. The data show that the shift is less obvious for *Cl. butyricum* and *Cl. butylicum*. The sequence is more pronounced when the results are calculated as the ratio, % acids/% alcohols. As is seen from Table 2.4 this index is markedly lower in bacterial cultures isolated from southern soils.

In all bacteria of the genus *Clostridium* which we studied, the nitrogen-fixing activity also changes in a regular mode. Southern cultures assimilate smaller amounts of N_2 per unit of utilized source of carbon than do the cultures isolated from soils of a colder climatic zone. The change is more pronounced with *Cl. pasteurianum* and *Cl. butyricum* than with *Cl. butylicum* and *Cl. acetobutylicum* (Table 2.5).

The important temperature characteristics of all the *Clostridium* species have been studied. Cultures of *Cl. pasteurianum* isolated from northern soils had an optimum temperature for growth of about 23 °C and a maximum of about 36 °C. For cultures of the same species from southern soils the optimum temperature was about 31 °C and the maximum about 45 °C. A similar picture was observed in cultures of *Cl. butyricum*. The cultures of *Cl. butylicum* and *Cl. acetobutylicum* of

Table 2.4. *Products of glucose fermentation by bacteria of the genus* Clostridium, *expressed as the ratio, % acids/% alcohol*

Climatic zone	*Cl. pasteurianum*	*Cl. butyricum*	*Cl. butylicum*	*Cl. acetobutylicum*
Northern	19.8	20.0	5.0	4.0
Moderate	8.0	16.5	4.0	3.8
Southern	4.5	15.0	2.5	2.5
Subtropical	3.2	14.7	2.0	2.1

Table 2.5. *Average nitrogen-fixing capacity of bacteria of the genus* Clostridium (*mg N/g glucose*)

Climatic zone	*Cl. pasteurianum*	*Cl. butyricum*	*Cl. butylicum*	*Cl. acetobutylicum*
Northern	8.5	9.5	3.7	3.3
Moderate	7.5	7.0	3.6	3.2
Southern	4.3	6.1	3.4	2.9
Subtropical	3.6	4.2	3.4	2.6

northern origin had optimum temperatures of about 30–35 °C and maxima of about 44 °C. In southern cultures of these bacteria the optimum temperatures for growth were about 36–40 °C and the maxima about 50 °C.

After pyrolysis, the cultures were studied by gas chromatography. The pyrograms of cultures of various origins differed somewhat from one another, suggesting a certain specificity in their composition. It should be noted that the pyrograms of *Cl. acetobutylicum* cultures were closer to one another than the pyrograms of strains of other anaerobes obtained from different soils.

From the results obtained, it can be concluded that *Clostridium* species include different ecological strains. A similar pattern was found earlier in *B. mycoides* (Mishustin, 1947).

A few remarks should be made about the activity of anaerobic nitrogen fixers in various soils. In our laboratory, T. A. Kalininskaya and V. Rao have conducted experiments on the fixation of $^{15}N_2$ in different soils, the fixer being included in the organic matter of the soil. In these experiments soils from different zones were taken and adjusted to contain 60 % moisture for all except rice and bog soils where the

Table 2.6. *Fixation of $^{15}N_2$ by different soils*

Soil	Excess atom % ^{15}N	Fixed nitrogen	
		mg/kg soil	kg/ha
Soddy-podzolic arable land	0.00–0.25	0.00–0.21	0.0–0.63
Grey forest arable land	0.06	0.78	2.35
Chernozem arable land	0.02–0.06	1.0–3.5	3.0–9.0
Chestnut arable land	0.00–0.03	0.0–1.2	0.0–3.6
Meadow-chernozem-like, planted to rice	0.21	11	33.0
Meadow-bog, virgin land	0.15	4.07	12.2

Table 2.7. *Effect of added cellulose on nitrogen fixation in flooded soils*

Soil	mg N fixed/g of utilized cellulose
Soddy-podzolic soils	1.5–2.0
Grey-forest soils	1.5–2.4
Chernozem soils	3.0–5.4
Meadow-chernozem soils planted to rice*	7.0–10.0

* The higher nitrogen-fixing capacity of flooded soils was also noted using the acetylene-reduction technique.

moisture was adjusted to 100 %. The soils were then exposed to $^{15}N_2$ for one month and the atom per cent excess ^{15}N label of the soil then calculated. The values were then extrapolated to a kilogram per hectare basis. Selected data (Table 2.6) show that the flooded meadow-bog soils and the rice fields accumulated considerably more nitrogen than the others. This observation can be explained on the basis that the nitrogen fixers get more catabolic substrates when anaerobic decomposition of the organic matter occurs than when complete oxidation is possible. Upon the application of cellulose to different soils, the efficiency of nitrogen fixation ($^{15}N_2$) was found to be higher in flooded soils thus confirming the above supposition. Some examples are given in Table 2.7.

References

Kaiser, P. (1961). *Etude de l'activité pectinolytique du sol et d'autres substrats naturel.* Imprimerie Barneoud: Laval.

Lazarev, N. M. (1949). Tipy organomineralnykh sistem razlichnykh pochv. *Trudy Inst. s. kh. microbiologii za 1941–1945. gg.*, **1**, 23–457.

Mishustin, E. N. (1947). *Ecological and geographical variability of soil bacteria.* USSR Acad. of Sciences: Moscow (in Russian).

Mishustin, E. N. & Yemtsev, V. T. (1973). Anaerobic nitrogen-fixing bacteria is USSR soils. *Soil Biol. Biochem.*, **5**, 97–107.

3. Nitrogen fixation in the rhizosphere of tropical grasses

JOHANNA DOBEREINER & J. M. DAY

Although it has been suggested for more than a decade that dinitrogen fixation of economic importance occurs under tropical grass cover (Parker, 1957; Moore, 1963; Jaiyebo & Moore, 1963; Dobereiner, 1961; 1966), the introduction of the acetylene-reduction method has helped to confirm these observations and to show the close plant–bacteria associations on which fixation depends (Rinaudo, 1970; Rinaudo, Balandreau & Dommergues, 1971; Yoshida & Ancajas, 1971; Dommergues, Balandreau, Rinaudo & Weinhard, 1973; Dobereiner, 1975; Dobereiner, Day & Dart, 1972*a*, *b*). Hardy & Holsten (1973) call these 'associative- or semi-symbiosis'. As most of these findings are confined to tropical grasses which, apart from rice, all possess the C_4-dicarboxylic acid photosynthetic pathway, it has been suggested that the more efficient solar energy conversion of these plants is connected with N_2 fixation on their roots (Balandreau & Villemin, 1973; Dobereiner *et al.*, 1972*a*).

The best known association is that of *Paspalum notatum* with *Azotobacter paspali*. The occurrence of these bacteria is restricted to the distribution of certain ecotypes of this grass (Dobereiner, 1970) and establishment of the bacteria in the rhizosphere takes weeks or months (Dobereiner & Campelo, 1971). In established associations, estimates of maximal N_2 fixation, by the acetylene-reduction method were 250 g N/ha/day (Dobereiner *et al.*, 1972*a*). Within hours, intact systems responded to light, indicating a rapid utilization by N_2-fixing bacteria of photosynthates exuded from the plant. Changes of oxygen tension in the gas phase above the soil from air to 0.04 atm O_2 or vice versa had little effect, but nitrogenase activity on roots removed from the soil was maximal at P_{O_2} around 0.04 atm and less than half under anaerobic conditions or in air. Most of the activity was localized on the roots and was not removed by vigorous washing in water (Dobereiner *et al.*, 1972*a*).

Relatively low root activities were reported initially for some other tropical grasses, including sugar cane, which showed relatively high activity in rhizosphere soil (Dobereiner *et al.*, 1972*b*). Yoshida &

Ancajas (1971) estimated a fixation rate of up to 50 kg N/ha in three months, most of which was fixed in the soil. In-situ studies of acetylene reduction in mixed savannas in the Ivory Coast revealed 10 to 25 nmoles C_2H_4/g roots/hour which correspond with a fixation rate of 7.2 to 16.8 g N fixed/ha/day (root weight 6000 kg/ha) while *Hyparrenia*-dominated savannas gave values of up to 100 nmoles/g compensating for nitrogen lost in fire and erosion in stable natural systems (Balandreau & Villemin, 1973).

The present paper reports high nitrogenase activities on grass roots indicating that several important cultivated forage grasses may obtain most of their nitrogen requirements by biological N_2 fixation. Additional data on the physiology of one such association are also presented.

Materials and methods

Field studies

Root samples for assay of nitrogenase activity were prepared in the following way: whole plants were collected from the field and roots shaken free from soil (rhizosphere soil) and immediately washed in water. Representative roots (about 1 g dry weight) were cut, distributed into 60 ml bottles which were closed with serum caps, evacuated three times and the gas phase replaced by a N_2/O_2 mixture of P_{O_2} 0.04 atm. Root samples were always collected in the afternoon, preincubated overnight without acetylene at P_{O_2} 0.04 atm and 5 % air, and 10 % C_2H_2 injected the next morning when the 2 to 5 hour rates of ethylene production were determined. Washing and overnight preincubation were routine procedures in these studies to avoid drying and to eliminate the 8 to 12 hour lag in acetylene-reduction reported previously (Dobereiner *et al.*, 1972*a*; Rinaudo *et al.*, 1971). All attempts to eliminate this lag, such as preparing roots under nitrogen or water, addition of phosphate buffer or carbon dioxide in various concentrations as well as different oxygen and acetylene concentrations (as suggested by Yoshida, 1971), were unsuccessful. In intact soil–plant systems, acetylene-reduction proceeds linearly without a lag (Dobereiner *et al.*, 1972*a*; Dobereiner & Dommergues, unpublished).

Greenhouse studies

Paspalum notatum was transplanted from the field into Leonard jars with vermiculite, watered with nitrogen-free nutrient solution (Norris, 1964), and grown for two months. Total nitrogen, in the vermiculite and plants at the beginning and at the end of the experiment were determined by the Kjeldahl method (semi-micro with a mercury catalyst).

Intact soil–plant cores of *P. notatum* were cut from the field into inverted Leonard jars (with bottleneck upwards), and placed into the lower part of the Leonard jar where a water level of 2 cm was maintained. For assay the bottleneck was closed with a suba seal and 50 ml C_2H_2 injected. Acetylene reduction was usually measured after 1 h, the suba seal then being removed. Several hours later the same cores could be used again. Preliminary determinations of rates enabled the jars to be sorted into replicates with similar activities thus reducing variability.

Sorghum vulgare was grown from non-sterile seeds for 25 days, in 25 ml test tubes containing sandy soil amended with PK (40 ppm phosphorus and 50 ppm potassium) and minor elements. At each sampling time, four replicate tubes of the intact soil–plant system were placed for 45 minutes into a dark incubator at 35 °C (to minimize algal fixation and equilibrate the temperature), then closed with serum caps. Acetylene (10 %) was then injected and the 1 hour rates of ethylene production determined at 35 °C in the dark.

Physiological studies with washed roots

Roots from the field, collected as described above, were preincubated overnight at P_{O_2} 0.04 atm. Carbon or nitrogen compounds were applied in 0.025 M phosphate buffer (pH 6.8) solution. Five ml were injected into each vial and distributed onto the roots by vigorous hand shaking. Acetylene (10 %) was injected immediately and the time course of ethylene production determined. Handling each set of replicates separately and storing gas samples in the syringes sealed by plunging into rubber bungs, permitted satisfactory timing of up to 6 replicates of 12 treatments at 20 minute intervals.

Pure culture studies

The organisms were grown in nitrogen-free mineral medium (Dobereiner, 1966) in shallow (1 cm) layer batch cultures at 35 °C or on the shaker. For experiments on the effect of added nitrogen compounds, 1 ml of these cultures was injected into 60 ml vials containing 1 ml of the same medium with the compound to be tested. Acetylene was injected immediately and the time course of ethylene production determined. For assays with carbon compounds, or when sequences of different nitrogen compounds were studied, the cells were centrifuged and washed twice in appropriate medium before assay. As in the root experiments, each set of replicates was prepared and timed separately.

Ethylene determinations

Ethylene was determined using a Perkin Elmer F-11 gas chromatograph with hydrogen flame ionization and a 2 m × 0.004 m Porapak-N column at 100 °C and a N_2 gas flow of 25 ml/min. Commercial acetylene was used and the ethylene impurities subtracted.

Results and discussion

Table 3.1 summarizes preliminary experiments of the occurrence of N_2 fixation on the roots of tropical grasses. The list includes most important forage grasses. The values obtained were very variable and higher fixation rates were obtained in moist soils (more than 50 % of field capacity) and during the hot summer months. More data are necessary to assess correlations with climatic conditions. Recent studies of Vlassak, Paul & Harris (1973) on nitrogenase activity of undisturbed soil cores from grasslands in Canada, showed highly significant correlations ($r = 0.8$) with soil moisture although overall rates were much lower (2 kg N/ha/season). A similar correlation was shown by Day, Harris, Dart & van Berkum (Chapter 5) for the more active system of temperate woodland floras. If values presented in Table 3.1 are transformed into kg N/ha (assuming an estimated 5000 kg roots/ha (Throughton, 1957) and the theoretical $C_2H_2:N_2$ conversion factor 3:1), maximum values of about 1 kg N/ha/day are obtained for roots and 0.15 kg for soils (2×10^6 kg soil/ha). These values, although exceptional, are much higher than have been estimated before.

Table 3.1 also shows that most of the nitrogenase activity is concentrated on the roots and little is found in the soil. Washing the roots in water does not remove the activity (Dobereiner *et al.*, 1972*a*); in fact, when relatively dry roots are washed immediately in water, activity is increased, presumably because it stops further drying and diminishes direct oxygen access (Dobereiner, 1975). Many earlier failures to observe nitrogen fixation in the rhizosphere might have been due to assaying the soil rather than the root (Kass, 1970).

The importance of seasonal variations which seems to vary with species is shown on Fig. 3.1. Roots of elephant grass (*Pennisetum purpureum*) were most active in the hot summer months during rapid growth of the

Table 3.1. *Nitrogenase activity on roots and in rhizosphere soil of tropical forage grasses**

Plant species	No. of sites examined**	nmoles C_2H_4/g dry roots/h	nmoles C_2H_4/g dry soil/h
Brachiaria mutica	6	156–730***	0
B. rugulosa (Tannergrass)	3	5–148	—
Hyparrhenia rufa	6	17–29	0–0.148
Digitaria decumbens	5	21–404***	0–0.349
Pennisetum purpureum	5	5–954***	0–0.085
Panicum maximum	5	20–299	0–0.148
Melinis minutiflora	3	13–41	0–0.187
Cynodon dactylon	2	17–269	0–0.068
Paspalum notatum ('Batatais')	6	2–283	0–0.330

* Minimum and maximum mean values (6 root or 3 soil samples). Individual values given by Neves *et al.* (1975). Acetylene-reduction rates (2 to 5 hour rates for roots and 6 to 12 hour rates for soil) measured on extracted roots as described.

** Areas of approximately 10 m², situated on various places of the Campus of the University or at the Experimental Station, Santa Monica, were designated 'sites'.

*** These maximum values were obtained in experiments with high phosphorus fertilization.

grass; after flowering began the activity decreased. This grass produces unexplained high nitrogen gains and does not respond to nitrogenous fertilizer in summer (Rocha & Aronovich, 1972) although improved growth was observed in the cooler winter months. Pangola grass (*Digitaria decumbens*) reacted differently to the seasonal changes; most nitrogenase activity was observed after establishment is complete. Activities of both species and of many others examined (including *P. notatum*) became insignificant when night temperatures below 15 °C became frequent. West (1970) showed that photosynthetic activity of Pangola and some other tropical grasses is impaired after cool nights due to mechanical injury of the chloroplasts by excessive starch accumulation which in warm nights is hydrolysed. Insufficient photosynthesis and lack of growth may also be responsible for cessation of nitrogen fixation. *Digitaria* cultivars have been bred with amylases which are more active in cool nights and which show some growth in winter (West, 1970). Studies of nitrogen fixation on these cultivars will be interesting.

Table 3.2 indicates further possibilities of plant breeding for increasing nitrogen fixation. The nitrogenase activities were relatively low because of the advanced stage of growth, but differences between cultivars of both *P. notatum* and *P. purpureum* are significant. The

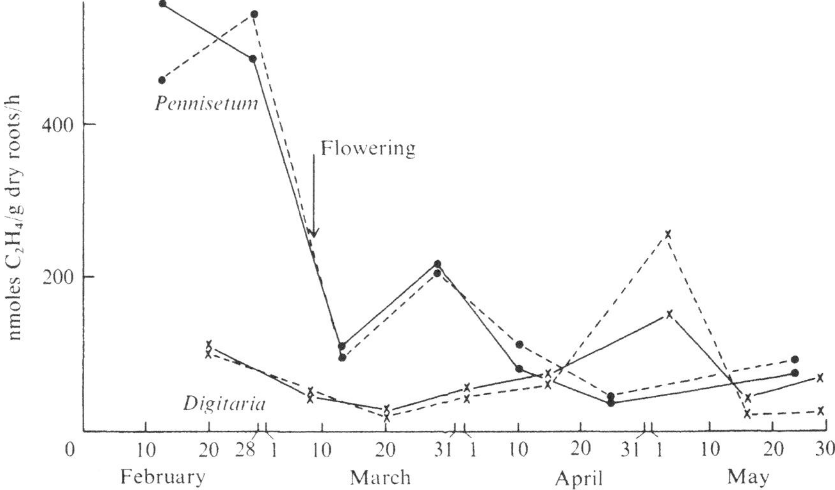

Fig. 3.1. Seasonal variation of nitrogenase activity on roots of *Pennisetum purpureum* and *Digitaria decumbens* grown in the field with (broken line) and without (solid line) nitrogen fertilizer (20 kg N/ha as NH_4NO_3 every two weeks, all applications two weeks before sampling). Points are means of 20 replicate measurements, four samples of five plots each.

Table 3.2. *Differences between cultivars and ecotypes of* P. notatum *and* P. purpureum *in relation to nitrogenase activity on their roots**

Paspalum notatum		*Pennisetum purpureum*	
Cultivar or ecotype	nmoles C_2H_4/ g roots/h	Cultivar	nmoles C_2H_4/ g roots/h
'Batatais' Km 47	39.8	Taiwan A-143*	33.6
'Batatais' Matao	21.0	Taiwan A-144**	60.2
Pensacola Florida	10.3	Puva Napier No. 1	15.8
Pensacola Argentine	12.2	Gigante de Pinda**	93.3
Tiften hybrid	11.4	Elefante de Pinda	33.2
Pensacola hybrid	6.7		
Diff. significant at	$p = 0.05$		$p = 0.01$

* Nitrogenase activity determined on extracted roots obtained in field experiments.
** Cultivars which had not started flowering while all others were in flower.

native 'batatais' types and some of the *P. purpureum* cultivars were more active than others where systematic breeding for increased response to mineral nitrogen fertilizer may have reduced the capacity for nitrogen fixation.

The challenge of characterizing optimal conditions in order to direct agricultural practices towards this goal is especially great in the tropics where the potential energy input by high light intensities and optimal temperatures and their efficient use by the C_4-dicarboxylic acid photosynthetic pathway of most grasses should lead to high rates of nitrogen fixation. High nitrogen fertilizer prices, and losses due to leaching make nitrogen input from rhizosphere nitrogen fixation more significant than in temperate zones. The large nitrogen gains under tropical grass cover such as those cited by Moore (1963) and Jaiyebo & Moore (1963) in Nigeria or those resulting in maize yield increases from 1100 to 4300 kg/ha under *Paspalum notatum* pastures in Southern USA (Beaty, 1973, personal communication) may be explained by rhizosphere nitrogen fixation.

Greenhouse studies with intact systems

Formerly nitrogen fixation by legumes or similar systems was measured as increase in total nitrogen in pot experiments. Because of the difficulty in establishing and maintaining the *P. notatum–A. paspali* association under artificial conditions (Machado & Dobereiner, 1969; Kass, 1970), plants of *P. notatum* were transplanted directly from the field into vermiculite and provided with a nitrogen-free nutrient solution. The nitrogen balances of these pots are presented in Table 3.3. Nitrogen gains comparable with those of legumes are shown. Plants grew well, were green and healthy, and the addition of 30 ppm nitrogen as NH_4NO_3 did not increase growth or percentage nitrogen content. If it is assumed that the 25 mg nitrogen fixed in the pots with dead plants were due to fixation by blue-green algae and photosynthetic bacteria, the grass association was responsible for the fixation of 84 mg nitrogen in two months.

Two other greenhouse experiments were carried out to complement the observations of the effects of seasonal changes by comparing daily fluctuations of nitrogenase activity on intact systems with the residual activity on washed roots collected in the afternoon, as in the field studies. For the first experiment, soil–plant cores of *P. notatum* were collected in the field and allowed to equilibrate for 24 hours in the greenhouse. The nitrogenase activity of 12 such cores was measured to group them in uniform replicates. Acetylene-reduction rates were measured over a period of one hour alternately on three different groups (four replicates each), every two or three hours; the bottlenecks were unstoppered after each gas sampling. In this way each set of four replicates had at least five hours in open bottles before the next one hour under

Table 3.3. *Nitrogen contents and nitrogen gains in the* Paspalum notatum–Azotobacter paspali *association grown in pot cultures* (mg N/pot)

	At transplant	Two months after transplant: Dead plants*	Two months after transplant: Growing plants**
Vermiculite	13.05	40.80	64.40***
Plants	5.47	3.28	63.87
Total	18.52	44.08	128.47
N_2 fixed	—	25.54	109.75

* Mean of 6 pots with dead plants.
** Mean of 33 pots with actively growing plants.
*** Included many fine roots which were impossible to separate from the vermiculite.

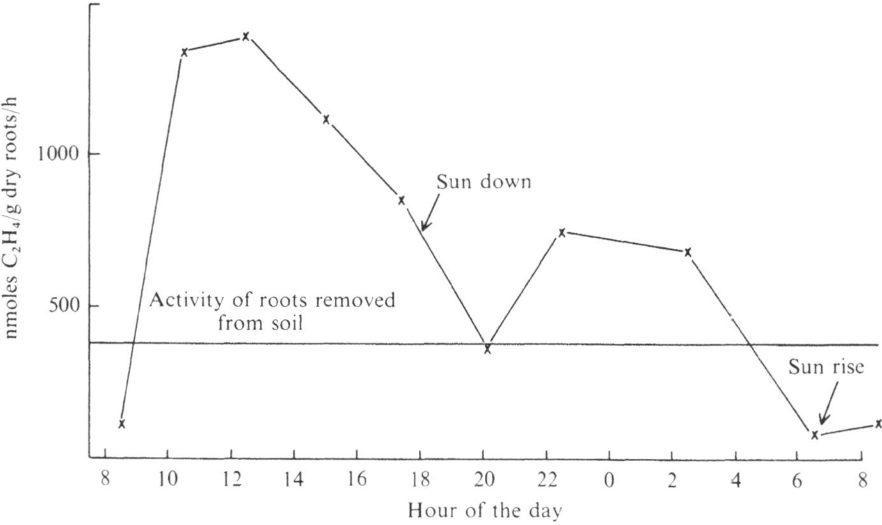

Fig. 3.2. Daily cycle of nitrogenase activity in intact soil plant cores with *Paspalum notatum* in comparison with the activity of the same roots extracted at 4 p.m., pre-incubated overnight at P_{O_2} 0.04 atm and acetylene reduction measured between 8 and 10 a.m. Points are means of four cores, soil activity being subtracted. The horizontal line represents the mean activity of extracted roots of all pots.

acetylene. Results are presented in Fig. 3.2. After nine such measurements during a 24 hour cycle, all roots were washed out the following afternoon and residual nitrogenase activity determined after pre-incubation overnight as described for the field assays. The mean value for all pots is shown for comparison in Fig. 3.2. The soil activity of each pot was also measured and subtracted from the values of the intact system. The results show a pronounced diurnal variation with one maximum at midday and another at night. The peak of activity at midday was more than three times higher, and that at night twice as high as the activity of washed roots. This residual activity is probably maintained by root reserves or by storage material within the bacteria. Sugar-grown *A. paspali* cells transferred into a sugar-free medium continued to show nitrogenase activity, after a short lag, for at least 12 hours. Only during the early morning hours was the activity of the intact system lower than the residual activity. Overall nitrogenase activity of washed roots showed only 60 % of the total daily activity of intact systems. Very similar two-peak nitrogenase activity curves were observed by Balandreau & Villemin (1973) and Balandreau & Dommergues (1973*b*) in in-situ measurements in Ivory Coast savannas. These authors suggest that solubilization of starch granules which accumulate in the chloroplasts of C_4 plants during the day, and which are metabolized during the night (West, 1970), may be responsible for the night peaks.

Fig. 3.3 shows results of a greenhouse experiment with sorghum grown in test tubes, which confirm the dependence of nitrogenase activity on plant photosynthesis. Rates of acetylene reduction were measured for one hour in four tubes. One week later the measurements were repeated with the same tubes. The first curve which was taken on a cloudy day reached its maximum at 3 p.m. while the second, which was taken on a sunny day, showed most activity at 1 p.m. The lower overall activity of the second sampling indicates that the plants may have been injured at the previous harvest. No night measurements were taken for the first curve but the second confirms the two-peak effect described above. Additional measurements on two sets of tubes which had their plant tops removed show more rapid increase of activity for two hours early in the morning, possibly because in intact plants, carbon substances are exported from roots to tops. This could also explain the low morning activity in the *Paspalum* experiment.

The pronounced and rapid effect of photosynthetic activity of the plant on nitrogenase activity on the roots indicates a closer relationship between higher plants and bacteria than can be explained by normal rhizosphere exudation. A close dependence on photosynthesis and the failure of washing to remove nitrogenase activity, indicates that the micro-organisms responsible for fixation are localized directly on the root, possibly within the *mucagel* layer as suggested for the

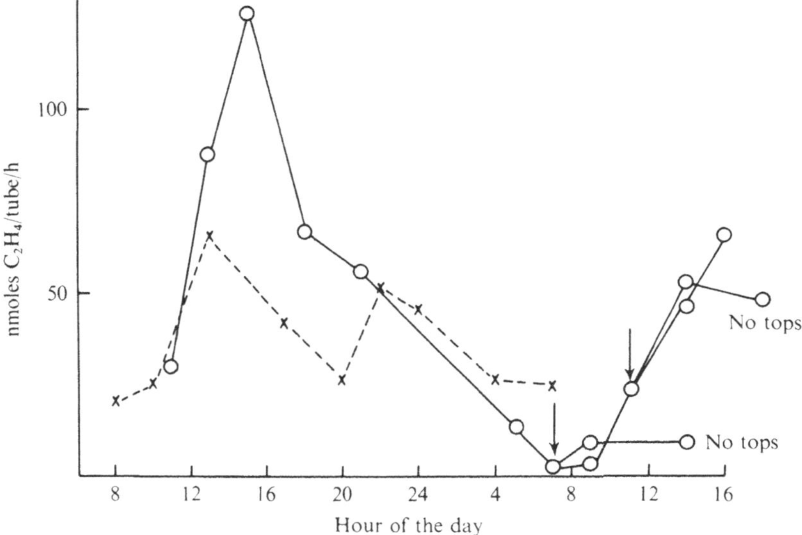

Fig. 3.3. Daily cycle of nitrogenase activity in intact soil plant systems of *Sorghum vulgare* in test tubes. Full line represents measurements on a cloudy day and dotted line measurements on a sunny day, one week later on the same systems. At arrows, on additional tubes, the plant tops were removed. Each point represents the mean of four replicate tubes.

Paspalum association (Dobereiner *et al.*, 1972*a*). The day–night variations resemble the curve presented by Bergersen (1970) for daily variations in nitrogenase activity of nodulated soybeans.

Physiological studies with washed roots

Laboratory experiments were carried out with washed *P. notatum* roots to obtain information about the availability of carbon substances for nitrogen fixation in well-illuminated plants. Figure 3.4 shows that nitrogenase activity of roots removed from the soil in the afternoon from normally illuminated plants (during a period of several sunny days), was three times higher and was increased less by the addition of sugars than that of roots from plants kept in the dark for 48 hours. This suggests that after sunny days, carbon substances are not the major limiting factor for nitrogen fixation in the rhizosphere. In a further experiment (Table 3.4) nitrogenase activity of *A. paspali* on roots, rhizosphere soil, and root surface soil, was compared at the time of harvest with that of the same system after 40 hours of incubation on nitrogen-free, 1 %-sucrose, agar medium to allow multiplication of the bacteria. *A. paspali* colonies formed on almost every grain of soil while

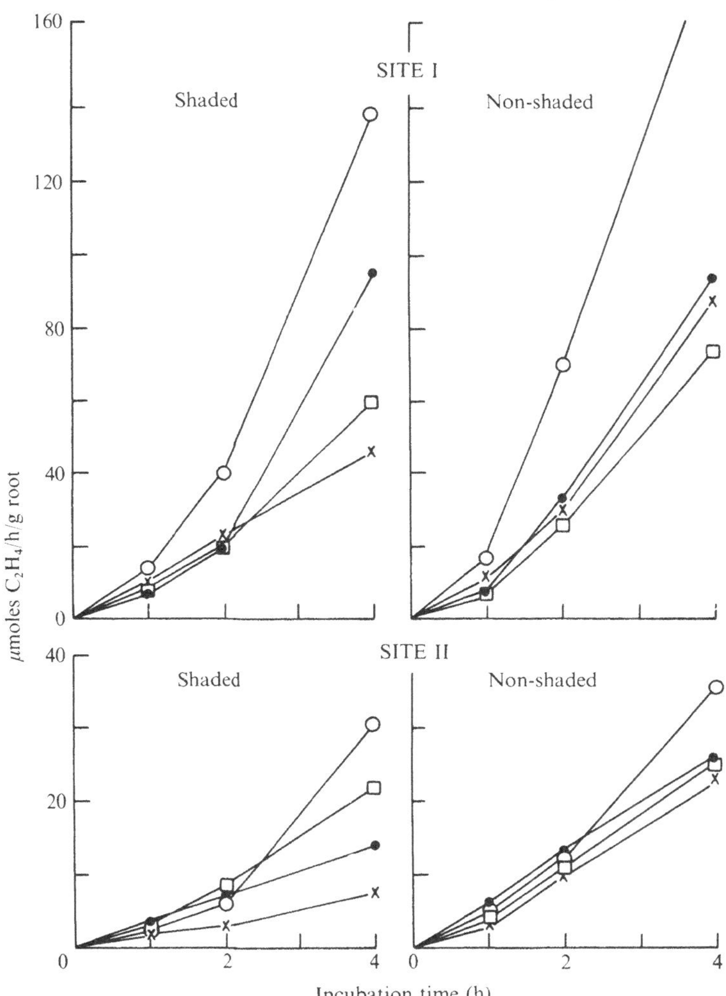

Fig. 3.4. Effect of sugar application on nitrogenase activity on roots of *P. notatum* extracted from the field (full sunlight or covered for 48 hours with a wooden box). Applications of sugar solutions (0.5 % in 0.025 M phosphate buffer pH 6.8) and acetylene at time zero. Points represent means of four replicates. ○, Sucrose; ●, arabinose; □, succinate; ×, control.

only a few root pieces produced colonies. Similar observations by Dobereiner *et al.* (1972*a*) also suggested that root *A. paspali* exposed to air do not multiply in contrast to soil organisms adapted to higher oxygen tensions which grew easily. For this reason, in the present experiment, three oxygen tensions were used for incubation and acetylene

Table 3.4. *Increases in nitrogenase activity after 40 h incubation on N-free sucrose medium of* P. notatum *roots, root surface or rhizosphere soil**

P_{O_2} during incubation and assay (atm.)		Initial nitrogenase activity (nmoles C_2H_4/g/h)	Activity after 40 h incubation on medium (nmoles C_2H_4/g/h)
Near zero	Rhizosphere soil	0.1	13 200
	Root surface soil	2.3	21 450
	Washed roots	103.6	2 046
0.04	Rhizosphere soil	0.3	33 000
	Root surface soil	11.5	30 800
	Washed roots	282.9	4 800
0.20	Rhizosphere soil	0.1	5 720
	Root surface soil	0.5	7 920
	Washed roots	79.5	1 210

* Twenty mg finely cut roots or sieved soil were scattered on nitrogen-free mineral salt sucrose agar slants in 60 ml bottles and incubated at 35 °C during 40 hours at the given P_{O_2}. Two hour acetylene-reduction rates were determined on samples of the original material and after the incubation. Almost every grain of soil formed a colony of *A. paspali* but very few root pieces did (means of 3 replicates).

reduction assays. Table 3.4 shows that rhizosphere soil nitrogenase activity was increased by a factor of 10^5 by incubation and root surface soil activity by 10^4 when root activity increased only 20 times. Oxygen tension, although affecting nitrogenase activity, as expected, did not alter this relationship. It seems therefore that other factors inhibit the multiplication of *A. paspali* on roots placed on culture medium.

To investigate further differences between in-vivo and in-vitro systems, the nitrogenase activities of pure cultures of *A. paspali* and those from washed roots were compared. Figure 3.5 shows that glycolate and acetate completely inhibited nitrogenase activity by cultures in 1.5 h but reduced that on roots only by 50 %. Citrate reduced culture activity to half but increased root activity after two hours, while succinate had no effect on cultures or roots. The inhibition of nitrogenase activity by glycolate may be related to the findings of Kurz & La Rue (1973) who observed no growth of *A. chroococcum* on nitrogen-free medium when glycolate was the carbon source, while cultures supplied with inorganic nitrogen salts grew on glycolate; acetate was used even by nitrogen-grown organisms. Test tube experiments with *A. vinelandii* by Peña & Dobereiner (unpublished) confirmed this but *A. paspali* growth was inhibited by glycolate and acetate even in the presence of mineral nitrogen. *Beijerinckia* spp. showed similar inhibition but two

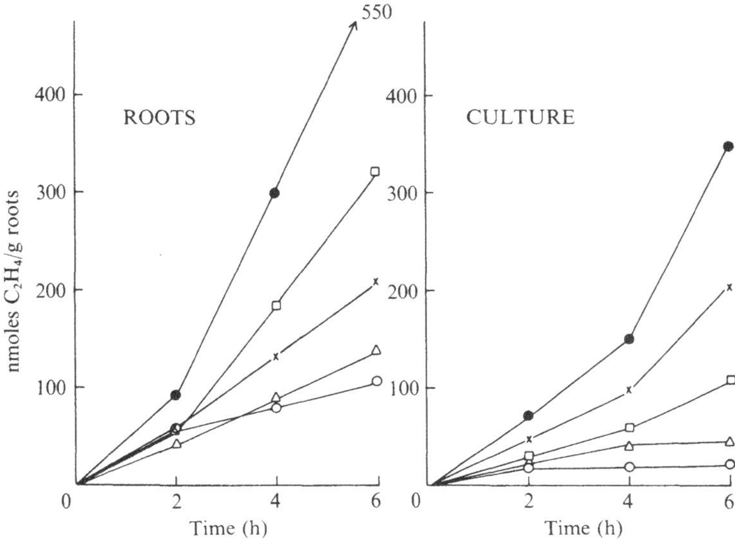

Fig. 3.5. Effects of organic acids on nitrogenase activity on *P. notatum* roots and pure cultures of *A. paspali*. Application of carbon compounds (0.5 % solutions in 0.025 M phosphate buffer, pH 6.8) and acetylene at time zero. For pure culture assays, 1 ml of washed cell suspensions (200×10^6 cells/ml) obtained from shallow-layer batch cultures was injected into 60 ml bottles at P_{O_2} 0.04 atm. containing 1 ml double strength substrate in medium (1 %), at the time of acetylene injection (time zero). All points are means of four replicates. ●, Sucrose; □, citrate; △, glycolate; ○, acetate; ×, control.

out of five *Derxia* strains grew on glycolate even in nitrogen-free medium.

The differential effect of nitrate is seen in more detail in Fig. 3.6. *A. paspali* nitrogenase remained unaffected by nitrate concentrations of up to 80 mM (unpublished) but inhibition of root activity by 10 mM NO_3^- was greater than inhibition by 10 mM NH_4^+. A possible explanation is that during NO_3^- uptake by roots there is a simultaneous efflux of NO_3^- out of the root (Morgan, Volk & Jackson, 1973). This efflux could be accompanied by traces of NO_2^- if the reduction in the roots was incomplete. Nitrite is very toxic to nitrogenase (Rigaud, Bergersen, Turner & Daniel, 1973). The inhibition of nitrogenase activity in *A. paspali* by NO_2^- is shown in Table 3.5. In the intact soil–plant system, reduction of nitrate to ammonia seems complete as the nitrate effect is similar to that of ammonia and shows up only two hours later (Fig. 3.6). The lack of a nitrate reductase in *A. paspali* was confirmed in a further experiment (Fig. 3.7) where it was shown that 10 mM NO_3^- did not affect nitrogenase activity of cells previously grown on N_2 or

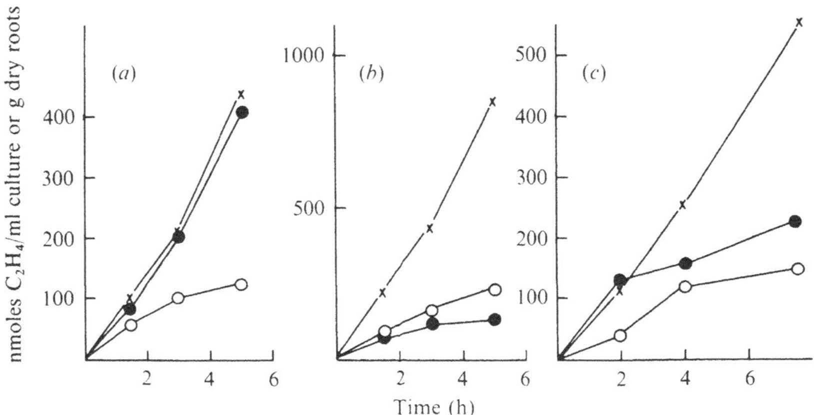

Fig. 3.6. Effects of ammonia and nitrate on nitrogenase activity of *A. paspali* in pure culture (*a*), on roots of *P. notatum* removed from soil (*b*) and intact systems with *P. notatum* (*c*). For pure culture assays, 1 ml of shallow layer batch culture (56.3×10^6 cells/ml) was injected into 60 ml bottles containing 1 ml of 20 mM NH_4^+ or NO_3^- in medium at the time of acetylene injection (time zero). To root samples, 5 ml of 10 mM NH_4^+ or NO_3^- in 0.025 M phosphate buffer (pH 6.8) and acetylene were applied at time zero. Intact soil–plant cores received 11.2 ppm N (soil basis) in solution at time zero. All nitrogen sources were $(NH_4)_2SO_4$ or KNO_3. All points are means of four replicates. ○, NH_4^+; ●, NO_3^-; ×, control.

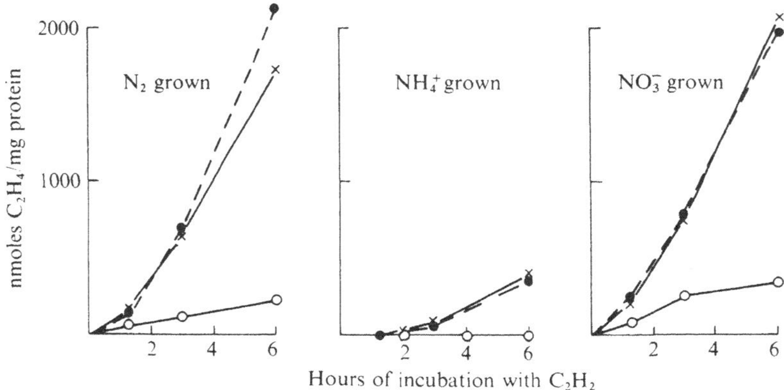

Fig. 3.7. Effects of ammonia and nitrate on nitrogenase activity of *A. paspali* grown in pure culture on N_2, ammonia or nitrate. One ml of washed cell suspensions, obtained from shake cultures grown with 10 mM $(NH_4)_2SO_4$ or KNO_3, or on N_2 (1.32, 0.44, 0.51 mg protein/ml resp.), was injected into 60 ml bottles with air, containing 1 ml of 20 mM NH_4^+ or NO_3^- in medium, at time zero. All points are means of three replicates. ○, NH_4^+; ●, NO_3^-; ×, control.

Table 3.5. *Effect of nitrite on growth and nitrogenase activity of* Azotobacter paspali

Concentration of NO_2^- (mM)	Cells grown on NO_2^-: Relative growth measured photometrically	Cells grown on NO_2^-: Ethylene production (nmol C_2H_4/ml)	Cells grown on N_2 with NO_2^- added at beginning of C_2H_2-reduction assay: Ethylene production (nmol C_2H_4/ml)
0.0	100*	74.3	67.8
0.1	89	66.3	68.0
0.2	58	45.7	65.1
0.5	24	18.4	61.2
1.0	4	0	31.6
2.5	0	0	0
5.0	0	0	0

* Control contained 90×10^6 cells/ml culture.

NO_3^-. Also nitrogenase in NH_4^+-grown cells was derepressed as rapidly in the presence as in the absence of NO_3^-. Green & Wilson (1953) observed a similar failure of a strain of *A. agilis* to utilize nitrate. This has been confirmed with a number of strains of *A. paspali* and this differentiates it from *A. chroococcum* which shows similar inhibition of nitrogenase activity and repression of nitrogenase synthesis by ammonia and nitrate (Drozd, Tubb & Postgate, 1972).

Studies on the effects of ammonia and nitrate on nitrogenase activity of three other tropical nitrogen-fixing bacteria revealed substantial differences between species (Peña & Dobereiner, 1974). In *Derxia gummosa* and *A. vinelandii* which apparently possess a very active nitrate reductase, nitrogenase activity stopped completely, three hours after application of 10 mM nitrate. In *Beijerinckia indica* nitrogenase activity continued, although at a slower rate, in the presence of 10 mM NO_3^- and in *B. fluminensis* nitrogenase activity was not affected by nitrate over a six hour period. In growth experiments, all three organisms, in contrast to *A. paspali*, showed at least partial repression of nitrogenase by 10 mM NO_3^-. These differences between tropical nitrogen-fixing bacteria and the classical *Azotobacter* spp. might be important in rhizosphere establishment and nitrogen fixation.

These observations might be of considerable ecological importance as the lack of a nitrate reductase enables these organisms to continue to fix nitrogen even at relatively high rates of nitrification in the soil, as long as the plants assimilate the nitrate without substantial exudation of nitrite or ammonia. In fact, when measurements of nitrogenase activity in the experiment with intact soil cores (Fig. 3.6*c*) were repeated after one week, the effect of nitrate had disappeared. Continued nitrogenase activity was also found after repeated small applications of nitrogen fertilizer (see Fig. 3.1) where neither *P. purpureum* nor *D. decumbens* nitrogenase activities were affected even after eight applications of 20 kg N/ha as NH_4NO_3, two weeks before sampling. Larger single applications are reported to inhibit nitrogenase activity on rice (Yoshida, 1971) and maize roots (Balandreau & Dommergues, 1973*a*).

Conclusion

Paspalum notatum has been shown to benefit appreciably from nitrogen fixation by associated micro-organisms. *Pennisetum* and many other tropical forage grasses have been shown to support similar levels of microbial fixation but to what extent the higher plant benefits from this awaits investigation using ^{15}N. Further study is also required to localize and identify the organisms responsible for this fixation and novel methods of culturing these nitrogen fixers must be developed to gain an insight into aspects of their biochemistry and physiology.

References

Balandreau, J. & Dommergues, Y. (1973*a*). Assaying nitrogenase (C_2H_2) activity in the field. *Bull. Ecol. Res. Comm.* (Stockholm), **17**, 247–54.

Balandreau, J. & Dommergues, Y. (1973*b*). Nycothermal variations in non-symbiotic nitrogen fixation in the rhizosphere. *Ann. Meet. Am. Soc. Microbiol.*, **73**, no. 7.

Balandreau, J. & Villemin, G. (1973). Fixation biologique de l'azote moléculaire en savanne de Lamto (Basse Cote d'Ivoire). Résultat préliminaires. *Revue Ecol. Biol. Sol*, **10**, 25–33.

Bergersen, F. J. (1970). The quantitative relationship between nitrogen fixation and the acetylene reduction assay. *Aust. J. biol. Sci.*, **23**, 1015–25.

Dobereiner, J. (1961). Nitrogen fixing bacteria of the genus *Beijerinckia Derx* in the rhizosphere of sugar cane. *Pl. Soil*, **14**, 211–17.

(1966). *Azotobacter paspali* n.sp. uma bacteria fixadora de nitrogenio na rhizofera de Paspalum. *Pesquisa Agropecuaria Brasileira*, **1**, 357–65.

(1970). Further research on *Azotobacter paspali* and its variety specific occurrence in the rhizosphere of *Paspalum notatum* Flugge. *Zentbl. Bakt., ParasitKde, Abt. II*, **124**, 224–30.

(1975). Nitrogen fixation in the rhizosphere of tropical forage grasses. *Global Impacts of Applied Microbiology*, IV. São Paulo, Brazil (in press).

Dobereiner, J. & Campelo, A. B. (1971). Non-symbiotic nitrogen-fixing bacteria in tropical soils. *Plant and Soil, Special Volume*, pp. 457–70.

Dobereiner, J., Day, J. M. & Dart, P. J. (1972*a*). Nitrogenase activity and oxygen sensitivity of the *Paspalum notatum–Azotobacter paspali* association. *J. gen. Microbiol.*, **71**, 103–16.

Dobereiner, J., Day, J. M. & Dart, P. J. (1972*b*). Nitrogenase activity in the rhizosphere of sugar cane and some other tropical grasses. *Pl. Soil*, **37**, 191–6.

Dommergues, Y., Balandreau, J., Rinaudo, G. & Weinhard, P. (1973). Non-symbiotic nitrogen fixation in the rhizosphere of rice, maize and different tropical grasses. *Soil Biol. Biochem.*, **5**, 83–9.

Drozd, J. W., Tubb, R. S. & Postgate, J. R. (1972). A chemostat study of the effect of fixed nitrogen sources on nitrogen fixation, membranes and free amino acids in *Azotobacter chroococcum*. *J. gen. Microbiol.*, **73**, 221–32.

Green, M. & Wilson, P. W. (1953). The utilisation of nitrate by the *Azotobacter*. *J. gen. Microbiol.*, **65**, 511–17.

Hardy, R. W. F. & Holsten, R. D. (1973). Global nitrogen cycling: pools, evolution, transformations, transfers, quantitations and research needs. In *The Aquatic Environment: Microbial transformations and water quality management implications* (eds. Ballantyne, R. K. and Guarraia, L. J.). Washington, D.C.

Jaiyebo, E. O. & Moore, A. W. (1963). Soil nitrogen accretion under covers in a tropical rain forest. *Nature, Lond.*, **197**, 317–18.

Kass, D. C. (1970). Sources of nitrogen in tropical environments with special reference to the relationship between *Paspalum notatum* Flugge and *Azotobacter paspali* Dobereiner. M.S. thesis. Cornell University.

Kurz, W. G. W. & LaRue, T. A. G. (1973). Metabolism of glycolic acid by *Azotobacter chroococcum* PRLH 62. *Can. J. Microbiol.*, **19**, 321–4.

Machado, W. C. & Dobereiner, J. (1969). Estudos complementares sobre a fisiologia de *Azotobacter paspali* e sua dependencia da planta. *Pesquisa Agropecuaria Brasileira*, **4**, 53–8.

Moore, A. V. (1963). Nitrogen fixation in a latosolic soil under grass. *Pl. Soil*, **19**, 127–38.

Morgan, M. A., Volk, R. J. & Jackson, W. A. (1973). Simultaneous influx and efflux of nitrate during uptake by perennial ryegrass. *Pl. Physiol.*, **51**, 267–72.

Neves, M. C. P., Day, J. M., Corneiro, A. M. & Dobereiner, J. (1975). Atividade da nitrogenase na rhizosfera de gramineas tropicais forrageiras. *Global Impacts of Applied Microbiology IV*, São Paulo, Brazil (in press).

Norris, D. O. (1964). *Legume bacteriology*. Commonwealth Bureau of Pastures and Field Crops, *Bulletin* no. **47**, 102–17.

Parker, C. A. (1957). Non-symbiotic nitrogen fixing bacteria in soil. III. Total nitrogen changes in a field soil. *J. Soil Sci.*, **8**, 48–59.

Peña, J. J. & Dobereiner, J. (1974). Efecto del nitrato y amonio en la actividad de la nitrogenasa de bacterias tropicales fijadoras de nitrogeno atmosferico. *Rev. lat-amer. Microbiol.*, **16**, 33–44.

Rigaud, J., Bergersen, F. J., Turner, G. L. & Daniel, R. M. (1973). Nitrate dependent anaerobic acetylene reduction and nitrogen fixation by soybean bacteroids. *J. gen. Microbiol.*, **77**, 137–44.

Rinaudo, G. (1970). Fixation biologique de l'azote dans trois types de sols de rizières de Côte d'Ivoire. *Thèse de Docteur Ingénieur*. Faculté des Sciences, Montpellier.

Rinaudo, G., Balandreau, J. & Dommergues, Y. (1971). Algal and bacterial non-symbiotic nitrogen fixation in paddy soils. *Plant and Soil, Special Volume*, pp. 471–9.

Rocha, G. L. da & Aronovich, S. (1972). Informe regional sobre problemas, atividades e programas recentes de desenvolvimento no campo dos pastos e plantas forrageiras. *Zootechnia, São Paulo*, **10**, 15–62.

Throughton, A. (1957). The underground organs of herbage grasses. *Commonwealth Bureau of Pastures and Field Crops, Bulletin no.* **44**, 1–163.

Vlassak, K., Paul, E. A. & Harris, R. E. (1973). Assessment of biological nitrogen fixation in grassland and associated sites. *Pl. Soil*, **38**, 637–49.

West, S. H. (1970). Biochemical mechanism of photosynthesis and growth depression in *Digitaria decumbens* when exposed to low temperatures. In *Proceedings 11th International Grasslands Congress, Australia*, pp. 514–17.

Yoshida, S. (1971). *Annual Report of the International Rice Research Institute*. IRRI: Los Baños, Philippines.

Yoshida, Y. & Ancajas, R. R. (1971). Nitrogen fixation by bacteria in the root zone of rice. *Proc. Soil Sci. Soc. America*, **35**, 156–7.

4. Nitrogen fixation in the rhizosphere of rice plants

J. BALANDREAU, G. RINAUDO, IBTISSAM FARES-HAMAD & Y. DOMMERGUES

The maintenance of soil fertility in irrigated paddy soils with no fertilizer addition has been attributed largely to the input of nitrogen by blue-green algae (e.g. Watanabe, Nishigaki & Konishi, 1951; Moore, 1966). However Ishizawa, Suzuki & Araragi (1970), Rinaudo (1970), Rinaudo, Balandreau & Dommergues (1971), Yoshida & Ancajas (1971; 1973) and Dommergues, Balandreau, Rinaudo & Weinhard (1973) showed that rhizosphere bacteria may also be responsible for significant nitrogen (N_2) fixation. The field studies reported here were not specifically designed to isolate the role of blue-green algae from that of rhizosphere bacteria; their main object was to evaluate as accurately as possible the actual N_2 fixation rate in paddy soils, to study its variations, and compare it to that of other ecosystems. Concurrently laboratory experiments were conducted in order to check N_2 fixation and to study the effects of some environmental factors such as air humidity, NH_4^+ content of the soil and diffusion of gases through the soil–plant system. Some preliminary investigations on rhizosphere bacteria were also carried out.

Materials and Methods

In-situ assays

The procedure was as described previously (Balandreau & Dommergues, 1973). Nitrogenase activity was measured 8–11 times per 24 h period, with four or five replications; each incubation was performed on a different plant.

Laboratory rice culture

Small culture-tubes containing soil were used, each of which contained 10 to 15 g of soil (dry weight). Rice seeds were surface sterilized with hydrogen peroxide, sterility was checked and, after germination, seeds were transferred to the small culture-tubes (one seed per tube), and were placed in a growth cabinet where environmental conditions were as follows: illumination 30000 lx, 14 h illumination per day; temperature 28 ± 3 °C; humidity 70–90 %. The rice soils were waterlogged.

Laboratory assays, long-term acetylene incubations

Experiments on the effect of NH_4^+ concentrations in soil on nitrogen fixation were carried out as follows: Samples of rice with attached soil were withdrawn from their culture-tubes, and placed in serum bottles which were evacuated and then flushed twice with argon; up to 10 % C_2H_2 was added, then the atmosphere was analysed for ethylene at regular intervals throughout a 24 h period. The observed maximum rate of acetylene reduction was used to estimate nitrogenase activity.

Laboratory assays, short-term acetylene incubations

Each rice seedling was covered by an inverted tube connected to the culture-tube by a piece of rubber tubing. One ml propane and 10 % overpressure C_2H_2 were added with a syringe. Propane was used as a tracer gas to allow the calculation of the total gas volume as in the in-situ assays. Incubations did not last more than three hours with two or three gas measurements.

Expression of nitrogen fixation. Results of laboratory assays were expressed as nmoles or μmoles C_2H_4/h/tube or C_2H_4/h/g dry soil. When possible, in-situ and laboratory results were related to root dry weight and expressed as nmoles or μmoles C_2H_4/h/g dry root; some in-situ results were extrapolated in terms of g N_2/day/ha or kg N_2/100 days/ha.

Soils and plant varieties

The main characteristics of the soils are summarized in Table 4.1. Two rice varieties were used: Taichung native 1 in the field experiment; IR8 in all laboratory experiments. Other plants studied in comparison with rice belonged to the following varieties: peanut: local variety; *Panicum maximum*: K 187; rye-grass: Viktoria.

Environmental conditions relating to field experiments

Paddy field. The investigated rice field located near Lamto (Ivory Coast) was made up of seedlings mixed with tillers originating from the previous crop. No fertilizer had been applied. Soil was saturated but not waterlogged. Algae could be seen but they were not abundant enough to form an algal layer. The assay was performed on a very hot sunny day. Air temperature was 45 °C and humidity 55 % at 3.30 p.m. (Fig. 4.1).

Peanut plot. This plot was located at Lamto (Ivory Coast); seedlings which were three weeks old, were densely nodulated. There were three shoots per cluster. The assay was performed on a hot day: temperature was above 40 °C at 2.00 p.m. Humidity was below 50 % around 4.00 p.m. A shower occurred at 4.30 p.m. (Fig. 4.1).

Panicum maximum. This plot was a part of an experimental grassland at Orstom station, Adiopodoumé (Ivory Coast). No manure had been applied for several months. The assay was performed on a cloudy, warm day (Fig. 4.1).

Rye-grass. This plot was a part of an experimental grassland at E.N.S.A.I.A. experimental station, La Bouzule, Eastern France. The assay was performed in May on a warm day; a shower occurred at 5.30 p.m., the night was very cool (Fig. 4.2).

Counts of N_2-fixing bacteria

The usual, most probable, number method was followed. The presence of aerobes was detected in tubes which had been incubated under air, using the acetylene-reduction test (Balandreau & Villemin, 1973). The presence of anaerobes was detected in tubes which had been incubated under nitrogen, by the change in colour of phenosafranin according to Brouzes, Mayfield & Knowles (1971).

Results

Estimation of N_2 fixation rates

Field estimations. Diurnal variations in N_2 fixation occurring in the paddy field were compared with those occurring in the two other tropical ecosystems in similar climatic conditions: the peanut and the *Panicum maximum* plot (Fig. 4.1.) Expressed on a dry root weight basis, maximum nitrogenase activities were respectively: 23.9, 35.0, 0.5 μmoles C_2H_4/h in the rice, peanut and *P. maximum* plots. Corresponding integrated daily values expressed in terms of nitrogen fixed on a per

Table 4.1. *Characteristics of the soils investigated*

	pH	C (%)	N (%)	C/N	Clay (%)
Paddy soils					
Djibelor (Sénégal)	4.3	3.8	0.31	12.2	33
Mas du Sauvage (France)	7.8	1.2	0.15	8.0	20
Lamto (Ivory Coast)	5.6	1.0	0.07	14.3	24
Other soils					
La Bouzule (France)	6.8	1.6	0.15	10.6	27
Lamto (Ivory Coast)	6.0	0.83	0.07	11.8	5

hectare basis have been calculated and published elsewhere (Balandreau, Millier & Dommergues, 1974); these values were respectively 32 kg, 26 kg and 0.8 kg N_2/100 days/ha.

Laboratory estimations. The pattern of diurnal variations in simulated day–night conditions (Fig. 4.3) was very similar to the field pattern (Fig. 4.1); however the maximum nitrogen fixation rate was 15 times lower: 1600 instead of 24000 nmoles C_2H_4/h/g dry root. This discrepancy was attributed mostly to the fact that illumination was less intense in the growth cabinet than in the field.

Laboratory studies of some factors affecting nitrogen fixation

Air humidity. Different humidity levels (92.5 %, 74.4 % and 43.7 %) were established in closed chambers by using saturated salt water solutions [$(NH_4)H_2PO_4$, NaCl and K_2CO_3, respectively]; a decrease in humidity from 92.5 % to 43.7 % induced a decrease of N_2 fixation rate from 1920 to 540 nmoles C_2H_4/h/g dry root (Fig. 4.4).

Ammonium levels in the soil. Applying increasing amounts of ammonium sulphate to the tubular soil at seeding time affected nitrogen fixation measured when the plants were 16 days old. With addition of up to 40 ppm, N_2 fixation was stimulated, but higher applications induced a marked decrease (Fig. 4.5). Estimations of ammonium content performed at the same time showed that all the ammonium added had been used up by the plants or immobilized by soil micro-organisms (Fig. 4.6).

Diffusion of gases into the soil–plant system. In order to check the role of rice in the diffusion of gases from the atmosphere to the soil, an

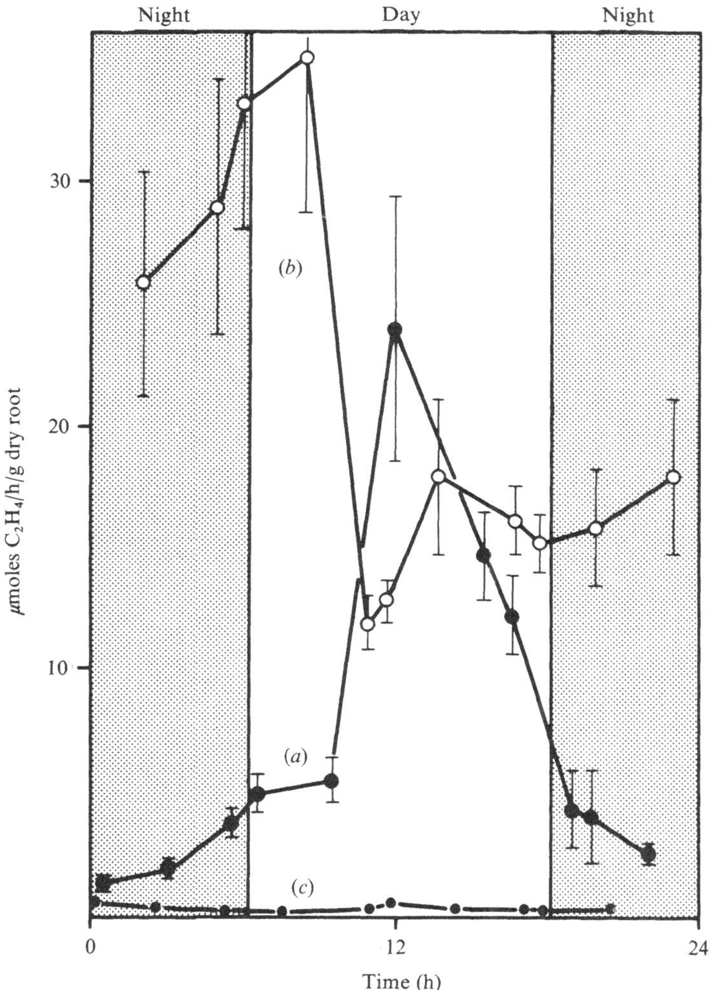

Fig. 4.1. Diurnal variations of nitrogenase activity in: (*a*) a rice field; (*b*) a peanut field; (*c*) a *Panicum maximum* field. Limits show standard error of means (from Balandreau *et al.*, 1974).

experiment was set up, using the waterlogged Mas du Sauvage soil, placed in U-shaped tubes. Each U-shaped soil column was 1.8 cm in diameter and approximately 27 cm long excluding the 1 cm water layer at each side. The following treatments were applied: (*a*) Both arms of the tube were planted with a germinated rice seed and the whole device was submitted to the regular day–night illumination; (*b*) both arms of the tube were planted with a rice seedling but the right arm was

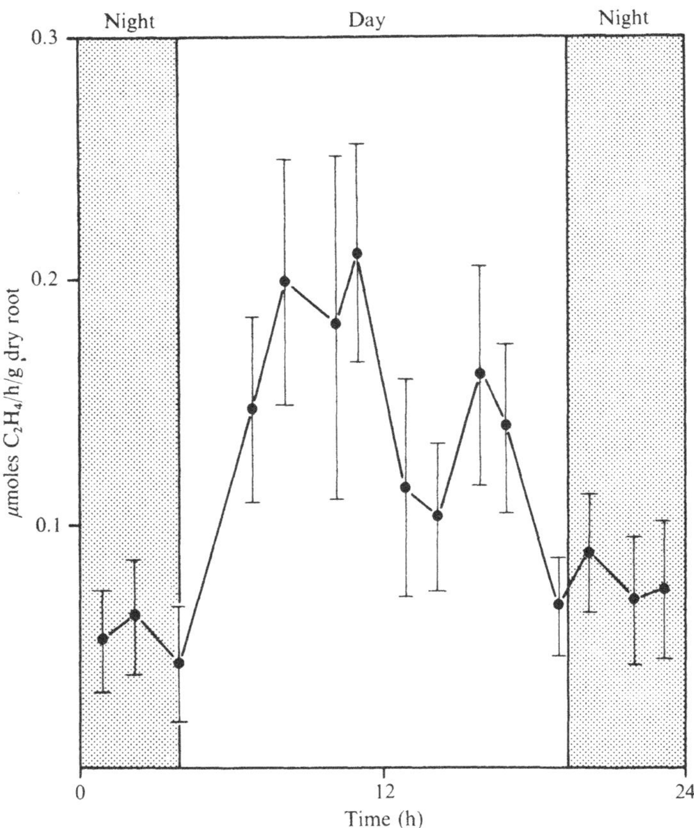

Fig. 4.2. Diurnal variations of nitrogenase activity in a rye-grass field. Limits show standard error of means.

maintained in the dark; (*c*) only the right arm of the tube was planted, the device being exposed to the regular day–night illumination; (*d*) neither arm of the tube was planted, the device being exposed to the regular day–night illumination.

When the plants were 21 days old each arm of the U-shaped tube was connected to a closed tube by a tight rubber ring through which 322 nmoles of acetylene were injected; any acetylene and ethylene formed and which had diffused through the system into the left arm were estimated at three intervals. Figure 4.7 shows that after 13 h the gas diffusion rate reached 0.06 nmoles/h in (*a*), only 0.01 nmole/h in (*b*) and was very low indeed in (*c*) (5 pmole/h). It was nil in (*d*) even after 48 h.

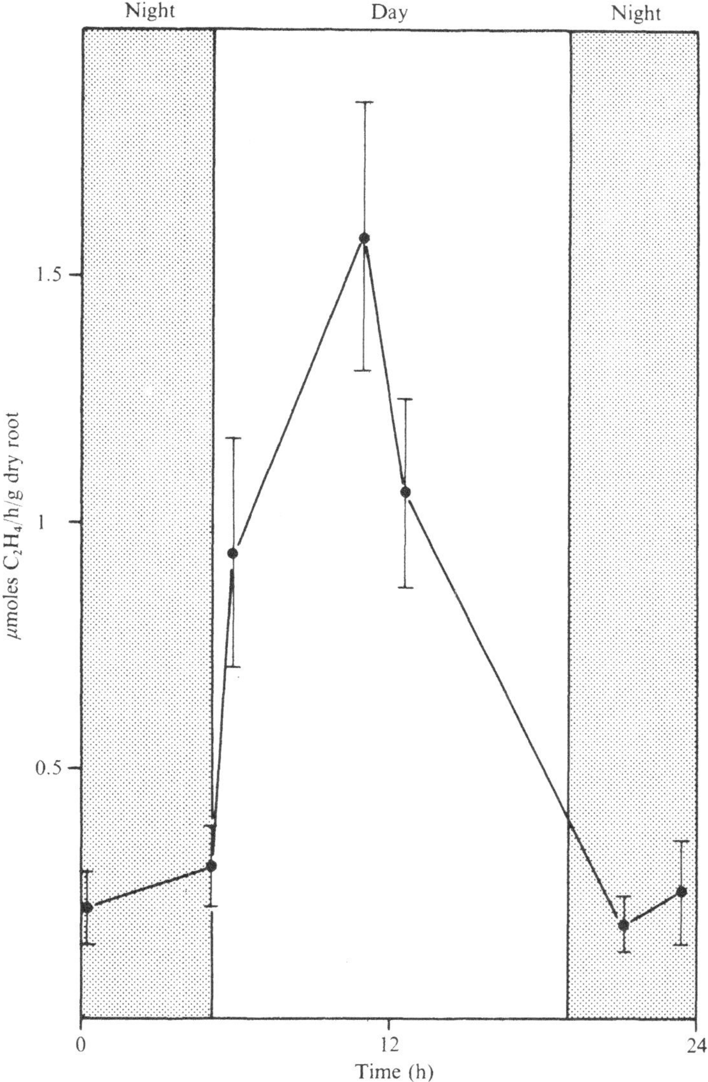

Fig. 4.3. Diurnal variations of nitrogenase activity in a soil–rice system placed in controlled conditions. Limits show standard error of means.

Bacterial microflora

The rhizosphere effect was evident especially in the case of aerobic bacteria (Table 4.2). Thus in non-rhizosphere soil aerobic bacteria counts were 20-fold less than in rhizosphere soil.

Table 4.2. *MPN counts of N_2 fixing bacteria in the rice rhizosphere (bacteria/g dry soil)*

	Aerobic	Anaerobic
Non-rhizosphere soil (control)	3 200 000	372 000
Rhizosphere soil	21 900 000	1 350 000
Rhizosphere soil + roots	61 500 000	472 600

MPN = most probable number.

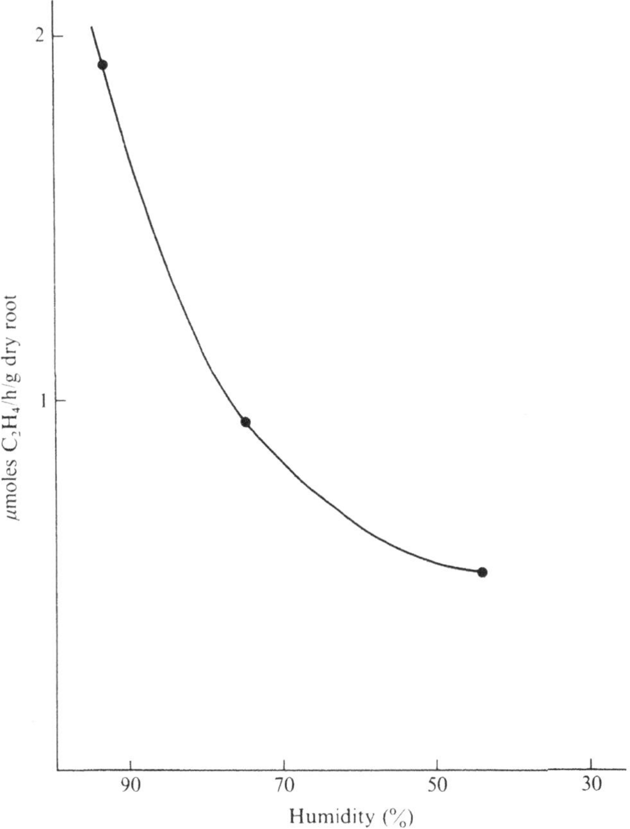

Fig. 4.4. Influence of air humidity on nitrogenase activity in the rhizosphere of rice grown in controlled conditions.

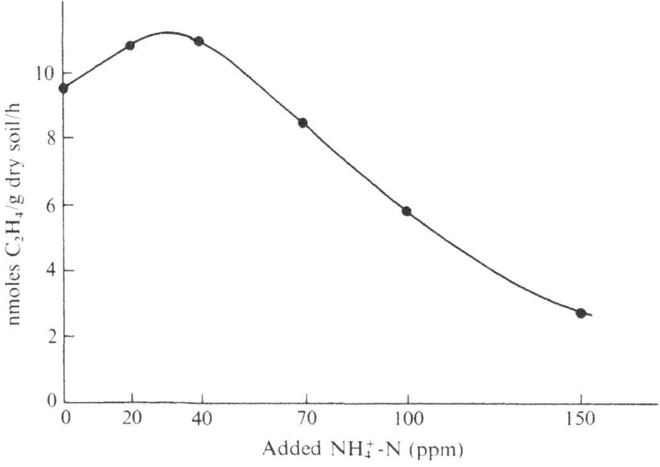

Fig. 4.5. Effect of ammonium–nitrogen additions on nitrogenase activity in the rice rhizosphere.

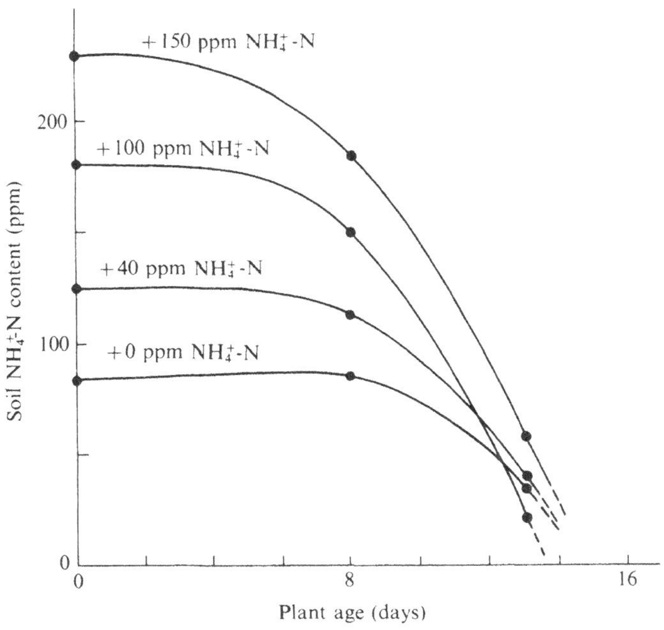

Fig. 4.6. Concentrations of ammonium–nitrogen in the rice rhizosphere at different times after the addition of different amounts of NH_4^+–N.

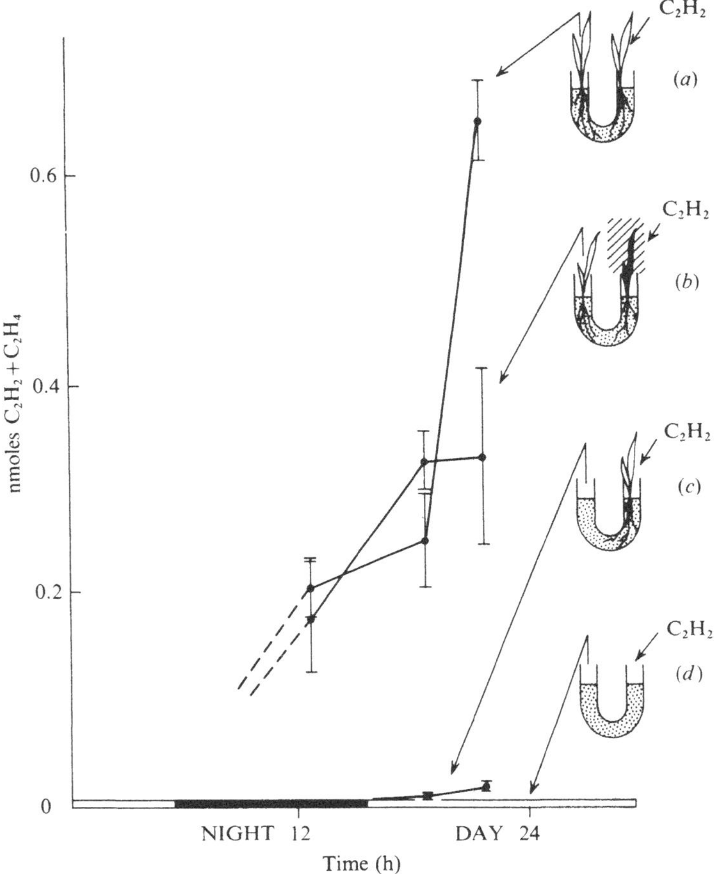

Fig. 4.7. Acetylene diffusion in the soil–plant (rice) system. Limits show standard error of means.

The following N_2-fixing bacteria were isolated from the rice rhizosphere using the highest positive dilution (acetylene test): *Azotobacter* sp., *Beijerinckia* sp., *Pseudomonas* sp., *Arthrobacter* sp. The *Beijerinckia* strain was isolated from the rhizosphere of rice growing in the Mas du Sauvage soil; it did not appear to be very different from *Beijerinckia fluminensis* (Dobereiner & Ruschel, 1958). These preliminary taxonomic studies give support to similar studies conducted in tropical-rice rhizosphere soil from which species of *Beijerinckia* and *Pseudomonas* were isolated (Yoshida, 1970).

Discussion and conclusion

N_2 fixation in paddy soils

Expressed on a per hectare basis, N_2 fixation in the paddy soil studied here was 32 kg N_2/100 days. This rate is lower than most rates previously reported (Moore, 1966; Yoshida, 1970; 1971), which ranged from 40 to 70 kg N_2/growth cycle. The relatively low rate measured here was possibly due to the low plant density in the field studied. Nevertheless, the most striking result was that paddy soil appeared to be a most efficient non-symbiotic, N_2-fixing ecosystem, since its daily integrated N_2 fixation was of the same order of magnitude as that of a symbiotic system (peanut).

Factors affecting N_2 fixation in paddy soils

According to present and previous studies (Rinaudo *et al.*, 1971; Dommergues *et al.*, 1973), N_2 fixation was shown to be affected by the following environmental factors: illumination; air humidity; soil type and especially NH_4^+ content. The response of rhizosphere N_2 fixation to illumination was unequivocally demonstrated in previous studies (e.g. Dommergues *et al.*, 1973). Obviously light acts upon exudation through increasing the rate of plant photosynthesis; light may also act upon nitrogen fixation in the rhizosphere by affecting the stomatal opening, hence the diffusion of gases especially nitrogen, through the plant. The acetylene-diffusion experiment (Fig. 4.6) substantiates this hypothesis since, when maintained in the dark, the rice plant allowed negligible diffusion of acetylene through the system, whereas diffusion was much better in plants submitted to regular illumination.

Nitrogen fixation appeared to be very sensitive to water stress induced by a decrease in air humidity (Fig. 4.4). This effect was attributed to the stomatal closure which reduced photosynthesis thus stopping exudation and which possibly induced a shortage of nitrogen in the rhizosphere, by impeding gas diffusion through the plant. Such a situation is thought to have occurred in the field experiment illustrated by Fig. 4.2 in which the low level of air humidity (less than 50 %) between 11.00 a.m. and 2.00 p.m. might have been responsible for the observed decrease of nitrogenase activity in the middle of the day. This water stress has also been detected in laboratory experiments, when plants are withdrawn from growth cabinets, where air humidity is usually higher than in the laboratory itself. In the case of irrigated rice fields air humidity is unlikely to be a limiting factor owing to the presence of a water layer on the soil.

The influence of the soil type was obvious and easily demonstrated (Rinaudo *et al.*, 1971). The effect of NH_4^+ content was more complex than anticipated. According to the results presented here (Figs. 4.5 and 4.6) no repression of nitrogenase synthesis occurred with additions of ammonium-nitrogen below 40 ppm; on the contrary, a slight increase was observed, which could be attributed to the increase of the exudate input into the soil. Since the 40 ppm (120 kg NH_4–N/ha) threshold for ammonium-nitrogen was a fairly high one, the risk of repression of nitrogenase synthesis could presumably be overlooked. Moreover it had to be stressed that, above the 40 ppm threshold, N_2 fixation was severely reduced even after the added ammonium-nitrogen had been used up, suggesting some transient modification of the microflora equilibrium.

Diffusion of N_2 through the soil–plant system

The experiment presented here showed clearly that the diffusion of acetylene through the plant–soil system occurred mostly through the plant. Although the properties of acetylene are very different from those of nitrogen, these results suggested that, if nitrogen diffused through the soil–rice system, such a diffusion occurred mainly through the plant tissues.

Diffusion of oxygen through the soil–plant system

On the one hand the occurrence of a P_{O_2} gradient in the soil might be assumed to favour N_2 fixation since any physiological group of bacteria involved in nitrogen fixation can find herein a micro-habitat where its own P_{O_2} requirement is met. On the other hand the rice rhizosphere may be the site of a P_{O_2} gradient more stable than that in the vicinity of other rhizospheres since it is mostly induced by a physical process, air diffusion. These assumptions may explain to some extent the outstanding activity of N_2-fixing bacteria in the rice rhizosphere.

Bacteria in the rice rhizosphere

The relative role of blue-green algae and rhizosphere bacteria in the process of N_2 fixation in paddy soils is difficult to assess, especially in the field, where algal populations are not evenly distributed. In some laboratory experiments, rhizosphere bacteria were significantly involved; indirect evidence of this is the finding of Dommergues *et al.* (1973) that, in a rice–soil system, nitrogenase activity was suppressed shortly after the plants had been decapitated. Even in the neutral soils studied here, such as Mas du Sauvage soil, *Azotobacter* counts were relatively low (approximately 7000/g dry soil); in contrast *Beijerinckia* sp., *Pseudomonas* sp., *Arthrobacter* sp., *Clostridium* sp. were much more abundant as shown by their isolation from higher dilutions (10^{-4} to 10^{-6}). Assuming that nitrogen fixation was achieved essentially by the most abundant strains, it can be inferred that, as far as aerobic fixation was concerned, *Azotobacter* played only a trivial role compared to that of other N_2-fixing aerobes. Although the occurrence of *Beijerinckia* in temperate soil has been reported infrequently (Becking, 1961), it may be that most investigators had overlooked it, because the sensitive acetylene-reduction technique was not then available.

References

Balandreau, J. & Dommergues, Y. (1973). Assaying nitrogenase (C_2H_2) activity in the field. *Bull. Ecol. Res. Comm., Stockholm*, **17**, 247–54.

Balandreau, J. & Villemin, G. (1973). Fixation biologique de l'azote moléculaire en savane de Lamto (Basse Côte d'Ivoire). Résultats préliminaires. *Rev. Ecol. Biol. Sol.*, **1**, 25–33.

Balandreau, J., Millier, C. & Dommergues, Y. (1974). Diurnal variations of nitrogenase activity in the field. *Appl. Microbiol.*, **27**, 662–5.

Becking, J. H. (1961). Studies on the nitrogen fixing bacteria of the genus *Beijerinckia. Pl. Soil*, **14**, 49–81 and 297–322.

Brouzes, R., Mayfield, C. I. & Knowles, R. (1971). Effect of oxygen partial pressure on nitrogen fixation and acetylene reduction in a sandy loam soil amended with glucose. *Plant and Soil, Special Volume*, 481–94.

Dobereiner, J. & Ruschel, A. P. (1958). Uma nova especie de *Beijerinckia. Revista de Biologia*, **1**, 261–72.

Dommergues, Y., Balandreau, J., Rinaudo, G. & Weinhard, P. (1973). Non-symbiotic nitrogen fixation in the rhizospheres of rice, maize and different tropical grasses. *Soil Biol. Biochem.*, **5**, 83–9.

Ishizawa, S., Suzuki, T. & Araragi, M. (1970). Trend of free-living nitrogen fixers in paddy soil. *Proc. Second Symp. N_2 fixation and Nitrogen cycle* (ed. Takahashi, H.), pp. 28–40.

Moore, A. W. (1966). Non symbiotic nitrogen fixation in soil and soil–plant systems. *Soils Fertil.*, **29**, 113–28.

Rinaudo, G. (1970). Fixation biologique de l'azote dans trois types de sols de rizière de Côte d'Ivoire. *Thèse Docteur Ingénieur, Fac. Sci., Montpellier.*

Rinaudo, G., Balandreau, J. & Dommergues, Y. (1971). Algal and bacterial non-symbiotic nitrogen fixation in paddy soils. *Plant and Soil, Special Volume*, 471–9.

Watanabe, A., Nishigaki, S. & Konishi, C. (1951). Effect of nitrogen fixing blue-green algae on the growth of rice plants. *Nature, Lond.*, **168**, 748–9.

Yoshida, T. (1970). *Soil Microbiology, Annual Report*, pp. 47–59. The International Rice Institute: Manila.

Yoshida, T. (1971). *Soil Microbiology, Annual Report*, pp. 23–6. The International Rice Institute: Manila.

Yoshida, T. & Ancajas, R. R. (1971). Nitrogen fixation by bacteria in the root zone of rice. *Soil Sci. Soc. Am. Proc.*, **35**, 156–7.

Yoshida, T. & Ancajas, R. R. (1973). The fixation of atmospheric nitrogen in the rice rhizosphere. *Soil Biol. Biochem.*, **5**, 153–5.

5. The Broadbalk experiment. An investigation of nitrogen gains from non-symbiotic nitrogen fixation

J. M. DAY, D. HARRIS, P. J. DART & P. VAN BERKUM

Between 1843 and 1856 Lawes and Gilbert started nine long-term classical experiments of which the one on Broadbalk Field is the most famous. Their main object was to measure the effects on crop yields of inorganic compounds containing nitrogen, phosphorus, potassium, sodium and magnesium, and compare them with farm yard manure. Weights of all produce harvested were recorded and samples analysed chemically. From these results they compiled a balance sheet for the major nutrients for each plot. Analysis of soil samples showed whether nutrients accumulated or diminished in the soil. Drains placed under each plot of the Broadbalk continuous wheat experiment in 1849 allowed losses of nutrients by leaching to be estimated. The design of the Broadbalk experiment remained largely unchanged from 1843 to 1968 when a crop rotation was introduced on 6 of the 10 sections; section 8 has also received no herbicides (*Rothamsted Experimental Station Report for 1968*, Pt 2).

Over this long period the nitrogen content of the soil remained remarkably constant (see Table 5.1). From 1885 to 1967 an average of 24 kg of nitrogen in the grain and straw was removed annually from plot 5, and the nitrogen balance showed a steady gain of 34 kg N/ha/annum (Jenkinson 1973).

Table 5.1. *Nitrogen percentage of Broadbalk soils 0–9 in.*

Plot	Manuring	1865	1944	1966
3	None	0.105	0.106	0.099
5	P, K, Na, Mg	0.107	0.105	0.107
7	N_2,* P, K, Na, Mg	0.117	0.121	0.115
22	F.Y.M.**	0.175	0.236	0.251

* $N_2 \equiv$ 96 kg N/ha/annum; ** farm yard manure.

Leguminous weeds are thought not to have contributed significantly to this nitrogen gain. Since 1955 they have been removed from part of the field by herbicide treatment and yields have not declined since.

In 1882 part of Broadbalk was enclosed, left unharvested and since then uncultivated. One part of this land (Broadbalk Wilderness) has developed into mixed woodland with a sparse ground flora, predominantly *Hedera helix* (ivy), with patches of *Mercurialis perennis* (dog's mercury) and isolated plants of *Heracleum sphondylium* (hogweed). Another part is 'stubbed' annually to remove woody species and now consists of coarse grasses with a mixed dicotyledonous flora of forty species (Brenchley & Adam, 1915). The vegetation cut down is removed. Large amounts of organic nitrogen have accumulated in both these areas of the Wilderness representing a net gain in the top 20 cm of soil of 2 tonnes N/ha since 1881, equivalent to 65 kg N/ha/annum in the 'wooded' and 55 kg N/ha/annum in the 'stubbed' area. Rainfall is estimated to account for 1.4 kg N/ha/annum, dry sorption of ammonia a maximum of 13 kg N/ha/annum and organic nitrogen in rain, dust and bird droppings 1.5 kg N/ha (Jenkinson, 1971). Legumes may have been important early in the experiment but nitrogen has continued to accumulate at the same rate since their disappearance in 1915 (Brenchley & Adam, 1915). As well as the assumed small losses from drainage or volatilisation, minimum annual gains from non-symbiotic nitrogen fixation are calculated to be 49 and 39 kg N/ha for the wooded and stubbed parts respectively. Broadbalk Wilderness had been limed prior to the experiment and the pH has remained above 7.

Geescroft Wilderness nearby has likewise developed since 1886 from an arable field. The soil type is similar but, because it had not been limed, the pH dropped from 7.1 to 4.5 and only 23 kg N/ha/annum has accumulated since, of which little needs to be attributed to biological nitrogen fixation (Jenkinson, 1973). We examined the nitrogenase activity in the soils of these and similar sites by an acetylene-reduction assay.

Materials and methods

Surface soil layers (to *c.* 1 cm deep), soil cores, soil cores with plants, leaf litter, and leaves were assayed from Broadbalk Arable plots 3, 5, 6, 9, 22 and the wooded and stubbed sections of Broadbalk Wilderness.

Because of the danger of causing permanent damage to the Wilderness by excessive sampling, the assay techniques were developed on a nearby site with a similar soil, pH 7.0–7.3, known as the Manor Wood.

Acetylene reduction was measured by gas chromatography (Dart, Day & Harris, 1972) using a 0.003 m diameter Durapak column (phenylisocyanate on Porasil C – Waters Associates) to separate ethane, ethylene and acetylene and a Varian model 2400 gas chromatograph coupled to a Vidar Autolab digital integrator 6300.

Acetylene reduction by crusts of blue-green algae was investigated by placing 2 cm diameter × *c.* 1 cm-deep surface soil cores in bottles closed with Suba Seals (W. Freeman & Co. Ltd, Staincross, Barnsley, England) or by in-situ assays using 15 cm diameter steel cores driven *c.* 10 cm into the soils and then sealed on top with perspex sheets fitted with a small Suba Seal for injection and sampling of gases (Froggatt *et al.*, 1973). Acetylene was injected and acetylene and ethylene assayed 30 and 90 minutes later, during which period acetylene loss and ethylene gain were almost linear.

Soil cores and soil cores with plants were obtained by driving 16 cm long × 11 cm diameter, 10-gauge, steel tubes into the test area and removing the tube and enclosed soil core. Each core, still in its tube, was transferred to a 5½ lb (*c.* 2.5 kg) confectionery jar (16 cm square, 3.7 l capacity – available from W.G. Glass Containers, Staines, Middlesex, England) fitted with a Suba Seal (size 140) retained by a black plastic cap (available from Johnsen & Jorgensen, Herringham Road, Charlton, London SE 7, England) drilled with holes to allow gas sampling through the Suba Seal, and its nitrogenase activity assayed. Activity is greatly decreased by removing the soil cores from the enclosing tube. Leaf litter and leaves were collected and assayed in bottles. Large volumes of acetylene, 500 ml per jar in the present experiments, can be injected conveniently using the assembly of the type in Fig. 5.1. The 'syringe' barrel is a 28 mm internal diameter glass tube fitted with a floating piston made by sticking together two plungers from 50 ml plastic, disposable hypodermic syringes. The syringe barrel is joined by rubber tubes to the tap assembly made by joining two-way-oblique glass taps.

The syringe barrel is first filled from an acetylene cylinder with the taps in the position shown, moving the plunger to the end of the barrel marked A. To deliver the acetylene the tap is moved to the position to empty, which shuts off the acetylene cylinder and opens the compressed air supply. This moves the plunger along the barrel to B expelling the acetylene through the tap and outlet which is connected to the incubation vessel. The volume of gas delivered can be varied by altering the length of the glass tube and to a lesser extent by altering the pressure

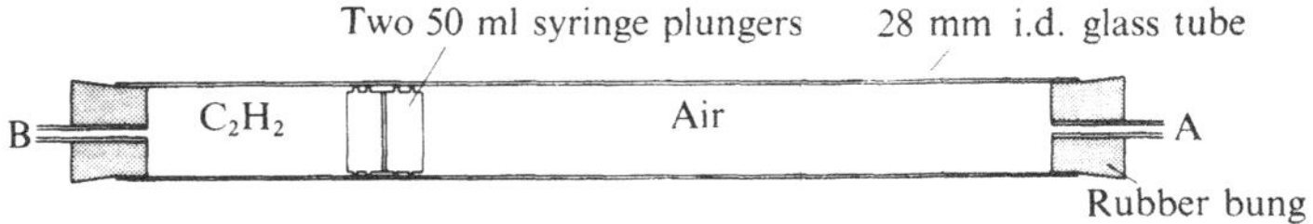

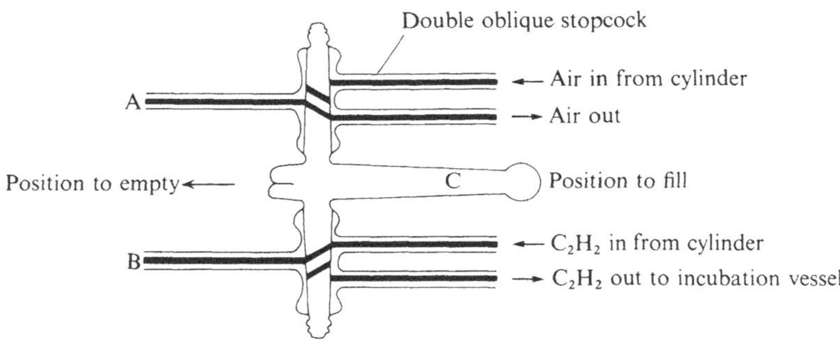

Fig. 5.1. Large capacity acetylene syringe. In operation, points A and B of the glass tube are joined by rubber tubing to points A and B respectively of the tap assembly.

of the acetylene cylinder (usually *c.* 0.75 kg/cm²). After injection of acetylene into the bottles, the gas phase can be returned to atmospheric pressure by bleeding excess gas through a hypodermic needle inserted through the Suba Seal for a short period.

Soils from a depth greater than 16 cm were sampled by driving 50 cm lengths of the 11 cm diameter steel tube into the soil, digging out the tube and sawing the pipe containing the soil into 16 cm lengths for assay. Broadbalk Wilderness was sampled with the mechanical core sampler to reduce disruption of the soil (Welbank & Williams, 1968), modified to take a 6.8 cm diameter plastic tube in place of the brass insert tube. Sixteen cm lengths of soil core were assayed in the plastic pipes as for the steel tubes.

There is a lag phase before the onset of rapid acetylene-reduction which varies with core size and soil moisture. This was accounted for by taking the first gas sample for assay 16–24 h after injection of the acetylene and subsequent samples a further 24–48 h later.

Soil moisture contents were determined on a *c.* 100 g, stone-free, subsample of the thoroughly mixed soil in the core. Soil pH was determined by suspending a 10 g subsample of soil from each core in 20 ml 0.01 M $CaCl_2$ and after 20 minutes measuring electrometrically.

Results

Development of the assay

Soil cores and leaf litter samples indicated that non-rhizosphere nitrogen fixation could account for only 4–5 kg N/ha/annum. A survey of nitrogenase activity of most of the 40 species occurring in the 'stubbed' part of the Wilderness showed that most activity was associated with the roots of *Heracleum sphondylium* (hogweed), *Anthriscus silvestris* (cow parsley), *Mercurialis perennis* (dog's mercury), *Rumex acetosa* (sorrel), and *Stachys sylvatica* (hedge woundwort) (Harris & Dart, 1972). To establish the contribution of these plants to the nitrogen economy of the wilderness a reliable in-situ assay system was desirable as transplanting plants in clods of soil gave large variations in activity.

An in-situ method for measuring nitrogenase activity was developed by Balandreau & Dommergues (1971). We drove 15 cm diameter steel tubes into the sward and sealed the top with a perspex sheet or a transparent polystyrene dome (Dart *et al.*, 1972), injected acetylene, and monitored gas losses by the disappearance of an added, known volume of propane standard. After a rapid initial loss of propane and acetylene the rate of loss was then extremely slow, indicating that acetylene was not rapidly entering the soil to any appreciable depth. This was confirmed by digging out similar, 60 cm deep, soil cores in tubes with 1 cm diameter holes drilled down one side of the tube 10 cm apart to enable sampling of the soil atmosphere at varying depths. The top of the tube was closed and acetylene injected into the headspace. The concentration of acetylene in the top 10 cm reached equilibrium in about 2 h but was very slow to migrate deeper into the core and on one occasion took 48 h to come to equilibrium at a depth of 50 cm in the wet, heavy clay soil. Assaying cores *in situ* was obviously impractical at our site and the assay technique detailed in the Methods section was then developed. Preliminary experiments showed that activity was greater in cores assayed in 10 cm diameter tubes than in 6.5 cm tubes, the 10 cm diameter tube being the largest to readily fit the available glass jars. Removing cores from the tubes decreased activity, and the 15 cm long cores taken had access to acetylene from both top and bottom so that the whole of the core rapidly equilibrated with the added acetylene.

Algal and rhizosphere activity on Broadbalk arable

The wheat on Broadbalk is sown in the autumn and stands the winter as young plants. Assays in the autumn, winter and spring of surface soil and soil cores containing wheat plants revealed very little nitrogenase activity in any samples before early June. Algal crusts develop and have nitrogenase activity only when the surface is moist. Their activity was related to the nitrogen status of the soils and activity for plot 9, receiving 192 kg N/ha/annum, was not appreciable until later in the season when the concentration of nitrogen in the surface soil had probably decreased.

Algal nitrogenase activities were determined during the summer of 1971 on 11 occasions on samples selected for presence and absence of a visible crust. The nitrogen fixation over the season was estimated using estimates of the algal cover, the moisture content of the surface soil, and the theoretical $3C_2H_2 \equiv 1\ N_2$ conversion factor. More nitrogen was fixed on the sections not treated with herbicides; plot 6 (given 48 kg N/ha without herbicides) was estimated to fix 28 kg N/ha/annum. Plots 3 and 5 (no nitrogen fertiliser) had less plant cover and the soil surface dried rapidly and estimated fixations ranged from 13–24 kg N/ha/annum. Sometimes nitrogenase activities were very high and, after heavy rain late in the season, estimated rates of nitrogen accumulation were as much as 1–2 kg N/ha/day (Froggatt *et al.*, 1973).

Soil cores (17 cm deep) containing wheat plants from Section 9 of plots 2*B*, 3, 5, 6 and 9 were assayed on three occasions in the summer of 1973; and soils from the non-planted areas of plots 2*B*, 3 and 5. The surface 1 cm of soil was removed to exclude any possible contributions from blue-green algae. Soils of plots 6 and 9, both receiving nitrogen fertiliser, were not included because nitrate levels differed between planted and non-planted areas in these plots (Nair & Talibudeen, 1973). Estimates of nitrogen fixation from acetylene-reduction activity are presented in Table 5.2. Activities were generally low and no significant differences were found between planted and non-planted areas of plots 2*B* and 3, but on plot 5 cores containing the wheat plants fixed significantly more than the unplanted area.

Soil moisture was low at the second sampling and nitrogenase activity was greatly decreased. Assuming that activity of the same order occurred on 100 days during the season, bacterial nitrogen fixation in the top 16 cm could account for only 2–3 kg N/ha/annum on these arable wheat fields and most of the nitrogen fixed can be attributed to blue-green algal crusts.

Table 5.2. *Nitrogen fixation by soil and wheat rhizosphere in cores from Broadbalk arable plots*

Plot no.		Sampling 1		Sampling 2		Sampling 3	
		N fixed	% Soil moisture	N fixed	% Soil moisture	N fixed	% Soil moisture
2	Wheat	17.1 ± 8.9*	18.7**	2.2* ± 0.9	15.5**	10.2 ± 4.3	21.4
F.Y.M.†	Soil	—	—	—	—	12.0 ± 3.0	24.1
3	Wheat	5.7 ± 2.0	15.0	2.7 ± 1.5	15.0	13.9 ± 3.2	17.7
No manure	Soil	—	—	—	—	13.0 ± 3.9	17.8
5	Wheat	18.7 ± 6.0	16.2	6.6 ± 1.8	15.6	44.8 ± 17	18.6
No N, P, K, Na, Mg	Soil	—	—	—	—	18.7 ± 9.2	19.5
6††	Wheat	27.8 ± 6.0	15.9	6.8 ± 3.0	15.6	13.8 ± 1.6	20.5
N_1, P, K, Na, Mg							

* g N fixed/ha/day, means of 4 replicates ± standard error. Cores were 16 cm deep in 11 cm diameter steel tubes.
** g H_2O/100 g oven dry soil.
† F.Y.M. ≡ Farm yard manure.
†† N_1 ≡ 48 kg N/ha/annum.

Nitrogenase activity associated with plant roots

Many of the preliminary investigations were performed in a recently cleared and replanted woodland site near to Broadbalk Wilderness known as the Manor Wood. The soil in this wood was disturbed prior to the replanting but the floristic composition is similar to Broadbalk although the soil is usually wetter possibly because there are fewer large trees.

Soil cores were taken from the centres of clumps of different plant species common on the Wilderness sites on three occasions in June and July 1973. High nitrogenase activities were obtained in all cores with similar variation between sampling dates for all species (Table 5.3). All cores had high activity on the last sampling taken the day after a heavy rain when the soils were exceptionally wet.

Preliminary sampling to 33 cm depth gave high activity in the 16–33 cm layer of soil under clumps of *Heracleum sphondylium* and *Nepeta glechoma* (Table 5.4, A). Further sampling up to 50 cm depth under *H. sphondylium*, *N. glechoma* and *Stachys sylvatica* gave highest nitrogenase activities in the 16–33 cm layer (Table 5.4, B). Although activity was less in the 33–50 cm layer, it still represented a significant nitrogen input. The mean activity of all six cores was 439 g N/ha/day

Table 5.3. *Nitrogen fixation in cores containing plants from the Manor Wood site*

Species	Sampling 1	Sampling 2 (g N fixed/ha/day)	Sampling 3
Nepeta glechoma (ground ivy)	264* ± 14	238 ± 14	658 ± 59
Stachys sylvatica (hedge woundwort)	303 ± 18	121 ± 10	447 ± 13
Heracleum sphondylium (hogweed)	122 ± 14	—	288 ± 16
Mercurialis perennis (dog's mercury)	99 ± 15	—	—

* Mean of 4 replicates ± standard error. Cores were 16 cm deep in 11 cm diameter steel tubes.

Table 5.4. *Variation in nitrogen fixation with depth in Manor Wood*

	Depth (cm)	*Heracleum*	*Nepeta* (g N/ha/day)	*Stachys*
		Sampling A		
	0–16	*263.1 ± 28.7	195.1 ± 16.9	—
	16–33	149.1 ± 24.7	96.9 ± 4.1	—
		Sampling B		
Core A**	0–16	184.1	171.1	160.5
	16–33	204.6	189.4	186.6
	33–50	166.5	135.0	135.4
	Total	555.2	495.5	482.5
Core B**	0–16	124.5	119.3	180.2
	16–33	149.9	146.4	160.4
	33–50	52.5	105.7	53.8
	Total	326.9	371.4	394.4
	Mean total	441.1	433.5	438.4

Mean of six cores = 438.7 ± 76.
* Mean of 4 cores ± standard error.
** Cores were taken in 11 cm diameter steel tubes.

with little difference between the plant species suggesting that some factor other than species was important in the variations between samples taken on different days. The high activities on the third sampling under *N. glechoma* (Table 5.3) extrapolate to a fixation of more than 1 kg N/ha/day in the 50 cm layer of soil.

Table 5.5. *Nitrogen fixation on the Wilderness site*

Vegetation	Sampling 1 N fixed	Sampling 2 N fixed	Sampling 2 Soil moisture	Sampling 3 N fixed	Sampling 3 Soil moisture
		Broadbalk stubbed pH 6.8–7.2			
Heracleum	81.4* ± 12.5	122 ± 11.8	28.0**	106 ± 10.1	31.6
Stachys	66.3 ± 3.9	112 ± 6.1	30.4	98 ± 8.1	29.7
Mixed	64.9 ± 4.8	238 ± 24.2	31.7	52.3 ± 8.6	25.8
		Broadbalk wooded pH 6.9–7.2			
Heracleum	29.1 ± 4.8	42.2 ± 3.1	21.4	32.2 ± 5.2	21.4
Hedera	24.0 ± 9.2	39.3 ± 8.4	21.3	11.3 ± 0.6	19.5
		Geescroft Wood pH 4.2–4.8			
Mercurialis	24.3 ± 2.4	—	—	—	—
Hedera	26.2 ± 2.2	49.5 ± 13.1	26.5	23.6 ± 1.9	26.7

* Mean of 4 replicates (g N fixed/ha/day) ± standard error. Cores were 16 cm deep in 11 cm diameter steel tubes.
** g H_2O/100 g oven dry soil.

Broadbalk Wilderness

Table 5.5 shows nitrogenase activities throughout mid-summer. The amounts of nitrogen fixed were generally lower than at the Manor Wood site but no samples were taken after heavy rain, and activity was probably restricted by low soil moisture. In the second sampling, one wet core fixed nearly 0.5 kg N/ha/day indicating the potential for very high fixation in wet conditions. At all samplings, cores from the 'wooded' section had consistently lower activity than cores containing the same plants from the 'stubbed' area. Activity was again closely correlated with soil moisture.

The variation of nitrogenase activity (associated with plant roots) with soil horizon was examined using a mechanical sampler (Welbank & Williams, 1968), modified to take cores in 6.8 × 17 cm diameter plastic tubes instead of the usual brass liners. The site was sampled to a depth of 75 cm the day after heavy rain following a previous dry period, and nitrogenase activities are shown in Table 5.6. The high activity in the top 7 cm compared with lower horizons may well be a response to differential wetting rather than a true indication of the relative abundance of nitrogen-fixing organisms. Nitrogenase activity below 41 cm was low and is probably of little significance at this site.

Nitrogenase activity under the dominant ground cover of *Hedera*

Table 5.6. *Variation in nitrogen fixation with depth in Broadbalk Wilderness stubbed*

Depth (cm)	(g N fixed/ ha/day)	(% H_2O)
0–7*	136.2**	35.1
7–24	33.3	21.2
24–41	18.6	17.9
41–58	13.4	24.0
58–75	14.4	26.8

* Cores taken in 6.8 cm diameter plastic tubes. ** Means of 4 cores.

helix at Geescroft Wilderness, pH 4.2–4.8, was similar to that of Broadbalk 'wooded' section, pH 6.8–7.2, showing that pH is not a major factor limiting root-associated nitrogen fixation.

The summer of 1973 was unusually dry and the area was not sampled whilst the soil was very wet. The results, though, suggest that fixation associated with plant roots could account for most of the nitrogen known to be entering the stubbed Wilderness.

Correlation between soil moisture and nitrogen fixed

Preliminary samples from Manor Wood with high nitrogenase activities were associated with wet soils. Subsequently all cores from all sites were subsampled after assay for soil moisture and pH. Figure 5.2 shows that nitrogenase activity was positively correlated with soil moisture ($\delta = 0.78$); nitrogenase increased exponentially with linear increase in soil moisture expressed as grams of water per 100 g dry soil.

Discussion

Significant nitrogen fixation occurs on the Broadbalk experiment. The nitrogen attributed to biological fixation in the 'stubbed' section of Broadbalk Wilderness (39 kg/ha/annum) (Jenkinson, 1971; 1973), is mainly fixed by organisms living in the rhizosphere of many genera of dicotyledonous weeds. Preliminary isolations suggest that several types of nitrogen-fixing organisms are associated with *Stachys sylvatica* and *Mercurialis perennis* roots, the most numerous belonging to the *Enterobacter–Klebsiella* group. Several other wood, roadside and hedgerow sites in North Hertfordshire containing similar species to Broadbalk Wilderness were sampled and all had similar levels of activity.

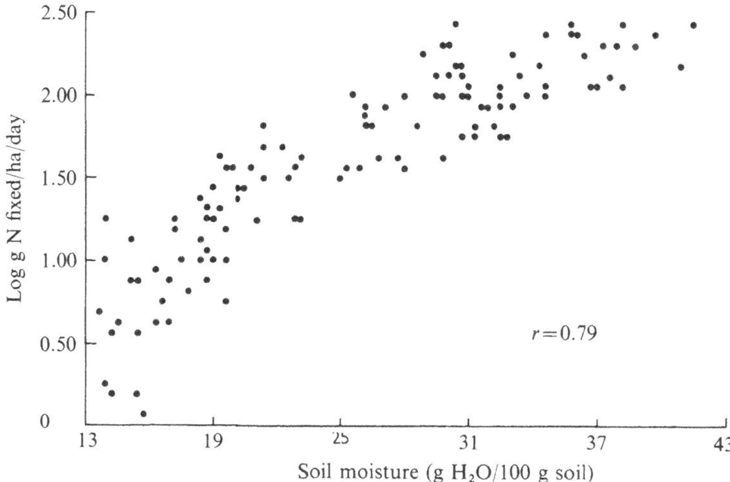

Fig. 5.2. Effect of soil moisture on nitrogen fixation by soil cores with plants, showing the highly significant and positive correlation.

Nitrogenase activity was highly and positively correlated with soil moisture levels. At wet sites such as the Manor Wood very high nitrogen fixation rates, up to 1 kg N/ha/day, were obtained. The 'wooded' sections of Broadbalk Wilderness and Geescroft Wilderness were rarely wet enough to support much fixation. Phyllosphere fixation on leaves and fixation associated with bark was insignificant for all the common tree species of the woodlands. The origin of the nitrogen (49 kg N/ha/annum) accumulated in the wooded section since 1882 is still a mystery. Perhaps very rapid fixation occurs in early spring before the trees are in leaf or alternatively the trees may be tapping nitrogen reserves deep in the subsoil or from outside the plot perimeter.

Vlassak, Paul & Harris (1973) also reported a similar correlation between soil moisture and nitrogen fixation in soil cores in grasslands. As soil moisture increases, the level of anaerobiosis in the soil crumbs and rhizosphere increases, and this affects nitrogenase activity rather than any effect of moisture *per se* on nitrogenase activity, provided that sufficient water is available to support microbial activity (Knowles, 1975). July and August of 1973 at Rothamsted were exceptionally dry with only 100 mm rainfall compared with the average of 130 mm over the past 125 years. Hence the fixation rates we report are likely to be low rather than exceptionally high. Differences in activities between species may be related to growth habit. *Mercurialis perennis* has most of its roots and rhizomes near the soil surface which is rarely wet enough for high activity. Deep rooting plants like *Heracleum sphondylium* can support considerable nitrogenase activity in the subsoil when the surface soil is dry.

The discrepancy between the present fixation rates and the much lower activities in prairie grasslands in Canada where only 2–3 kg N/ha is fixed annually (Vlassak *et al.*, 1973) may be partly due to lower soil temperatures in Canada or to differences in the state of equilibrium of the vegetation of the two systems. The prairie grasslands have been left undisturbed for centuries and have perhaps reached a state where the soil nitrogen content is high, and nitrogen is efficiently cycled within the ecosystem, whereas Broadbalk is still accumulating nitrogen at the same high rate as at the beginning of the experiment. Broadbalk Wilderness was left to revert to natural vegetation relatively recently after many years of agricultural use, probably going back to Roman times.

Broadbalk arable field had very little bacterial nitrogen (N_2) fixation, probably because under normal agricultural practice such a soil is rarely wet enough to support significant rhizosphere fixation. Activity would perhaps be greater in heavy soils, or under non-tillage wheat or in subsoils under agricultural crops. Most of the N_2 fixation on the Broadbalk wheat experiment was associated with blue-green algal crusts which developed late in the growing season.

Rhizosphere-stimulated N_2 fixation is widespread. Significant activity is also associated with the marine angiosperms *Thalassia*, *Syringodium*, *Diplanthera*, and *Zostera* (Patriquin & Knowles, 1972), *Paspalum notatum–Azotobacter paspali* association (Dobereiner, Day & Dart, 1972*a*), *Zea mays* (Hauke-Pacewiczowa, Balandreau & Dommergues, 1970; Rinaudo, Balandreau & Dommergues, 1971; Raju, Evans & Seidler, 1972), sugar cane associated with *Beijerinckia* (Dobereiner, *et al.*, 1972*b*), *Carex* sp. (Porter & Grable, 1969), rice (Yoshida & Ancajas, 1973) and various tropical grasses (Weinhard, Balandreau, Rinaudo & Dommergues, 1971; Dart *et al.*, 1973: Dobereiner & Day, Chapter 3).

It is not known how directly and to what extent plants benefit from non-symbiotic fixation as ^{15}N studies of the amount and rate of transfer of the fixed nitrogen to the host plant are lacking. *Paspalum notatum* plants transplanted from the field into Leonard jars benefited substantially, gaining 85 mg N/jar in two months, and the growth rates of plants relying on rhizosphere fixation were similar to those fed combined nitrogen (Dobereiner & Day, Chapter 3). Some of this fixed nitrogen may be lost again by denitrification as this is favoured by the same conditions as nitrogen fixation, i.e. high soil temperature and moisture content, and abundant root exudation.

References

Balandreau, J. & Dommergues, Y. (1971). Mésure *in situ* de l'activité nitrogénasique. *C. r. hebd. Séanc. Acad. Sci., ser. D*, **273**, 2020–3.

Brenchley, W. E. & Adam, H. (1915). Recolonisation of cultivated land allowed to revert to natural conditions. *J. Ecol.* **3**, 193–210.

Dart, P. J., Day, J. M. & Harris, D. (1972). Assay of nitrogenase activity by acetylene reduction. In *Use of isotopes for study of fertiliser utilisation by legume crops*. International Atomic Energy Agency Technical Report No. 149, pp. 85–100.

Dart, P. J., Harris, D. & Day, J. M. (1973). Nitrogen fixation associated with the roots of tropical grasses. *Report of the Rothamsted Experimental Station for 1972*, Pt 2, p. 87.

Dobereiner, J., Day, J. M. & Dart, P. J. (1972*a*). Nitrogenase activity and oxygen sensitivity of the *Paspalum notatum–Azotobacter paspali* association. *J. gen. Microbiol.* **71**, 103–16.

Dobereiner, J., Day, J. M. & Dart, P. J. (1972*b*). Nitrogenase activity in the rhizosphere of sugar cane and some other tropical grasses. *Pl. Soil*, **37**, 191–6.

Froggatt, P. J., Keay, P. J., Witty, J. F., Dart, P. J. & Day, J. M. (1973). Algal nitrogen fixation in Rothamsted fields. *Report of the Rothamsted Experimental Station for 1972*, Pt 2, p. 87.

Harris, D. & Dart, P. J. (1973). Nitrogenase activity in the rhizosphere of *Stachys sylvatica* and some other dicotyledonous plants. *Soil Biol. Biochem.*, **5**, 277–9.

Hauke-Pacewiczowa, T., Balandreau, J. & Dommergues, Y. (1970). Fixation microbienne de l'azote dans un sol salin Tunisien. *Soil Biol. Biochem.*, **2**, 47–53.

Jenkinson, D. S. (1971). The accumulation of organic matter in soil left uncultivated. *Report of the Rothamsted Experimental Station for 1970*, Pt 2, pp. 113–37.

Jenkinson, D. S. (1973). Organic matter and nitrogen in soils of the Rothamsted classical experiments. *J. Sci. Food Agric.*, **24**, 1149–50.

Knowles, R. (1975). The significance of asymbiotic dinitrogen fixation by bacteria. In *Dinitrogen* (*N_2*) *Fixation* (ed. Hardy, R. W. F.), vol. II, Pt A. Wiley–Interscience: New York (in press).

Nair, P. K. R. & Talibudeen, O. (1973). Dynamics of K and NO_3 concentrations in the root zone of winter wheat at Broadbalk using specific ion electrodes. *J. agric. Science, Camb.*, **81**, 327–37.

Patriquin, D. & Knowles, R. (1972). Nitrogen fixation in the rhizosphere of marine angiosperms. *Marine Biol.*, **16**, 49–58.

Porter, L. K. & Grable, A. R. (1969). Fixation of atmospheric nitrogen by non-legumes in wet mountain meadows. *Agron. J.*, **61**, 521–3.

Raju, P. N., Evans, H. J. & Seidler, R. J. (1972). An asymbiotic nitrogen fixing bacterium from the root environment of corn. *Proc. Nat. Acad. Sci.*, **69**, 3474–8.

Rinaudo, G., Balandreau, J. & Dommergues, Y. (1971). Algal and bacterial non-symbiotic nitrogen fixation in paddy soils. *Plant and Soil, Special Volume*, 471–9.

Vlassak, K., Paul, E. A. & Harris, R. E. (1973). Assessment of biological nitrogen fixation in grassland and associated sites. *Pl. Soil*, **38**, 637–49.

Weinhard, P., Balandreau, J., Rinaudo, G. & Dommergues, Y. (1971). Fixation non-symbiotique de l'azote dans la rhizosphère de quelques non-légumineuses tropicales. *Revue Ecol. Biol. Sol*, **8**, 367–73.

Welbank, P. J. & Williams, E. D. (1968). Root growth of a barley crop estimated by sampling with portable powered soil-coring equipment. *J. appl. Ecol.*, **5**, 477–81.

Yoshida, Y. & Ancajas, R. R. (1971). Nitrogen fixation by bacteria in the root zone of rice. *Proc. Soil Sci. Soc. America*, **35**, 156–7.

Yoshida, Y. & Ancajas, R. R. (1973). Nitrogen-fixing activity in upland and flooded rice fields. *Proc. Soil Sci. Soc. America*, **37**, 42–6.

6. Nitrogen fixation in the phyllosphere

JAKOBA RUINEN

This paper considers the salient features of the phyllosphere as a habitat for free-living, nitrogen-fixing microbes. Particular attention is paid to the interrelationships of microbes with living plant cover in natural habitats but it is hoped that such studies will serve as a base for future research on the interrelations between microbes and crop plants.

The search for energy-supplying nutrients for free-living nitrogen fixers in the soil has recently shifted from dead and decaying plant material to the living plant. Information on root exudates has stimulated research in this direction, and fixation in the rhizosphere has been recorded recently (Dobereiner & Campelo, 1971; Dobereiner, Day & Dart, 1972). Although the available evidence strongly points to the phyllosphere as a suitable site for nitrogen fixation, this milieu has, so far, received less attention than the rhizosphere. Nitrogen-fixing organisms often inhabit the leaf surface in great numbers and reports from all over the world are available, mostly from the humid tropics, where the first studies were carried out, and also from the subtropics and temperate zones (Ruinen, 1974). The available data indicate that nitrogen-fixing agents with widely overlapping physiological potentialities occur in the phyllosphere together with numerous non-fixing saprophytes, often conspicuously covering the leaf surface. They comprise the autotrophic Cyanophyceae, either free-living or in symbiosis with lichens and mosses, and heterotrophic bacteria (Table 6.1). Common non-nitrogen-fixing organisms are included in this table for completeness. Plates 6.1 and 6.2 show different examples of such phyllosphere populations. The mere abundance and diversity of the microbial community is an indication of an extremely rich environment in which water enhances its spread and growth.

Studies to date have been rather haphazard. Much available information has been discarded by concentrating on the search for the well-known azotobacters and beijerinckias using orthodox methods and media. Thus potential nitrogen fixers with other growth requirements, but present in this environment, have been missed. Negative accounts do not constitute proof, therefore, that these agents are absent.

Table 6.1. *Common micro-organisms observed in the phyllosphere*

		Australia	Great Britain	Italy	Hawaii	India	Indonesia	Ivory Coast	The Netherlands	New Zealand	Nigeria	Central and South America*	Surinam	USSR
Azotobacter spp.	+	15	—	10x	15	2, 12	10x	—	10x	—	9	5, 6	10b	—
Beijerinckia spp.	+	15	—	10x	15	2, 12	10a	—	—	—	9	5, 6	10b	—
Derxia	+	—	—	—	—	—	10x	10x	—	—	—	—	—	—
Agrobacterium		—	—	—	—	—	10x	10d	—	—	—	—	—	14
Pseudomonadaceae		—	—	—	—	2	10d	—	1	11	—	6	1, 10b	13, 14
Pseudomonas	+	—	—	—	—	—	10x	10d	—	—	—	—	—	—
Xanthomonas		—	—	—	—	—	10x	10x	1	11	—	—	1	—
Azotomonas	+	—	—	—	—	—	10x	—	—	—	—	—	—	—
Mycoplana rubra		—	—	—	—	—	3, 10a	—	—	—	—	—	—	—
Spirillum		—	—	—	—	—	10x	10d	1	—	—	—	10b	—
Myconostoc		—	—	—	—	—	10x	10d	1	—	—	—	—	—
Achromobacteriaceae		—	8	—	—	—	10x	—	1	11	—	—	1	—
Flavobacterium	+	—	8	10x	—	—	10x	10x	1	11	—	—	—	14
Enterobacteriaceae		—	—	—	—	2	10x	10d	1	11	—	—	1	—
Klebsiella	+	—	—	—	—	—	10x	10d	1	—	—	—	1	—
Aerobacter		—	—	—	—	—	10x	10d	1	11	—	6	1	—
Micrococcaceae		—	—	10x	—	—	10x	—	—	11	—	—	—	14
Corynebacteriaceae		—	—	—	—	2	10x	10d	1	—	—	—	1, 10x	—
Bacillus spp.	+	—	—	10x	—	2	10x	10d	1	11	—	6	1, 10x	14
Clostridium	+	—	—	—	—	—	—	10d	—	—	—	—	10b	14
Mycobacteriaceae		—	—	—	—	—	10x	10d	1	11	—	—	—	14
Actinomycetaceae														
Nocardia		—	—	—	—	—	10x	10d	1	—	—	—	1, 10x	—
Cyanophyceae	+	—	—	—	—	—	10x	—	—	—	—	5, 6	10b	—
Chlorophyceae		—	—	—	—	—	10x	—	—	—	—	6	10b	—
Fungi–Yeasts		—	—	10x	—	—	10x	10d	—	4, 11	—	5, 6, 7	10b, c	13, 14
Lichens	+	—	—	—	—	—	10x	—	—	—	—	5, 6, 7	10b	—
Bryophyta		—	—	—	—	—	10x	—	—	—	—	5, 6	10b	—
Protozoa		—	—	—	—	—	10x	—	—	—	—	—	10b	—

+ Nitrogen-fixing capacity demonstrated by laboratory experiments.
* Including Colombia, Panama, Peru and Puerto Rico.

1, Bessems (1973); 2, Bhat, Limaye & Vasantharajan (1971); 3, De Vries & Derx (1951); 4, Di Menna (1959*a*); 5, Edmisten (1970); 6, Harrelson (1969); 7, Hutton & Rasmussen (1970); 8, Jones (1970); 9, Meiklejohn (1962); 10, Ruinen a (1956), b (1961), c (1963), d (1970; 1971), x (1974); 11, Stout (1961; 1964); 12, Vasantharajan & Bhat (1968); 13, Voznyankovskaya (1959; 1962); 14, Voznyankovskaya & Khudyakov (1960); 15, unpublished or personal communication.

Plate 6.1. (*a*) Collodion film with embedded elements of the phyllosphere population on a mature *Citrus* leaf from Surinam. The growth has been partly ripped off. In the upper right-hand corner the impression of fungi and bacteria may be seen. In the lower centre a patch of algal growth on top of the *Azotobacter* layer. More to the right *Trentepohlia* sp. (from Ruinen, 1961). ×525. (*b*) Microbial mucus from the sheath of *Tripsacum laxum* Nash. Note yeast capsules invaded by bacteria and degeneration of the yeast cells (from Ruinen, 1970). ×1800.

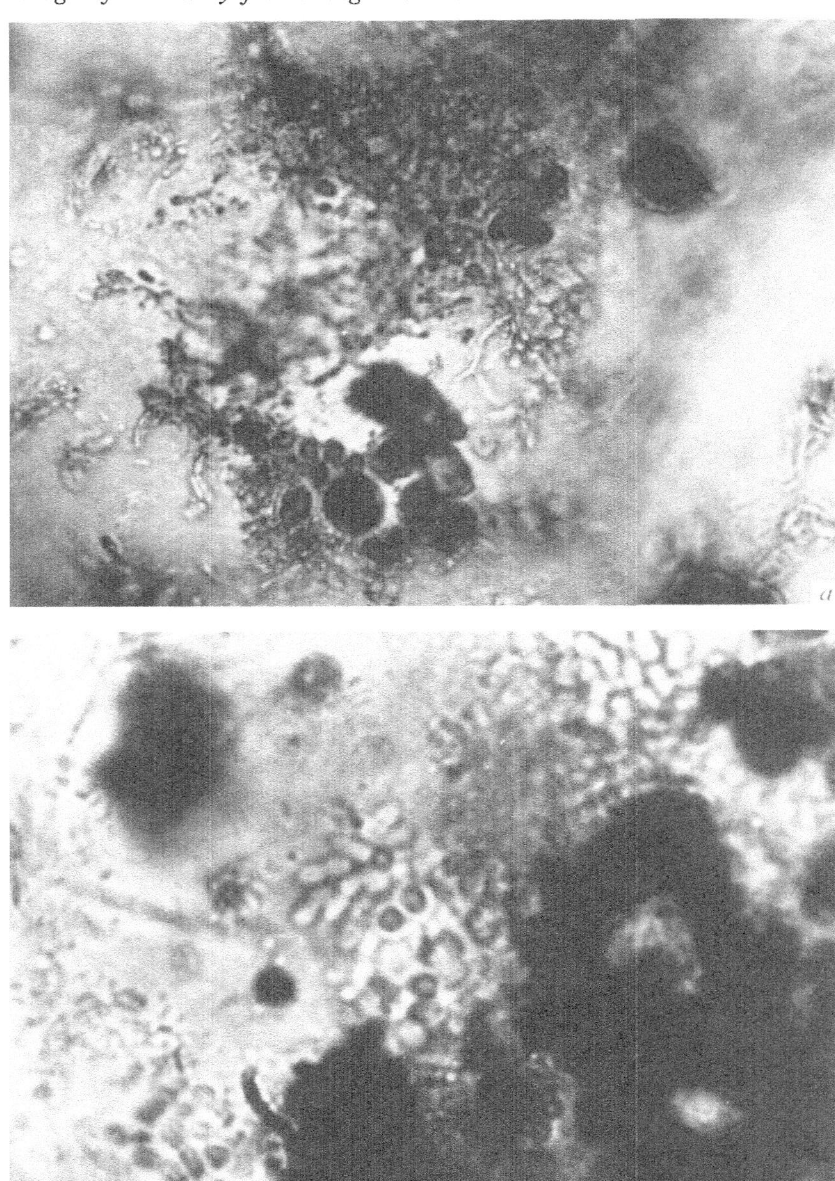

Plate 6.2. (*a*) Surface cut of leaf epidermis of *Pellionia* sp. grown in a tropical greenhouse of the Leiden Botanical Garden: *Azotobacter* and *Beijerinckia* spp., yeasts and algae, besides other bacteria. × 1000. (*b*) Surface cut of *Potamogeton* sp. growing in a pond at Wageningen, with covering by *Azotobacter vinelandii* (from Quispel, 1974). × 2000.

Quantitative data on population density are widely divergent, and should be considered in relation to plant species and the environmental conditions of the habitat. Values ranging from 10 to 20×10^6 nitrogen-fixing bacteria/cm^2 of leaf surface derived from direct counts on citrus and cacao in Surinam do not seem high if compared with plate counts of 3 to 15×10^6 propagules/ml of dew collected from the same plants and counted on nitrogen-free agar (Ruinen, 1961; 1965). The numbers of *Azotobacter* and *Beijerinckia* cells on mulberry leaves cultivated in the open in India (1.8 to $4.8 \times 10^6/cm^2$) fall within the same range (Vasantharajan & Bhat, 1968). Plate counts of oligonitrophilic bacteria in sheath water collected from the lumen between stem and leaf-sheath of *Tripsacum laxum* growing in the Ivory Coast (Ruinen, 1970; 1971) and those reported from Surinam (Bessems, 1973) fall within the same range, viz up to $10^9/cm^2$. The widely different species composition of the population strongly suggests different ecological conditions.

There is little information on the ecological conditions of the various habitats studied except in a few instances, for example in grassland and forest studies in New Zealand (Stout, 1961; 1964; Di Menna, 1959*a*, *b*) and in the research on the nitrogen economy of a tropical forest integrated into the program on energy cycling at the USEAC Center in Puerto Rico (Odum & Pigeon, 1970).

Similar arguments apply to the controversial reports of actual fixation rates in the phyllosphere. There is now irrefutable evidence obtained by Kjeldahl, ^{15}N and acetylene-reduction techniques that the rates are high under optimum conditions of light, humidity, temperature and nitrogen deficiency. Then bacteria and blue-green algae provide appreciable amounts of combined nitrogen to the ecosystem (Edmisten, 1970; Edmisten & Harrelson, 1969; Edmisten & Kline, 1968; Harrelson, 1969; Jones, 1970; Ruinen, 1965; 1974). For example Harrelson's acetylene-reduction figures for epiphyllae scraped from forest leaves were five times higher than those of the scraped leaves. Edmisten and Kline found ^{15}N labelling of 9.67 and 0.70–5.56 atom % in the epiphyllae and scraped leaves respectively, after 48 h exposure to an atmosphere containing 5 % ^{15}N. The ^{15}N labelling in the non-exposed controls was one third of these amounts. It should be noted that a complete removal of surface growth is hardly possible, and that the fixed nitrogen is rapidly assimilated by the leaves. These data agree well with earlier observations (Ruinen, 1965). However, poor or complete lack of nitrogen fixation in the phyllosphere has also been reported (Becking, 1975; Bessems, 1973) but in these cases no data on the ecological conditions were presented.

The phyllosphere is intrinsically an aquatic environment where microbes with widely divergent nutritional requirements are linked in a food chain – rather a food web, which starts and ends in the leaf. In between, the conditions are continuously modified by the different components of the biocoenosis. The nitrogen-fixers have a clear advantage over other organisms being independent of combined nitrogen and not highly selective in their carbon requirements. Thus they grow and fix nitrogen under conditions inhibitory to the development of other colonizers in the ecosystem and possibly profiting from the reduced oxygen tension caused by the rich populations. The picture is complicated further by the diurnal rhythm in the physical environment affecting all processes in the ecosystem and, with it, the flow of metabolites.

As Gates (1968) points out, in any ecosystem the circumstances are shaped by the total sum of organisms, the environment and the processes of interaction between and within all its parts. The controlling systems in the phyllosphere are best illustrated in a much simplified diagram of a plant in its environment, or as an ecological unit. It is seen (Fig. 6.1) that the leaf surface is an environment *par excellence* for microbial growth in which nitrogen fixers may find a selective habitat because the requisites for development and performance are present, viz water, steady supply of minerals, energy-providing nutrients, a low nitrogen level, and a suitable temperature.

In the three compartments the plant is placed in the centre. In the root zone bounded by the rhizosphere and at the top by the phyllosphere, it makes contact with the soil and the atmosphere, respectively. The phyllosphere and rhizosphere form a continuum, the phytosphere, interconnected, but preserving their specific identity by contact with the adjoining compartments. The whole is affected by radiation which influences the environmental factors: light, temperature and humidity in a diurnal and seasonal rhythm, resulting in physiological conditions which vary widely. Geographic, climatic and topographic zonation are further important determinants of activity.

The central position of the plant, and more particularly of the foliage, is self-evident. The latter is the accumulator, transformer and distributor of energy, atmospheric gases and nitrogen compounds. It is also a trapping and pumping device for water and water-soluble matter between the soil, atmosphere and phyllosphere (see Pate, 1973).

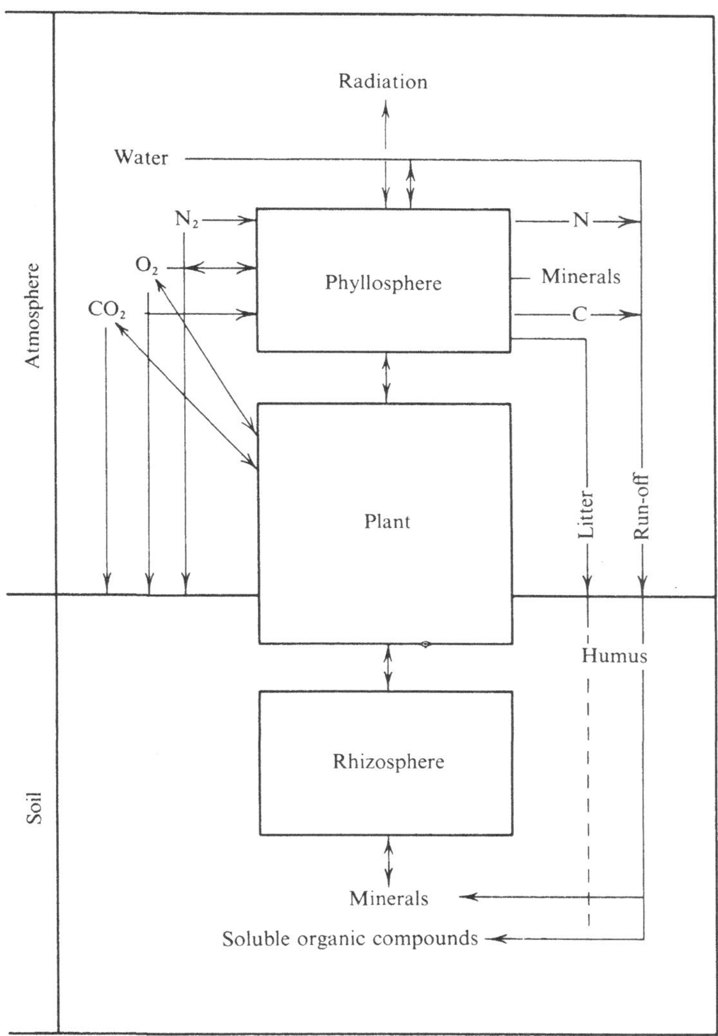

Fig. 6.1. Phyllosphere interrelationships in the compartments of the ecosystem (from Quispel, 1974).

In terrestrial habitats the water is provided as rain, mist and dew. Dry spells on the leaf surface during insolation alternate with nightly wetting by condensation of atmospheric water. On exposed surfaces such as open grassland or the forest canopy, wetting periods may last from evening until late in the morning (more than 15 h). Plant structures and the epiphyllic growths enhance the catchment of water and the leaching of substances from the plants.

In the leachates, essential elements, macro- and micro-nutrients, and organic metabolites are all to be found. The last include mainly sugars and organic acids, but also vitamins, auxins, alkaloids and phenolic compounds. In fact, all substances accumulated in excess of the requirements for plant growth are quickly and easily leached out until an equilibrium between internal and external milieus has resulted (see Tukey, 1970; 1971*a*, *b*). Thus, particular substances will occur in the phyllosphere at times of nutritional imbalance within the growing plant, which in turn is a reflection of soil conditions and the age of the leaves. Thus, any deficiency or excess will show in both leaves and the composition of the exudates. Moreover, conditions of diurnal stress in the canopy caused by fluctuations in the physical environment will result in a cycling within the ecosystem by alternating flows inside and outside the plant, often in short pulses (Gates, 1968; Pate, 1971).

Obviously the proximity of the phyllosphere to the photosynthetic and transforming centre, and the ready release of the latter's products is the one important factor for the development and functioning of a heterotrophic community on the leaf surface. Multiplication and subsequent partial die-back follow diurnal rhythms and cause a selection of organisms best adapted to the temporary changes in the environment. Through the intensive oxygen consumption by the developing aerobic organisms during the water phase, the oxygen supply falls drastically and creates suitable conditions for other microbes tolerating, or requiring, particular P_{O_2} levels for subsistence or performance in an apparently aerobic environment. This offers a part explanation of the presence of anaerobic microbes in the phyllosphere (Table 6.1).

Some of these points may be extended to tropical situations. The major limiting nutrient factor in most tropical soils is shortage of available combined nitrogen and microelements (Date, 1973). Also, high photosynthetic rates by the vegetation create conditions of permanent nitrogen deficiency in the leaves. Then photosynthate is leached into the phyllosphere. The resulting high C/N ratio enhances conditions for effective colonization by oligonitrophilic microbes. This might go some way towards explaining the earlier findings of nitrogen-fixing organisms on the leaf surface of tropical vegetation (Meiklejohn, 1962; Ruinen, 1956).

Evaluation of the amount in leaf drip collected over longer periods evidently underestimates leached carbohydrate because of microbial consumption. Schweizer (1941) reports up to 500 ppm in freshly collected dew and rain drip from trees in Java, and Ruinen (1965) found 10 to 112 ppm from different plant species in Surinam. She also found considerable nitrogen fixation in this dew. The concentrations in the lumen of the leaf sheaths of grasses (up to more than 4000 ppm) are even higher (Bessems, 1973; Ruinen, 1970).

Tukey (1970) reports high leaching rates of carbohydrate under constant wetting (equivalent to 6 % of the dry weight in bean leaves within 24 hours). More spectacular are the figures of Bessems (1973) for carbohydrates in the sheath water of *Tripsacum laxum*, growing in Surinam. After thoroughly washing the lumen with 10 l of sterile water, Bessems observed 500 ppm after 15 minutes. After a second rinse immediately afterwards, the concentrations increased to more than 3000 ppm in 45 min. Sampling was carried out in bright weather on leaves of different ages. Under these conditions active photosynthesis took place in all green parts of the plants and transpiration enhanced transport through the canopy. Effects of bacterial consumption were precluded by the high dilution rates at the start of the experiment and the short exposure time.

Ruinen (1970; 1971), later corroborated by Bessems (1973) in Wageningen and Surinam, and by Ruinen (unpublished) in Indonesia, observed a conspicuous microbial slime in the sheath water of grasses growing in the Ivory Coast, thus pointing to an ample carbohydrate supply (Plate 6.2). An analysis of this sheath water of *Tripsacum laxum* revealed concentrations which were related to the topophysis, viz the age of the leaves, and a shift in population density and composition (Fig. 6.2). It is clear that this sink of leached metabolites maintains the outward flow of these substances which competes with internal sinks such as developing organs within the plant.

Comparison of the upper graph (Fig. 6.2*a*) for sugar concentrations with the lower one for numbers of six groups of microbes shows: (1) high carbohydrate concentrations occur at the top of the culm (the internal sink), increasing with maturity and with the inherent photosynthesizing area of the expanding leaves and translocation from the lower tiers; (2) this sink effect is also found in the laterals which are conspicuous by the depletion of carbohydrate in the enveloping leaf sheath (8, 10, 12); (3) the competitive microbial sink is evident in that microbial population density is inversely related to the carbohydrate concentration during, and after maturation of the leaves; (4) there is

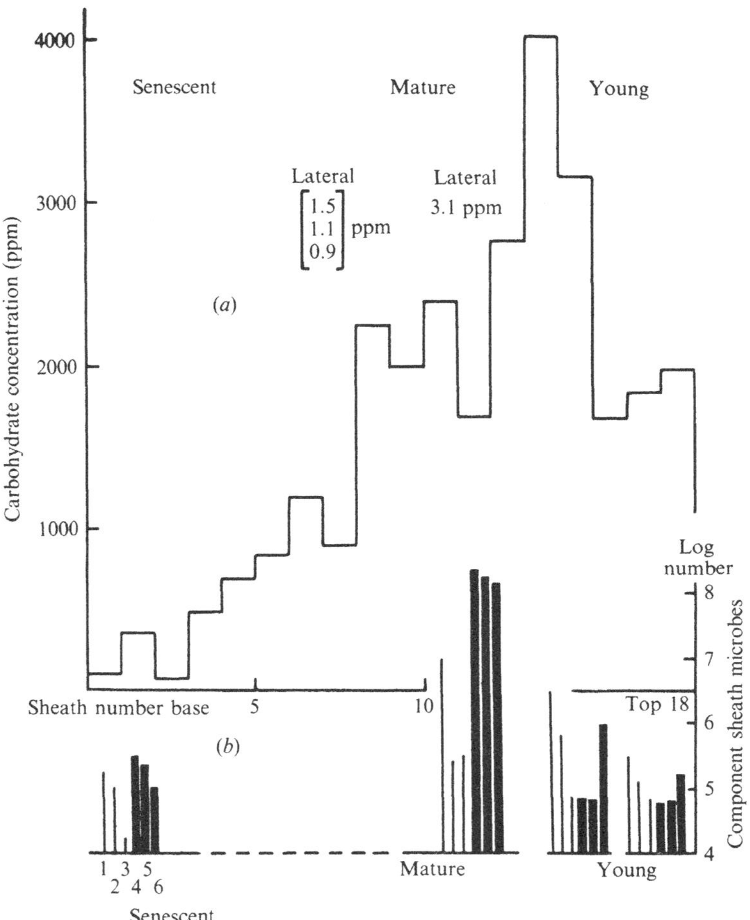

Fig. 6.2. Relation between topophysis and sugar concentration of the moisture contained within successive leaf sheaths of *Tripsacum laxum* Nash grown in the Ivory Coast (*a*) and shift in numbers of 6 components of the leaf sheath population (*b*). Top: funnel of furled leaves; base: partly dead leaf. (1) Mycobacteriaceae; (2) Yeasts; (3) *Myconostoc* spp.; (4) *Spirillum lipoferum*; (5) *Pseudomonas ambigua*; (6) *Klebsiella* sp. Nos. 4–6 are nitrogen fixers (from Quispel 1974).

a shift from the dominance of oligonitrophiles (Mycobacteriaceae, *Myconostoc*, and yeasts) to that of potential nitrogen fixers (*Pseudomonas ambigua*, *Klebsiella* sp. and *Spirillum lipoferum*); (5) the dominance of the latter group is preserved in the lower tiers in reduced numbers as available substrate becomes gradually depleted.

These findings suggest that the phyllosphere is a permanent base for invasion of the rhizosphere and upper soil layers by potential nitrogen fixers which are transported downwards by stemflow and throughfall, together with the excesses of the leachates, as a primary nutrient medium. They also suggest enhanced fixation at these sites.

The nitrogen-fixing capacity of many of the isolates from the phyllosphere has been ascertained in pure and in mixed culture. Dense suspensions of *Azotobacter*, *Pseudomonas ambigua* or *Klebsiella* cells in standing culture in the stationary phase or even on the decline due to lack of substrate, resume nitrogen fixation immediately when glucose is supplied (Ruinen, unpublished). In the field as well as in the laboratory, experiments on crowding of the micro-organisms suggest that this may result in some form of respiratory protection which enhances nitrogen fixation and results in a more economical use of substrate.

A few more additional points may be made on topics such as incompatability of host and nitrogen-fixing agents, symbiotrophy, competition and antibiosis between the components of the population, the specific functions of the components, the limitations in the release and the nature of the substrate, release and cycling of the fixation products, effects of irradiation and pollution, etc. (Ruinen, 1974; see also Preece & Dickinson, 1971). They all tend to complicate the evaluation of non-symbiotic nitrogen fixation in the phyllosphere and its contribution to the build-up of the biomass in natural vegetation. The main difficulty is that the release of the nitrogenous products from the phyllosphere outside the plant inevitably causes losses of available plant nutrient during mineralization of the bacterial protein. Therefore, only a part of the fixed amounts will reach the plant body by foliar and mycorrhizal uptake or by absorption by the roots, while that at the foliar level is most directly available and most economical. Thus increases in biomass scarcely reflect the potentialities of the nitrogen-fixing micro-organisms living on the leaf surface.

However variable, the accumulation of nitrogen may be considerable. In addition to the examples already mentioned, there are the experiments of Vasantharajan & Bhat (1968) who found enhanced shoot and root growth, with significant increases of dry weight and nitrogen content in *Azotobacter*-inoculated sterile mulberry seedlings, compared to the untreated controls. Moore (1963) found in pot cultures of *Eleusine*, in which subsoil effects were excluded, that there was an increase in the plant–soil system corresponding to 110 to 145 kg/ha/4 months, which appeared to be the combined effect of fixation in the rhizosphere

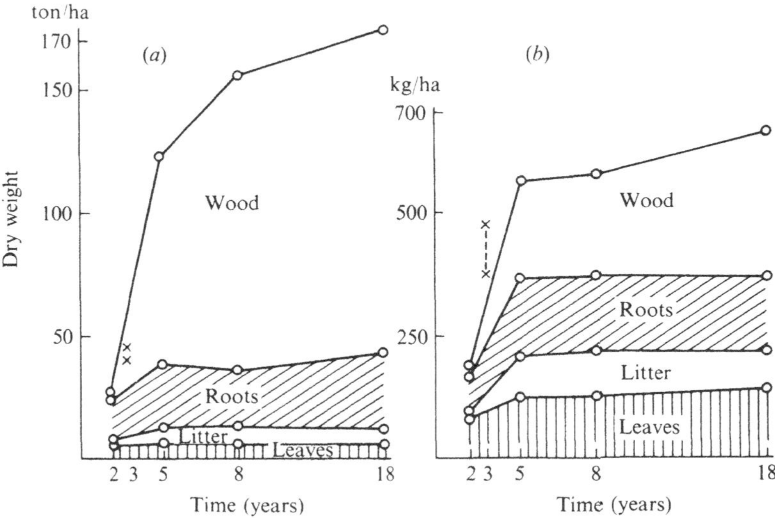

Fig. 6.3. (*a*) Increases in dry matter content of the biomass of a secondary mixed forest in Yangambi (Zaïre) during 18 years of fallow growth; (*b*) increases in nitrogen content. × Dry weight, ×----×. nitrogen content in biomass of three-year-old grass monoculture (drawn from the data of Bartholomew, Meijer & Laudelout, 1953).

and phyllosphere by free-living, nitrogen-fixing bacteria. Parker (1957) showed a nitrogen accumulation under wheat in West Australia of approximately 24 and 110 kg/ha/yr in plant and soil.

Figure 6.3, constructed after the data of Bartholomew, Meijer & Laudelout (1953), contains the distribution of organic matter in an 18-year-old mixed forest fallow in Zaïre following two years of food crop. The last food crop was a mixed cassava–banana plantation; the forest contained leguminous trees. Graph 6.3*a* shows the dry matter estimates in the different parts of the biomass; graph *b* the corresponding nitrogen levels. There is an apparently constant level of living leaf in the closed canopy. There is also a slight increase in the litter fraction during the first five years due to the shift from herbaceous to woody vegetation in the shrub phase of the forest regrowth, combined with decomposition equilibrating on the forest floor compared with the spectacular immobilization in roots and wood which amount to about 20 times the leaf biomass. There is evidence that the leaves are the production centre in the ecosystem. Data for the nitrogen content of the fractions suggest that the foliage is an accumulator and distributor of nitrogenous compounds. The figures for grass monocultures in the same region compare well with the early forest regrowth. It may be noted that the accretion is quickest in the early years of fallow growth.

The diagram does not contain the contribution to soil nitrogen by leaf-fall and run-off, or the losses by volatilization. For the mixed forest region discussed above, increases of soil nitrogen are mentioned ranging from 1000 to 1500 kg/ha in the top soil, and up to 500 kg/ha in the 25 to 100 cm layer during the 18-year fallow period. For a five-year-old, non-leguminous forest regrowth in Nigeria an increase of 585 kg/ha was mentioned for the top 40 cm layer (Jaiyebo & Moore, 1963). The data for different climatic zones do, however, vary widely (400 to 600 kg/ha/yr) depending on vegetation composition and age, seasonal influences, soil, and collecting procedures. These values are counterbalanced by equally variable values for losses by leaching, erosion, volatilization and animal consumption, ranging from 25 to 250 kg/ha/yr (see Bullock, 1963; Date, 1973; Moore, 1966).

It is thus evident that Moore's (1966) dictum holds the directive for future research, that is, the contribution of free-living nitrogen fixers to the nitrogen economy of a vegetation cannot be evaluated until a complete balance sheet is obtained.

References

Bartholomew, W. V., Meijer, J. & Laudelout, H. (1953). Mineral nutrient immobilization under forest and grass fallow in the Yangambi (Belgian Congo) region. *Publications de l'Institut National d'Etudes Agronomiques Congo Belge, Série Scientifique*, **57**, 1–27.

Becking, J. H. (1975). Nitrogen fixation in some natural ecosystems in Indonesia. In *Symbiotic Nitrogen Fixation in Plants* (ed. Nutman, P. S.), pp. 539–50. Cambridge University Press.

Bessems, E. P. M. (1973). Nitrogen fixation in the phyllosphere of Gramineae. *Agric. Res. Rep.*, **786**, 1–68.

Bhat, J. V., Limaye, K. S. & Vasantharajan, V. N. (1971). The effect of the leaf surface microflora on the growth and root exudation of plants. In *Ecology of the Leaf Surface Micro-organisms* (eds. Preece, T. F. and Dickinson, C. H.), Section v, pp. 581–95. Academic Press. London, New York.

Bullock, J. A. (1973). Preliminary observation on litter accumulation and decomposition in Malayan forests. In *Proceedings of the Pre-Congress Conference of the Pacific Science Association* of the Symposium: Planned utilization of the low-land Tropical Forest, Cipajung 1971 (ed. Partoatmodjo, Soeratno), pp. 193–206. Biotrop, National Biological Institute: Bogor.

Date, R. A. (1973). Nitrogen, a major limitation in the productivity of natural communities, crops and pastures in the Pacific area. *Soil Biol. Biochem.*, **5**, 5–18.

De Vries, J. M. & Derx, H. G. (1951). On the occurrence of *Mycoplana rubra* and its identity with *Protaminobacter rubrum. Annls Bogorienses*, **1**, 53–60.

Di Menna, M. E. (1959*a*). Yeasts from the leaves of pasture plants. *N.Z. J. agric. Res.*, **2**, 394–405.

Di Menna, M. E. (1959*b*). Some physiological characters of yeasts from soils and allied habitats. *J. gen. Microbiol.*, **20**, 13–23.

Dobereiner, J. & Campelo, A. B. (1971). Non-symbiotic nitrogen-fixing bacteria in tropical soils. *Plant and Soil, Special Volume*, 457–70.

Dobereiner, J., Day, J. M. & Dart, P. J. (1972). Nitrogenase activity and oxygen sensitivity of the *Paspalum notatum–Azotobacter paspali* association. *J. gen. Microbiol.*, **71**, 103–6.

Edmisten, J. (1970). Preliminary studies on the nitrogen budget of a tropical rain forest. In *A Tropical Rain Forest* (eds. Odum, H. T. and Pigeon, R. F.), pp. 211–15. Office of Information Services US Atomic Energy Commission: Springfield, Virginia.

Edmisten, J. & Harrelson, M. A. (1969). Nitrogen fixation in the epiphyllae at El Verde. In *The Rain Forest Project Annual Report*, FY-1969, pp. 136–8. USAEC Report PRNC-**129**, Puerto Rico Nuclear Center: Puerto Rico.

Edmisten, J. & Kline, R. J. (1968). Nitrogen fixation by epiphyllae. In *The Rain Forest Project Annual Report*, FY-1968, pp. 141–3. USAEC Report PRNC-**119**, Puerto Rico Nuclear Center: Puerto Rico.

Gates, D. M. (1968). Towards understanding ecosystems. *Adv. ecol. Res.* **5**, 1–34.

Harrelson, M. A. (1969). Tropical epiphyllae and nitrogen fixation. Ph.D. Thesis, University of Georgia.

Hutton, R. S. & Rasmussen, R. A. (1970). Microbiological observations in a tropical forest. In *A Tropical Rain Forest* (eds. Odum, H. T. and Pigeon, R. F.), pp. 43–56. Office of Information Services US Atomic Energy Commission: Springfield, Virginia.

Jaiyebo, E. O. & Moore, A. W. (1963). Soil nitrogen accretion under different covers in a tropical rain forest environment. *Nature, Lond.*, **197**, 317–18.

Jones, K. (1970). Nitrogen fixation in the Douglas fir, *Pseudotsuga douglasii. Ann. Bot.*, **34**, 239–44.

Meiklejohn, J. (1962). Microbiology of the nitrogen cycle in some Ghana soils. *Emp. J. exp. Agric.*, **30**, 115–26.

Moore, A. W. (1963). Nitrogen fixation in latosolic soil under grass. *Pl. Soil*, **19**, 127–38.

Moore, A. W. (1966). Non-symbiotic nitrogen fixation in soil and soil–plant systems. *Soils and Fertilizers*, **29**, 113–28.

Odum, H. T. & Pigeon, R. F. (eds.) (1970). *A Tropical Rain Forest.* Office of Information Services, US Atomic Energy Commission: Springfield, Virginia.

Parker, C. A. (1957). Nonsymbiotic nitrogen fixing bacteria in soil. III. Total nitrogen changes in a field soil. *J. Soil Sci.*, **8**, 48–59.

Pate, J. S. (1971). Movement of nitrogenous solutes in plants. In *N-15 in Soil Plant Studies.* Proceedings of a meeting on recent developments in the use of N-15 in soil–plant studies, Sofia 1969 (ed. IAEA), Vienna 1971, pp. 165–87.

Pate, J. S. (1973). Uptake, assimilation and transport of nitrogen compounds by plants. *Soil Biol. Biochem.*, **5**, 109–19.

Preece, T. F. & Dickinson, C. H. (eds.) (1971). *Ecology of Leaf Surface Micro-organisms.* Academic Press: London, New York.

Quispel, A. (ed.) (1974). *The Biology of Nitrogen Fixation.* North-Holland: Amsterdam.

Ruinen, J. (1956). Occurrence of *Beijerinckia* species in the 'phyllosphere'. *Nature, Lond.*, **177**, 220–1.

(1961). The phyllosphere. I. An ecologically neglected milieu. *Pl. Soil*, **15**, 81–109.

(1963). II. Yeasts from the phyllosphere. *Antonie van Leeuwenhoek*, **29**, 425–38.

(1965). III. Nitrogen fixation in the phyllosphere. *Pl. Soil*, **22**, 375–94.

(1970). V. The grass sheath, a habitat for nitrogen-fixing micro-organisms. *Pl. Soil*, **33**, 661–71.

(1971). The grass sheath as a site for nitrogen fixation. In *Ecology of Leaf Surface Micro-organisms* (eds. Preece, T. F. and Dickinson, C. H.), pp. 567–79. Academic Press: London, New York.

(1974). Nitrogen fixation in the phyllosphere. In *The Biology of Nitrogen Fixation* (ed. Quispel, A.), pp. 121–69. North-Holland: Amsterdam.

Schweizer, J. (1941). Wat halen een koffie- en rubberaanplant uit den grond en wat geven zij aan den grond terug. *Verslag Vergadering van Proefstations Personeel*, **28**, 221–38.

Stout, J. D. (1961). A bacterial survey of some New Zealand forest lands, grasslands and peats. *N.Z. J. agric. Res.*, **4**, 1–30.

(1964). Bacterial populations of some grazed pastures in Hawke's Bay. *N.Z. J. Agric. Res.*, **7**, 91–117.

Tukey, H. B. (1970). The leaching of substances from plants. *Ann. Rev. Pl. Physiol.*, **21**, 305–24.

(1971*a*). Leaching of metabolites from foliage and its implications in the tropical rain forest. In *A Tropical Rain Forest* (eds. Odum, H. T. and Pigeon, R. F.), pp. 155–60. Office of Information Services, US Atomic Energy Commission: Springfield, Virginia.

(1971*b*). Leaching of substances from plants. In *Ecology of Leaf Surface Micro-organisms* (eds. Preece, T. F. and Dickinson, C. H.), pp. 67–80. Academic Press: London, New York.

Vasantharajan, V. N. & Bhat, J. V. (1968). Interrelations of micro-organisms and mulberry. II. Phyllosphere microflora and nitrogen fixation in leaf and root surfaces. *Pl. Soil*, **28**, 258–67.

Voznyakovskaya, Y. M. (1959). A new species of epiphytic micro-organism *Pseudomonas epiphytica* n.sp. *Mikrobiologiya, USSR*, **28**, 840–3.
Voznyakovskaya, Y. M. (1962). Epiphytic yeasts. *Mikrobiologiya, USSR*, **31**, 504–10.
Voznyakovskaya, Y. M. & Khudyakov, Y. P. (1960). Species composition of the epiphytic microflora of living plants. *Mikrobiologiya, USSR*, **29**, 97–103.

7. Associative growth of nitrogen-fixing bacteria with other micro-organisms

V. JENSEN & ESTHER HOLM

It is always difficult to draw conclusions from pure culture studies in the laboratory as to what will happen under natural conditions. With regard to non-symbiotic nitrogen fixation, it has been claimed that considerably larger amounts of nitrogen can be fixed when the nitrogen fixers grow in association with other micro-organisms, as they do in nature, than when they are grown in pure culture. Consequently non-symbiotic nitrogen fixation might be a more important process in nature than pure culture studies suggest.

The authors' attention was drawn to this problem in connection with examinations of a non-identified nitrogen-fixing bacterium (probably related to *Derxia*), which was isolated some years ago from a forest soil (Petersen & Holm, 1964). In preliminary experiments with this bacterium, only small amounts of nitrogen (and sometimes no nitrogen at all) were fixed by pure cultures, whereas contaminated cultures sometimes gave much higher yields. It was also observed that fungal colonies occurring as random air-contaminants in Petri dish cultures on nitrogen-deficient agar media, strongly stimulated bacterial growth surrounding the colonies. It was decided, therefore, to make a more detailed study of the growth of this organism in association with other micro-organisms, especially fungi, and of the effects of the associative growth on the amounts of nitrogen fixed.

Review of previous literature

Beijerinck & van Delden (1902) first observed that *Azotobacter chroococcum* fixed more nitrogen when grown in association with other bacteria, e.g. strains of *Agrobacterium*, *Aerobacter* and *Clostridium*, than it did in pure culture. Later similar observations have been made by Richards (1939), Lind & Wilson (1942), Lal & Achari (1953), Parker (1955), Bouisset & Breuillaud (1958), Gadgil & Bhide (1960), and Fedorov & Savkina (1961) for associations of *Azotobacter* with strains of *Achromobacter*, *Aerobacter*, *Bacillus*, *Chromobacterium*, *Clostridium*, *Pseudomonas*, *Rhizobium*, various actinomycetes and a number of unidentified soil bacteria.

Bucksteeg (1936) was probably the first to study associations of *Azotobacter* with cellulolytic bacteria, and this line of research was continued by Jensen (1940; 1941*a*) and by various Russian workers (see Fedorov, 1960, and Rubenchik, 1963, for references). It was demonstrated that such associations could grow and fix nitrogen on cellulose media, whereas *Azotobacter* was unable to grow in pure culture.

A beneficial effect of fungi on nitrogen fixation by *Azotobacter* was first observed by Löhnis & Westermann (1909), who found that the presence of *Aureobasidium pullulans* stimulated nitrogen fixation by the somewhat doubtful species *Azotobacter vitreum*. A stimulatory effect of *A. pullulans* and of two species of *Saccharomyces* on *Azotobacter* was also observed in experiments mentioned by Fedorov (1960), and Ruinen found a similar effect of *Rhodotorula* sp. on *Azotobacter* (unpublished results mentioned by Mulder, Lie & Woldendorp, 1969). Bucherer (1933) and Vartiovaara (1938) demonstrated that unavailable carbon sources such as cellulose can be made available to *Azotobacter* by fungal as well as by bacterial activity.

Hills (1916) found a very slight increase in nitrogen fixation by *Azotobacter* in soil cultures in the presence of protozoa, and both Nasir (1923) and Cutler & Bal (1926) found generally a higher nitrogen fixation in the presence of protozoa than in their absence. Similar results are presented by Fedorowa-Winogradowa & Gurfein (1928) in experiments with soil amoebae, and by Hirai & Hino (1928) for soil ciliates. Hervey & Greaves (1941) found a stimulatory effect of both living and heat-killed cells of *Colpoda maupasii*, and Nikulyuk (cited by Rubenchik, 1963) found that the presence of protozoa caused a decrease in cell number but an increase in the amount of nitrogen fixed in *Azotobacter* cultures. Recently these relationships have been reinvestigated by Darbyshire (1972*a*, *b*), who found that *Colpoda steini* stimulated *Azotobacter*, but only at temperatures which were suboptimal for nitrogen fixation.

Jensen (1941*b*) noted that nitrogen-fixing clostridia, like *Azotobacter*, can develop on cellulose media in association with various cellulolytic bacteria and fungi, and similar observations are mentioned by Fedorov (1960) and Rubenchik (1963). Associations of *C. pasteurianum* with oligonitrophilic soil bacteria are described by Fedorov & Kalininskaya (1959). Emtsev (1960) found that the presence of *Bacillus closter-*

oides in cultures of *C. pasteurianum* enhanced growth and nitrogen fixation.

The mutualistic relationship between clostridia (especially *C. pasteurianum*) and *Azotobacter* is well known, and according to Rubenchik (1963) filtrates of young *Azotobacter* cultures also have a stimulatory effect on *C. pasteurianum*. Rice & Paul (1972) have described experiments in which decomposition products of other soil bacteria are utilized by nitrogen-fixing clostridia. Line & Loutit (1973) have recently drawn attention to the fact that clostridia can fix nitrogen and reduce acetylene when occurring as contaminants in aerobically incubated cultures of other bacteria. The same observation has been made in the authors' laboratory, where it was found that certain *Bacillus* species could grow vigorously on nitrogen-deficient agar plates under aerobic conditions when contaminated with clostridia (Z. Tyle, private communication).

Associations involving *Beijerinckia* have been considered only by Dommergues & Mutaftschiev (1965). They observed that nitrogen fixation in cultures of both *B. indica* and *B. fluminensis* was strongly stimulated by the presence of *Lipomyces starkeyi*, and also by the addition of filtrates of *Lipomyces* cultures. *Beijerinckia* and *Lipomyces* often occur together in tropical environments.

According to Fedorov & Kalininskaya (1959) mixed cultures of oligonitrophilic soil bacteria often fix nitrogen, even if none of the implicated bacteria can fix nitrogen in pure culture. The same authors (1961*a*) isolated a new bacterial species (later named *Mycobacterium flavum*), which was able to fix appreciable amounts of nitrogen in mixed cultures with species of *Escherichia*, *Pseudomonas* or *Flavobacterium*, whereas only a little nitrogen was fixed in pure culture. Later it was found that *Mycobacterium flavum* could grow well in a medium with ethanol, or organic acids, as carbon source, but that nitrogen fixation was still improved in the presence of other bacteria (Fedorov & Kalininskaya, 1961*b*).

Kalininskaya (1967*a*) suggested the term 'facultative symbiotrophic nitrogen fixers' for micro-organisms with a limited ability for nitrogen fixation in pure culture, but which fix nitrogen readily in mixed cultures with other micro-organisms. In subsequent papers (Kalininskaya, 1967*b*, *c*) it was demonstrated that such nitrogen-fixing bacterial associations are widespread in the most diverse types of soil.

Nitrogen-fixing bacteria of the *Klebsiella–Aerobacter* type were isolated from the rhizosphere by Evans, Campbell & Hill (1972), and it was found that in pure culture these bacteria could fix nitrogen only anaerobically, whereas they could also fix nitrogen under aerobic conditions in mixed cultures with *Rhizobium japonicum*. Bacteria of the same type were isolated by Seidler, Aho, Raju & Evans (1972) from decaying wood, where they apparently grew in a mutualistic association with the wood-decomposing fungi.

Associations involving phototrophic nitrogen-fixing bacteria have been studied especially by Japanese workers (Okuda, Yamaguchi & Kobayashi, 1961; Okuda & Kobayashi, 1961; 1963; Kobayashi, Katayama & Okuda, 1965; Katayama, Kobayashi & Okuda, 1965). It was found that *Rhodopseudomonas capsulatus* could fix more nitrogen when grown in association with *Azotobacter vinelandii*, *Bacillus subtilis* or *Bacillus megaterium* than it could in pure culture under the same conditions.

Finally it may be mentioned that associations of nitrogen-fixing blue-green algae with various bacteria have also been studied, and it has been observed that the presence of bacterial contaminants in algal cultures often greatly stimulates nitrogen fixation (Bunt, 1961; Bjälfve, 1962; Laloraya & Mitra, 1964).

Materials and methods

Bacteria

Two strains of the nitrogen-fixing bacterium labelled N-61 and N-63 were studied. Their properties which are almost identical for both organisms are as follows: Rods, often irregular, non-motile, encapsulated, Gram-negative to Gram-variable, yellow-pigmented, strictly aerobic. There is no fermentation of carbohydrates, no reduction of nitrate, no hydrolysis of gelatin, casein, starch or lipids, no production of indole or hydrogen sulphide. Urease is produced. The pH optimum is about 6 to 8; the temperature optimum about 25 to 30 °C (Plate 7.1).

Media

The basal medium (a modified Burk medium) contained the following in g/l: KH_2PO_4, 0.12; K_2HPO_4, 0.8; $MgSO_4.7H_2O$, 0.2; $CaSO_4.2H_2O$, 0.1; $Fe_2(SO_4)_3$, 0.01; Na_2MoO_4, 0.005; yeast extract, 0.01. Agar was added in concentrations of 2.0 % to solid media and 0.1 % to liquid media. Unless otherwise stated, $(NH_4)_2SO_4$ was added in a concentration of 0.05 g/l and 1.0 % glucose, autoclaved in the medium, was used as carbon source.

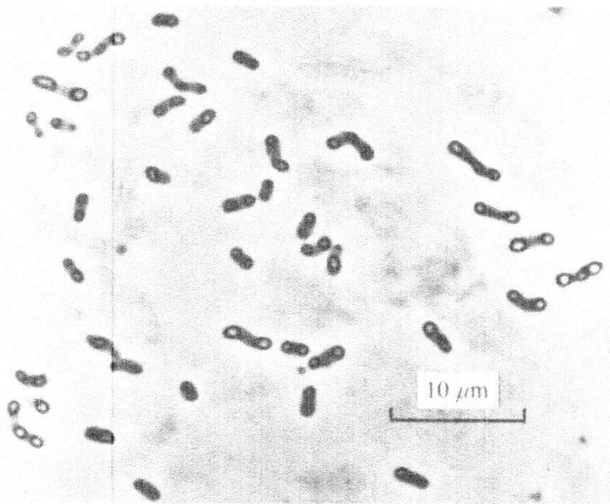

Plate 7.1. Bacterium N-63, grown on nitrogen-deficient medium. Phase contrast.

Cultivation methods

Cultures for nitrogen determinations were grown in 20 ml liquid medium in 100 ml Erlenmeyer flasks, closed with metal caps. Cultures for acetylene-reduction tests were grown on agar slopes in culture tubes (18 mm diameter).

Chemical analyses

In cases where residual glucose was to be determined, filter-sterilized glucose was added aseptically to the basal medium. The determinations were made enzymatically by means of the GOD-Perid reagent (Boehringer Mannheim GMBH).

Total nitrogen was determined by a semi-micro Kjeldahl procedure, using a mercury catalyst.

Acetylene-reduction tests

Before testing, the culture tubes were closed with Suba-seals and 5 ml of acetylene was injected into each tube. The ethylene produced was measured by gas chromatography after incubation times from 2 to 6 hours.

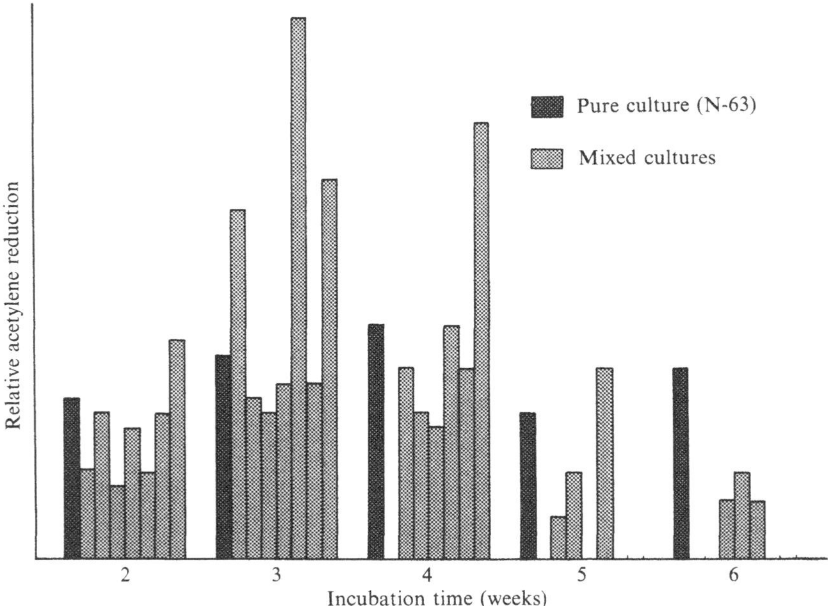

Fig. 7.1. Comparison of nitrogen-fixing activity in pure and mixed cultures, measured by acetylene-reduction tests after varying incubation periods (from 2 to 6 weeks) at 25 °C.

The eight columns represent from left to right: (1) N-63, pure culture; (2) N-63 mixed with *Penicillium* sp.; (3) N-63 mixed with *Aureobasidium pullulans*; (4) N-63 mixed with *Cladosporium herbarum*; (5) N-63 mixed with *Alternaria* sp.; (6) N-63 mixed with *Mucor hiemalis*; (7) N-63 mixed with *Lipomyces starkeyi*; and (8) N-63 mixed with another bacterium (N-668, a small, Gram-negative, oligonitrophilic rod).

Results

In preliminary experiments a number of moulds and yeasts were tested for a possible influence on nitrogen fixation when grown in mixed cultures with the two bacteria, N-61 and N-63. The activity of nitrogen fixation was examined both by Kjeldahl analyses and by acetylene-reduction tests (Fig. 7.1). The results were not very consistent, and it was found in subsequent experiments that changes in the growth environment may alter the stimulatory or antagonistic effect of a certain fungus. Apparently the stimulatory effect of the associated growth depends upon a rather delicate balance between the two organisms, which can easily shift in one direction or the other.

It was possible, however, to distinguish between (1) fungi, which usually had a stimulatory effect: *Mucor hiemalis*, *Mortierella ramanniana*, *Cladosporium herbarum*, *Penicillium* sp., *Alternaria* sp.; (2) fungi, which usually had little or no influence on nitrogen fixation: *Chaetomium* sp., *Aureobasidium pullulans, Candida curvata, Torulopsis aeria, Lipomyces starkeyi*; and (3) fungi, which usually had an inhibitory effect: *Penicillium vermiculatum, Stachybotrys* sp., *Saccharomyces cerevisiae, Saccharomyces pastorianus, Rhodotorula mucilaginosa.* A single bacterial strain was also tested. This was a small, Gram-negative, oligonitrophilic rod, which had been isolated from the same enrichment culture, as N-63. It usually showed a stimulatory effect (Fig. 7.1).

Table 7.1. *Influence of initial nitrogen content on nitrogen fixation in pure cultures of N-63 and in cultures mixed with various fungi. The cultures were incubated for two weeks at 25 °C*

	Nitrogen fixation (mg N/100 ml medium)							
	Without $CaCO_3$				With $CaCO_3$ added			
$(NH_4)_2SO_4$ g/l	0	0.05	0.25	1.00	0	0.05	0.25	1.00
Initial nitrogen content (mg/100 ml)	0.45	1.56	6.17	22.84	0.45	1.56	6.17	22.84
N-63, pure culture	4.96	8.85	3.25	0.21	2.42	5.25	2.08	0.12
N-63 mixed with *Penicillium* sp.	8.81	13.02	0.20	0.41	—	—	—	—
N-63 mixed with *Alternaria* sp.	4.46	21.03	0.05	0.15	5.39	6.82	6.52	0.45
N-63 mixed with *Cladosporium herbarum*	4.85	17.05	0.14	0	5.12	6.54	6.75	1.32
N-63 mixed with *Mortierella ramanniana*	3.87	3.39	3.35	0	4.57	5.73	2.21	0.95
N-63 mixed with *Mucor hiemalis*	—	14.61	7.16	0	5.08	7.85	6.06	0
N-63 mixed with *Aureobasidium pullulans*	3.35	3.98	2.37	0	4.43	4.93	2.55	0.08
N-63 mixed with *Torulopsis aeria*	3.69	3.59	5.47	0	—	—	—	—
N-63 mixed with *Lipomyces starkeyi*	3.73	5.30	0.68	0	—	—	—	—

The initial nitrogen content of the medium had a strong influence on the development of both pure and mixed cultures (Table 7.1). In pure culture the bacteria develop very slowly and sometimes fail to grow at all if the initial nitrogen content is too low. Most efficient nitrogen fixation is obtained generally with 10 to 20 mg/l of combined nitrogen. If the initial nitrogen concentration is above this, nitrogen fixation decreases, and practically ceases with 200 mg/l of combined nitrogen.

In mixed cultures the equilibrium between the two organisms appeared to be determined largely by the nitrogen content of the

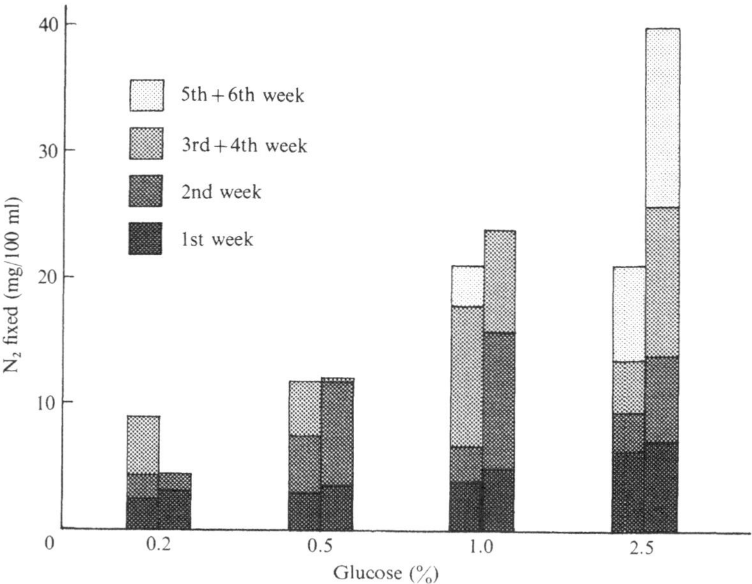

Fig. 7.2. Influence of varying glucose concentrations on nitrogen fixation in pure and mixed cultures.
Left column: Pure culture of N-63.
Right column: Mixed culture of N-63 and *Alternaria* sp.
Nitrogen determinations were made on individual cultures, incubated for 1, 2, 4 and 6 weeks respectively at 25 °C.

medium. At the lowest nitrogen level the fungi showed very slight growth and had little effect on nitrogen fixation. At the highest nitrogen levels, on the other hand, the fungi usually dominated and often depressed nitrogen fixation. *Penicillium*, *Alternaria* and *Cladosporium* had a strong stimulatory effect on nitrogen fixation in the presence of 15 mg/l of combined nitrogen but there was a clear inhibitory effect with 60 mg/l. This inhibition was due mainly to acid production and could be neutralized by the addition of calcium carbonate, but in most other cases nitrogen fixation was lower in presence of calcium carbonate than in its absence.

The influence of varying glucose concentrations on nitrogen fixation was also examined in pure cultures of N-63 and in mixed cultures of N-63 and *Alternaria* sp. (Fig. 7.2). The fungus had a slightly stimulatory effect at all concentrations during the first week, but at the lowest concentration the mixed culture stopped fixing after two weeks, and the final amount of nitrogen was twice as high in the pure culture as in the mixed culture. At 0.5 % glucose the final amounts of nitrogen fixed

Table 7.2. *Nitrogen fixation in relation to the amounts of glucose decomposed in pure cultures of N-63 and in cultures mixed with various fungi. Individual cultures were incubated for 1, 2, 3, 4, 5, 6 and 7 weeks respectively, at 25 °C, and then analysed for total nitrogen and residual glucose. Initial glucose content in the medium 975 mg/100 ml*

	N-63 Pure culture			N-63 mixed with *Mucor hiemalis*			N-63 mixed with *Alternaria* sp.			N-63 mixed with *Cladosporium herbarum*		
	Glucose decomp.	Nitrogen fixed		Glucose decomp.	Nitrogen fixed		Glucose decomp.	Nitrogen fixed		Glucose decomp.	Nitrogen fixed	
Incubation period (weeks)	(mg/ 100 ml)	(mg/ 100 ml)	(mg/ g gluc.)	(mg/ 100 ml)	(mg/ 100 ml)	(mg/ g gluc.)	(mg/ 100 ml)	(mg/ 100 ml)	(mg/ g gluc.)	(mg/ 100 ml)	(mg/ 100 ml)	(mg/ g gluc.)
1	98	1.84	18.8	125	2.64	21.1	120	2.36	19.7	95	2.15	22.6
2	173	4.38	25.3	300	6.20	20.7	238	6.16	25.9	178	5.46	30.7
3	333	7.80	23.4	570	8.00	14.0	458	10.07	22.0	313	7.97	25.5
4	538	11.70	21.7	863	17.69	20.5	780	16.03	20.6	533	—	—
5	718	15.01	20.9	975	18.92	19.4	975	17.81	18.3	775	17.47	22.5
6	905	18.45	20.4	975	18.89	19.4	975	17.42	17.9	975	20.96	21.5
7	975	20.16	20.7	975	18.74	19.2	975	17.00	17.4	975	20.82	21.4

were almost identical. At 1.0 % glucose, the mixed culture stopped fixing after four weeks, but although the pure culture continued to fix nitrogen during weeks 5 to 6, it did not reach the level of the mixed culture. At 2.5 % glucose both cultures continued their activity throughout the six week period, with the activity of the mixed culture being highest until finally it contained almost twice as much nitrogen as the pure culture.

The efficiency of nitrogen fixation using filter-sterilized glucose was studied in cultures in which total nitrogen and residual glucose was measured after varying incubation periods (Table 7.2). Under these conditions the pure cultures grew very slowly, and at 25 °C the glucose was not completely decomposed until after seven weeks. The mixed cultures showed faster growth and contained the highest amounts of nitrogen during the first weeks. During the last two weeks, however, the pure cultures reached the same level. The efficiency of nitrogen fixation in pure cultures was a little above 20 mg/g glucose decomposed. It was slightly lower in the associations with *Mucor hiemalis* and *Alternaria* sp. and slightly higher with *Cladosporium herbarum*. The stimulating effect of the fungi seems to be mainly on the growth rate, there being little effect on the efficiency of nitrogen fixation.

In experiments using autoclaved glucose, growth was considerably faster than in cultures growing on filter-sterilized glucose (Fig. 7.3), but the total amount of nitrogen fixed was slightly lower. The curves show an initial lag for media with filter-sterilized glucose, whereas growth in media with autoclaved glucose followed a rectilinear curve from the very beginning. Furthermore, the curves indicate that the stimulatory effects of autoclaving and the stimulation caused by the presence of fungi are independent.

Several other carbon sources were tested. In media with glycerol the development of both pure and mixed cultures was almost identical to the development on filter-sterilized glucose. Sodium pyruvate was utilized well at the start, but growth soon stopped in both pure and mixed cultures as the medium became alkaline. On starch media the pure cultures were unable to grow because of lack of amylase, but some growth and nitrogenase activity occurred in the mixed cultures.

The effect of aeration on nitrogen fixation was examined by comparing the amounts of nitrogen fixed in static cultures and in shake cultures (Table 7.3). Increased aeration inhibited nitrogen fixation almost completely in the pure cultures, whereas some growth and nitrogen fixation occurred in the presence of fungi, especially *Penicillium* sp. Apparently

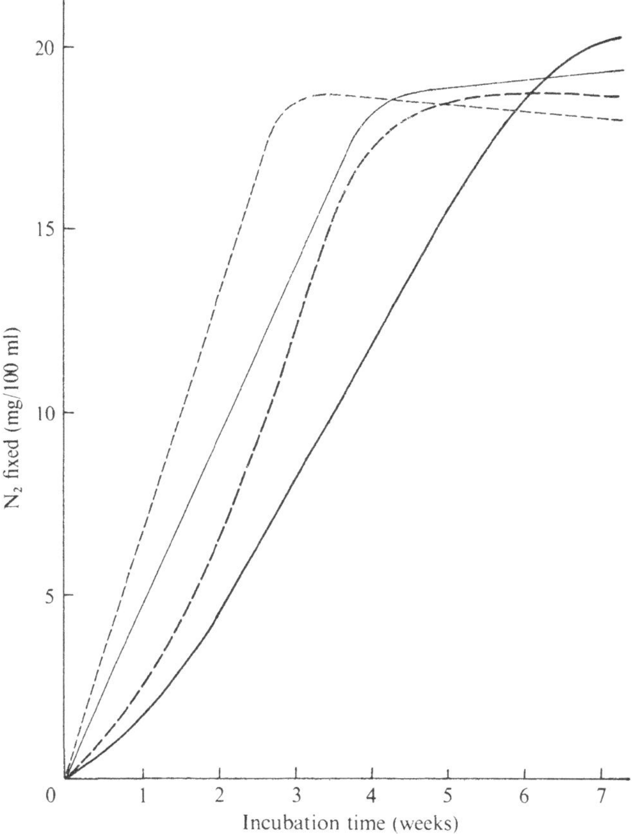

Fig. 7.3. Comparison of nitrogen fixation by pure and mixed cultures in media with filter-sterilized or autoclaved glucose as carbon source.
—— N-63, pure culture, filter-sterilized glucose.
- - - N-63 mixed with *Mucor hiemalis*, filter-sterilized glucose.
—— N-63, pure culture, autoclaved glucose.
- - - N-63 mixed with *Mucor hiemalis*, autoclaved glucose.

the oxygen consumption of the fungi was able to counteract to some degree the effect of the increased oxygen tension. In another experiment an attempt was made to vary the oxygen availability by varying the surface/volume ratio of the cultures (Table 7.4). However, it was not possible in this way to obtain a high enough oxygen tension in the medium to suppress nitrogen fixation. A depth of a few millimetres seems to be enough for the bacterium to establish a sufficiently reduced oxygen tension for nitrogen fixation to take place freely. There was no clear correlation between the stimulatory effect of the fungi and the

Table 7.3. *Comparison of nitrogen fixation in static cultures and cultures incubated on a continuous shaker. Incubation for 2 weeks at 25 °C*

	Static cultures (mg N/100 ml)	Shaken cultures (mg N/100 ml)	Shaken cultures (% Reduction)
N-63, pure culture	7.31	0.03	99.6
N-63 mixed with *Penicillium* sp.	7.75	1.97	74.6
N-63 mixed with *Alternaria* sp.	13.15	0.85	93.5

Table 7.4. *Nitrogen fixation by pure and mixed cultures in relation to the surface/volume ratio of the growth medium. Nitrogen fixation of the mixed cultures expressed as percentage of nitrogen fixation of the corresponding pure cultures*

		Nitrogen fixation			
Incubation period (weeks)	Surface/volume ratio (cm^2/cm^3)	N-63 pure culture (mg N/100 ml)	N-63 mixed with *Mucor hiemalis*	N-63 mixed with *Alternaria* sp.	N-63 mixed with *Penicillium* sp.
			(% of pure culture)		
3	2.51	14.50	131	146	152
	1.19	9.26	216	206	215
	0.54	9.28	167	193	193
	0.11	2.18	213	228	284
5	2.51	20.15	91	117	101
	1.19	17.87	113	124	118
	0.54	17.05	98	122	114
	0.11	4.90	159	161	188

surface/volume ratio, except that stimulation generally was strongest under conditions where pure cultures showed the slowest growth.

The effect of incubation temperature on the relationships between N-63 and various fungi was also examined (Fig. 7.4). It was found that nitrogen fixation by pure cultures increased with increasing temperature throughout the range examined (18–32 °C), whereas the final amounts of nitrogen fixed by the mixed cultures were highest at 18 °C or 25 °C. After three weeks, the stimulatory effect of the fungi was strongest at 25 °C, where the mixed cultures had fixed about twice as much nitrogen as the pure culture. After five weeks, the effect was strongest at 18 °C, where the mixed cultures had fixed from 127 % to 168 % more nitrogen than the pure culture. Superficially it might

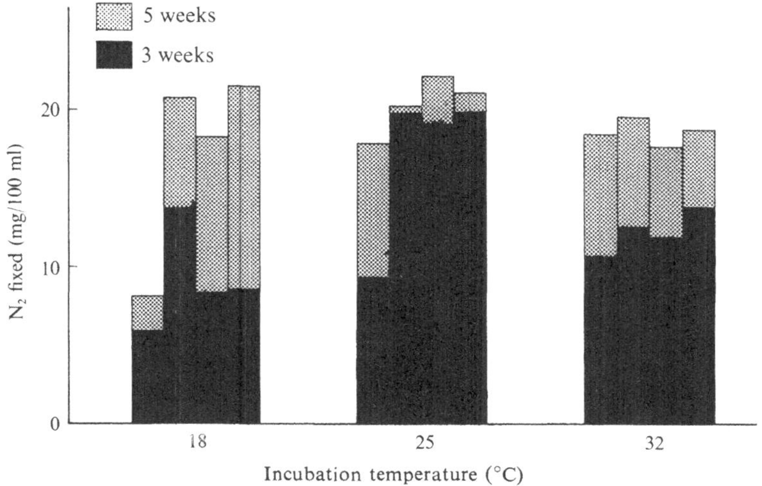

Fig. 7.4. Influence of incubation temperature on nitrogen fixation by pure and mixed cultures.

The four columns represent from left to right: (1) N-63, pure culture; (2) N-63 mixed with *Mucor hiemalis*; (3) N-63 mixed with *Alternaria* sp.; (4) N-63 mixed with *Penicillium* sp.

appear as if the temperature optimum was lower in the presence of fungi than in their absence. However, the results can also be explained by assuming a general increase in growth rate in the presence of fungi throughout the temperature range. If the experiment had continued over a longer incubation period, the pure cultures might possibly have reached the level of the mixed cultures at all three temperatures.

The observations of concentric zones of increased growth around accidental fungal colonies on Petri dish cultures of N-61 and N-63 suggest production of diffusible growth-promoting substances by the fungi. In order to test this hypothesis a number of experiments were made with N-63 growing in cultures, where half of the normal medium had been replaced by either filter-sterilized or autoclaved cultures of fungi, grown on the same medium for one week (Fig. 7.5). Initially the addition of both filtrates and autoclaved cultures caused a slight depression of nitrogen fixation, but later the effect was reversed in the case of three of the four fungi tested (*M. hiemalis*, *Alternaria* sp., *Penicillium* sp.). The stimulatory effects after 20 and 30 days incubation were rather similar to those caused by the presence of the living fungi, and both positive and negative effects were stronger in experiments with autoclaved cultures than with filtrates. No obvious explanation

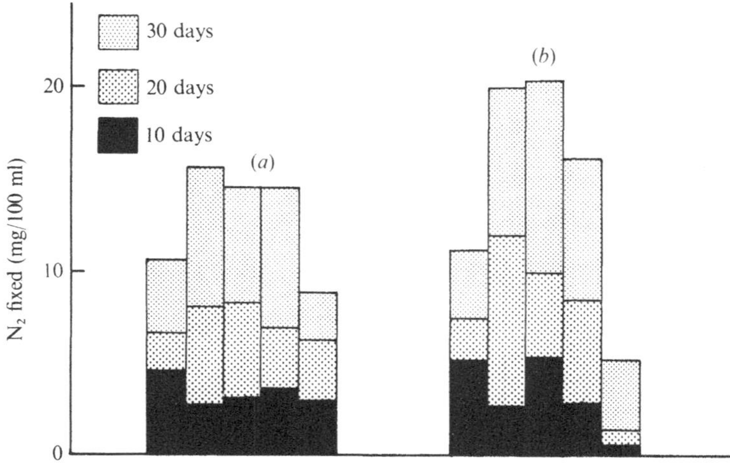

Fig. 7.5. Influence of addition of filter-sterilized (*a*) or autoclaved (*b*) fungal cultures on nitrogen fixation by pure cultures of strain N-63.

The five columns represent from left to right first an untreated control (1) and then four cultures with addition of filter-sterilized or autoclaved cultures of the following fungi: *Mucor hiemalis* (2), *Alternaria* sp. (3), *Penicillium* sp. (4), and *Cladosporium herbarum* (5).

The three different shades on the columns represent nitrogen fixation after incubation periods of 10, 20 and 30 days respectively, at 25 °C.

was found for the persistently negative effect of *C. herbarum*. Parallel experiments were made with measurements of nitrogenase activity by acetylene reduction, and a slight stimulation was observed in some cultures, but the results were generally less consistent than those based on total nitrogen determinations.

In a final series of experiments the distribution of nitrogen between cells and medium was examined at various intervals after inoculation in order to see whether the concentration of nitrogen in the medium was kept at a lower level in mixed than in pure cultures. This might be expected, as the fungi apparently were supplied with nitrogen from the bacteria, but no significant difference between pure and mixed cultures in this respect could be detected. The initial nitrogen content of 10–15 mg/l of medium decreased slightly during the first 10 days and then increased gradually again to 30 to 40 mg/l medium after 30 days.

Discussion and conclusions

The mutualistic relationship between cellulolytic micro-organisms and non-cellulolytic nitrogen-fixing bacteria is easily understood (Bucherer,

1933; Bucksteeg, 1936; Vartiovaara, 1938; Jensen, 1940, 1941*a*, *b*; Rubenchik, 1963), and several analogous relationships have been described, where an associative micro-organism partly decomposes an otherwise unavailable carbon source and thereby renders it available to the nitrogen fixer (Richards, 1939; Fedorov & Kalininskaya, 1961*a*, *b*; Katayama *et al.*, 1965; Kobayashi *et al.*, 1965; Kalininskaya, 1967*c*). This process was also demonstrated in the present investigation in experiments where N-63 was cultivated with various fungi on starch media.

The associative micro-organisms may improve the growth medium in other ways. Lind & Wilson (1942) found that a *Bacillus* sp. could improve growth conditions for *Azotobacter* by decomposing iron humates so that iron was made available. Other essential metals, e.g. molybdenum, occurring in suboptimal concentrations may be liberated in a similar manner by the decomposition of organic substances.

Organic growth factors may also be produced. In cases where living and dead cells, or culture filtrates, are effective in stimulating nitrogen fixation, the most natural explanation is the production of necessary or stimulatory growth factors by the associative organisms (Hervey & Greaves, 1941; Fedorov & Kalininskaya, 1959; 1961*a*, *b*; Dommergues & Mutaftschiev, 1965; Sandrak, 1971). In the case of blue-green algae the effect of contaminating bacteria may be due to production of substances with hormone properties, e.g. indole-3-acetic acid (Bunt, 1961). Convincing evidence of the presence of growth-promoting substances in certain fungal cultures has also been presented in the present study. In most cases the observed stimulation of growth and nitrogen fixation can be accounted for in this way, but no attempt has been made as yet to isolate or identify the active substances.

Growth conditions for the nitrogen-fixing micro-organism may also be improved by the removal of toxic substances, which were either present originally in the growth environment or were produced as by-products during growth of the nitrogen-fixing bacterium. Special attention has been paid to oxygen. The relationship is quite obvious in connection with the anaerobic or facultatively anaerobic nitrogen fixers, which are active only at low redox potentials (Jensen, 1941*b*; Rubenchik, 1963; Evans *et al.*, 1972; Line & Loutit, 1973). However, a high oxygen tension may also be inhibitory to nitrogen fixation by strictly aerobic micro-organisms, and it has been shown in some cases that the observed stimulation is due to removal of excess oxygen

(Mulder *et al.*, 1969; Darbyshire, 1972*a*, *b*). In the present experiments continuous shaking during incubation completely inhibited nitrogen fixation by pure cultures of N-63, but some growth and nitrogen fixation was possible in shaken cultures in the presence of fungi, which apparently were able to reduce the toxic effect of free oxygen to some extent.

Removal of other inhibitory substances, e.g. accumulated metabolic products, dead cells, etc., has been mentioned as an explanation of a stimulatory effect (Rubenchik, 1963) but not supported by experimental evidence. Mulder *et al.* (1969) suggested removal of excreted nitrogen compounds that otherwise would have reduced nitrogen fixation, as another possible explanation, but direct experimental evidence also seems to be lacking here. In the experiments with varying concentrations of the energy substrate (Fig. 7.2) we found that nitrogen fixation by the pure culture was nearly the same with 2.5 % glucose in the medium as with 1 %, whereas the mixed culture could fix much larger amounts of nitrogen with 2.5 % glucose. It seems that nitrogen fixation by pure cultures is suppressed when the nitrogen content reaches about 200–250 mg/l, whereas the mixed culture can fix more nitrogen possibly because excess nitrogen is assimilated by the fungus. However, this theory is not supported by the observed distribution of nitrogen between cells and medium in the cultures.

Cutler & Bal (1926) explained the increased nitrogen fixation by *Azotobacter* in the presence of protozoa by assuming that the efficiency of the *Azotobacter* cells was maintained for a longer time, when old and non-viable cells were constantly phagocytosed by the protozoa. However, Darbyshire (1972*a*, *b*) did not observe any delay in the normal development of the *Azotobacter* culture in the presence of protozoa, and the proportion of viable cells was not affected. In the present study the stimulatory effect was not due to a prolongation of the bacterial growth period. On the contrary this period was shortened as the growth rate increased in mixed culture.

The relationships between incubation temperature and stimulatory effect of associative organisms found in the present investigations (Fig. 7.4) are rather similar to the results presented by Darbyshire (1972*a*). The differences between pure and mixed cultures are largest at the start and diminish gradually. At high temperatures such a difference is less noticeable and the stimulatory effect is most easily observed at lower temperatures, where the mixed culture remains more active over a longer period.

We thank Alice Stengade Andersen, Hanne Kristiansen and Dorte Rasmussen for carrying out the numerous Kjeldahl analyses, and Ulla Hartmann-Petersen and Jens Wolstrup for assistance with the acetylene-reduction tests.

References

Beijerinck, M. W. & Delden, A. van (1902). Ueber die Assimilation des freien Stickstoffs durch Bakterien. *Zentrbl. Bakt. ParasitKde Abt. II*, **9**, 3–43.

Bjälfve, G. (1962). Nitrogen fixation in cultures of algae and other micro-organisms. *Physiologia Pl.*, **15**, 122–9.

Bouisset, L. & Breuillaud, J. (1958). Etude bacteriologique de germes oligonitrophiles isolés du sol. *Bull. Soc. Hist. Nat. Toulouse*, **93**, 479–82.

Bucherer, H. (1933). Experimenteller Beitrag zur Frage der ernährungs-biologischen Wechselbeziehungen zwischen Bakterien und Pilzen. *Zentrbl. Bakt., ParasitKde, Abt. II*, **89**, 273–83.

Bucksteeg, W. (1936). Zur Frage der symbiontischen Beziehungen zwischen zellulosezersetzenden und stickstoffbindenden Bakterien. *Zentrbl. Bakt., ParasitKde, Abt. II*, **95**, 1–24.

Bunt, J. S. (1961). Blue-green algae, growth, *Nature, Lond.*, **192**, 1274–5.

Cutler, D. W. & Bal, D. V. (1926). Influence of protozoa on the process of nitrogen fixation by *Azotobacter chroococcum. Ann. appl. Biol.* **13**, 516–34.

Darbyshire, J. F. (1972*a*). Nitrogen fixation by *Azotobacter chroococcum* in the presence of *Colpoda steini*. – I. The influence of temperature. *Soil Biol. Biochem.*, **4**, 359–69.

Darbyshire, J. F. (1972*b*). Nitrogen fixation by *Azotobacter chroococcum* in the presence of *Colpoda steini*. – II. The influence of agitation. *Soil Biol. Biochem.*, **4**, 371–6.

Dommergues, Y. & Mutaftschiev, S. (1965). Fixation synergique de l'azote atmosphérique dans les sols tropicaux. *Annls Inst. Pasteur*, **109**, *Suppl.* no. 3, 112–20.

Emtsev, V. T. (1960). Symbiotic relationships between *Clostridium* and *Bacillus closteroides*. *Mikrobiologiya*, **29**, 529–35.

Evans, H. J., Campbell, N. E. R. & Hill, S. (1972). Asymbiotic nitrogen-fixing bacteria from the surfaces of nodules and roots of legumes. *Can. J. Microbiol.*, **18**, 13–21.

Fedorov, M. V. (1960). *Biologische Bindung des atmosphärischen Stickstoffs.* VEB Deutscher Verlag der Wissenschaften: Berlin.

Fedorov, M. V. & Kalininskaya, T. A. (1959). Nitrogen-fixing activity of mixed cultures of oligonitrophile microorganisms. *Mikrobiologiya*, **28**, 323–30.

(1961*a*). A new species of nitrogen-fixing Mycobacterium and its physiological properties. *Mikrobiologiya* **30**, 9–14.

(1961*b*). Relationship of a nitrogen-fixing Mycobacterium (*Mycobacterium* sp. 301) to various sources of carbon and supplementary growth factors. *Mikrobiologiya*, **30**, 833–40.

Fedorov, M. V. & Savkina, E. A. (1961). Interrelationships between *Azoto-*

bacter and typical rhizosphere bacteria of corn. *Mikrobiologiya*, **29**, 862–7.

Fedorowa-Winogradowa, T. & Gurfein, L. N. (1928). Beiträge zur Frage der Wirkung des Bodenamöben auf das Wachstum und die Entwicklung des *Azotobacter chroococcum* unter Versuchsbedingungen auf sterilem Boden. *Zentrbl. Bakt. ParasitKde, Abt. II*, **74**, 14–22.

Gadgil, P. D. & Bhide, V. P. (1960). Nitrogen fixation by *Azotobacter* in association with some associated soil microorganisms. *Proc. natn. Inst. Sci. India*, **26***B*, 60–3.

Hervey, R. J. & Greaves, J. E. (1941). Nitrogen fixation by *Azotobacter chroococcum* in the presence of soil protozoa. *Soil Sci.* **51**, 85–100.

Hills, T. L. (1916). The relation of protozoa to certain groups of soil bacteria. *J. Bacteriol.*, **1**, 423–33.

Hirai, H. & Hino, I. (1928). Influence of soil protozoa on nitrogen fixation by *Azotobacter*. *Proc. Papers First Int. Congr. Soil Sci.*, **3**, 160–5.

Jensen, H. L. (1940). Nitrogen fixation and cellulose decomposition by soil microorganisms. I. Aerobic cellulose-decomposers in association with *Azotobacter*. *Proc. Linnean Soc. New South Wales*, **65**, 543–56.

(1941*a*). Nitrogen fixation and cellulose decomposition by soil microorganisms. II. The association between *Azotobacter* and facultative-aerobic cellulose-decomposers. *Proc. Linnean Soc. New South Wales*, **66**, 89–106.

(1941*b*). Nitrogen fixation and cellulose decomposition by soil microorganisms. III. *Clostridium butyricum* in association with aerobic cellulose-decomposers. *Proc. Linnean Soc. New South Wales*, **66**, 239–49.

Kalininskaya, T. A. (1967*a*). Methods of isolating and culturing nitrogen-fixing bacterial associations. *Mikrobiologiya*, **36**, 282–5.

(1967*b*). Quantitative evaluation of facultatively symbiotrophic nitrogen fixers. *Mikrobiologiya*, **36**, 436–8.

(1967*c*). Utilization of different carbon sources by nitrogen-fixing bacterial associations. *Mikrobiologiya*, **36**, 520–4.

Katayama, T., Kobayashi, M. & Okuda, A. (1965). Nitrogen-fixing microorganisms in paddy soils. XIV. Nitrogen fixation in mixed culture of photosynthetic bacteria (*R. capsulatus*) with other heterotrophic bacteria (4), association with *B. megaterium*. *Soil Sci. Pl. Nutr.*, **11**, 78–83.

Kobayashi, M., Katayama, T. & Okuda, A. (1965). Nitrogen-fixing microorganisms in paddy soils. XIII. Nitrogen fixation in mixed culture of photosynthetic bacteria (*R. capsulatus*) with other heterotrophic bacteria (3), association with *B. subtilis*. *Soil Sci. Pl. Nutr.*, **11**, 74–77.

Lal, A. & Achari, T. K. T. (1953). Microbiological studies. VII. Studies on *Azotobacter* and *Actinomycetes* in relation to nitrogen fixation and cellulose utilization. *Proc. natn. Acad. Sci. India*, **23***B*, 137–49.

Laloraya, V. K. & Mitra, A. K. (1964). Fixation of elementary nitrogen by *Scytonema hoffmanni* Ag. ex Born. et Flah. and *Fischerella muscicola* (Thuret) Gomont in pure and unialgal culture. *Current Sci.*, **33**, 619–20.

Lind, C. J. & Wilson, P. W. (1942). Nitrogen fixation by *Azotobacter* in association with other bacteria. *Soil Sci.*, **54**, 105–11.

Line, M. A. & Loutit, M. W. (1973). Nitrogen-fixation by mixed cultures of aerobic and anaerobic micro-organisms in an aerobic environment. *J. gen. Microbiol.*, **74**, 179–80.

Löhnis, F. & Westermann, T. (1909). Ueber stickstofffixierende Bakterien. *Zentrbl. Bakt. ParasitKde, Abt. II*, **22**, 234–54.

Mulder, E. G., Lie, T. A. & Woldendorp, J. W. (1969). Biology and soil fertility. In *Natural Resources Research IX, Soil Biology*, pp. 163–208. UNESCO.

Nasir, S. M. (1923). Some preliminary investigations on the relationship of protozoa to soil fertility with special reference to nitrogen fixation. *Ann. appl. Biol.*, **10**, 122–33.

Okuda, A. & Kobayashi, M. (1961). Production of slime substance in mixed cultures of *Rhodopseudomonas capsulatus* and *Azotobacter vinelandii. Nature, Lond.*, **192**, 1207–8.

Okuda, A. & Kobayashi, M. (1963). Symbiotic relationship between *Rhodopseudomonas capsulatus* and *Azotobacter vinelandii. Mikrobiologiya*, **32**, 936–45.

Okuda, A., Yamaguchi, M. & Kobayashi, M. (1961). Nitrogen-fixing micro-organisms in paddy soils. VII. Products in mixed culture of *Rhodopseudomonas capsulatus* with *Azotobacter vinelandii. Soil Sci. Pl. Nutr.*, **7**, 115–18.

Parker, C. A. (1955). Non-symbiotic nitrogen-fixing bacteria in soil. II. Studies on *Azotobacter. Aust. J. Agric. Res.*, **6**, 388–97.

Petersen, E. J. & Holm, E. (1964). On nitrogen fixation in Danish deciduous forests. *Royal Veterinary and Agricultural College, Copenhagen, Yearbook 1964*, 209–26.

Rice, W. A. & Paul, E. A. (1972). The organisms and biological processes involved in asymbiotic nitrogen fixation in waterlogged soil amended with straw. *Can. J. Microbiol.*, **18**, 715–23.

Richards, E. H. (1939). Note on the effect of temperature on a mixed culture of two organisms in symbiotic relation. *J. Agric. Sci.* **29**, 302–5.

Rubenchik, L. I. (1963). Azotobacter *and its use in agriculture.* Israel Program for Scientific Translations: Jerusalem.

Sandrak, N. A. (1971). Effect of extracellular *Azotobacter* secretions on other microorganisms. *Mikrobiologiya*, **40**, 603–7.

Seidler, R. J., Aho, P. E., Raju, P. N. & Evans, H. J. (1972). Nitrogen fixation by bacterial isolates from decay in living white fir trees (*Abies concolor* (Gord. and Glend.) Lindl.). *J. gen Microbiol.*, **73**, 413–16.

Vartiovaara, U. (1938). The associative growth of cellulose-decomposing fungi and nitrogen-fixing bacteria. *J. scient. agric Soc. Finland*, **10**, 241–64.

8. Nitrogen fixation associated with vascular aquatic macrophytes

W. S. SILVER & A. JUMP

The widespread occurrence of free-living, nitrogen-fixing bacteria and blue-green algae in a variety of habitats has been known for many years but it is only since the development of the acetylene-reduction method (see Hardy, Burns & Holsten, 1973) that extensive assessments have been made of nitrogen fixation in natural materials (Knowles, 1975; Mague, 1975) rather than with cell masses of isolated nitrogen-fixing organisms. Such studies have revealed that ecologically significant accretions of nitrogen may be found in diverse habitats including the rhizosphere of terrestrial plants (Dobereiner, Day & Dart, 1972; Dommergues, Balandreau, Rinaudo & Weinhard, 1973), the surface of sea grasses (Goering & Parker, 1972), and the rhizosphere of marine (Patriquin & Knowles, 1972) and freshwater angiosperms (Bristow, 1974). The latter was of special interest to us as rhizosphere related nitrogen fixation might provide sufficient nitrogen to facilitate the growth of submerged, emergent and floating freshwater macrophytes, many of which pose serious problems to the proper use and management of Florida's vast waterways.

The object of this investigation was to determine whether the extensive nitrogen fixation by rhizosphere microbes, which has been reported for temperate aquatic macrophytes (Bristow, 1974), also occurred in aquatic angiosperms characteristic of subtropical ponds, lakes and rivers.

Methods

The plant species investigated were the submerged *Hydrilla verticillata* (Florida elodea), the marginal *Hydrocotyle umbellata* (pennywort) and *Cyperus tetragonus* (water sedge), and the floating *Eichornia crassipes* (water hyacinth). Plants were gently removed from the lake bottom with their roots intact, agitated to remove large soil particles and mud, and transported to the laboratory in plastic buckets containing the habitat water. Healthy root tissue was cut from the plant with a razor blade, washed thoroughly in tap water, blotted to remove excess water, and 3 g fresh weight of tissue was added to 125 ml Erhlenmeyer flasks. The flasks were fitted with a rubber stopper through which a 5 cm-long glass tube passed which was capped tightly with a 14 mm rubber serum stopper.

The flasks contained the plant tissue, distilled water and appropriate supplements (glucose, antibiotics) in a total volume of 15 ml. All solutions were degassed with argon prior to use. The stoppered flasks were purged with argon flowing through 19 gauge syringe needles for 2 min. Sufficient gas was removed to allow the injection of acetylene to give a final concentration of 0.1 atm.

In some experiments aerobic flasks with 0.2 atm acetylene were assayed. In all experiments control flasks without acetylene were included to assess the extent of endogenous ethylene production (negligible). Flasks were incubated at 28 °C.

Samples, each of 0.2 ml, were withdrawn at appropriate times with a 1 ml plastic, gas-tight syringe and analyzed for ethylene production using a Varian gas chromatograph (Model 1400) equipped with a 5 ft × 1 in. (*c.* 1.52 m × 25.4 mm) stainless-steel column packed with 100–200 mesh Porapak N (Waters Associates) and operated at 90 °C. Under these conditions the retention times for acetylene and ethylene were 6 min and 4 min respectively. The molar concentration of ethylene produced was calculated from an appropriate standard curve. At the termination of each experiment the dry weight of the plant tissue was determined in order to express the results on a dry weight basis.

Results and discussion

Preliminary experiments with *Hydrilla* roots revealed that tissue gassed with acetylene aerobically formed little ethylene, with or without glucose supplementation, and that under anaerobic conditions glucose stimulated ethylene production markedly (Fig. 8.1*a*, curves 1, 2). Epiphytic bacteria rather than plant tissue were responsible for the activity, as evidenced by the fact that the addition of streptomycin and penicillin abolished ethylene production. During the second day of incubation the liquid in some of the flasks containing plant species became turbid, possibly because of outgrowth of antibiotic-resistant mutants (Fig. 8.1*c*, curve 3).

Acetylene reduction was not unique to *Hydrilla* roots (Fig. 8.1*b–d*), because the roots of the other plant species tested showed a similar activity. It is of interest that the water hyacinth, which typically has a

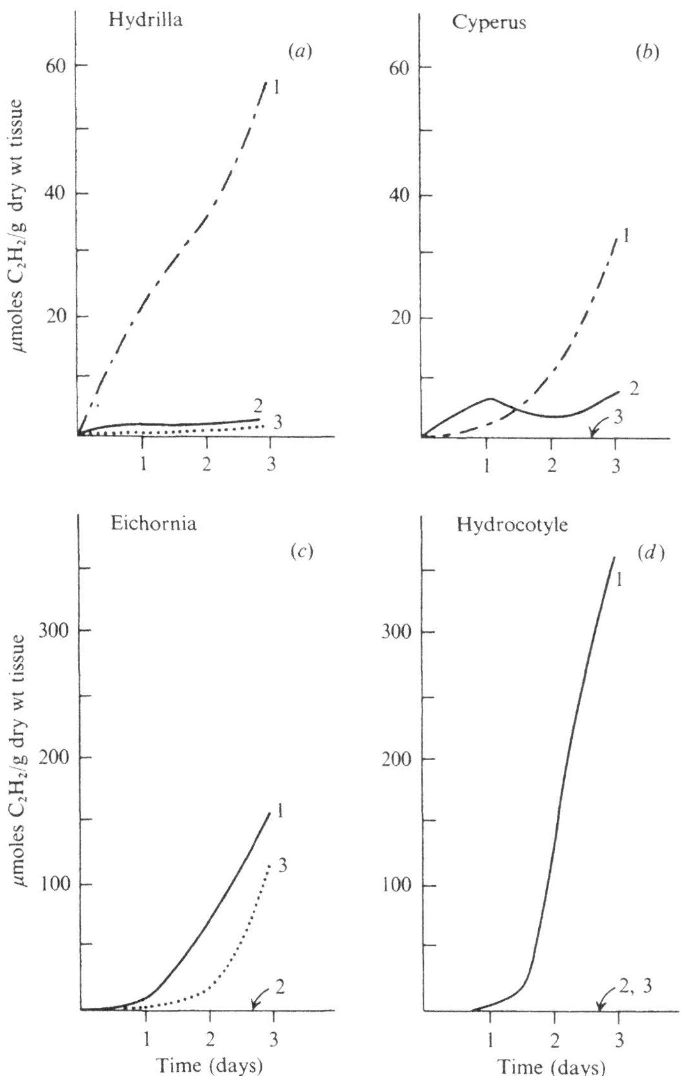

Fig. 8.1. Acetylene reduction to ethylene catalyzed by roots of aquatic angiosperms. Analyses described in test. Total gas volume 130 ml: total liquid volume, 15 ml. Curve 1: 1 ml of 1.0 M glucose; curve 2: no supplement; curve 3: 1 ml of 1.0 M glucose, 100 μg streptomycin sulfate and 100 units penicillin G per ml final concentration.

floating habit and therefore a relatively aerobic root environment, was also active under anaerobic conditions. This may be a reflection of the plant collection site which was very shallow with an extremely dense stand in which the levels of dissolved oxygen at the roots was only 0.4 ppm. In other experiments not reported here hyacinth roots collected from plants growing in a rapidly flowing river reduced acetylene aerobically. Thus, it appears that the optimal P_{O_2} for acetylene reduction is a function of the oxygen tension of the roots in the stand.

The factor common to these plant types (including the water hyacinth in dense stands) is the anaerobic nature of the root growth zone. The sediment is typically black, rich in detritus and has a marked odor of hydrogen sulfide when disturbed. Furthermore, bottom sediments not supplemented with glucose and acetylene evolve methane. This is similar to the habitat which typifies *Typha* and *Glyceria* (Bristow, Cardenas, Fullerton & Sierra, 1972) genera in which rhizosphere associated nitrogen-fixation has been established (Bristow, 1974). Acetylene-reduction is not merely the result of sediment nitrogen-fixers because the presence of roots greatly increased the reduction. Although accurate estimates of the population of nitrogen-fixers have not yet been made, on plating of the washings of freshly collected roots of water sedge which were subjected to vigorous agitation in a vortex mixer mucoid colonies of facultative nitrogen-fixers developed which covered the surface of nitrogen-free mineral agar.

We do not have sufficient data at this time to make estimates of how much of the nitrogen requirement of the plants may be obtained via N_2 fixed in the rhizosphere. However, for *Typha* this can represent 10–20 % of its nitrogen requirement (Bristow, 1974). Access to nitrogen in the form of N_2 may be important in the phosphate rich waters typical of the region studied, because nitrogen is the limiting nutrient. We have noted lush growth of *Hydrilla* in non-eutrophic lakes whose waters contain only traces of ammonium and nitrate ions, and few blue-green algae. In *Hydrilla*, leaf epiphytes show negligible acetylene-reducing activity in contrast to the leaf epiphytes of some terrestrial plants (Ruinen, 1970).

These preliminary results suggest that, in waters where nitrogen is the factor limiting plant growth, rhizosphere associated N_2-fixation may be of considerable significance. This should also apply to coastal waters which are known to be nitrogen limited (Ryther & Dunstan, 1971). Indeed, we have noted high rates of acetylene reduction associated with mangrove roots and rhizomes (Silver & Pukatzki, unpublished observations). The presence of N_2-fixing epiphytes on the mangrove rhizome may be an important factor in the reclaimation of sandy shores where mangrove stands are difficult to establish (Savage, 1972).

The capable technical assistance of J. A. Illingworth is gratefully acknowledged. This work was supported by a research contract with the Florida Department of Natural Resources and the Southwest Florida Water Management District.

References

Bristow, J. M. (1974). Nitrogen fixation in the rhizosphere of freshwater angiosperms. *Can. J. Bot.*, **52**, 217–21.

Bristow, J. M., Cardenas, J., Fullerton, T. M. & Sierra, F. (1972). *Aquatic Weeds*. International Plant Protection Center, Oregon State University: Corvallis.

Dobereiner, J., Day, J. M. & Dart, P. J. (1972). Nitrogenase activity and oxygen sensitivity of the *Paspalum notatum–Azobacter paspali* association. *J. gen. Microbiol.*, **71**, 103–16.

Dommergues, Y., Balandreau, J., Rinaudo, G. & Weinhard, P. (1973). Non-symbiotic nitrogen fixation in the rhizospheres of rice, maize and different tropical grasses. *Soil Biol. Biochem.*, **5**, 83–9.

Goering, J. J. & Parker, P. L. (1972). Nitrogen fixation by epiphytes on sea grasses. *Limnol. Oceanogr.*, **15**, 320–3.

Hardy, R. W. F., Burns, R. C. & Holsten, R. D. (1973). Applications of the acetylene–ethylene assay for measurement of nitrogen fixation. *Soil Biol. Biochem.*, **5**, 47–81.

Knowles, R. (1975). The significance of asymbiotic dinitrogen fixation by bacteria. In *Dinitrogen* (N_2) *Fixation* (ed. Hardy, R. W. F.). Wiley-Interscience: New York (in press).

Mague, T. H. (1975). Ecological aspects of dinitrogen fixation by blue-green algae. In *Dinitrogen* (N_2) *Fixation* (ed. Hardy, R. W. F.). Wiley-Interscience: New York (in press).

Patriquin, D. & Knowles, R. (1972). Nitrogen fixation in the rhizosphere of marine angiosperms. *Marine Biol.*, **16**, 49–58.

Ruinen, J. (1970). The phyllosphere. v. The grass sheath, a habitat for nitrogen-fixing micro-organisms. *Pl. Soil.* **33**, 661–71.

Ryther, J. H. & Dunstan, W. M. (1971). Nitrogen, phosphorus and eutrophication in the coastal marine environment. *Science*, **171**, 1008–13.

Savage, T. (1972). *Florida mangroves as shoreline stabilizers*. Professional Papers Series, No. **19**. Marine Research Laboratory, Fla. Dept. Nat. Res.: St. Petersburg, Florida.

PART II

Nitrogen fixation by free-living, blue-green algae

9. Nitrogen assimilation and metabolism in blue-green algae

W. D. P. STEWART, A. HAYSTEAD
& M. W. N. DHARMAWARDENE

During the period of the International Biological Programme there has been a great increase in our understanding of nitrogen fixation by blue-green algae. In particular new groups of nitrogen-fixing algae have been discovered, the enzyme has been localised, many of its features characterised and there has been a substantial increase in our understanding of the ecology of nitrogen fixation, much of this being due to the introduction of the acetylene-reduction technique. These findings, particularly the more physiological and biochemical aspects, are detailed and referenced in a recent review (Stewart, 1973) while up-to-date information on the ecological role of these algae in various parts of the world is provided in the papers subsequent to this one. In this paper we would like to bridge the gap between physiology and biochemistry, and ecology, by considering, at the molecular level, the ways in which combined nitrogen, the one ecological factor which specifically affects nitrogen fixation, exerts its effect. We also wish to consider possible ways in which blue-green algae regulate their metabolism under nitrogen-fixing conditions so that ammonia, the end product of nitrogen fixation, does not inhibit the nitrogen-fixing process. Information on this aspect is relevant to current agricultural practice in paddy soils and elsewhere where nitrogen-fixing, blue-green algae occur naturally but where increasing use is being made of synthetic nitrogen fertiliser.

Materials and methods

Algae

Anabaena cylindrica was obtained from the Culture Collection of Algae and Protozoa, Cambridge, and was used in pure culture.

Culture and growth of the algae

The algae were grown, free of combined nitrogen, unless otherwise stated, in the medium of Allen & Arnon (1955), in small batch culture, in continuous culture, or in mass culture. Batch culture material was grown in 200 ml aliquots of culture medium in 500 ml conical flasks on a reciprocal shaker at 80 revolutions/min. The light intensity was continuous at 3000 lux and the temperature was 27 °C. Cultures in exponential phase of growth were always used for experimentation unless otherwise stated. Continuous cultures of the algae were grown as described by Lyne & Stewart (1973), and mass cultures were grown as described by Haystead & Stewart (1972).

Chlorophyll a *determinations*

Chlorophyll *a* was extracted in methanol and was estimated using a specific absorption coefficient at 663 nm of 78.74 l/g/cm (Meeks & Castenholz, 1971).

Preparation of nitrogen-fixing cell-free extracts and assay of nitrogenase in vitro

The procedures followed are detailed in Haystead, Robinson & Stewart (1970).

Preparation of heterocyst and vegetative cell fractions

Five litres of exponentially growing cells were harvested by centrifugation, and the cell paste was washed twice in HEPES buffer (0.05 M, pH 7.2) containing 10 mM $MgCl_2$ and 1 mM dithiothreitol. It was then resuspended in fresh HEPES buffer and incubated with lysozyme (0.5 mg/ml) at 25 °C for 30 min. The filaments were then resuspended in buffer and passed through a Yeda Press (Scientific Instruments, Rehovot, Israel) at 2000 p.s.i. until the vegetative cells, but not the heterocysts were clearly ruptured. The suspension was then centrifuged at 2500 × ***g*** for 30 min and the supernatant which contained the ruptured vegetative cell material was assayed for enzyme activity. The 2500 × ***g*** for 30 min pellet, which contained the heterocysts, was resuspended in fresh buffer, washed and repelleted two further times, and the final pellet which contained virtually pure heterocysts was subjected to French pressure cell treatment at 16000 p.s.i. This broke the heterocysts and the extract was then assayed for enzymic activity.

Acetylene-reduction assays

The method of Stewart, Fitzgerald & Burris (1967) was used.

Nitrogen measurements

Total nitrogen was measured by the method of Burris & Wilson (1957) and nitrate-nitrogen was measured using the phenol–disulphonic acid method (Hora & Webber, 1960).

Free intracellular ammonia and amino acids were extracted from intact algal filaments in 70 % (v/v) ethanol at room temperature for 15 min. The extracted material was then evaporated to dryness on a rotary evaporator below 35 °C and the residue taken up in citrate buffer (0.2 M Na citrate; pH 2.20). The amino acids and ammonia were separated by ion exchange chromatography using a Bio-Cal model BC-100 amino acid analyser (Bio-Cal Instruments Ltd, St Albans, Herts.). A sodium citrate, single-column, buffer system or a lithium citrate buffer system (Benson, Gordon & Patterson, 1967) was used as required. The amino acids and ammonia were detected using ninhydrin reagent and were identified by comparing their elution times with those of known standards (Spackman, Stein & Moore, 1958). The amino acid concentrations in the extracts were calculated from the recorder tracings by measuring the peak areas and referring to standard curves and internal standards.

The incorporation of ^{15}N into various amino acids in *Anabaena cylindrica* was examined by supplying intact algae with $^{15}N_2$ and after suitable incubation periods extracting the amino acids and separating these using the amino acid analyser as described above. Each amino acid collected was then assayed for ^{15}N enrichment using a Zeiss N015 atomic emission spectrophotometer and the techniques of Akkermans (1971). The $^{15}N_2$ was obtained from Prochem Ltd, South Croydon, England. Further details are given in the Results section.

^{14}C studies

The incorporation of ^{14}C-label from fumarate into various amino acids was measured by adding 50 mM [2,3-^{14}C]fumarate (Radiochemical Centre, Amersham, Bucks.) to cell-free extracts of *Anabaena cylindrica*, prepared by French pressure cell treatment at 16000 p.s.i. The incubation period was 15 min at 30 °C. The free amino acids were then extracted, separated and quantified as described above. Each amino acid was then assayed for ^{14}C-label by liquid scintillation counting using a Tracerlab Corumatic 200 scintillation counter and Triton X100 scintillation fluid (Patterson & Greene, 1965). Further details are given in the Results section.

Extraction and assay of intracellular ATP pool

The extraction procedure, with minor modifications, was that of Cole, Wimpenny & Hughes (1967). A 6 ml sample of algal suspension was injected rapidly with 1.5 ml of ice cold 20 % (w/v) $HClO_4$ and extracted with shaking for 10 min. The solution was then neutralised with KOH (1 M), centrifuged at $6000 \times \boldsymbol{g} \times 10$ min at 1 °C and the ATP in the supernatant assayed immediately. A delay of approximately 7 sec occurred between sampling and treating with $HClO_4$ but experiments showed that delays of up to 30–45 sec had no effect on the amount of ATP measured in the final assay. ATP was measured by a modification of the firefly-luciferin method of McElroy (1963). To 0.9 ml of HEPES/Mg^{2+} buffer (0.025 M, pH 7.4) at room temperature was added 80 μl of filtered firefly-lantern extract in arsenate–magnesium buffer (Sigma FLE-250) which had been kept at 2 °C for 24 h after reconstitution. The mixture was shaken for exactly 90 sec and disintegrations counted immediately using a Tracerlab Corumatic 200 liquid scintillation counter. The first 10 sec count was proportional to the ATP in the sample. ATP standards, in the range 2–25 pmoles, were assayed at the same time.

Gases

All gases, except acetylene and ethylene (British Oxygen Company) were obtained pre-mixed from Air Products Ltd.

Chemicals

The chemicals used were the highest grade commercially available from Sigma, London; British Drug Houses, Poole; Boehringer Corporation (London) Ltd and LKB Instruments, South Croydon, Surrey.

Results

The effects of combined nitrogen on nitrogen fixation

Nitrogen-fixing blue-green algae, like other nitrogen-fixing plants, assimilate both inorganic nitrogen and N_2, and when the former is available it may be assimilated preferentially and nitrogen fixation may be inhibited. The effects of combined nitrogen on nitrogen fixation vary, however, depending on the quantity and type of combined nitrogen supplied, on the length of the incubation period in the presence of combined nitrogen, and on the physiological state of the algae. There is, in fact, a confusing literature on the subject. We decided therefore to reinvestigate the effect of nitrate and ammonium-nitrogen on nitrogenase activity by *Anabaena cylindrica* using both batch and continuous cultures.

Table 9.1. *Effect of various levels of nitrate-nitrogen, ammonium-nitrogen and sodium chloride on in-vivo nitrogenase activity of* Anabaena cylindrica

Nitrogen concentration (mg/l)	NaCl concentration (mmoles/l)	% Residual nitrogenase activity		
		NO_3-N	NH_4-N	NaCl
0	0	100	100	100
50	3.5	100	100	100
100	7.0	100	100	100
150	10.5	100	100	100
200	14.0	98.2	99.5	100
250	17.5	96.1	94.2	96.0
300	21.0	62.0	65.0	64.0
400	28.0	10.2	52.1	15.4
500	35.0	2.0	47.0	1.4
750	52.5	0.0	0.0	0.0

One ml samples of algae in the concentrations of salt indicated in the table were incubated in 7.0 ml serum bottles at 25 °C and 3000 lx in the presence of 20 % C_2H_2, and C_2H_4 production measured after 30 min. The 100 % activity is 6.5 nmoles C_2H_4 per sample per 30 min. Each value is the mean of triplicates.

Effect of nitrate-nitrogen and ammonium-nitrogen on the activity of preformed nitrogenase. Data on the effect of various levels of NO_3^- and NH_4^+ on the in-vivo activity of preformed nitrogenase are presented in Table 9.1. These show that there is no significant inhibition of acetylene reduction at nitrogen concentrations at, or below, 250 mg/l but that higher concentrations cause an inhibition. However, Table 9.1 also shows that an identical inhibitory effect can be obtained when the nitrogen salt is replaced by an equimolar concentration of NaCl. These results suggest that physiological levels of combined nitrogen do not have an immediate effect on preformed nitrogenase activity, and that the immediate inhibition noted at very high nitrogen levels is a non-specific salt effect. These results thus agree with those of Kennedy (1970) who found with legumes that KCl, $(NH_4)_2SO_4$ and K_2SO_4 at high concentration inhibited nitrogenase activity in a non-specific manner. The data showing no inhibition of acetylene reduction at concentrations up to 250 mg N/l also support the earlier data of Stewart *et al.* (1968)

Table 9.2. *Effect of nitrate-nitrogen, ammonium-nitrogen and sodium chloride on in-vitro nitrogenase activity in crude extracts of* Anabaena cylindrica

NO_3-N or NH_4-N concentration (mg/l)	NaCl concentration (mmoles/l)	% Residual nitrogenase activity		
		NO_3-N	NH_4-N	NaCl
0	0	100	100	100
10	0.7	104	101	97
20	1.4	99	99	97
40	2.8	97	98	94
60	4.2	97	97	97
80	5.6	95	96	94
100	7.0	92	91	93

The extracts were prepared and assayed as described by Haystead & Stewart (1972). The 100 % activity is 2.9 nmoles C_2H_4/mg protein/min. Each value is the mean of quadruplicate determinations.

and Ohmori & Hattori (1972) who observed no effect of NO_3^- or NH_4^+ on the activity of preformed nitrogenase in short-term experiments.

The effects of NO_3^-, NH_4^+ and equimolar NaCl on in-vitro nitrogenase activity are shown in Table 9.2. It is seen that there is no specific inhibition of acetylene reduction even in the presence of 100 mg/l of combined nitrogen. It is possible that an inhibition of activity by combined nitrogen would have been noted if N_2 fixation rather than acetylene reduction had been measured (it was not possible to measure N_2 fixation in our crude extracts because of the presence of a vigorous ATP-dependent aminating mechanism) (see Bothe, 1970, and below). However, the fact that combined nitrogen has no effect on in-vitro N_2 reduction by extracts of heterotrophic bacteria (see Carnahan, Mortenson, Mower & Castle, 1960; Dalton & Mortenson, 1972) argues against this possibility.

The effect of nitrate-nitrogen and ammonium-nitrogen on nitrogenase synthesis. The effect of NO_3^- and NH_4^+ on the activity of preformed nitrogenase was studied in continuous culture experiments. The results (Fig. 9.1) show that when 150 mg/l of nitrate-nitrogen are added to steady-state cultures of *Anabaena cylindrica* nitrogenase activity decreases over a 24 h period and then levels off at a rate which is approximately 80 % of that obtained with N_2-grown cells.

On the other hand when ammonium-nitrogen is added to algae growing in continuous culture on N_2 the pattern is quite different

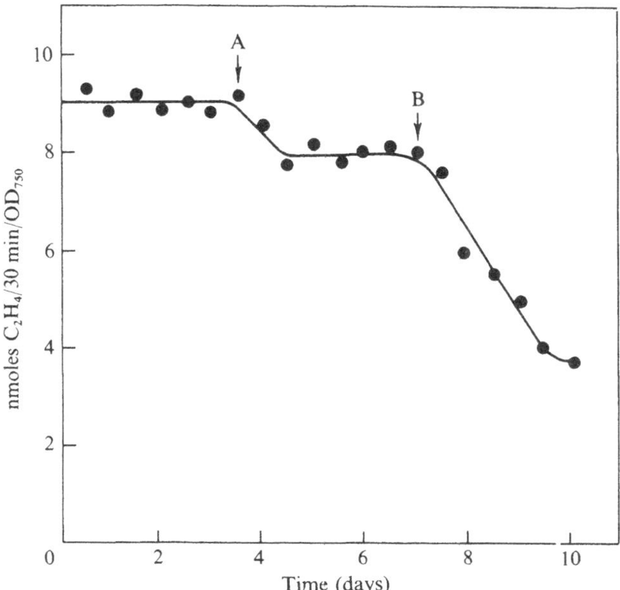

Fig. 9.1. Effect of nitrate-nitrogen on nitrogenase activity (acetylene reduction) by *Anabaena cylindrica* grown in continuous culture. The alga was grown initially on N_2 and at time A 150 mg/l of nitrate-nitrogen were added; at time B the culture was allowed to grow as a batch culture. The temperature was 26 °C, the light intensity 3000 lx and the mean generation time of the alga approximately 22 h. Each value is the mean of triplicate determinations.

(Fig. 9.2). On the addition of NH_4^+ there is a marked decrease in acetylene reduction with activity being inhibited almost completely within 48 h. When the ammonium-nitrogen is removed nitrogenase activity returns to its original level within 36–48 h. As the mean doubling time of the alga is 22 h, much of the loss of activity is due probably to a simple dilution out of the enzyme with growth, but it is possible, from the rapid drop-off in activity that there is also a rapid turnover of nitrogenase protein without resynthesis. Bone (1971*a*, *b*) and Shah, Davis & Brill (1972) have evidence of nitrogenase turnover in *A. flos-aquae* and in the heterotrophic *Klebsiella* respectively. It is thus clear that the effects of NO_3^- and NH_4^+ on preformed nitrogenase are quite different in that the inhibitory effect of nitrate-nitrogen on nitrogenase synthesis is much less than that of ammonium-nitrogen.

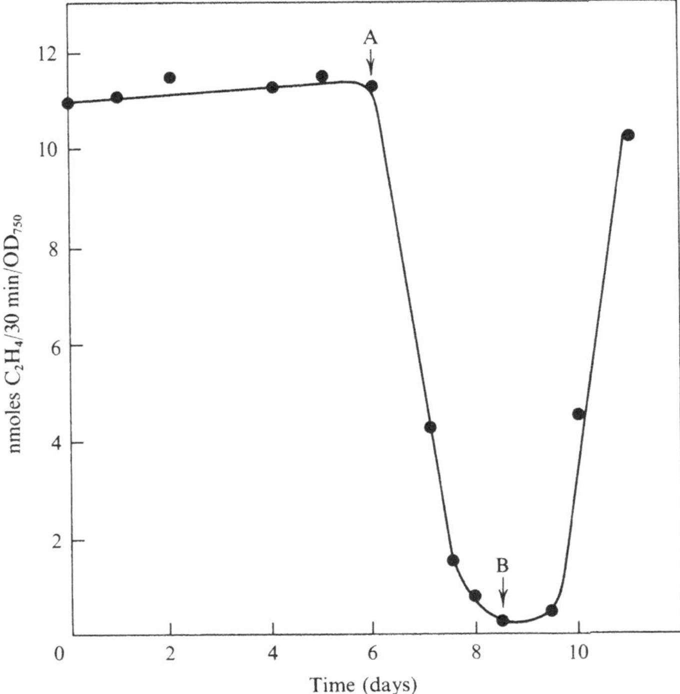

Fig. 9.2. Effect of ammonium-nitrogen on nitrogenase activity (acetylene reduction) by *Anabaena cylindrica* grown in continuous culture. The alga was grown initially on N_2 and at time A 50 mg/l of ammonium-nitrogen were added; at time B the medium was changed back to ammonium-free medium. The temperature was 26 °C, the light intensity 3000 lx and the mean generation time of the alga approximately 22 h. Each value is the mean of triplicate determinations.

The relationship between the intracellular pool of ammonia and nitrogenase activity in intact filaments of Anabaena cylindrica. The above data show that the addition of exogenous ammonium-nitrogen to continuous cultures of *Anabaena cylindrica* inhibits nitrogenase synthesis more than does nitrate-nitrogen, and there is ample evidence (Smith & Evans, 1970; Haystead & Stewart, 1972) that cells growing under nitrogen-depleted conditions, if metabolically active, develop high rates of nitrogenase activity. It thus appears likely that nitrogenase synthesis is regulated, directly or indirectly, by the intracellular level of combined nitrogen, probably NH_4^+, and we checked therefore the relationship between nitrogenase activity and the intracellular pool of ammonia. Typical data obtained are presented in Fig. 9.3. It is seen that during the first four days growth on N_2 the rate

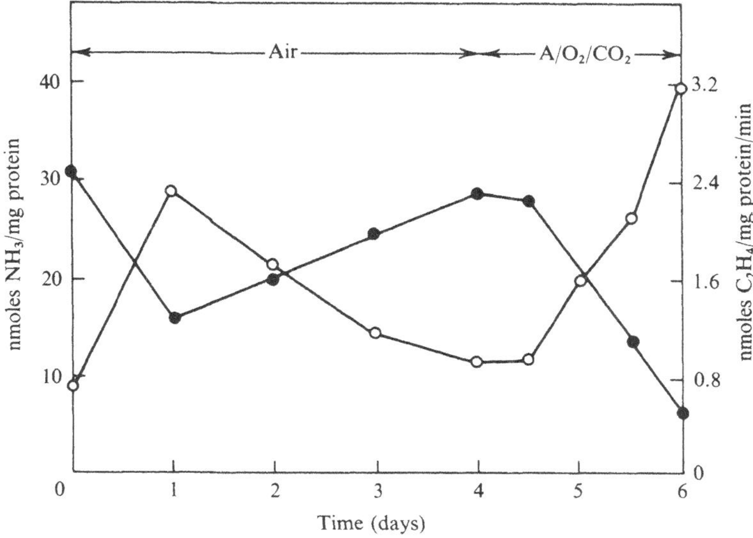

Fig. 9.3. Nitrogenase activity (acetylene reduction) (○) and the intracellular pool of ammonia (●) in *Anabaena cylindrica* filaments grown in batch culture in air and in the absence of N_2 ($A/O_2/CO_2$:79.96/20.00/0.04) in medium free of combined nitrogen. The temperature was 26 °C, the light intensity 3000 lx. Each value is the mean of duplicate determinations.

of acetylene reduction increases then subsequently falls off as is typical of nitrogenase activity in batch culture (Stewart *et al.*, 1968; Stewart & Lex, 1970; Neilson, Rippka & Kunisawa, 1971) and that in general there is an inverse relationship between nitrogenase activity and the size of the intracellular pool of ammonia. This relationship is maintained when, during days 4–6, the algae are placed under conditions of nitrogen starvation; then the rate of nitrogenase activity increases as the ammonia pool decreases.

Effect of ammonium-nitrogen on intracellular ATP pool levels. High concentrations of ammonium-nitrogen are known to uncouple phosphorylation in a variety of plant species (see e.g. McGarty & Coleman, 1969) and it is possible than an effect of NH_4^+ on nitrogenase in *Anabaena cylindrica* could be due to its uncoupling of photophosphorylation, which in turn results in a shortage of ATP for nitrogenase. We therefore investigated the effect of NH_4^+ on both the rate of nitrogenase

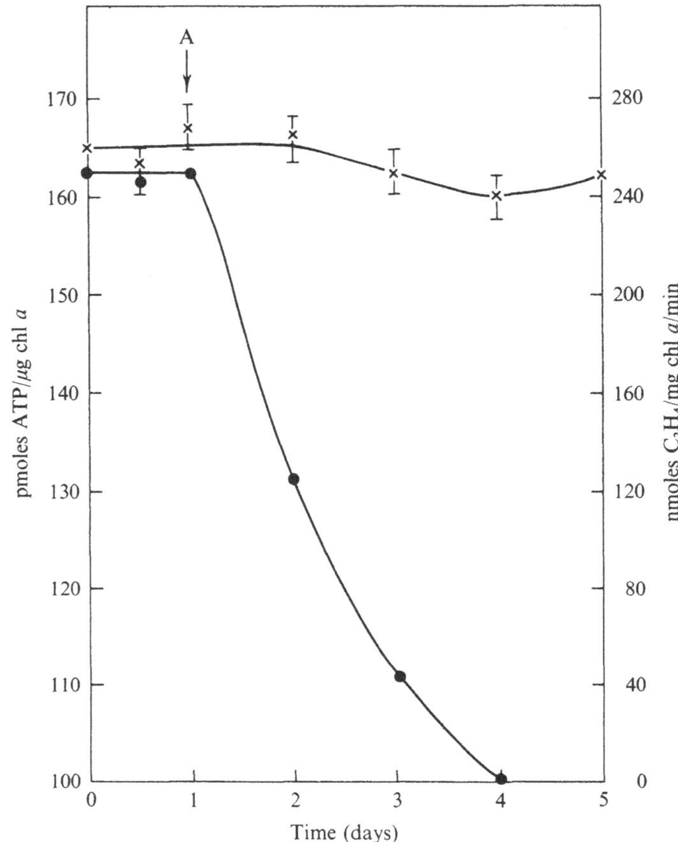

Fig. 9.4. Nitrogenase activity (acetylene reduction) (●) and the intracellular pool of ATP (×) in *Anabaena cylindrica* filaments grown in continuous culture on N_2 and ammonium-nitrogen (50 mg N/l). The algae were grown on N_2 under steady-state conditions and at time A ammonium-nitrogen was added to the culture. The temperature was 25 °C, the light intensity 4000 lx and the mean generation time of the alga 22 h. Each value is the mean of triplicate determinations.

activity and the intracellular ATP pool of *Anabaena cylindrica* growing in continuous culture. The results obtained are shown in Fig. 9.4. It is seen that when 3.6 mM NH_4^+ (50 mg/l of ammonium-nitrogen) are supplied to *Anabaena cylindrica* growing in continuous culture there is a rapid decrease in the acetylene-reducing capacity of the culture, but the cellular ATP pool does not decrease significantly. We conclude therefore that shortage of ATP is probably not responsible for decreases in nitrogenase activity such as those observed in Fig. 9.4.

Nitrogen reserves in blue-green algae

Evidence has been presented above that nitrogenase synthesis in blue-green algae is inhibited directly or indirectly by NH_4^+. It is thus important for continuing nitrogenase activity that ammonia, the end-product of dinitrogen reduction, is not allowed to accumulate. The ways by which the blue-green algae overcome this problem, and yet maintain a supply of fixed nitrogen for metabolism under nitrogen-limiting conditions are at least two-fold. First, the algae possess a variety of ammonia-assimilating mechanisms, and second, they contain two major nitrogen reserves into which excess nitrogen is channelled. This section deals with these reserves; the following section considers ammonia-assimilating enzymes.

The two major nitrogen storage products in blue-green algae are phycocyanin and structured granules (cyanophycin granules). Phycocyanin, and the closely related allophycocyanin, are phycobiliproteins which, being composed of conjugated pyrrole rings, are rich in nitrogen. Although phycocyanin acts as an accessory pigment in photosynthesis, it also serves as a readily mobilised nitrogen reserve which is used up under conditions of nitrogen-deficiency and which reappears rapidly when nitrogen becomes available (see Allen & Smith, 1969; Stewart & Lex, 1970; Van Gorkom & Donze, 1971). Structured granules are readily observed under the electron microscope as dense osmiophilic bodies which accumulate, particularly at the end of the log phase (Simon, 1971; 1973) in nitrogen-sufficient cells. According to Simon (1971) these granules, which have a molecular weight of 25000–100000, are co-polymers of L-aspartic acid and L-arginine in a 1:1 molar ratio. Although structured granules serve as nitrogen reserves they are mobilised less readily than phycocyanin under conditions of nitrogen-deficiency.

Ammonia assimilation and metabolism

The enzymic mechanisms of ammonia assimilation and metabolism have not, until now, been investigated in any detail in blue-green algae. Early work by Magee & Burris (1954) showed that the highest ^{15}N label from $^{15}N_2$ occurred in glutamic acid over a 90 min incubation period. However, Hoare, Hoare & Moore (1967) reported that they could not find glutamic dehydrogenase (GDH) in various non-nitrogen-fixing algae and commented on the apparent anomaly of their findings, although others subsequently detected low levels of GDH in various blue-green algae (Pearce, Leach & Carr, 1969; Scott & Fay, 1972; Neilson & Doudoroff, 1973). We decided to check for the presence of GDH and other ammonia-assimilating enzymes in *Anabaena cylindrica* (Haystead, Dharmawardene & Stewart, 1973).

Table 9.3. *Coenzyme specificity of glutamic dehydrogenase and alanine dehydrogenase of nitrogen-fixing* Anabaena cylindrica

Enzyme	Coenzyme	Activity (nmoles coenzyme oxidised/ sample/min)
Glutamic dehydrogenase	NADPH	2.0
	NADH	0.1
Alanine dehydrogenase	NADH	16.0
	NADPH	0.0

The enzymes were extracted and assayed as described by Haystead *et al.* (1973). Each value is the mean of triplicate determinations.

Glutamic dehydrogenase (GDH). Assays for GDH in crude extracts ($2500 \times g \times 15$ min supernatant) of *Anabaena cylindrica* were carried out using a modification of the method of Strecker (1955) (see Haystead *et al.*, 1973). Low but consistent activity, which was NADPH dependent, was noted. There was negligible activity with NADH (Table 9.3). The NADPH specificity is similar to that noted by Pearce *et al.* (1969) and by Neilson & Doudoroff (1973) although NADH-specific GDH is present in heterocysts of *Anabaena variabilis* according to Scott & Fay (1972). We were unable to show an oxidative deamination of glutamate in the reverse reaction.

Alanine dehydrogenase (ADH). Tests for ADH show that an NADH-specific enzyme is present in *Anabaena cylindrica*, just as it is in many of the algae examined by Neilson & Doudoroff (1973). We did not find (Table 9.3) any NADPH-specific activity. In our work on ADH we recognised the possibility that our results could be due perhaps to GDH, which has been shown from studies on e.g. bovine liver to dissociate in dilute extracts to give a monomeric form of GDH with ADH activity (Frieden, 1963). This was checked by measuring ADH and GDH activity at different protein concentrations, as well as in the presence of added ADP, which is known to increase the aggregation of the monomeric to the polymeric form of GDH. We obtained no evidence from such studies that the GDH activities of our *Anabaena cylindrica* extracts were due to a modified form of GDH rather than to ADH. This is also backed up by the fact that our GDH activity was NADPH specific whereas the ADH activity was NADH specific.

Table 9.4. *Partial purification of glutamine synthetase from* Anabaena cylindrica

Treatment	Biosynthetic activity (ΔO.D.$_{660}$/ mg protein/ min)
Crude broken cell suspension	0.068
$40000 \times g \times 30$ min supernatant	0.088
Sephadex G25 eluate	0.109
Protamine-SO_4 supernatant	0.263
NH_4Cl supernatant	0.485

The reaction mixture was a modification of that of Shapiro & Stadtman (1970) and was as follows, in a final volume of 1.0 ml: ATP, 10 μmoles; Mg^{2+}, 50 μmoles; NH_4Cl, 50 μmoles; sodium glutamate, 100 μmoles; HEPES buffer, pH 7.2, 50 μmoles. Assays were carried out at 37 °C for 30 min. (After Dharmawardene *et al.*, 1973.)

Glutamine synthetase (GS). The presence of GS in a variety of blue-green algae has been demonstrated in our laboratory (Dharmawardene *et al.*, 1972) and its main features characterised (Dharmawardene, Haystead & Stewart, 1973). Partially purified preparations show a high specific activity and the enzyme can be demonstrated readily using both biosynthetic (Table 9.4) and transferase reactions. It has a divalent cation requirement satisfied by Mg^{2+}, Mn^{2+}, Ca^{2+}, Co^{2+} and Zn^{2+}, but not by Cu^{2+} or Ba^{2+}; its pH optimum with Mn^{2+} is 7.2 and its temperature optimum is near 37 °C.

The enzyme appears to be regulated at several levels. First, activity is low in NH_4^+ grown cultures and high in nitrogen-depleted cultures suggesting that the level of combined nitrogen may regulate enzyme synthesis and/or activity. Second, GS activity decreases markedly at Mn/ATP ratios greater and less than 1 which suggests a role for this ion and possibly the total nucleoside triphosphate level (see below) in regulating GS activity. Third, there is evidence that the enzyme is subject to feed-back inhibition by potential products of glutamine metabolism. Evidence for the presence or absence of these regulatory

Table 9.5. *Effect of various feedback inhibitors on the biosynthetic activity of glutamine synthetase from log-phase and stationary-phase cultures of* Anabaena cylindrica

Inhibitor	% Inhibition	
	Log-phase culture	Stationary-phase culture
Glycine	8–10	8–12
Alanine	8–10	8–10
Glucosamine-6-phosphate	5–8	5–8
AMP	75–80	75–80
ADP	15–20	10–15
CTP	10–12	10–12
Glycine + alanine + glucosamine-6-phosphate + AMP + ADP + CTP	80–100	80–100
Glycine + alanine + AMP + ADP + CTP	80–85	80–90

Partially purified enzyme (see Table 9.4) was used in the standard biosynthetic assay. The results are expressed as per cent inhibition of biosynthetic activity by 5 mM concentrations of each inhibitor.

mechanisms is detailed in Dharmawardene *et al.* (1973) and additional data on feed-back inhibition are presented in Table 9.5. The latter show that alanine, glycine, adenosine monophosphate, adenosine diphosphate, cytidine triphosphate and glucosamine-6-phosphate are inhibitory to GS activity in extracts from both log-phase and stationary-phase cultures. The cumulative nature of the feedback inhibition noted in the presence of AMP, CTP, glycine and alanine is shown by the kinetic data presented in Fig. 9.5. It is seen that curves obtained with AMP alone, AMP + glycine, or alanine, or ATP, contrast with the case of glucosamine-6-phosphate which enhances the inhibitory effect of AMP, i.e. there is an independent, cumulative, allosteric inhibition by alanine, glycine and CTP, while glucosamine-6-phosphate shows synergism with AMP. In all cases AMP is the most potent inhibitor of GS. Hubbard & Stadtman (1967) observed a similar situation in *B. licheniformis* where AMP acted as the key inhibitor and was synergistic with histidine and glutamine. It could not be determined with certainty whether *A. cylindrica* GS showed an adenylylation system because although changes in the ratio of transferase activity to biosynthetic activity were noted, other studies (see Dharmawardene *et al.*, 1973) provided no positive evidence for an adenylylating system of the *E. coli* type.

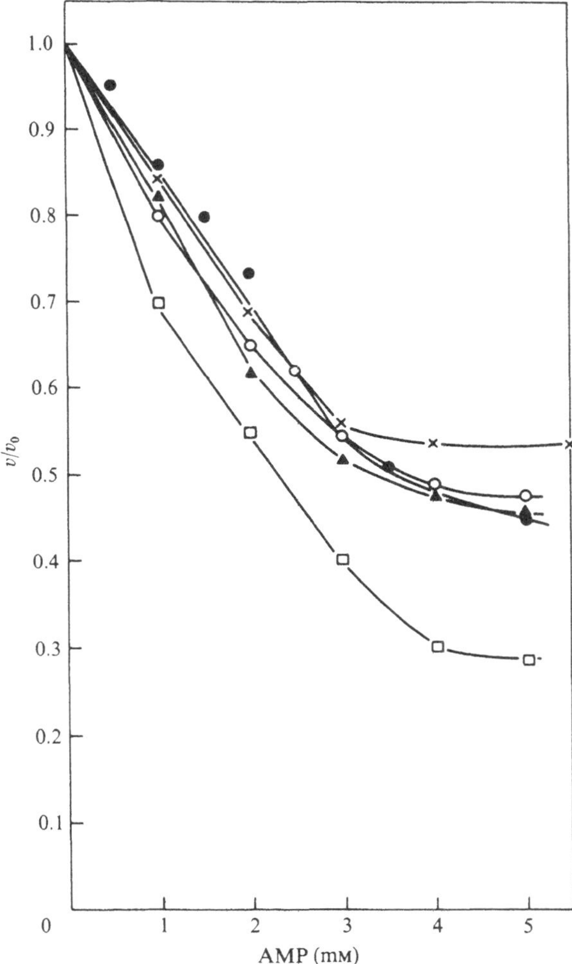

Fig. 9.5. Inhibition of *Anabaena cylindrica* glutamine synthetase activity by different levels of AMP in the presence and absence of other inhibitors. The residual activity with 5 mM concentrations of each single compound used was taken as 1.0 and the data plotted as v/v_0 against AMP concentration where v = residual activity and v_0 = full activity. ▲, AMP alone; ×, AMP+5 mM glycine; ●, AMP+5 mM alanine; ○, AMP+5 mM cytidine triphosphate; □, AMP+5 mM glucosamine-6-phosphate. Each value is the mean of quadruplicate determinations.

Aspartase, aspartic acid dehydrogenase and asparagine synthetase. In spectrophotometric assays for ASDH (Haystead *et al.*, 1973) we obtained a value of 7.6 nmoles NADH oxidised/mg protein/min using crude extracts of *A. cylindrica* ($2500 \times \boldsymbol{g} \times 15$ min supernatant). Activity with NADPH was barely detectable. However, as there is a rapid

Table 9.6. *The aspartate synthesising capacity of extracts of* Anabaena cylindrica

Additions	Concentration of ^{14}C in aspartate as % of original ^{14}C added as fumarate
[^{14}C]fumarate	0
[^{14}C]fumarate + NH_3	0
[^{14}C]fumarate, NH_3 NADP/NADPH (1:1)	0.8
[^{14}C]fumarate, NH_3 NAD/NADH (1:1)	1.2
[^{14}C]fumarate, NH_3 NADP/NADPH + fumarase + malate dehydrogenase	1.7
[^{14}C]fumarate and glutamate	1.9

50 mM (2,3-^{14}C, 0.1 μCi) fumarate was added to Sephadex G25 treated extracts of *Anabaena cylindrica* with or without ammonia in the presence of various co-factors shown and ^{14}C-labelling into aspartate determined (see text for further details).

decarboxylation of C_4 acids in such extracts these data cannot be taken as unequivocal evidence of ASDH activity. We therefore investigated the ability of crude *Anabaena* extracts to synthesise aspartate by adding 50 mM, [2,3-^{14}C]fumarate and measuring incorporation of the ^{14}C-label at 30 °C after 15 min. The results are summarised in Table 9.6. No ^{14}C-label was found in aspartate after the addition of fumarate, with or without ammonia, indicating that aspartase was absent, or inactive under our conditions. On the addition of pyridine nucleotides, fumarate, fumarase and malate dehydrogenase, 1.7 % of the total ^{14}C-label in fumarate was found in aspartate suggesting the passage of label from fumarate into malate and oxaloacetate and then into aspartate via ASDH. On the replacement of NH_4^+ by glutamate (25 mM) on the other hand, 1.9 % of the ^{14}C-label appeared in aspartate suggesting that in-vivo synthesis of aspartate occurs rapidly via transaminase-mediated reactions as well. Under all conditions used in these experiments at least 40 % of the added label was recovered in alanine with negligible amounts in glutamate and glutamine (Fig. 9.6). There was no evidence for the presence of asparagine synthetase (see Haystead *et al.*, 1973).

Other enzymes. In addition to the above enzymes there is also present in *Anabaena cylindrica* a variety of other ammonia-incorporating enzymes. These include glutamate-dependent transaminases such as GOT (glutamate oxaloacetate amino transferase) and GPT (glutamate pyruvate

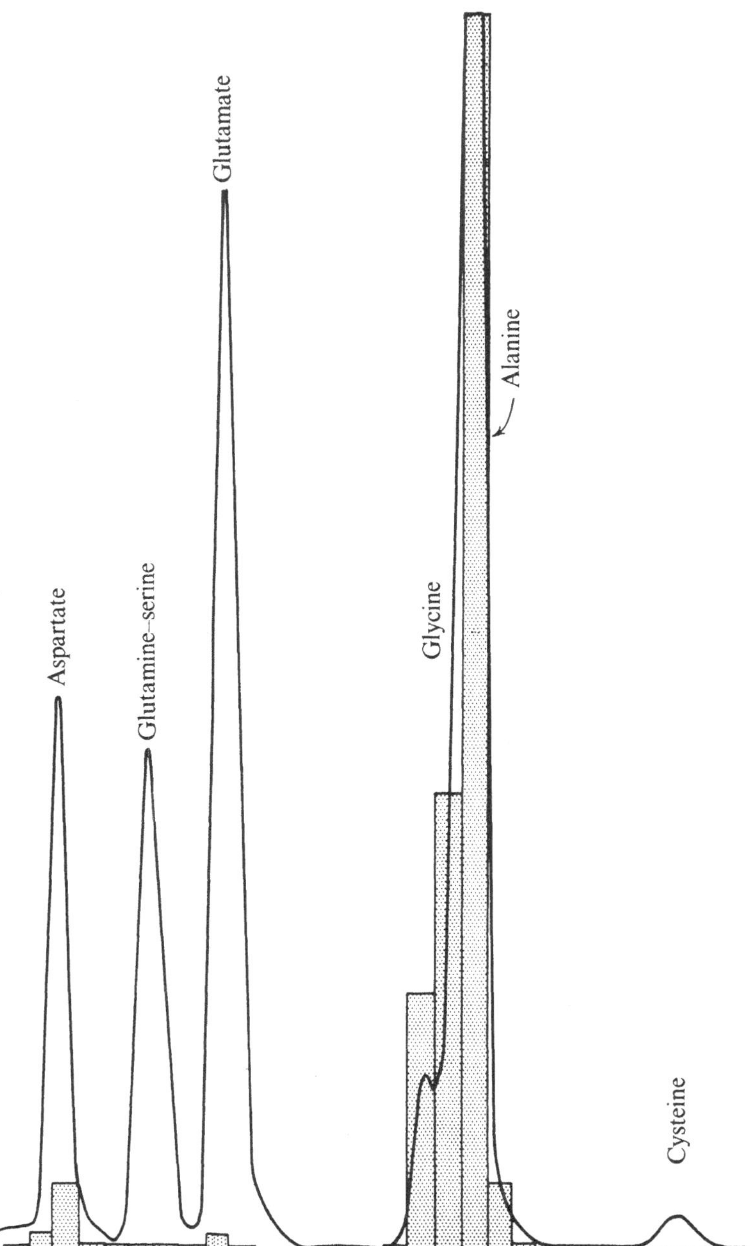

Fig. 9.6. Relative soluble (70 % ethanol) levels of various amino acids, and their ^{14}C label (cpm) 15 min after the addition of 50 mM, [2,3-^{14}C]fumarate (0.1 μCi) to crude extracts of N_2-grown *Anabaena cylindrica*. Shaded histograms indicate ^{14}C label.

amino transferase), as well as others which result in the synthesis of leucine, valine, *iso*leucine (see Haystead *et al.*, 1973), glycine and serine (see Codd & Stewart, 1973). The former three enzymes have also been reported in the non-nitrogen-fixing alga *Anabaena variabilis* by Hood & Carr (1972). Carbamyl phosphate synthetase which may be important in citrulline formation is also present in *Anabaena cylindrica* (Haystead *et al.*, 1973). The presence of glutaminase in our extracts is of particular interest as it suggests that the intracellular levels of glutamine could be regulated to some extent by the opposing effects of GS and glutaminase. The discovery of glutaminase renders equivocal our earlier report of the presence of glutamate (amide): 2-oxoglutarate aminotransferase (GOGAT) in *Anabaena cylindrica*, because the NADPH oxidation observed in the presence of glutamine and 2-oxoglutarate could conceivably have been due to the action of GDH on NH_3 produced on the de-amidation of glutamine by glutaminase. Neilson & Doudoroff (1973) and Brown, Burn & Johnson (1973) have failed to detect GOGAT in several blue-green algae. It may be noted that the intracellular free amino acid pool patterns of *A. cylindrica* (Dharmawardene *et al.*, 1972) are characteristic of organisms that possess GOGAT (Brown & Stanley, 1972; Brown *et al.*, 1973).

Michaelis constants for GDH, ADH and GS. The above data show clearly that there is no shortage of ammonia-assimilating mechanisms in nitrogen-fixing algae, and in an attempt to elucidate which might be the more important of these in removing NH_4^+ from, and around, nitrogenase, tests were carried out to determine the $K_m^{app.}$ for NH_4^+ for three of the possible primary ammonia-assimilating enzymes: GDH, ADH and GS. The rationale to this study was that an enzyme with a low K_m for NH_4^+ is likely to be particularly efficient in scavenging low concentrations of NH_4^+ and may thus ensure that ammonia does not accumulate in sufficient quantity to inhibit nitrogenase synthesis. Double reciprocal plots of GS, ADH and GDH activity at varying levels of ammonium-nitrogen are presented in Figs. 9.7–9.9. The $K_m^{app.}$ values for NH_4^+ from these data are respectively 1 mM, 5 mM and 12.5 mM. The values for GS are obtained consistently, but although, as shown in Fig. 9.8, ADH can have a low $K_m^{app.}$ this varies with the pH, the higher the pH within the range 7–9 the lower the K_m (Rowell and Stewart, unpublished). Thus the possible importance of ADH as an ammonia-scavenging enzyme will depend on the pH level of the subcellular environment in which it occurs.

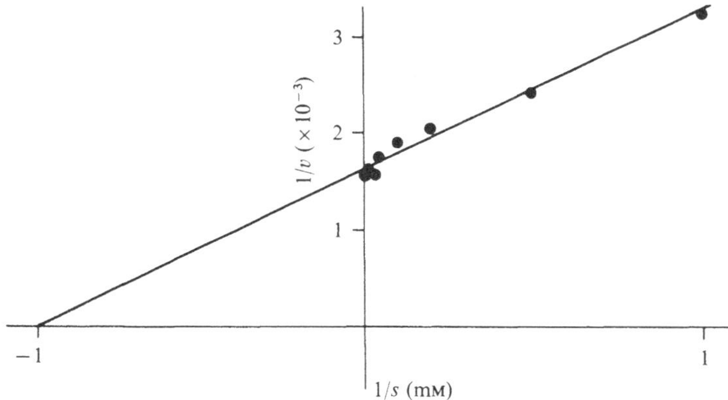

Fig. 9.7. Double reciprocal plot of glutamine synthetase of N_2-grown *Anabaena cylindrica* at varying ammonium-nitrogen levels. Each value is the mean of quadruplicate determinations.

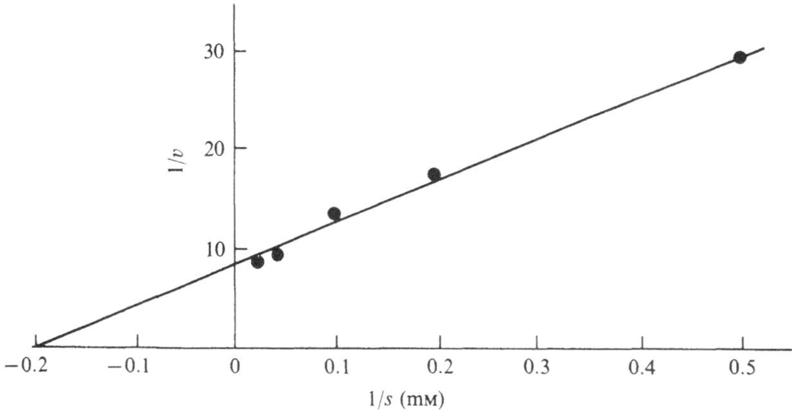

Fig. 9.8. Double reciprocal plot of alanine dehydrogenase of N_2-grown *Anabaena cylindrica* at varying levels of ammonium-nitrogen. Each value is the mean of triplicate determinations.

Levels of activity of various ammonia-assimilating enzymes in heterocysts and vegetative cells of A. cylindrica

The relative abundance of GS, ADH, GDH, GOT and GPT in heterocysts and vegetative cells of *Anabaena cylindrica* is shown in Table 9.7. It is seen that on a protein basis GS activity is almost twice as high in the heterocysts as in the vegetative cells while this is not so for ADH. GDH activity is low but of the same order both in heterocysts

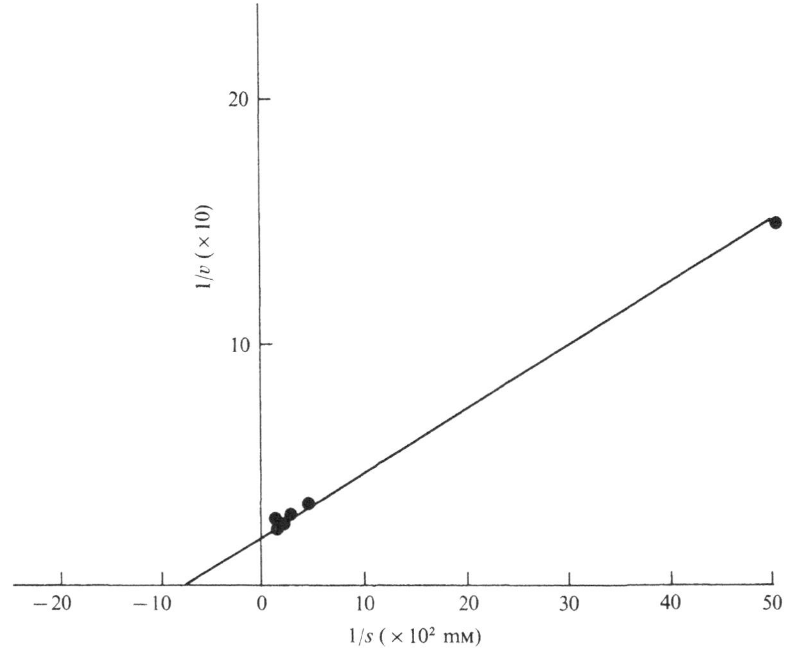

Fig. 9.9. Double reciprocal plot of glutamic dehydrogenase of N_2-grown *Anabaena cylindrica* at varying levels of ammonium-nitrogen. Each value is the mean of triplicate determinations.

Table 9.7. *Ammonia assimilating and metabolising enzymes extracted from vegetative cell and heterocyst fractions of N_2-fixing* Anabaena cylindrica

Enzyme	Specific activity in the heterocyst fraction	Specific activity in the vegetative cell fraction
GS	102 nmoles of P_i released	42 nmoles of P_i released
ADH (NADH)	12.8 nmoles of NADH oxidised	15.1 nmoles of NADH oxidised
GDH (NADPH)	2.0 nmoles of NADH oxidised	1.8 nmoles of NADH oxidised
GOT	59 nmoles of oxaloacetate formed	42 nmoles of oxaloacetate formed
GPT	3.8 nmoles of pyruvate formed	1.5 nmoles of pyruvate formed

Specific activities are expressed as nmoles of substrate utilized or product formed per mg protein per min.

and vegetative cells, while the activities of GOT, and GPT, are higher in the heterocysts than in the vegetative cells. These results are considered in more detail later.

Kinetics of incorporation of ^{15}N into various amino acids after exposure of intact filaments of A. cylindrica *to $^{15}N_2$*

The experimental set-up used in this study is shown in Plate 9.1. The algae were grown on N_2 in continuous culture in 5.0 l batches in reservoir A and when required for experimentation 250 ml were drawn over into a 500 ml capacity reaction vessel (B). There the algal suspension was stirred magnetically and illuminated continuously at 3000 lx. The suspension was then bubbled with $A/O_2/CO_2$ (79.96/20.00/0.04, v/v) and evacuated to remove the N_2 in air. Subsequently 50 % N_2 (enriched with 90 atom % ^{15}N) was introduced and the balance added as $A/O_2/CO_2$ (29.96/50.00/20.00/0.04, v/v). After the desired exposure period to ^{15}N, the suspension was blown on to a Whatman GF/C filter via outlet C and after filtration (which was completed within 15 sec), the filter pad plus alga was rapidly extracted with 80 % ethanol. The amino acids present in the extract were then separated using a Bio-Cal 100 amino acid analyser and analysed for ^{15}N enrichment using a Zeiss N015 atomic emission spectrophotometer as detailed in the Materials and Methods section.

Data obtained after exposure of the algae for 1, 2, 5 and 10 min are presented in Fig. 9.10. These show that within 2 min of introducing $^{15}N_2$, the highest ^{15}N label is found in the amide group of glutamine, followed closely by alanine and glutamate. This pattern is maintained for incubation periods under $^{15}N_2$ of at least 10 min. Furthermore, the amino group of glutamine has a significantly lower ^{15}N label than the amide group.

Discussion

The aim of the early part of the studies described in this paper was to help clarify a rather confused area of research on nitrogen fixation by blue-green algae, namely the effect of combined nitrogen on nitrogen fixation. The results presented in this paper indicate that ammonia, the end-product of dinitrogen reduction, is an important regulator of nitrogen fixation which, when present in high concentration, inhibits nitrogenase synthesis and that increased nitrogenase synthesis occurs

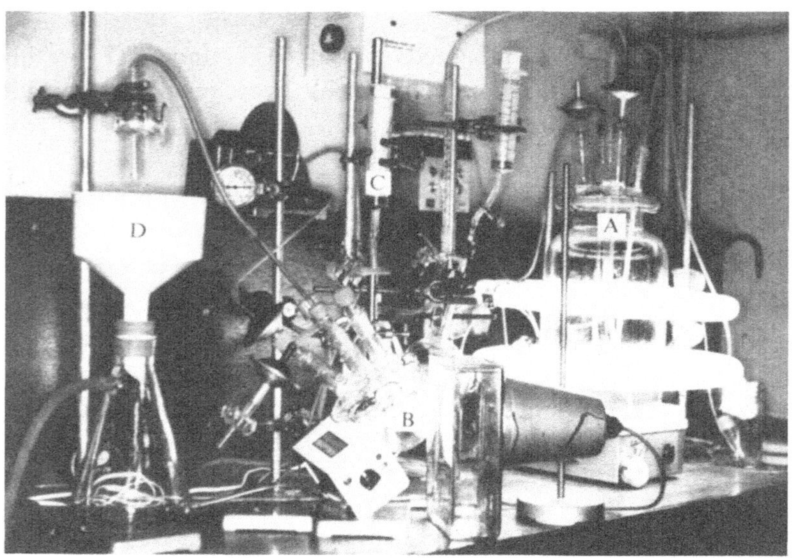

Plate 9.1. Experimental arrangement used in ^{15}N uptake studies. The algae were grown axenically in semi-continuous culture in vessel A and 250 ml aliquots siphoned when required into reaction vessel B which was fitted with a magnetic stirrer, gassing system and light source. Air was removed by flushing with argon/CO_2 (99.96/0.04 %) and $^{15}N_2$ (25 %) introduced by syringe (C). At the end of the exposure period to $^{15}N_2$ the algal suspension was transferred by tube to filter funnel D and filtration was completed within 15 sec. The algae were then extracted in 80 % ethanol.

when the ammonia concentration is low. Our results furthermore provide evidence that the critical concentration regulating nitrogenase synthesis in blue-green algae is the size of the intracellular pool of ammonia, or its product, rather than the level in the external medium. They also indicate that the regulatory effect of exogenous nitrate on nitrogen fixation is not direct, but that it modifies the size of the intracellular pool of ammonia, which in turn exerts its effect, directly or indirectly, on nitrogenase synthesis.

These results thus support the earlier supposition (Stewart *et al.*, 1968; Bone, 1971*a*, *b*) that ammonia inhibits the synthesis of new nitrogenase while allowing the continuing functioning of existing nitrogenase. Our data showing a general inverse correlation between the size of the intracellular pool of ammonia and nitrogenase activity accords with the hypothesis of Neilson *et al.* (1971), Thomas & David (1971) and Ohmori & Hattori (1972), all based on indirect evidence that the intracellular level of ammonia probably regulates nitrogenase synthesis. It is not necessarily implied from these results that ammonia

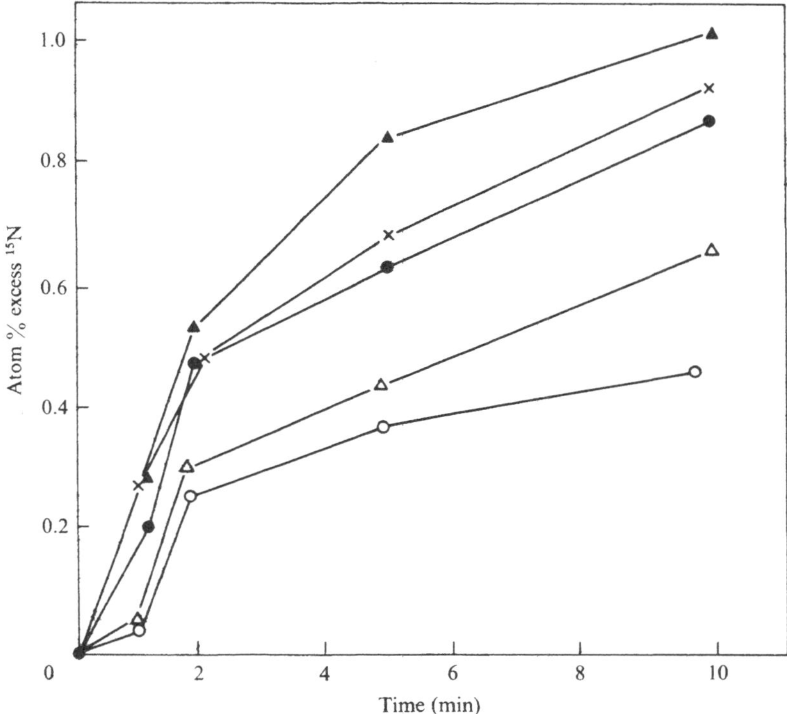

Fig. 9.10. ^{15}N incorporation into various free amino acids after exposing intact filaments of *Anabaena cylindrica* to $^{15}N_2$ for 1, 2, 5 and 10 min. ▲, Amide-nitrogen of glutamine; ●, glutamate; ×, alanine; △, amino-nitrogen of glutamine; ○, aspartate.

rather than some compound derived from it is the actual inhibitor of nitrogenase activity. Indeed data which suggest that ammonia *per se* is not the inhibitor have been obtained recently by Stewart & Rowell (1975). They showed that on adding the analogue L-methionine-DL-sulphoximine (1 μM), which inhibits glutamine synthetase activity, to nitrogen-fixing cultures of *Anabaena cylindrica*, nitrogenase continued to be synthesised and remained active even although there was a build-up both of intracellular and extracellular ammonia.

There are at least three major ways in which the algae may overcome the problem of repression of nitrogenase synthesis by ammonia, and these are not mutually exclusive: (1) there may be subcellular compartmentation of the site of nitrogenase synthesis and the site of its activity; (2) the ammonia may be transported rapidly from the nitrogen-fixing site and at sites removed from the loci of nitrogenase synthesis may be assimilated into general cell metabolism; (3) there may be present a very efficient ammonia-scavenging mechanism which at the site of nitrogenase activity assimilates the ammonia into a non-inhibitory form(s) which then becomes available for general cell metabolism at, or remote from the site of nitrogenase activity. It should be noted that such mechanisms are only partially successful in that although the algae continue to produce nitrogenase while fixing nitrogen, growth of the algae under nitrogen-depleted conditions results in increased nitrogenase activity (Smith & Evans, 1970; Haystead & Stewart, 1972), indicative of a further derepression of nitrogenase synthesis, compared with air-grown cells.

To date we have no evidence for subcellular compartmentation of the sites of nitrogenase synthesis. However the kinetics of ^{15}N incorporation which show differential ^{15}N labelling of glutamic acid and the amide and amino groups of glutamine suggest that there may be at least two pools of glutamate involved in the subsequent metabolism of newly-fixed ammonia. There is also good evidence for a variety of ammonia-assimilating enzymes in *Anabaena cylindrica* although one must be cautious in extrapolating in-vitro enzymic data to in-vivo situations.

Collectively, the above data provide some pointers to the relative merits of various pathways of ammonia assimilation in *Anabaena cylindrica*. First, the K_m data suggest that GS with its low K_m (1 mM) could act as an ammonia-scavenging enzyme more efficiently than GDH (12.5 mM) and ADH (> 5 mM). (The intracellular NH_4^+ concentration in filaments of *Anabaena cylindrica* is 1.5 mM.) It is always possible, of course, that these K_m values are of limited significance if, as is likely, subcellular compartmentation of ammonia, or other substrates occurs. Second, the high levels of detectable GS and ADH compared with GDH suggest that in nitrogen-fixing *Anabaena cylindrica* GDH is less important than it is in other plants where it is the key ammonia-assimilating enzyme (see McKee, 1962). Third, the finding that the GS/ADH ratio is higher in the nitrogen-fixing heterocysts than in the vegetative cells suggests that GS may be more important than ADH in the general metabolism of nitrogen there. Fourth, the ^{15}N and enzymic data suggest that transaminase-mediated reactions are also of major importance in nitrogen-fixing *Anabaena cylindrica*.

On the basis of these data we propose, as a model for future research,

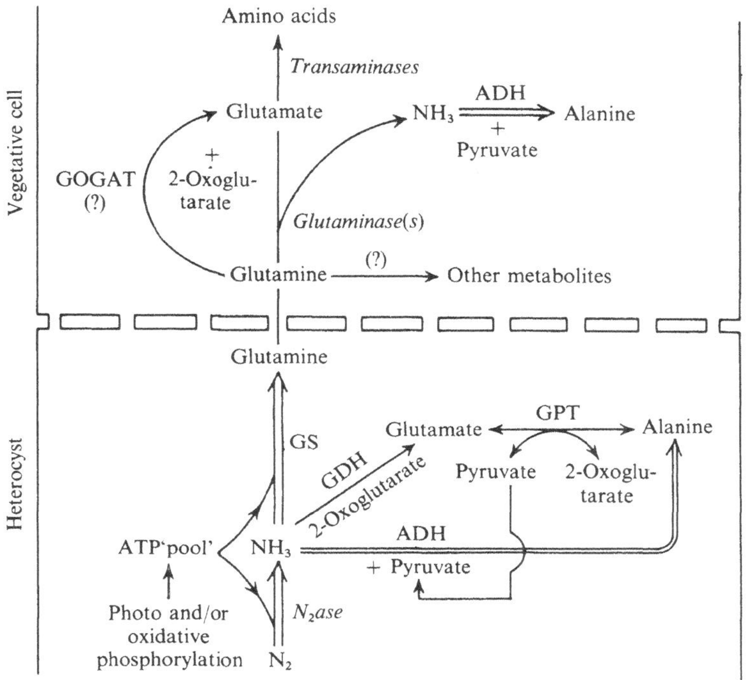

Fig. 9.11. Possible interrelations of nitrogenase and some ammonia-assimilating enzymes in *Anabaena cylindrica* (see also text).

the scheme outlined in Fig. 9.11. In this scheme we have indicated that the newly-fixed ammonia is removed rapidly from the nitrogen-fixing site as glutamine, but, although not shown, it could also be alanine, aspartate, or free ammonia (free ammonia is a major extracellular product of symbiotic blue-green algae; Stewart, Rodgers and Rowlands, unpublished) so we would not wish to be categorical here. The release of ammonia from glutamine could be brought about perhaps by glutaminase, GOGAT, or transaminases, with glutamate-mediated reactions, transaminases and ADH functioning both at, and remote from, the site of nitrogenase. Such a scheme could account for the two pools of glutamate suggested by the ^{15}N kinetic data, and it is also possible that the glutamate could be recycled to GS at the site of nitrogenase activity although this is not shown in the figure. The nitrogen-fixing site could be considered as the heterocyst and the site remote from the nitrogen-fixing site as the vegetative cell (see Fig. 9.11) although the scheme could apply equally to subcellular compartmentation within one cell.

It should also be appreciated that this scheme has several imponderables. For example the presence of subcellular pools could mean that enzymes such as GDH may be more important in removing ammonia than say GS. Also the evidence for glutamate dehydrogenase is equivocal (Dharmawardene *et al.*, 1972; 1973; Neilson & Doudoroff, 1972; Brown *et al.*, 1973) and GS is an ATP requiring enzyme. It is possible, and indeed probable, therefore that in nitrogen-fixing algae enzymes such as GS and ADH play roles other than in the scavenging of ammonia. For example GS is an important regulatory enzyme in various heterotrophic bacteria (see Prusiner & Stadtman, 1973) and this could conceivably be its major role in nitrogen-fixing algae such as *Anabaena cylindrica*. These and other points will be resolved by further research.

Finally, it is worth considering the environmental implications of the above findings. There is now evidence from the other papers in Part III of this volume, and elsewhere (Stewart, 1975) that a wide variety of blue-green algae fix nitrogen, and that in areas such as rice paddy soils natural populations of these algae fix about 30–50 kg N/ha/ann (see Venkataraman, 1973; Chapter 14). There is, in addition, assimilation of this fixed nitrogen by the rice plant after autolysis of the algae and/or the release of algal extracellular products. The results presented here suggest that the addition of fertiliser-nitrogen to natural populations of such blue-green algae is likely to depress nitrogen fixation and that of the two types of fertiliser-nitrogen the depressive effect of ammonium-nitrogen is likely to be more severe than the effect of nitrate-nitrogen. In fact substantial amounts of nitrogen could still be fixed in the presence of nitrate-nitrogen whereas in the presence of ammonium-nitrogen activity would probably be inhibited completely. At first sight this suggests that the addition of nitrate fertiliser would be most beneficial. However this is questionable because rapid rates of denitrification are likely to occur in waterlogged paddy soils making its addition inappropriate and expensive. On the other hand the beneficial effect of added ammonium-fertiliser as a source of nitrogen is likely to be exerted only in soils where more than 50 kg/ha/ann can be added. Through the agency of IBP we have established the role and contribution of blue-green algae in areas such as the rice paddy field ecosystem. What is required now are more detailed studies on the economic and agronomic merits of algal fertilisers relative to the more conventional types of synthetic nitrogen fertilisers presently available.

This work was made possible through grants to W.D.P.S. from the Science Research Council, the Natural Environment Research Council, the Agricultural Research Council and the Royal Society.

References

Akkermans, A. D. L. (1971). Nitrogen fixation and nodulation of *Alnus* and *Hippophaë* under natural conditions. Doctoral thesis, University of Leiden.

Allen, M. B. & Arnon, D. I. (1955). Studies on nitrogen-fixing blue-green algae. I. Growth and nitrogen fixation by *Anabaena cylindrica* Lemm. *Pl. Physiol.*, **30**, 366–72.

Allen, M. M. & Smith, A. J. (1969). Nitrogen chlorosis in blue-green algae. *Arch. Mikrobiol.*, **69**, 114–20.

Benson, J. V., Gordon, M. J. & Patterson, J. A. (1967). Accelerated chromatographic analysis of amino acids in physiological fluids containing glutamine and asparagine. *Analyt. Biochem.*, **18**, 228–40.

Bone, D. H. (1971*a*). Nitrogenase activity and nitrogen assimilation in *Anabaena flos-aquae* growing in continuous culture. *Arch. Mikrobiol.*, **80**, 234–41.

Bone, D. H. (1971*b*). Kinetics of synthesis of nitrogenase in batch and continuous culture of *Anabaena flos-aquae*. *Arch. Mikrobiol.*, **80**, 242–51.

Bothe, H. (1970). Photosynthetische Stickstoffixierung mit einem zellfreien Extrakt aus der Blaualge *Anabaena cylindrica*. *Ber. dt. bot. Ges.*, **83**, 421–32.

Brown, C. M., Burn, V. J. & Johnson, B. (1973). Presence of glutamate synthetase in fission yeasts. *Nature, Lond.*, **246**, 115–16.

Brown, C. M. & Stanley, S. O. (1972). Environment-mediated changes in the cellular content of the 'pool' constituents and their associated changes in cell physiology. *J. appl. Chem. Biotech.*, **22**, 363–90.

Burris, R. H. & Wilson, P. W. (1957). Methods for measurement of nitrogen fixation. In *Meth. Enzym.*, **4**, 355–65.

Carnahan, J. E., Mortenson, L. E., Mower, H. F. & Castle, J. E. (1960). Nitrogen fixation in cell-free extracts of *Clostridium pasteurianum*. *Biochim. biophys. Acta*, **42**, 530–35.

Codd, G. A. & Stewart, W. D. P. (1973). Pathways of glycollate metabolism in the blue-green alga *Anabaena cylindrica*. *Arch. Mikrobiol.*, **94**, 11–28.

Cole, H. A., Wimpenny, J. W. T. & Hughes, D. W. (1967). The ATP pool in *Escherichia coli*. 1. Measurement of the pool using a modified luciferase assay. *Biochim. biophys. Acta*, **143**, 445–53.

Dalton, H. & Mortenson, L. E. (1972). Dinitrogen (N_2) fixation (with a biochemical emphasis). *Bacteriol. Rev.*, **36**, 231–60.

Dharmawardene, M. W. N., Haystead, A. & Stewart, W. D. P. (1973). Glutamine synthetase of the nitrogen-fixing alga *Anabaena cylindrica*. *Arch. Mikrobiol.*, **90**, 281–95.

Dharmawardene, M. W. N., Stewart, W. D. P. & Stanley, S. O. (1972). Nitrogenase activity, amino acid pool patterns and amination in blue-green algae. *Planta, Berlin*, **108**, 133–45.

Frieden, C. (1963). Different structural forms of reversibly dissociated glutamic dehydrogenase: Relation between enzymic activity and molecular weight. *Biochem. biophys. Res. Comm.*, **10**, 410–15.

Haystead, A., Dharmawardene, M. W. N. & Stewart, W. D. P. (1973). Ammonia assimilation in a nitrogen-fixing blue-green alga. *Pl. Sci. Lett.*, **1**, 439–45.

Haystead, A., Robinson, R. & Stewart, W. D. P. (1970). Nitrogenase activity in extracts of heterocystous and non-heterocystous blue-green algae. *Arch. Mikrobiol.*, **75**, 235–43.

Haystead, A. & Stewart, W. D. P. (1972). Characteristics of the nitrogenase system from the blue-green alga *Anabaena cylindrica*. *Arch. Mikrobiol.*, **82**, 326–36.

Hoare, D. S., Hoare, S. L. & Moore, R. B. (1967). The photoassimilation of organic compounds of autotrophic blue-green algae. *J. gen. Microbiol.*, **49**, 351–70.

Hood, W. & Carr, N. G. (1972). Branched chain amino acid biosynthesis in blue-green algae. *J. gen. Microbiol.*, **73**, 317–26.

Hora, F. B. & Webber, P. J. (1960). Determination of NO_3^- using phenol 2,4 disulphonic acid. *The Analyst*, **85**, 567.

Hubbard, J. S. & Stadtman, E. R. (1967). Regulations of glutamine synthetase. v. Partial purification and properties of glutamine synthetase from *Bacillus licheniformis*. *J. Bact.*, **94**, 1007–15.

Kennedy, I. R. (1970). Kinetics of acetylene and cyanide reduction by the nitrogen-fixing system of *Rhizobium lupini*. *Biochim. biophys. Acta*, **222**, 135–44.

Lyne, R. L. & Stewart, W. D. P. (1973). Emerson enhancement of carbon fixation but not of acetylene reduction (nitrogenase activity) in *Anabaena cylindrica*. *Planta, Berlin*, **109**, 27–38.

McCarty, R. E. & Coleman, C. H. (1969). The uncoupling of photophosphorylation by valinomycin and ammonium chloride. *J. biol. Chem.*, **244**, 4292–8.

McElroy, W. D. (1963). Determination of adenosine-5-triphosphate. In *Methods of Enzymatic Analysis* (ed. Bergmeyer, H. U.), pp. 565–6. Academic Press: New York, London.

McKee, H. S. (1962). *Nitrogen Metabolism in Plants*. Clarendon Press: Oxford.

Magee, W. E. & Burris, R. H. (1954). Fixation of N_2 and utilization of combined nitrogen by *Nostoc muscorum*. *Am. J. Bot.*, **11**, 777–82.

Meeks, J. C. & Castenholz, R. W. (1971). Growth and photosynthesis in an extreme thermophile, *Synechococcus lividus* (Cyanophyta). *Arch. Mikrobiol.*, **78**, 25–41.

Neilson, A. H. & Doudoroff, M. (1976). Ammonia assimilation in blue-green algae. *Arch. Mikrobiol.*, **89**, 15–22.

Neilson, A. H., Rippka, R. & Kunisawa, R. (1971). Heterocyst formation and nitrogenase synthesis in *Anabaena* spp.: a kinetic study. *Arch. Mikrobiol.*, **79**, 164–75.

Ohmori, M. & Hattori, A. (1972). Induction of nitrate and nitrite reductases in *Anabaena cylindrica*. *Pl. Cell Physiol., Tokyo*, **11**, 873–8.

Patterson, M. S. & Greene, R. C. (1965). Measurement of low energy β-emitters in aqueous solution by liquid scintillation counting of emulsions. *Analyt. Chem.*, **37**, 854–7.

Pearce, J., Leach, C. K. & Carr, N. G. (1969). The incomplete tricarboxylic acid cycle in the blue-green alga *Anabaena variabilis. J. gen. Microbiol.*, **54**, 371–78.

Prusiner, S. & Stadtman, E. R. (eds.) (1973). *The Enzymes of Glutamine Metabolism.* Academic Press: New York.

Scott, W. E. & Fay, P. (1972). Phosphorylation and amination in heterocysts of *Anabaena variabilis. Br. phycol. J.*, **7**, 283–4.

Shah, V. K., Davis, L. C. & Brill, W. J. (1972). Nitrogenase, I. Repression and derepression of the iron–molybdenum and iron proteins of nitrogenase in *Azotobacter vinelandii. Biochim. biophys. Acta*, **256**, 498–511.

Shapiro, B. M. & Stadtman, E. R. (1970). Glutamine synthetase (*Escherichia coli*) biosynthetic assay, P_i measurement, transferase assay, γ-glutamyl hydroxamate measured. In *Methods in Enzymology* (eds. Tabor, H. and Tabor, C. U.), vol. **17**A, pp. 910–22. Academic Press: London, New York.

Simon, R. D. (1971). Cyanophycin granules from the blue-green alga *Anabaena cylindrica*: a reserve material consisting of copolymers of aspartic acid and arginine. *Proc. natn. Acad. Sci., USA*, **68**, 265–7.

Simon, R. D. (1973). Measurement of the cyanophycin granule polypeptide contained in the blue-green alga *Anabaena cylindrica. J. Bact.*, **114**, 1213–16.

Smith, R. V. & Evans, M. C. W. (1970). Soluble nitrogenase from vegetative cells of the blue-green alga *Anabaena cylindrica. Nature, Lond.*, **225**, 1253–4.

Spackman, D. H., Stein, W. H. & Moore, S. (1958). Automatic recording apparatus for use in the chromatography of amino acids. *Analyt. Chem.*, **30**, 1190–206.

Stewart, W. D. P. (1973). Nitrogen fixation by photosynthetic microorganisms. *Ann. Rev. Microbiol.*, **27**, 283–316.

Stewart, W. D. P. (1975). Blue-green Algae. In *Dinitrogen (N_2) fixation* (ed. Hardy, R. W. F.), vol. **2**. Wiley-Interscience: New York (in press).

Stewart, W. D. P., Fitzgerald, G. P. & Burris, R. H. (1967). *In situ* studies on N_2 fixation using the acetylene reduction technique. *Proc. natn. Acad. Sci., USA*, **58**, 2071–78.

Stewart, W. D. P., Fitzgerald, G. P. & Burris, R. H. (1968). Acetylene reduction by nitrogen-fixing blue-green algae. *Arch. Mikrobiol.*, **62**, 336–48.

Stewart, W. D. P. & Lex, M. (1970). Nitrogenase activity in the blue-green alga, *Plectonema boryanum* strain 594. *Arch. Mikrobiol.*, **73**, 250–60.

Stewart, W. D. P. & Rowell, P. (1975). Effects of L-methionine-DL-sulphoximine on the assimilation of newly fixed NH_3, acetylene reduction and heterocyst formation in *Anabaena cylindrica. Biochem. Biophys. Res. Comm.* **65**, 846–56.

Strecker, J. H. (1955). Glutamic dehydrogenase from liver. *Meth. Enzym.*, **2**, 220–1.

Thomas, J. & David, K. A. V. (1971). Studies on the physiology of heterocyst production in the nitrogen-fixing blue-green alga *Anabaena* sp. L-31 in continuous culture. *J. gen. Microbiol.*, **66**, 127–31.
Van Gorkom, H. J. & Donze, M. (1971). Localization of nitrogen fixation in *Anabaena. Nature, Lond.*, **234**, 231–2.
Venkataraman, G. S. (1973). *Algal Biofertilizers and Rice Cultivation.* Today and Tomorrow's Printers & Publishers: New Delhi.

10. The physiology of nitrogen fixation by a *Gloeocapsa* sp.

J. R. GALLON, W. G. W. KURZ & T. A. LARUE

Gloeocapsa is the only unicellular blue-green alga which so far has been demonstrated to fix atmospheric dinitrogen (Rippka, Neilson, Kunisawa & Cohen-Bazire, 1971). Although a soil organism it may be grown in a liquid medium and it was by using such cultures that Wyatt & Silvey (1969) first showed that cells of *Gloeocapsa* sp. LB795 were able to grow in nitrogen-free medium and were also able to reduce acetylene to ethylene.

Gloeocapsa sp. LB795, although a unialgal culture, is contaminated with bacteria which live in the extensive slime capsule surrounding the organism. Wyatt & Silvey showed that nitrogen fixation by their cultures was not the result of bacterial contamination, and this has been confirmed by the demonstration that the bacterial contaminant of *Gloeocapsa* sp. LB795 cannot fix nitrogen itself (Gallon, LaRue & Kurz, unpublished). Moreover, an axenic culture of *Gloeocapsa* sp. LB795 has been isolated by Stanier, Kunisawa, Mandel & Cohen-Bazire (1971), and this organism, *Gloeocapsa* sp. 27152, was also able to reduce acetylene and in most respects behaved in culture in an identical manner to *Gloeocapsa* sp. LB795.

Unicellular blue-green algae, like *Gloeocapsa*, of the order Chroococcales are evolutionarily much more primitive organisms than the other Cyanophyceae, and recently, Stanier *et al.* (1971) have suggested that they may be more properly classified as a major group of bacteria than as algae, even though they possess a photosynthetic mechanism like that of higher plants, in which water is the ultimate electron donor, and oxygen is evolved.

This photosynthetic evolution of oxygen presents a special and unique problem to the nitrogenases of blue-green algae; all of which so far isolated are irreversibly inactivated *in vitro* by exposure to very low levels of oxygen (Haystead, Robinson & Stewart, 1970; Smith & Evans, 1971). Despite this, these enzymes function *in vivo*, and must therefore, in the intact organism, be protected from inactivation by photosynthetically produced oxygen.

Many filamentous blue-green algae possess heterocysts. These algae, of which *Anabaena* is an example, may, under aerobic conditions, fix at least some of their nitrogen in these specialised cells (Fay, Stewart, Walsby & Fogg, 1968; Stewart, Haystead & Pearson, 1969; Weare & Benemann, 1973). Heterocysts have a low intracellular concentration of oxygen and since they do not seem to possess a functional Photosystem II of photosynthesis (Donze, Haveman & Schiereck, 1972), they are unable to evolve oxygen photosynthetically. They therefore provide a suitable site for the oxygen-sensitive nitrogenase in these organisms, and represent one means of protecting the enzyme from oxygen inactivation. Other filamentous blue-green algae such as *Plectonema*, which do not possess heterocysts, can only fix nitrogen under microaerophilic conditions (Stewart & Lex, 1970) and therefore appear to have poor protection of their nitrogenase *in vivo*.

On the other hand, *Gloeocapsa* grows and fixes nitrogen aerobically. Because of this, and because the photosynthetic apparatus and the nitrogen-fixing enzymes are present in a single, undifferentiated cell, *Gloeocapsa* provides a unique tool with which to study the relationships between photosynthesis and nitrogen fixation.

Nitrogen fixation by *Gloeocapsa*

The nitrogenase of Gloeocapsa

Broken-cell preparations of *Gloeocapsa* sp. LB795 reduced acetylene when provided with an ATP generating system and sodium dithionite as a source of reducing power (Gallon, LaRue & Kurz, 1972). The activity of these preparations however was not sufficient to demonstrate unequivocally the reduction of $^{15}N_2$.

Gloeocapsa nitrogenase was inhibited by 0.01 atm of carbon monoxide, and in the absence of a suitable alternative substrate it catalysed the ATP-dependent evolution of hydrogen from H^+ (Gallon *et al.*, 1972). In its physical and kinetic properties, the enzyme was similar in most respects to the nitrogenases isolated from *Anabaena* (Smith & Evans, 1971) and *Plectonema* (Haystead *et al.*, 1970) despite the phylogenetic differences between *Gloeocapsa* and these filamentous blue-green algae. Unlike these nitrogenases however, the *Gloeocapsa* enzyme was as stable to storage at 0 °C as at room temperature, although activity was lost rapidly under both conditions, with a half-life of 8.7 h (Gallon *et al.*, 1972). However, *Gloeocapsa* nitrogenase was stable to storage in liquid nitrogen for up to 3 months.

Gloeocapsa nitrogenase was sedimented by centrifugation at $10000 \times g$

and therefore appeared to be particulate. In this respect too the enzyme differed from the nitrogenases of *Anabaena* and *Plectonema*. All efforts to obtain a soluble preparation of *Gloeocapsa* nitrogenase failed, and so no purification of the enzyme was attempted.

In its kinetic properties, *Gloeocapsa* nitrogenase also resembled the enzymes from the bacteria *Azotobacter vinelandii* (Burns, 1969; Hardy, Holsten, Jackson & Burns, 1968) and *Clostridium pasteurianum* (Hardy *et al.*, 1968). In its particulate nature and its relative stability at 0 °C and at room temperature, *Gloeocapsa* nitrogenase further resembled crude preparations of *Azotobacter* nitrogenase (Bulen & LeComte, 1966) and also the nitrogenase from the photosynthetic bacterium *Chromatium* (Winter & Ober, 1973). However, unlike particulate preparations of these two bacterial enzymes, *Gloeocapsa* nitrogenase was irreversibly inactivated by oxygen *in vitro* (Table 10.1).

Table 10.1. *The effect of oxygen on acetylene reduction by broken cell preparations of* Gloeocapsa

Partial pressure of O_2 (atm)	C_2H_4 formed (nmol/min/mg protein)	Inhibition (%)
0	0.573	0
0.0005	0.586	0
0.001	0.571	0
0.005	0.458	20
0.01	0.109	81
0.05	0.034	94
0.1	0	100

For conditions of assay, see Gallon *et al.* (1972).

Source of reducing power

Sodium dithionite was used as an artificial source of reducing power for *Gloeocapsa* nitrogenase during the determinations of the kinetic parameters of the enzyme. However, the natural source of this reducing power for the nitrogenases of blue-green algae is ferredoxin (Bothe, 1970; Smith, Noy & Evans, 1971), and the *Gloeocapsa* enzyme seemed to be no exception. A crude preparation of *Gloeocapsa* ferredoxin, when reduced itself, was able to supply reducing power for the reduction of acetylene by broken cell preparations of *Gloeocapsa*. Spinach ferredoxin at the same concentration could replace *Gloeocapsa* ferredoxin, but with only 23 % of the efficiency of the latter (Gallon, Kurz & LaRue, 1973).

Table 10.2. *The effect of DCMU or darkness on acetylene reduction and photosynthesis by intact cells of* Gloeocapsa (*from Gallon* et al., *1973, reproduced by permission of the National Research Council of Canada from the* Canadian Journal of Microbiology, **19**, *1973*)

	Concn of DCMU (μM)	C_2H_4 formed (nmol/min/ mg protein)	Inhibition (%)	Photo-synthetic O_2 evolved (nmol/min/ mg protein)	Inhibition (%)
Light	0	3.77	0	21.4	0
	1	3.45	9	0	100
	10	3.08	18	0	100
Dark	0	1.85	51	0	100

Cells were removed from cultures when nitrogenase activity was greatest and were assayed under an atmosphere of air–C_2H_2 (99:1) as described by Gallon *et al.* (1972).

In blue-green algae, as in higher plants, the generation of reduced ferredoxin is achieved *in vivo* by the photosynthetic flow of electrons from water. Theoretically then, nitrogen fixation by *Gloeocapsa* may be supported directly by reductant generated photosynthetically in the light. However, as 3-(3,4-dichlorophenyl)-1,1-dimethylurea (DCMU), an inhibitor of Photosystem II, had very little effect on the ability of whole cells of *Gloeocapsa* to reduce acetylene at concentrations which inhibited their oxygen evolution completely (Table 10.2), *Gloeocapsa* nitrogenase would seem not to be entirely dependent upon photosynthesis as a source of reducing power. Assuming that DCMU at these concentrations had, by its action on Photosystem II, completely blocked electron transport between water and ferredoxin, then acetylene reduction must have been supported either by electrons inserted into the photosynthetic electron transport chain at a point subsequent to the inhibited Photosystem II, in a manner analogous to the artificial ascorbate-2,6-dichlorophenol indophenol (DCPIP) system, or by an entirely non-photosynthetic means of generating reduced ferredoxin.

Cells of *Gloeocapsa*, grown under continuous illumination of 2500 lx, were able to reduce acetylene in the dark at 50 % of their rate in the light (Table 10.2). This reduction was linear with respect to time for up to 4 h after the cells were removed from the light, and was not altered by the addition of 10 μM DCMU (Gallon *et al.*, 1973). Thus, *Gloeocapsa* nitrogenase was able to reduce acetylene in the absence of reductant produced directly by photosynthesis.

Photosynthetic oxygen evolution by *Gloeocapsa* was completely inhibited in the dark, and also in the light in the presence of 10 μM DCMU (Table 10.2). Acetylene reduction was however about 30% greater under the latter conditions than in the dark, which strongly suggested that a light driven reaction, insensitive to DCMU, was responsible for part of the acetylene reduced in the light in the presence of DCMU. This could mean that the earlier assumption that photosynthetic electron flow through Photosystem II was completely blocked by 10 μM DCMU was incorrect, or it could indicate that electrons were inserted into the electron transport chain in the manner suggested above.

Three possible means of generating reduced ferredoxin to support nitrogen fixation by *Gloeocapsa* therefore exist. They are:

1. Conventional photosynthetic electron transport, with water as the ultimate electron donor and involving both photosystems.
2. A light-dependent reduction of ferredoxin which may not involve Photosystem II, but which probably does involve the subsequent portion of the photosynthetic electron transport chain.
3. An entirely non-photosynthetic generation of reduced ferredoxin.

This last possibility, the only one capable of supporting acetylene reduction in the dark, was investigated using broken cell preparations of *Gloeocapsa* nitrogenase (Gallon *et al.*, 1973). In the presence of a system generating ATP, a reduced ferredoxin preparation from *Gloeocapsa* supported acetylene reduction by *Gloeocapsa* nitrogenase. The ferredoxin was itself reduced by a system consisting of *iso*citrate, *iso*citrate dehydrogenase ($NADP^+$), $NADP^+$, and ferredoxin-$NADP^+$ reductase. As the enzymes used in this latter system were also prepared from extracts of *Gloeocapsa*, the intact organism appeared to possess all the necessary components to generate reduced ferredoxin in this way. The oxidation of *iso*citrate by *iso*citrate dehydrogenase ($NADP^+$) may, then, serve in *Gloeocapsa* as a source of reducing power enabling nitrogenase to function in the dark.

A search for alternative substrates revealed that the oxidation of malate by malate dehydrogenase ($NADP^+$) could also supply reducing power to ferredoxin and thus support acetylene reduction by *Gloeocapsa* nitrogenase. It was noted that irrespective of whether the substrate was *iso*citrate or malate, ferredoxin was an essential component of the system generating reductant, but no absolute requirement for $NADP^+$ could be demonstrated (Gallon *et al.*, 1973). NAD^+ could not replace $NADP^+$ however.

Both *iso*citrate dehydrogenase ($NADP^+$) and malate dehydrogenase

($NADP^+$) were found in cells of *Gloeocapsa* fixing nitrogen, so the oxidation of either *iso*citrate or malate, or perhaps of both, may be of physiological importance in supplying reducing power to *Gloeocapsa* nitrogenase in the dark.

Photosynthesis and nitrogen fixation

Studies on aerobically grown cells

Although in the dark *iso*citrate and malate may provide reducing power to *Gloeocapsa* nitrogenase, in the light the photosynthetic reduction of ferredoxin is probably of major importance. Photosynthesis by *Gloeocapsa* resembles that of higher plants in that water is the ultimate electron donor and oxygen is evolved. Whether or not photosynthesis supplies reducing power directly to *Gloeocapsa* nitrogenase in the light, this associated evolution of oxygen makes necessary some form of protection for the oxygen-sensitive, nitrogen-fixing enzyme.

Since *Gloeocapsa* is a unicellular organism, its nitrogenase is present in a cell evolving oxygen and not in specialised cells such as heterocysts. In this organism, therefore, protection of nitrogenase from damage by oxygen is not achieved by the differentiation of nitrogen-fixing cells. However, *Gloeocapsa* nitrogenase is associated with cell fragments into which oxygen may be unable to diffuse, so it might be expected that the enzyme is protected from inactivation in this way, as may be the case with *Azotobacter* nitrogenase (Oppenheim & Marcus, 1970). This is not so either; even though particulate, *Gloeocapsa* nitrogenase was irreversibly inactivated *in vitro* by exposure to oxygen (Table 10.1) and furthermore, because the ability of whole cells of *Gloeocapsa* to reduce acetylene was decreased in an atmosphere containing more than 0.2 atm of oxygen (Table 10.3), oxygen also had an inhibitory effect on the enzyme *in vivo*, although this inhibition was reversible.

Cells of *Gloeocapsa* showed maximum nitrogenase activity when the concentration of oxygen in the atmosphere was about 0.1 atm. When exposed to an atmosphere containing 0.2 atm of oxygen, they reduced acetylene at 80 % of their optimum rate. However, even though the enzyme does not function most efficiently in cells cultured under an atmosphere of air, that *Gloeocapsa* can grow and fix nitrogen aerobically at all indicates that its nitrogenase is sufficiently active under these conditions (and in cells evolving oxygen) to provide enough fixed nitrogen to support the organism's growth. Some protection of the

Table 10.3. *The effect of varying concentrations of oxygen in the atmosphere above cells of* Gloeocapsa *on their ability to fix nitrogen*

Partial pressure of components of gas phase (atm)				
N_2	Ar	O_2	C_2H_4 formed (nmol/min/10^8 cells)	Relative activity
0.60	0.40	0	0.016	94
0.60	0.35	0.05	0.020	118
0.60	0.30	0.10	0.021	124
0.60	0.20	0.20	0.017	100
0.60	0.10	0.30	0.013	77
0.60	0	0.40	0.006	34

Cells were removed from aerobically grown cultures of *Gloeocapsa* when nitrogenase activity was greatest and assayed under a light intensity of 2500 lx as described by Gallon *et al.* (1972).

enzyme from damage by oxygen, particularly that produced photosynthetically, must therefore exist *in vivo.*

The relationship between photosynthetic oxygen evolution and nitrogenase was examined during the growth of batch cultures of *Gloeocapsa*, grown under air with a constant illumination of 2500 lx. For the first seven days after inoculation, the cell population of those cultures increased logarithmically, with an average doubling time of 65 h. After this, cell division slowed down and eventually stopped about ten days after inoculation.

Because the cell is the basic unit of both nitrogen fixation and photosynthesis, these activities were calculated in terms of a constant cell number, 10^8 cells.

Nitrogenase activity per 10^8 cells, as determined by acetylene reduction, usually reached a maximum in cultures of *Gloeocapsa* six days after inoculation, but was always greatest between days four and eight (Fig. 10.1). Nitrogenase activity was therefore highest during the period of maximum cell division, and by the time cell division ceased, nitrogenase activity had fallen to half of its maximum activity.

Oxygen evolution per 10^8 cells remained remarkably constant at a relatively low level between five and eight days after inoculation of *Gloeocapsa* cultures (Fig. 10.2) but after day eight, it increased rapidly. During the period of maximum nitrogenase activity, therefore, photosynthetic oxygen evolution was maintained at a constant low level and it always increased dramatically as nitrogenase activity fell. From this, it appeared that a temporal separation between maximum nitrogenase

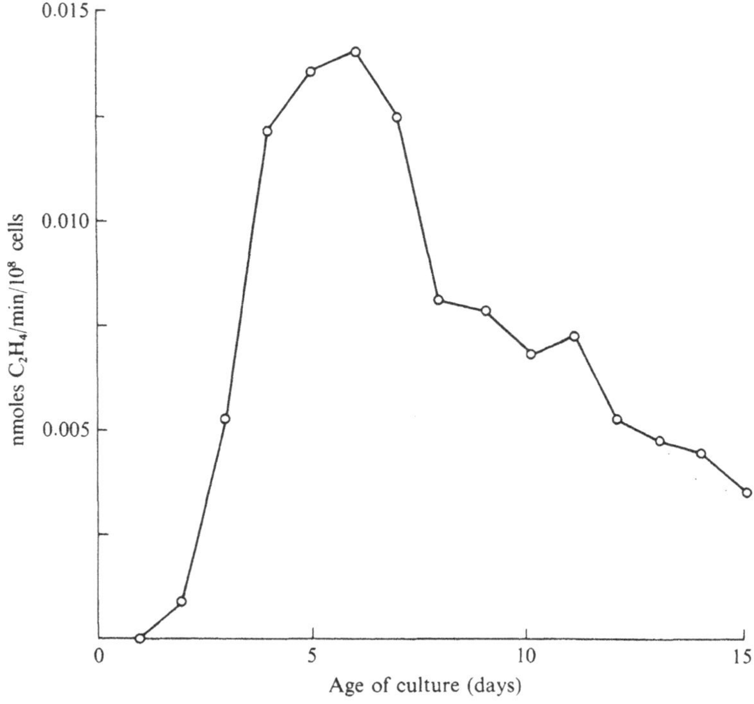

Fig. 10.1. Relationship between nitrogenase activity and age of a culture of *Gloeocapsa*.

activity and maximum photosynthetic oxygen evolution, protected *Gloeocapsa* nitrogenase from inactivation by photosynthetically produced oxygen.

During the logarithmic phase of growth, cultures of *Gloeocapsa* appeared bright green in colour; later they appeared blue-green. The colour change usually occurred when the culture was seven or eight days old and coincided with the start of increased oxygen production.

Blue-green algae contain two main classes of photosynthetic pigment, the green pigment chlorophyll *a* and the phycobilins of which phycocyanin (blue) is quantitatively the most important in *Gloeocapsa*. Spectrophotometric analysis of the pigment composition of cells of *Gloeocapsa* during their growth in batch cultures showed that both chlorophyll *a* and phycocyanin were present only in very small amounts during the first seven days after inoculation, but after this, their levels increased rapidly (Fig. 10.3). In this respect, the pigment content per 10^8 cells reflected the photosynthetic oxygen production per 10^8 cells

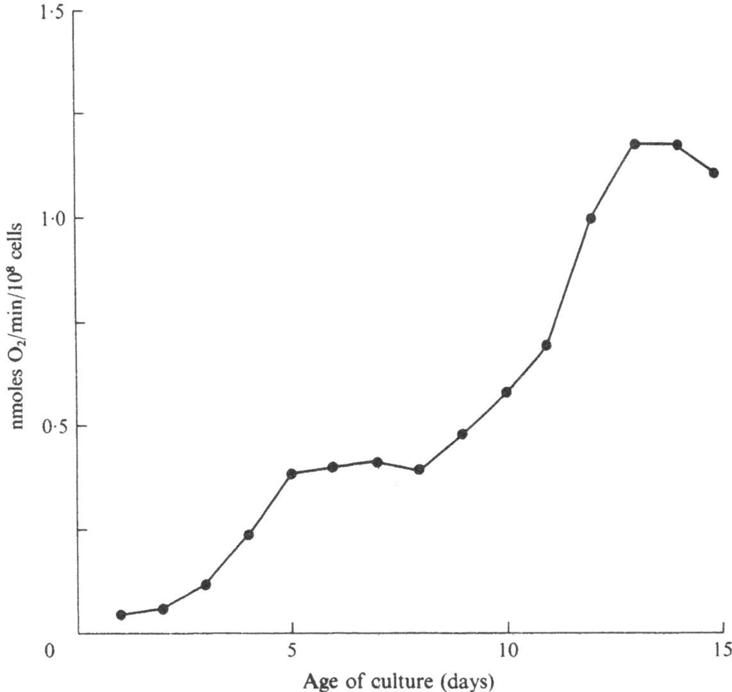

Fig. 10.2. Relationship between photosynthetic oxygen evolution and age of a culture of *Gloeocapsa*.

at the same stage of growth (Fig. 10.2). When the intracellular concentration of the photosynthetic pigments was low, oxygen evolution was also relatively low, and an increase in the pigment concentration was followed by an increase in photosynthetic oxygen evolution.

Furthermore, the ratio of chlorophyll *a* to phycocyanin in cells of *Gloeocapsa* was relatively high three days after inoculation but decreased rapidly during the following six days (Fig. 10.4). As well as containing only a low concentration of photosynthetic pigments therefore, rapidly dividing cells of *Gloeocapsa* were also more deficient than older cells in phycocyanin relative to chlorophyll *a*.

Phycocyanin is a protein, and there is evidence that it may be used as an emergency reserve of nitrogen (Allen & Smith, 1969). Hence, for the first three days after inoculation, cells of *Gloeocapsa* might use this protein as a source of fixed nitrogen, since nitrogenase activity was low over this period. However, young cells of *Gloeocapsa* contained low levels of both photosynthetic pigments, not just phycocyanin. This

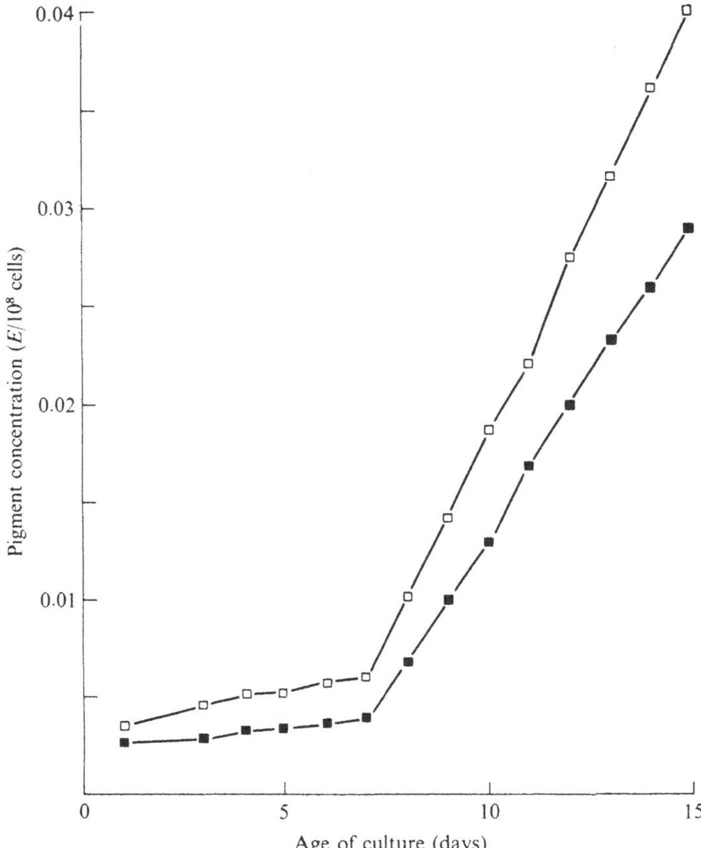

Fig. 10.3. Relationship between the age of a culture of *Gloeocapsa* and the cellular concentration of the photosynthetic pigments; chlorophyll *a* (□) and phycocyanin (■) were determined spectrophotometrically at 675 nm and 615 nm respectively.

may mean that the simultaneous loss of chlorophyll *a* and phycocyanin serves a function other than to supply fixed nitrogen.

Furthermore, while the dramatic loss of both photosynthetic pigments occurred within 24 h of inoculation of a new culture of *Gloeocapsa*, the change in the ratio of chlorophyll *a* to phycocyanin occurred between one and three days after inoculation and seemed to be as much associated with chlorophyll *a* synthesis as phycocyanin utilisation.

The phycobilin pigments are mainly associated with Photosystem II, and therefore also with the photoproduction of oxygen from water, so a decrease in the levels of phycocyanin relative to chlorophyll *a* could

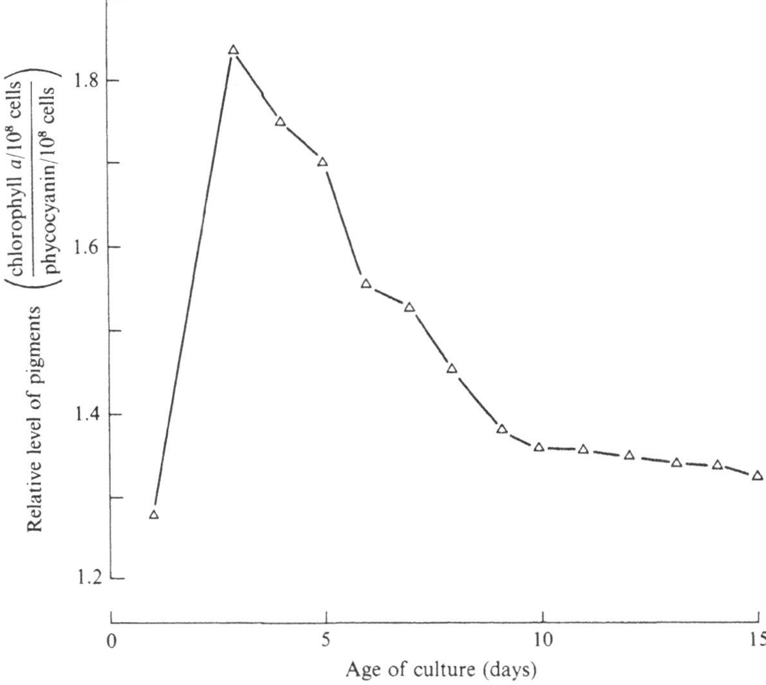

Fig. 10.4. Relationship between the age of a culture of *Gloeocapsa* and the relative concentrations of chlorophyll *a* and phycocyanin.

be an important factor in reducing photosynthetic oxygen evolution. The low level of oxygen production observed in cells of *Gloeocapsa*, during the period of maximum nitrogenase activity may thus be a reflection of, not only the low total levels of the photosynthetic pigments in the cells, but also a relatively high ratio of chlorophyll *a* to phycocyanin.

A similar relationship between the photosynthetic pigments and nitrogenase activity has been reported in the heterocysts of *Anabaena* (Thomas, 1972; David & Thomas, 1970). When nitrogenase activity in these cells was greatest, they were deficient in the pigments of Photosystem II. As the cells aged, the levels of phycocyanin and phycoerythrin increased and nitrogenase activity was progressively inhibited. It has also been noticed that whole filaments of *Anabaena flos-aquae*, grown under conditions when few of them possess heterocysts (Kurz & LaRue, 1971), appear bright green when they are fixing nitrogen at the maximum rate.

Table 10.4. *The effect of the concentration of oxygen in the gas mixture used to sparge growing cultures of* Gloeocapsa *on their ability to fix nitrogen*

Partial pressure of components of gas phase (atm)		C_2H_4 formed over 10 days by 10^8 cells (μmol)	Relative activity
N_2	O_2		
1.00	0	4.135	49
0.95	0.05	9.239	109
0.90	0.10	9.826	116
0.80	0.20	8.476	100
0.70	0.30	6.323	75

The gas mixture in each case contained a carefully controlled level of 3×10^{-4} atm CO_2. The ethylene formed was calculated from the activity of the culture over the four days before and the six days after maximum nitrogenase activity.

Studies on cells grown under artificial atmospheres

While heterocystous blue-green algae may protect their nitrogenase from the effects of exposure to oxygen by a spatial separation of nitrogen fixation and photosynthetic oxygen evolution, *Gloeocapsa* achieves this by a temporal separation of the maximum activities of the two processes. However, the concentration of oxygen within cells of *Gloeocapsa* is not entirely dependent upon that produced photosynthetically, it may also be dependent upon the concentration of dissolved oxygen in the culture medium which in turn depends upon the oxygen content of the atmosphere to which the culture medium is exposed.

Cells of *Gloeocapsa* sp. LB795, grown under standard laboratory conditions, but sparged with the gas mixtures shown in Table 10.4 showed exactly the same pattern of nitrogenase activity with respect to age, as cultures sparged with air. In these experiments, maximum nitrogenase activity always occurred when photosynthetic oxygen evolution was at a relatively low level and this relationship was entirely independent of the oxygen content of the sparging gas mixture over the range examined. The total amount of nitrogen fixed by a constant cell number of these cultures (expressed in terms of acetylene reduction) however, was dependent upon the oxygen content of the sparging gas mixture, being maximal at 0.1 atm (Table 10.4).

The concentration of oxygen within cells of *Gloeocapsa* depends upon the diffusion of oxygen into and out of the cells themselves. That this takes place is indicated by the effect of oxygen on acetylene reduction by whole cells of *Gloeocapsa* (Table 10.3). Furthermore, except in the vicinity of the thylakoid membranes (the site of photosynthesis in *Gloeocapsa*), the concentration of oxygen in the cell may be expected to approximate to the concentration of dissolved oxygen in the growth medium.

The results in Table 10.4 indicate that although the concentration of oxygen in the medium did have an effect on nitrogen fixation by growing cultures of *Gloeocapsa*, there was an inverse relationship between nitrogenase activity and photosynthetic oxygen evolution which was entirely independent of the oxygen composition of the medium. This suggests that the activity of nitrogenase is more closely affected by photosynthetic oxygen evolution than by the oxygen content of the atmosphere, which in turn suggests that the site of nitrogenase activity is close to that of photosynthetic oxygen evolution.

As *Gloeocapsa* nitrogenase is particulate (Gallon *et al.*, 1972), it is possible that the enzyme is associated with the thylakoid membranes themselves and consequently that reductant generated by conventional photosynthesis may be used directly to fix nitrogen under normal conditions of growth. Although, in the experiments described, photosynthetic oxygen evolution was low when nitrogenase activity was greatest, sufficient reductant could nevertheless have been supplied by photosynthesis to support nitrogen fixation at the levels observed.

J.R.G. gratefully acknowledges the award of a National Research Council of Canada Postdoctoral Research Fellowship from 1970–1972 and a Science Research Council Postdoctoral Fellowship at the Botany School, South Parks Road, Oxford, from 1972–1973.

The skilled technical assistance of Mr C. S. Mallard and Mr K. B. Chatson is also gratefully acknowledged.

References

Allen, M. M. & Smith, A. J. (1969). Nitrogen chlorosis in blue-green algae. *Arch. Mikrobiol.*, **69**, 114–20.

Bothe, H. (1970). Photosynthetische Stickstoffixierung mit einem Zellfreien Extract aus der Blaualge *Anabaena cylindrica. Ber. dt. bot. Ges.*, **83**, 421–32.

Bulen, W. A. & LeComte, J. R. (1966). The nitrogenase system from *Azotobacter*: two enzyme requirements for N_2 reduction, ATP dependent H_2 evolution and ATP hydrolysis. *Proc. natn. Acad. Sci. USA*, **56**, 979–86.

Burns, R. C. (1969). The nitrogenase system from *Azotobacter*. Activation energy and divalent cation requirement. *Biochim. biophys. Acta*, **171**, 253–9.

David, K. A. V. & Thomas, J. (1970). Nitrogenase activity and its relation to Photosystems I and II of photosynthesis in heterocysts of a blue-green alga. In *Proceedings of the Department of Atomic Energy Symposium on Radiation and Radioisotopes in Soil Studies and Plant Nutrition*, Bangalore, Dec. 1970, pp. 435–43.

Donze, M., Haveman, J. & Schiereck, P. (1972). Absence of photosystem 2 in heterocysts of the blue-green alga, *Anabaena*. *Biochim. biophys. Acta*, **256**, 157–61.

Fay, P., Stewart, W. D. P., Walsby, A. E. & Fogg, G. E. (1968). Is the heterocyst the site of nitrogen fixation in blue-green algae? *Nature, Lond.*, **220**, 810–12.

Gallon, J. R., Kurz, W. G. W. & LaRue, T. A. (1973). Isocitrate supported nitrogenase activity in *Gloeocapsa* sp. LB795. *Can. J. Microbiol.*, **19**, 461–5.

Gallon, J. R., LaRue, T. A. & Kurz, W. G. W. (1972). Characteristics of nitrogenase activity in broken cell preparations of the blue-green alga *Gloeocapsa* sp. LB795. *Can. J. Microbiol.*, **18**, 327–32.

Hardy, R. W. F., Holsten, R. D., Jackson, E. K. & Burns, R. C. (1968). The acetylene–ethylene assay for nitrogen fixation: laboratory and field evaluation. *Pl. Physiol.*, **43**, 1185–1207.

Haystead, A., Robinson, R. & Stewart, W. D. P. (1970). Nitrogenase activity in extracts of heterocystous and non-heterocystous blue-green algae. *Arch. Mikrobiol.*, **74**, 235–43.

Kurz, W. G. W. & LaRue, T. A. (1971). Nitrogenase in *Anabaena flos-aquae* filaments lacking heterocysts. *Naturwissenschaften*, **58**, 417.

Oppenheim, J. & Marcus, L. (1970). Correlation of ultrastructure in *Azotobacter vinelandii* with nitrogen source for growth. *J. Bact.*, **101**, 286–91.

Rippka, R., Neilson, A., Kunisawa, R. & Cohen-Bazire, G. (1971). Nitrogen fixation by unicellular blue-green algae. *Arch. Mikrobiol.*, **76**, 341–8.

Smith, R. V. & Evans, M. C. W. (1971). Nitrogenase activity in cell-free extracts of the blue-green alga, *Anabaena cylindrica*. *J. Bact.*, **105**, 913–17.

Smith, R. V., Noy, R. J. & Evans, M. C. W. (1971). Physiological electron donor systems to the nitrogenase of the blue-green alga *Anabaena cylindrica*. *Biochim. biophys. Acta*, **253**, 104–9.

Stanier, R. Y., Kunisawa, R., Mandel, M. & Cohen-Bazire, G. (1971). Purification and properties of unicellular blue-green algae (order Chroococcales). *Bact. Rev.*, **35**, 171–205.

Stewart, W. D. P., Haystead, A. & Pearson, H. W. (1969). Nitrogenase activity in heterocysts of blue-green algae. *Nature, Lond.*, **224**, 226–8.

Stewart, W. D. P. & Lex, M. (1970). Nitrogenase activity in the blue-green alga, *Plectonema boryanum* strain 594. *Arch. Mikrobiol.*, **73**, 250–60.

Thomas, J. (1972). Relationship between age of culture and occurrence of the pigments of Photosystem II of photosynthesis in heterocysts of a blue-green alga. *J. Bact.*, **110**, 92–5.

Weare, N. M. & Benemann, J. R. (1973). Nitrogen fixation by *Anabaena cylindrica*. I. Localization of nitrogen fixation in the heterocysts. *Arch. Mikrobiol.*, **90**, 323–32.

Winter, H. C. & Ober, J. A. (1973). Isolation of particulate nitrogenase from *Chromatium* strain D. *Pl. Cell Physiol.*, **14**, 769–73.

Wyatt, J. T. & Silvey, J. K. G. (1969). Nitrogen fixation by *Gloeocapsa*. *Science*, **165**, 908–9.

11. Nitrogen fixation by blue-green algae in polar and subpolar regions

VERA ALEXANDER

This paper will deal with nitrogen fixation by blue-green algae in cold-dominated regions. Nitrogen fixation by blue-green algae is a particularly interesting phenomenon because these are the only algae in which nitrogen fixation occurs, and because their efficiency is such that where they occur, their contribution of nitrogen to the system is often spectacular. This is especially true for warmer environments, such as hot springs, tropical and subtropical regions and temperate lakes following stratification in spring. However, I suggest that even in very cold environments these organisms contribute a highly significant portion of the nitrogen budget, in terms of income.

Until recently, polar and subpolar regions have been among the biologically least understood areas of the world. However, interest in the nutrient regimes of cold-dominated environments has been increasing, partly as a result of human utilization of these areas. In the case of nitrogen fixation, much of the recently available information for the Arctic and Antarctic has resulted from the International Biological Programme. The rationale for including nitrogen fixation work is clear. Nitrogen supply is known to limit biological productivity in some systems, and in this respect nitrogen input through nitrogen fixation becomes highly significant. In cold environments, where decomposition rates are low, such *new* nitrogen input is perhaps more important than in many other regions. Besides nitrogen fixation, rain represents the only major nitrogen input into many of these areas.

The subject of nitrogen fixation by blue-green algae has been reviewed in detail by Fogg & Stewart (1965) and by Stewart (1970; 1973). It will suffice here to point out that as a result of a current surge of interest in the ecological role of nitrogen-fixing, blue-green algae, these organisms are turning out to be far more significant on a global scale than had previously been acknowledged. Early northern studies were largely restricted to the aquatic environment. For example, low rates of nitrogen fixation were reported in lakes of Afognak Island, Alaska (Dugdale,

Dugdale, Neess & Goering, 1959; Dugdale & Guillard, 1966), although this represents a relatively warm, wet, maritime climate rather than a continental cold region. In interior Alaska, Billaud (1968) found very high rates of nitrogen fixation in a subarctic lake near Fairbanks during an *Anabaena* bloom. The earliest positive result from true arctic tundra in Alaska was the detection of nitrogen fixation in a tundra pond (Paul's Pond) near Barrow by Dugdale & Toetz (1961). Here, *Nostoc* sp. on the pond bottom appeared to be responsible. In each of the studies mentioned here, blue-green algae were the nitrogen-fixing agents. In the case of southern polar regions, Holm-Hansen (1963; 1964) found large populations of blue-green algae in Antarctica, and showed that *Nostoc commune* was an abundant form which fixed nitrogen in impure culture. These observations, then, provide the background against which the wealth of recently acquired information can be examined. Although clearly blue-green algae reach their maximum development in warmer regions, they have also been shown of major importance to soil development in antarctic soils (Boyd, Staley & Boyd, 1966). Similarly, Novichkova-Ivanova (1972) states that algae of the family Nostocales are typical of the tundra climax stage (moss–lichen–polar desert) and as a rule are absent during the primary stages in tundra vegetation. Crusts of *Nostoc commune* are apparently widespread in the spotted tundra of Taimyr (Tichomirov, 1957), and fourteen species of nitrogen-fixing, blue-green algae are listed for the tundra and arctic desert in the Soviet Union. The indications are that blue-green algae are extremely widely distributed in the far northern and southern regions, and that they are tolerant of extreme conditions of temperature.

Recent studies of nitrogen fixation in polar and subpolar regions

An interesting generalization has emerged from the nitrogen fixation work carried out during the past few years under the auspices of the International Biological Programme, in that blue-green algae, either free-living or as symbionts in lichens, have been attributed a major role as nitrogen fixers in the majority of tundra sites. This is true for Barrow (Alexander & Schell, 1973*a*, *b*; Schell & Alexander, 1973) and for Stordalen (Granhall & Selander, 1973); at these sites blue-green algae were important nitrogen-fixing agents. At Kevo, lichens with blue-green algae as symbionts are the most important nitrogen-fixing organisms (Kallio, Suhonen & Kallio, 1972; Kallio, 1973; Kallio & Kallio, 1974), and the same holds for Hardangervidda (Torsvik, 1972).

Lichens are also very important at Barrow and at the Alaskan alpine tundra sites. At only one circumpolar site was major nitrogen-fixing activity by bacteria apparent – Devon Island, Canada (Stutz, 1973). Here, there is a strange exception to the usual situation, since *Nostoc commune* was very abundant, but was considered to be not very active in fixing nitrogen.

Initial work on Signy Island, South Orkney Islands, was carried out by Fogg & Stewart (1968). *Nostoc commune*, *Collema pulposum* and *Stereocaulon* sp. were found sufficiently active as nitrogen fixers to have ecological significance. Horne (1972) estimated biomass and activities for these organisms, and found that *Nostoc* ranged from 28 to 4.5 mg/m^2, whereas *Collema* varied between 13 and 2.9 mg/m^2. Their estimate for fixation by *Nostoc* and *Collema* together in favorable sites was a maximum of 2.1 mg N/m^2/yr. Freshwater phytoplankton and benthos were found to be insignificant compared with algae in wet flushes and soils. This is very similar to Barrow, Alaska, where high rates of nitrogen fixation are attributable to blue-green algae and lichens, which occur primarily in wet, marshy areas which are terrestrial rather than aquatic. Although nitrogen-fixing, blue-green algae do at times occur in permanent ponds, either as mats or balls of *Nostoc* sp., these are relatively uncommon, whereas *Nostoc* sp., either free-living, or living in, or on, moss, is characteristic of the terrestrial environment around Barrow. The nitrogen fixation input is considerably higher than found in Signy Island, varying between 23 and 180 mg N/m^2/yr, depending on the terrain. In contrast, in spite of a *Nostoc commune* biomass of 39 mg/m^2 in favorable habitats, with 17 mg/m^2 in mesic meadows and 9 mg/m^2 in wet meadows, annual nitrogen fixation input by these algae was estimated at a mere 2 μg N/m^2 for Devon Island. *Peltigera aphthosa* was also present and actively fixing nitrogen, but was considered unimportant (Stutz, 1973).

Three principal sites were used for the United States Tundra Biome work on nitrogen fixation. The major area was at Barrow, and additional sites in northern alpine tundra were located at Eagle Summit and in the Alaska Range (Fig. 11.1). Initial observations at Barrow were carried out on a relatively dry meadow, and here acetylene-reduction rates were appreciable but not spectacular. However, during the first year's work at the alpine tundra site in the Alaska range, we observed that nitrogen fixation occurred in wet moss, and upon checking this at Barrow, found that here, too, wet moss almost always has high acetylene-reduction activity. At the Alaska Range alpine tundra site,

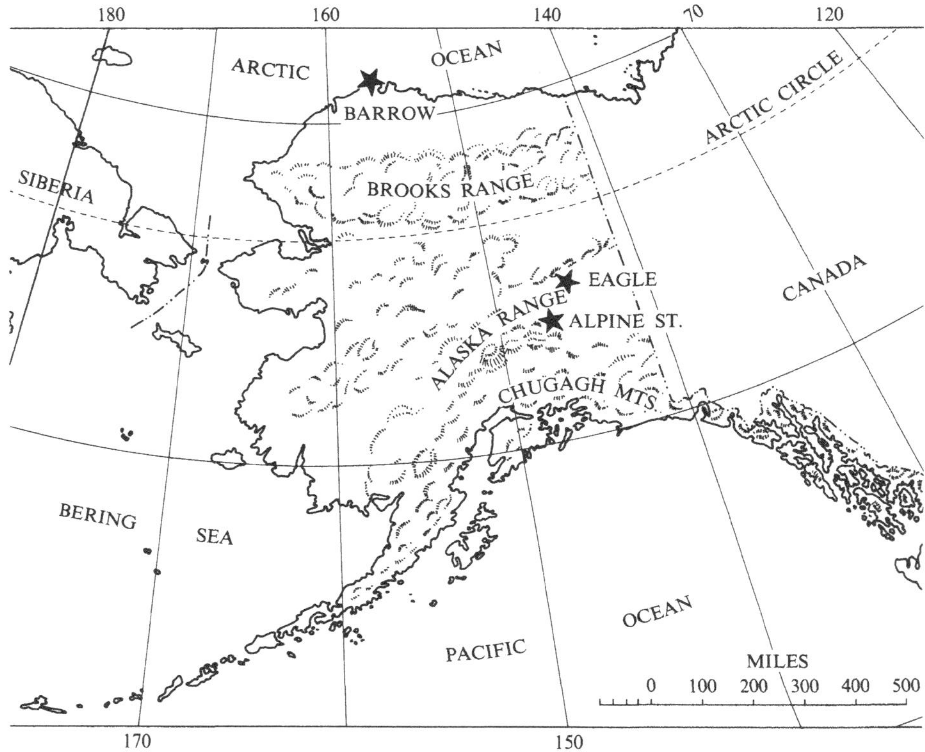

Fig. 11.1. Map showing the Alaskan nitrogen fixation sites.

small ponds also had sufficiently high blue-green algae populations associated with all plants so that high acetylene-reduction rates were obtained with all samples taken. A lichen acetylene-reduction survey was carried out at the Alaska Range site, and several active forms were found, in particular *Peltigera aphthosa*, *Nephroma arcticum* and *Stereocaulon* sp. Although a legume (*Oxytropis nigrescens*) and a *Dryas* sp. (*Dryas octopetala*) showed some activity, the nitrogen-fixing, free-living and symbiotic blue-green algae were by far the most obvious and active nitrogen fixers here. The same turned out to be true for Eagle Summit in 1971. Here, *Oxytropis nigrescens* was the only non-blue-green algal nitrogen fixer of significance, but its distribution was not very widespread compared with damp mossy areas, when fixation rates of 0.6 μg N/m^2/h were measured.

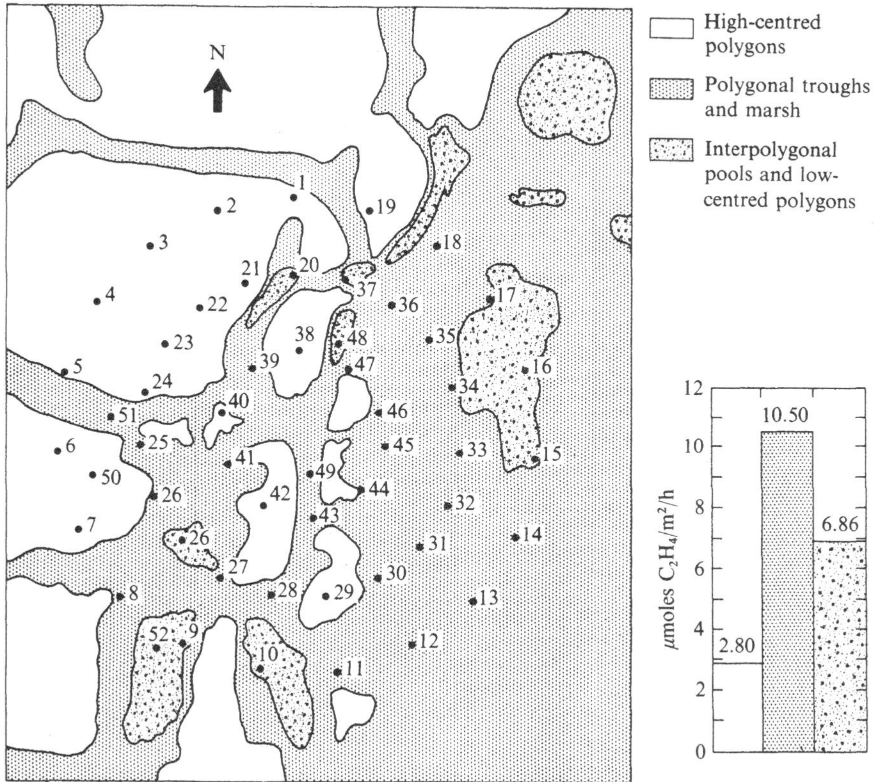

Fig. 11.2. Map of Site 12 stations, Barrow, with mean 1971 acetylene-reduction experimental results (from Schell & Alexander, 1973).

It is difficult to make quantitative estimates of algae associated with moss, and we feel that our biomass estimates from Barrow probably do not include most of the *Nostoc* present and therefore prefer not to present this information. Examination with epifluorescent microscopy has proved the best way to observe intercellular and epiphytic algae on moss. However, due to these difficulties we found it impossible to estimate fixation based on biomass and efficiency information. Our approach was more direct; in order to quantify the nitrogen input into tundra as well as to assess the contribution of the moss community, we established a nitrogen fixation intensive site at Barrow (site 12). Here, a concentric series of 53 stations was laid out and each station has been sampled for nitrogenase activity at regular intervals throughout three summer seasons (Fig. 11.2). The acetylene-reduction

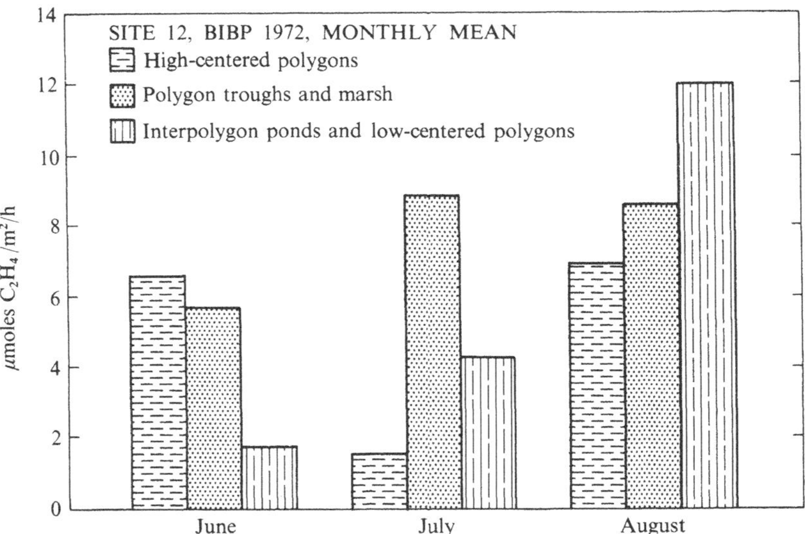

Fig. 11.3. 1972 acetylene-reduction results, Site 12.

values shown here represent the results of the first summer's work. The second year's data are shown in Fig. 11.3, a somewhat different representation. Here, seasonal trends in the various microenvironments are shown. The seasonal pattern appears to depend on the moisture conditions at the particular site. For a dry station, rates decline with the season as further drying progresses. However, in a wet area rates tend to increase with season as temperatures increase and some drying occurs. Similar results were obtained on a second site at Barrow (Fig. 11.4). Site 4 was selected by the microbiology group as an intensive site to study moisture gradients effects. Close examination of the wet marshy areas shows *Nostoc* sp. to be very abundant in, around, and on the moss. In some cases the algae float freely in shallow pools overlying the moss. Other blue-green algae are present in these communities, but are generally much less abundant than the dominant *Nostoc*.

The nitrogen fixation pattern described here is very similar to that found by Granhall & Basilier (1973). These authors found large numbers of blue-green algae associated with moss either as epiphytes or as intercellular organisms, and these communities were associated with high acetylene-reduction rates. In common with Barrow, *Drepanocladus* and *Sphagnum* mosses were the most important in such associations, although in Barrow we found acetylene-reduction activity in connection with other genera also, specifically *Campylium stellatum* and

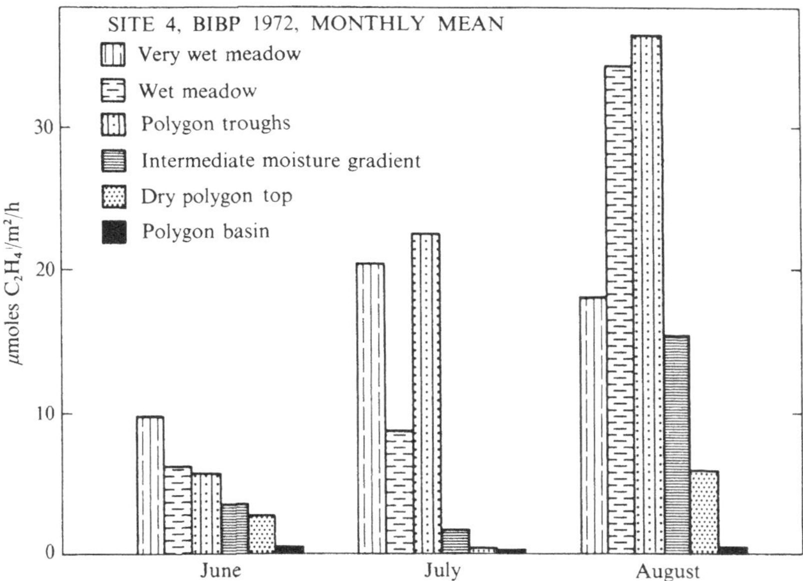

Fig. 11.4. Site 4 acetylene-reduction results, Barrow, 1972.

Calliergon giganteum. Rates of acetylene reduction measured in connection with such communities are included in Table 11.1. There appears little question that in regions where such damp, mossy environments occur, nitrogen fixation by blue-green algae is highly significant. At Kevo, Finland, very little nitrogen fixation by free-living, blue-green algae occurs, because the soil pH is too low. However, even here blue-green algae occur in *Sphagnum* and could be important on a very local basis (S. Kallio, personal communication).

Nitrogen fixation by blue-green algae in lichen symbiosis is a major input at several of the northern tundra biome sites. This is certainly true for Kevo, where several genera are important, but in particular *Stereocaulon paschale* and *Nephroma arcticum* are important, with *Stereocaulon alpinum*, *Peltigera* sp. and *Solorina crocea* also contributing. At Barrow, the principal nitrogen-fixing lichens are *Peltigera aphthosa*, *Peltigera scabrosa*, *Nephroma arcticum* and a *Lobaria* sp.; *Stereocaulon tomentosum* also occurs. Here lichens are less abundant than free-living, blue-green algae, but are very active in acetylene reduction, and are probably extremely important on a local basis.

At Stordalen, Norway, *Solorina crocea*, *Stereocaulon paschale* and a third unidentified lichen (possibly *Nephroma*) were important.

Table 11.1. *Nitrogen fixation by blue-green algae in cold-dominated systems*

Site	(nmoles C_2H_4/g dw/h)	(mg N/m^2/yr)	Reference
Hardangervidda			
1. Stigstuv			
(*a*) lichen heath	—	100	Torsvik (1971)
(*b*) wet meadow	—	250	
(*c*) dry meadow	—	100	
2. Birchwood (Maurset)	—	100	
3. Snow bed (Finse)	—	200	
Devon Island			
Nostoc commune	6200.0–53 000	0.002	
Soil-algae	—	0.014	Stutz (1973)
Abisko			
1. Stordalen mire			
Mosses with free-living algae	11–42	160	Granhall & Selander (1973)
Mosses with intracellular algae	3.9–89	9 400	
Mosses with epiphytic algae	170–2700	—	
Mosses without algae	1.3–3.3	—	
Free-living, blue-green algae	—	11 500	
Barrow			
Nostoc commune	394–452 (max. rates)	—	
Moss–algae (*Campylium stellatum*,	—	115	Alexander & Schell (1973*a*)
Calliergon giganteum)	11 (5 °C) – 61 (20 °C)	—	
Drepanocladus revolvens	4.7 (6 °C) – 21.5 (20 °C)	—	
Eagle Summit			
Nostoc sp.	206	—	
Sphagnum–algae	60 nmoles/m^2/h	—	
Alaska Range			
Anabaena sp.	178	—	
Nostoc sp. mat	374	—	
Nostoc sp.	57	—	
Signy Island			
Nostoc commune (4 sites)	—	1.8, 0.3, 0, 0.08	Horne (1972)
Nostoc commune	2.36 μg N/mg/day	—	Fogg & Stewart (1968)

Acetylene-reduction data for lichens from all these sites are synthesized in Table 11.2.

Factors determining the occurrence of nitrogen-fixing blue-green algae in polar and subpolar regions

Possibly the most important factor affecting the distribution of nitrogen-fixing, blue-green algae is the soil pH. Several workers have reported

Table 11.2. *Symbiotic nitrogen fixation by blue-green algae in lichens*

Site	(nmoles C_2H_4/g dw/h)	(mg N/m^2/yr)	Reference
Hardangervidda			
1. Stodalen mire			
(*a*) Finse Snow Bed	—	—	Torsvik (1971)
Stereocaluon paschale	—	100	
Solorina crocea	—	1100	
Unidentified lichen	—	200	
(*b*) Lichen heath (Stigstuv)			
Stereocaulon paschale	—	100	
Solorina crocea	—	500	
Unidentified lichen	—	200	
2. 1972 data	—	—	Torsvik (1972)
Solorina crocea	460	8400	
Stereocaulon paschale	—	500	
Unknown lichen (*Nephroma*?)	85	1100	
Devon Island			
Peltigera aphthosa	5100	—	Stutz (1973)
Abisko			
(*a*) Stordalen mire	—	—	Granhall *et al.* (1972)
Peltigera malacea	490, 140	—	
Lobaria linita	60	—	
Peltigera polydactyla	1	—	
Peltigera scabrosa	1400	—	
Nephroma parile	+	—	
Stereocaulon tomentosum	+	—	
Barrow			
Peltigera aphthosa	96.6–24480	—	Alexander & Schell (1973*a*)
Peltigera scabrosa	706	—	
Alpine tundra (Alaska)			
Peltigera aphthosa	42–125	—	Alexander & Schell (1973*a*)
Nephroma arcticum	22–183	—	
Kevo			
1. Birch forest	(mg N/g dw/yr)		
Nephroma arcticum	16.9	93.5	Kallio & Kallio (1974)
Stereocaulon paschale	15.6	42.1	
2. Pine forest			
Nephroma arcticum	12.1	—	
Stereocaulon paschale	9.0	349.2	
3. Low alpine heath			
Nephroma arcticum	6.2	3.1	
Stereocaulon paschale	4.2	135.7	
Signy Island			
Stereocaulon sp. cephalodia	0.2 μg N/mg N/day	—	Fogg & Stewart (1968)
Collema pulposum	1.02–0.05	—	

that blue-green algae do not occur in soils below pH 5. This is the reason given for the absence of blue-green algae at Kevo. Granhall & Henriksson (1969) have studied the distribution of nitrogen-fixing, blue-green algae in Swedish soils, and found that between pH 3 and 5, no blue-green algae were observed, whereas in alkaline soils blue-green algae were abundant, with nitrogen-fixing species present in over 75 % of investigated soils. Similarly, Dooley & Houghton (1973) have observed in Glenamoy, Ireland, that pH and soil moisture appear to operate together as the critical factors determining the distribution of blue-green algae. No blue-green algae were found at a pH lower than 5.4. Granhall & Basilier (1973) found that blue-green algal nitrogen fixation showed a strong maximum at pH 6.8 in laboratory experiments, although in field work little correlation of rates with local pH was found. Moisture appears to be the major factor at Signy Island. In Barrow, the distribution of blue-green algae appears to be related to that of moisture, and, although large mats of the algae are found in very dried out condition, these have resulted from prior inundation, and are dormant while in the dry state. Similarly, at Signy Island, blue-green algae are present primarily in wet depressions. In the case of lichens, distribution is less determined by moisture, and lichens are apparently extremely resistant to desiccation, and may recover completely following extensive dry periods. This particular ability, namely recovery of nitrogen-fixing capacity following desiccation, was studied by Henriksson (1971). Lichen nitrogen fixation is appreciable at Barrow sites even where algal fixation is precluded by the low moisture.

Factors controlling rates of nitrogen fixation by blue-green algae in polar and subpolar regions

Ecological studies on factors influencing rates of nitrogen fixation have been carried out in much of the arctic and antarctic work discussed here. Moisture and temperature appear to be the major factors which affect nitrogen fixation rates. Temperature influences rates radically on a year-round basis, since very few blue-green algae, either free-living or symbiotic, are able to fix nitrogen effectively below 0 °C. However, Kallio *et al.* (1972) found that nitrogen fixation was detectable at −5 °C in *Solorina crocea.* This is the lowest temperature recorded for nitrogen fixation, and this *Solorina* species would appear to be the best-adapted lichen to arctic conditions. Granhall & Henriksson (1969) have reported that appreciable nitrogen fixation can occur in blue-green algae at temperatures as low as 0 °C. In our work at Barrow, we have found only very low rates of fixation at 0 °C in *Nostoc* (Alexander & Schell, 1973*b*), and optimum temperatures for nitrogen fixation for *Nostoc*, *Nostoc*-moss association and *Peltigera aphthosa* were 18 °C to 20 °C. Kallio *et al.* (1972) have found that the temperature optimum for *Nephroma arcticum* and *Solorina crocea* was 15 °C, although the exact pattern of response also depended on light intensity. For *Stereocaulon paschale*, the optimum temperature for nitrogen fixation was near 20 °C, although maximum photosynthesis takes place around 0 °C. Moisture also enters into this relationship, and maximum nitrogen fixation at 20 °C is attained when the moisture content is 500 % of the dry weight of the thallus (Kallio, 1973). Preliminary results from Barrow suggest that saturation of the nitrogen-fixing process in *Peltigera aphthosa* takes place at somewhat lower water content (250 %). Granhall & Selander (1972) found that light was the major factor limiting nitrogen fixation in moss–algae communities at Stordalen mire. For lichens, Kallio *et al.* (1972) and Kallio (1973) have found a rather complex relationship of nitrogen fixation with light which involves oxygen tension also. Under natural conditions where oxygen levels are about 20 %, light is believed to be relatively unimportant as a controlling factor.

The final important factor to be considered is oxygen. Alexander & Schell (1973*a*) found that both *Nostoc* and *Peltigera aphthosa* responded to reduction in oxygen tensions with an increase in nitrogenase activity, with a maximum under anaerobic conditions. These experiments were carried out in the light. The response curves were such that at normal aerobic oxygen tensions, fixation by *Nostoc* is 50 % of its maximum and that of *Peltigera* is considerably reduced. The possibility exists that low oxygen tensions in the moss layer may be of considerable importance in allowing high rates of nitrogen fixation.

Diurnal studies have supported the conclusions on major factors influencing nitrogen fixation rates. Granhall & Basilier (1973) found that their rates on a 24 hour basis followed the light curve closely. At Barrow, a diurnal study of nitrogen fixation by *Peltigera aphthosa* showed a marked cycle which followed fluctuations in light and temperature, but again appeared more closely correlated with light. Kallio & Kallio (1974) have found that diurnal cycles in the Kevo sites varied depending on the specific environment. In open lichen heath, the diurnal cycle followed moisture, with a decline at mid-day when conditions became driest. However, in the birch and pine forest sites, nitrogenase activity increased during the warmest and lightest part of the day.

Discussion

Overall, it appears from the work described here that nitrogen fixation by free-living or symbiotic blue-green algae is extremely important in terms of the total nitrogen fixation in polar and subpolar regions. The major factors influencing rates appear to be pH (which largely determines the occurrence of free-living, blue-green algae), moisture, light and temperature, with oxygen also playing a role. The rates attributable to these organisms are somewhat lower in polar regions than in more temperate areas but rates in subarctic sites compare favorably. As an example of temperate rates, Henriksson (1971) found fixation to be 0.4 to 4.4 g $N/m^2/yr$ at Lakeside meadow (Sunnersta), attributable to *Nostoc*, *Anabaena*, *Cylindrospermum* and *Calothrix*, and even higher at field sites with *Nostoc* bulbs present (Bergsbrunna) (1.5–5.1 g $N/m^2/yr$). The Stordalen values exceed these considerably. In some environments, nitrogen fixation by blue-green algae is the major nitrogen input. This appears to be true for Barrow, Alaska, and probably also for the mossy regions of Stordalen site.

In polar regions, it is logical that nitrogenase activity, which is highly temperature sensitive, should occur at maximum rates in the zone of greatest impact of solar heating, at or slightly above the ground surface. Dark-colored mosses absorb heat, and provide a particularly warm environment. This, coupled with a relatively abundant energy source in the form of light, enables such algal fixation to proceed at rates sufficiently high to be of ecological importance. The buffering capacity associated with mosses may also be a significant factor. In addition, moisture conditions are often greatly improved in moss tufts compared with adjacent areas. A rough calculation suggests that this nitrogen fixation associated with moss may be adjusted to supply all the nitrogen required by the moss. Assuming a daily growth rate of 50 mg/m^2 for moss, and a nitrogen content of 1.5 % for moss tissue, then the nitrogen requirement would be 0.75 mg $N/m^2/day$. Nitrogen fixation rates in mossy areas are as high as twice this value (Alexander & Schell, 1973*a*).

Algal nitrogen fixation may be of special significance because the products become available to the biological system quite rapidly, whereas with plants like *Dryas* sp. the newly-fixed nitrogen is primarily retained by the plant for its own use. In lichens, the fungal partner benefits, and in the case of the moss–algae system the moss possibly obtains much of the nitrogen. We do not, at present, know how long the nitrogen is retained by these plants. In the case of free-living algae, nitrogen is undoubtedly released as has been shown in other systems. Preliminary experiments using $^{15}N_2$ in the presence of *Nostoc* and *Eriophorum* suggest that the *Eriophorum* obtains newly-fixed nitrogen rather rapidly, as Stewart (1967) noted elsewhere.

Much of the US IBP Tundra Biome work discussed in this paper was carried out as a joint project with Dr D. M. Schell. We are especially appreciative of the field and laboratory assistance of Miss Margaret Billington. The author expresses sincere thanks to the many circumpolar tundra workers who so freely provided their data for the purpose of synthesis.

References

Alexander, V. & Schell, D. M. (1973*a*). Nitrogen fixation in arctic and alpine tundra. *US Tundra Biome Data Report* 73–10, 54 pp.

Alexander, V. & Schell, D. M. (1973*b*). Seasonal and spatial variation of nitrogen fixation in the Barrow, Alaska, Tundra. *Arctic and Alpine Res.*, **5**, 77–88.

Billaud, V. A. (1968). Nitrogen fixation and the utilization of other nitrogen sources in a subarctic lake. *J. Fish. Res. Bd Canada*, **25**, 2101–10.

Boyd, W. L., Staley, J. T. & Boyd, J. W. (1966). Ecology of soil microorganisms of Antarctica. In *Antarctic soils and soil forming processes* (ed. Tedrow, J. C. F.), *Antarctic Res. Ser.* **8**, 125–59.

Dooley, F. & Houghton, J. A. (1973). The nitrogen-fixing capacities and the occurrence of blue-green algae in peat soils. *Br. phycol. J.*, **8**, 289–93.

Dugdale, R. C. & Guillard, R. R. L. (1966). Nitrogen fixation in lakes on Afognak Island. In *Final Report to the Arctic Institute of North America. Nutrition of algae in subarctic lakes* (ed. Dugdale, R. C.). Arctic Institute of North America subcontract ONR-276, 25 pp.

Dugdale, R. C., Dugdale, V. A., Neess, J. & Goering, J. (1959). Nitrogen fixation in lakes. *Science*, **130**, 859–60.

Dugdale, R. C. & Toetz, D. (1961). Sources of nitrogen for arctic Alaska lakes. In *Final report of investigations of the nitrogen cycle in Alaska lakes* (ed. Dugdale, R. C.). Arctic Institute of North America, subcontract ONR-253, 21 pp.

Fogg, G. E. & Stewart, W. D. P. (1965). Nitrogen fixation in blue-green algae. *Sci. Progr.*, **53**, 191–201.

Fogg, G. E. & Stewart, W. D. P. (1968). *In situ* determinations of biological nitrogen fixation in Antarctica. *Br. Antarct. Surv. Bull.*, **15**, 39–46.

Granhall, U. & Basilier, K. (1973). Nitrogen fixation in tundra moss communities. In *IBP Swedish Tundra Biome Project Technical Report*, no. **14**, 174–90.

Granhall, U. & Henriksson, E. (1969). Nitrogen-fixing blue-green algae in Swedish soils. *Oikos*, **20**, 175–8.

Granhall, U. & Selander, H. (1973). Nitrogen fixation in a subarctic mire. *Oikos*, **24**, 8–15.

Granhall, U., Selander, H. & Forss, M. (1972). Nitrogen fixation. In *IBP Swedish Tundra Biome Project Technical Report*, no. **9**. *Progress Report, 1971* (ed. Sonneson, M.), pp. 32–3.

Henriksson, E. (1971). Algal fixation in temperate regions. *Plant and Soil, Special Volume*, pp. 415–19.

Holm-Hansen, O. (1963). Algae: nitrogen fixation by antarctic species. *Science*, **139**, 1059–60.

Holm-Hansen, O. (1964). Isolation and culture of terrestrial and fresh-water algae of Antarctica. *Phycologia*, **4**, 43–51.

Horne, A. J. (1972). The ecology of nitrogen fixation on Signy Island, South Orkney Islands. *Br. Antarct. Surv. Bull.*, **27**, 1–18.

Kallio, S. (1973). The ecology of nitrogen fixation in *Stereocaulon paschale. Rep. Kevo. Subarct. Sta.*, **10**, 34–42.

Kallio, P. & Kallio, S. (1974). Nitrogen fixation in subarctic ecosystems in *Nephroma arcticum* and *Stereocaulon paschale. Rev. Kevo. Subarct. Res. Stat.* (in press).

Kallio, P., Suhonen, S. & Kallio, H. (1972). The ecology of nitrogen fixation in *Nephroma arcticum* and *Solorina crocea. Rep. Kevo. Subarctic Res. Stat.*, **9**, 7–14.

Schell, D. M. & Alexander, V. (1973). Nitrogen fixation in Arctic coastal tundras in relation to vegetation and moss relief. *Arctic*, **26**, 120–7.

Stewart, W. D. P. (1967). Nitrogen turnover in marine and brackish habitats. II: Use of ^{15}N in measuring nitrogen fixation in the field. *Ann. Bot., N.S.*, **31**, 385–407.

(1970). Algal fixation of atmospheric nitrogen. *Pl. Soil*, **32**, 555–88.

(1973). Nitrogen fixation by photosynthetic micro-organisms. *Ann. Rev. Microbiol.*, **27**, 283–316.

Stutz, R. C. (1973). *Nitrogen fixation in a high arctic ecosystem.* Ph.D. dissertation. University of Alberta, 62 pp.

Novichokova-Ivanova, L. N. (1972). Soil and aerial algae of polar deserts and arctic tundra. In *International Biological Programme Tundra Biome proceedings of international meeting on the biological productivity of tundra* (ed. Wielgolaski, F. E. and Rosswall, T.), pp. 261–5.

Tichomirov, B. A. (1957). Dynamics of vegetation on arctic sport tundra. *Bot. Zhur.*, **42**, II.

Torsvik, V. L. (1970, 1971 and 1972). *Norwegian IBP Tundra Biome Annual Reports.*

12. Nitrogen fixation by blue-green algae in temperate soils

ULF GRANHALL

Nitrogen-fixing, blue-green algae occur frequently in temperate soils either as free-living soil algae, algal crusts, epiphytes or as symbionts (mainly in lichens). The distribution and role of nitrogen-fixing soil algae in temperate regions have been investigated recently in Russia (Gollerbakh & Shtina, 1969; Shtina, 1969; Shtina & Roizin, 1966), northern and western Europe (Granhall & Henriksson, 1969; Henriksson, 1971; Henriksson, Englund, Héden & Was, 1972; Stewart, 1967*a*, *b*) and North America (Jurgensen & Davey, 1968; Mayland, McIntosh & Fuller, 1966; Shields & Durrell, 1964). Soil moisture, pH, mineral nutrients and the level of combined nitrogen are probably the major factors which determine the distribution of algae in the different soil types and Shtina (1969) has used the species of blue-green algae which occur in particular soils as indicators of soil type.

Results and discussion

The relationship between soil type and the occurrence of blue-green algae in Swedish soils is shown in Table 12.1. It is seen that 47 % of the soils contained potential nitrogen-fixing algae. A rather similar situation has been found in Scottish soils (Stewart, personal communication). The sandy soils sampled in this study were located near lakes or sea shores and were moist, low in organic matter, and often acid. They contrast with the sand-dune slacks investigated by Stewart (1965; 1967) on the east coast of England which have a higher pH and are rich in *Nostoc* (Stewart, 1967*b*). The moraine soils were poor in humus, were acid or neutral, and were temporarily quite dry. The absence of nitrogen-fixing species here is not readily explained, especially as other blue-green algae occurred. The clay soils generally showed a high humidity and most were neutral or alkaline. These soils develop abundant growths of nitrogen-fixing algae. The calcareous soils were alkaline, rich in mineral nutrients including phosphates, and nitrogen-fixing algae occurred in over 90 % of samples investigated.

The acidity of forest soils probably explains the low numbers of blue-

Table 12.1. *Relationship between soil type and occurrence of algae*

Soil characteristics	No. of soils	Soils with blue-green algae (%)	Soils with N_2-fixing algae (%)	Occurrence of heterocystous filaments (%)	Total no. of genera	No. of N_2-fixing genera
Sand	6	50	17	0	4	1
Moraine	7	57	0	0	4	0
Clay	16	100	81	31	13	7
Calcareous	15	100	93	40	10	4
Forest soils (humus)	20	25	10	0	3	1
Total	64	66	47	17	17	7

green algae found there. Jurgensen & Davey (1968) and Kallio, Suhonen & Kallio (1972) also found a general lack of nitrogen-fixing algae in acid forest soils in the United States and in tundra soils respectively. However, Jurgensen & Davey (1968) did find algae abundant (up to 100000 algae/g of surface crusts) in certain forest-tree nurseries. These populations decreased rapidly with depth. Nitrogen-fixing *Calothrix braunii* occurs as crusts in 15-year-old stands of Scots pine at Jädraås, 200 km north of Stockholm (Granhall, unpublished) and these crusts fix 20–70 μg N/m^2 crust surface/h at 6.5 °C and a soil pH of 4.4. This suggests that when forest stands are opened through logging operations, or fire, they may become favourable habitats for algal development because of altered soil properties and improved light conditions.

Of the various environmental conditions affecting the development of nitrogen-fixing, blue-green algae in temperate regions, soil humidity is probably the most important. In Sweden nitrogen-fixing algae occur twice as frequently in moist soils as they do in dry soils (Granhall & Henriksson, 1969). Waterlogging has a still greater effect on their proliferation, probably because of lowered oxygen levels (Granhall & Selander, 1973). Blue-green algae occur also in arid areas (Cameron & Blank, 1966; Mayland *et al.*, 1966; Shields & Durrell, 1964; Stewart, 1969; 1970) in the form of algal crusts. The reason for this is probably that they lack competition from other organisms and are relatively well protected against desiccation by aggregating with fungal mycelia and soil particles. Then they become important sources of organic matter and combined nitrogen, and also improve infiltration, reduce erosion and aid seedling establishment (Mayland *et al.*, 1966). The moisture content of these crusts correlates well with their nitrogen fixation rates. Algal crusts from a desert in Arizona (Mayland *et al.*, 1966) fix 10 to 16 mg N/m² crust surface/day (equivalent to 10.9 kg N/ha/yr) under cycling wet–dry and continuous wet conditions respectively. This compares with an in-situ fixation rate by algal crusts on a subarctic *Dryas* heath in northern Sweden (Abisko) of 11 mg N/m² crust surface/day at 12 °C, pH 7.3 and semi-moist conditions (Granhall & Selander, 1973).

Another important factor is soil pH. Blue-green algae do not occur in soils below pH 4.4 but are frequent at higher pH values. For example, nitrogen-fixing species occurred in over 75 % of the investigated alkaline soils in Sweden (Granhall & Henriksson, 1969). *Nostoc* and *Anabaena* were the commonest genera and showed the greatest pH tolerance (Granhall, 1970*a*). Laboratory experiments with an isolated soil alga (*Nostoc punctiforme*) demonstrated that nitrogenase activity occurred in the pH range 5.0–10.5 with an optimum at 7.6.

The light dependence of the nitrogenase activity of two *Nostoc* species was studied by Granhall (1970*a*). Reduction of acetylene occurred at light intensities of 83 lx but not at 18 lx and the light saturation level was 3000–4000 lx. The temperature optimum was 20–22 °C. The general nitrogen-fixing capacity of unialgal cultures isolated from Swedish soils (see Table 12.1) was of the same magnitude as pure cultures (1–4 nmoles C_2H_4/mg protein/min) (see Table 12.2) except for a strain of *A. variabilis* which was infected by cyanophages (Granhall & Hofsten, 1969) and one heterocystless *Pseudanabaena* that did not reduce acetylene. No algal sample kept in the dark for 24 h reduced acetylene.

Data on in-situ rates of fixation of nitrogen(C_2H_2) in Swedish soils (mainly clay soils near Uppsala) are presented in Table 12.3. Nitrogen-(C_2H_2) fixation by blue-green algae (0–2 cm layer) varied between 0.06 and 25.7 mg N/m²/h in the light and the main nitrogen-fixing genera were *Anabaena*, *Cylindrospermum* and *Nostoc* (Granhall & Henriksson, 1969). Clay soils are among the most suitable habitats for blue-green soil algae and the environmental factors, e.g. moisture, temperature, pH and light conditions, were nearly optimal during incubation. No difference was observed between samples incubated under aerobic or ambient gas phases. Heterotrophic fixation by bacteria, as judged from samples kept in the dark for 24 h, was sometimes observed and varied between 0.09 and 1.2 mg N/m²/h. Henriksson *et al.* (1972) have reported values from 0.01 to 4.5 mg N/m²/h for aerobic algal fixation in other clay soils near Uppsala. Stewart (1967*b*) reports values of 2.5 g N/m²/yr

Table 12.2. *Acetylene reduction of isolated unialgal cultures, compared with two pure cultures*

Alga	Acetylene reduction (nmoles C_2H_4/ mg protein/ min)
Anabaena sp.	2.87
A. variabilis strain B	2.09
A. variabilis strain S	0.22
Nodularia spumigena	1.28
Nostoc commune	1.00
N. muscorum	3.36
N. punctiforme	3.87
Nostoc sp.	1.56
Pseudanabaena catenata	0.00
Anabaena inaequalis (pure)	3.65
A. cylindrica (pure)	3.50

All figures represent the mean of triplicate determinations.

for algae in sand-dune slacks. The nitrogen fixed may be assimilated subsequently by associated higher plants (Stewart, 1967).

Results of tests carried out on some other soils under simulated, optimized field conditions are presented in Table 12.4 (Granhall 1970*b*). The soils were homogenized and kept moist with distilled water or Bristol's nitrogen-free culture medium (Henriksson, 1951). The samples were incubated in a light chamber (20.5 °C, 2500 lx, 16 h light: 8 h dark) for six weeks and the upper 0.5 cm tested for acetylene reduction. No significant change in pH was observed during incubation. Initial numbers of soil algae were determined. Nitrogen-fixing species were identified on cover slips placed on the surface and examined under the microscope (Granhall & Henriksson, 1969). In three cases out of four, fixation increased considerably on the addition of the nitrogen-free inorganic salts (Table 12.4).

Finally, it may be noted that blue-green algae occur epiphytically or intracellularly associated with *Sphagnum* and other mosses in temperate and subarctic bog mosses. Such associations were reported from Great Britain (Stewart, 1966), Alaska (Schell & Alexander, 1973) and temperate and subarctic regions of Sweden (Granhall & Selander, 1973; Granhall & Basilier, 1973). Investigations over several years at the Swedish IBP site at Stordalen (Abisko) have shown that mosses within wet minerotrophic depressions associated with only heterotrophic

Table 12.3. *In-situ nitrogenase activity in various Swedish soils*

Site	Soil characteristics	Moisture content (%fw)	pH	Mean incub. temp (°C)	Mean incub. light intensity (lx)	Mean† $N_2(C_2H_2)$ fixation (mg $N/m^2/h$)	Main nitrogen-fixing organisms
Bergsbrunna I	Bare, clay	Waterlogged	7.0	18	> 3.000	25.7	*Nostoc*
Bergsbrunna I	Bare, clay	3.6 %	7.0	22	> 3.000	0.3	*Nostoc*
Bergsbrunna I	Bare, clay	3.6 %	7.0	22	0*	0.0	*Nostoc*
Bergsbrunna II	Bare, clay	Waterlogged	7.1	26	10.000	11.0	*Nost.*, *Cylindrospermum*
Bergsbrunna II	Bare, clay	Waterlogged	7.1	16**	11.800	1.3	*Nost.*, *Cylindrospermum*
Funbo I	Bare, clay	27.6 %	7.9	25	3.000	9.6	*Anabaena*
Funbo I	Bare, clay	27.6 %	7.9	25	0*	0.1	Heterotrophic bacteria
Funbo II	Bare, clay	38.0 %	7.9	25	2.300	4.8	*Anabaena*
Berthåga	Grass, clay	Moist	7.1	22	7.000	1.4	*Nostoc*
Berthåga	Grass, clay	Moist	7.1	22	0*	0.0	*Nostoc*
Lurbo	Bare, clay	Moist	—	14	> 3.000	1.3	Heterocystous blue-green algae
Hågaån	Bare, clay	Moist	6.6	22	0*	1.2	Heterotrophic bacteria
Hågaån	Bare, clay	Moist	6.6	12	4.000	0.4	Heterotrophic bacteria
Rosenlund I	Bare, clay	48.6 %	6.5	15.5	1.700	0.6	Heterotrophic bacteria
Rosenlund II	Mosses, clay	32.8 %	6.5	18	1.000	0.1	
Årsta	Bare, hydromorphous	Waterlogged	7.4	24	3.300	0.2	Heterotrophic bacteria
Årsta	Bare, hydromorphous	Waterlogged	7.4	24	0*	0.1	Heterotrophic bacteria
Skarholmen	Grass, clay	Semi-moist	—	12	> 3.000	0.0	None

* Samples kept in the dark for 24 h.
** After one freeze–thaw cycle.
† Mean of 10 samples.

Table 12.4. *Nitrogenase activity in various Swedish soils under simulated in-situ conditions*

Sampling site	Soil characteristics	Initial no. of soil algae (cells/g dry weight of soil)	pH	Mean $N_2(C_2H_2)$ fixation* (mg N/m^2/h): Distilled water	Mean $N_2(C_2H_2)$ fixation* (mg N/m^2/h): Nitrogen-free medium	Main nitrogen-fixing organisms
Fyris	Hydromorphous	800	6.1	0.91	3.40	*Anabaena*
Hammarby I	Organogenic	300	5.2	0.69	0.09	*Anabaena*
Ultuna	Grassland, clay/loam	9300	6.0	0.36	0.86	*Cylindrospermum, Nostoc*
Hammarby II	Arable land, clay/loam	3100	5.4	0.07	0.59	*Anabaena, Cylindrospermum*
Kungshamn	Peat	400	4.7	0.00	0.00	None
Lunsen	Podsol	9000	4.6	0.00	0.00	None

* The theoretical 3:1 ratio for C_2H_2/N_2 reduction was used. Mean of triplicate determinations.

bacteria fix $N_2(C_2H_2)$ at the rate of 12–22 ng N/g dry weight of moss/h, mosses with mainly intracellular algae fix 36–830 ng N/g dry weight of moss/h, mosses surrounded by mainly free-living algae 102–395 ng/g dry weight of moss/h, and mosses with mainly epiphytic algae 0.6–25 μg N/g dry weight of moss/h (Granhall & Selander, 1973). The mean annual fixation rates were estimated as 0.05, 0.2, 0.2 and 9.4 g N/m^2/yr respectively. Schell & Alexander (1973) have reported values of 62 μg N/m^2/h for wet moss–algal associations in arctic coastal areas. The highly efficient nitrogen fixation by these epiphytic algae is probably responsible for the higher plant production in the minerotrophic parts of the mire. The fixation rates under ambient gas conditions are directly related to global radiation. *Nostoc* spp. are, as in Alaska (Schell & Alexander, 1973), the most common algae. In this ecosystem the pH (H_2O) values were remarkably low (4.5 to 5.6) for algal fixation, but no correlation was found in the field between pH and fixation within this range. Laboratory experiments with epiphytic moss-associated algae also showed little change in nitrogenase activity within the pH range observed in the field but there was a distinct pH-optimum at 6.8 which is the pH value for stored mire water. Light saturation of nitrogenase activity in the laboratory was not achieved even at 14000 lx.

In such ecosystems the more reduced oxygen tensions and increased carbon dioxide content towards the bottom of the wet depressions

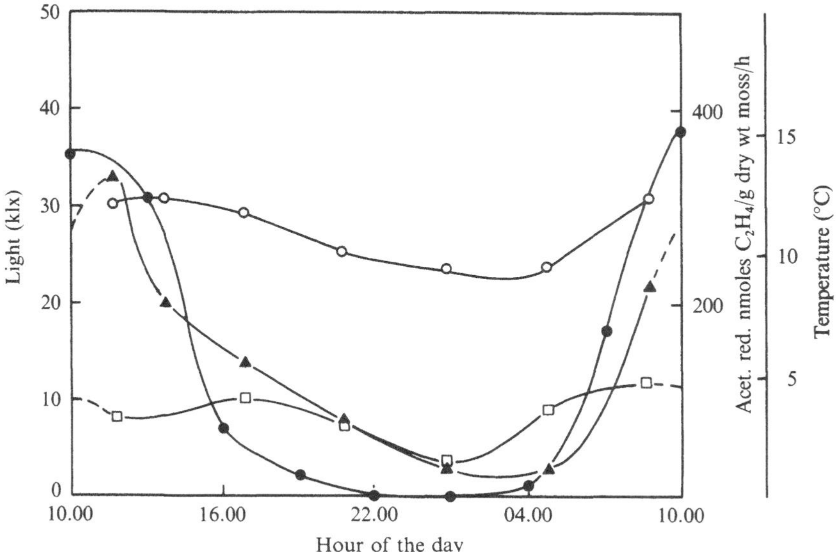

Fig. 12.1. The 24 h cycle of algal nitrogen fixation on floating *Sphagnum riparium* as indicated by the acetylene-reduction technique. The global radiation and temperature variations for the same period are also shown. ○, Temperature; ●, light; ▲, acetylene reduction in light; □, acetylene reduction in dark.

favoured algal fixation wherever light conditions were not significantly altered. Preliminary results indicate that the temperature optimum is 15 °C. This agrees with the findings of adaptation by other subarctic and antarctic algae, including lichens, to low temperatures (Fogg & Stewart, 1968; Kallio *et al.*, 1972). In the Antarctic, Fogg & Stewart (1968) noted a high Q_{10} value (4–6) for nitrogenase activity. A 24 h cycle of nitrogen fixation on floating *Sphagnum riparium* is shown in Fig. 12.1 (Granhall & Basilier, 1973). This shows maximum activity at noon and a close relation between global radiation and nitrogen fixation rates. Microbial methane production during the day was found to be closely related to nitrogen fixation (Svensson, 1973). Seasonal variation studies (unpublished) also show that a decrease in nitrogen fixation rates occurs from July to August and that this is correlated with a decrease in total radiation values.

Symbiotic blue-green algae in lichens are of considerable importance in both temperate and subarctic areas and are considered elsewhere (Granhall, 1973; Granhall & Selander, 1973; Henriksson, 1951; Hitch & Stewart, 1973; Kallio *et al.*, 1972).

In summary therefore nitrogen-fixing, blue-green algae are often found in many temperate regions and acetylene-reduction studies show that far from being unimportant they play a significant role in the input of new nitrogen into many such ecosystems.

I wish to thank Dr Elisabeth Henriksson and Mr Björn Nilsson for help with soil analyses. This work was made possible through grants from the Swedish Natural Research Council.

References

Cameron, R. E. & Blank, G. B. (1966). Desert algae: soil crusts and diaphanous substrate as algal habitats. *Jet Propulsion lab., Pasadena, Tech. Rep.* no. 32–971, pp. 1–41. Pasadena, USA.

Fogg, G. E. & Stewart, W. D. P. (1968). *In situ* determinations of biological nitrogen fixation in Antarctica. *Br. Antarct. Surv. Bull.*, **15**, 39–46.

Gollerbakh, M. M. & Shtina, E. A. (1969). *Pochvennye Vodorosli* (Soil Algae). Idz. 'Nauka' Leningrad, 228 pp.

Granhall, U. (1970*a*). Acetylene reduction by blue-green algae isolated from Swedish soils. *Oikos*, **21**, 330–2.

(1970*b*). IBP-recommendations for the use of the acetylene reduction technique in estimation of the nitrogen fixation in the field. Free-living microorganisms. *Norden*, **5**, 17–19.

(1973). Nitrogenase activity in lichens. In *Progress Report 1972. IBP Swedish Tundra Project, Tech. Rep.* (ed. Sonesson, M.), **14**, 191–4.

Granhall, U. & Henriksson, E. (1969). Nitrogen-fixing blue-green algae in Swedish soils. *Oikos*, **20**, 175–8.

Granhall, U. & v. Hofsten, A. (1969). The ultrastructure of a cyanophage attack on *Anabaena variabilis*. *Physiol. Pl.*, **22**, 713–22.

Granhall, U. & Basilier, K. (1973). Nitrogen fixation in tundra moss communities. In *Progress Report 1972, IBP Swedish Tundra Project, Tech. Rep.* (ed. Sonesson, M.), **14**, 174–90.

Granhall, U. & Selander, H. (1973). Nitrogen fixation in a subarctic mire. *Oikos*, **24**, 8–15.

Henriksson, E. (1951). Nitrogen fixation by a bacteria-free symbiotic *Nostoc* strain isolated from *Collema*. *Physiol. Pl.*, **4**, 542–5.

Henriksson, E. (1971). Algal nitrogen fixation in temperate regions. *Plant and Soil, Special Volume*, pp. 415–19.

Henriksson, E., Englund, B., Hedén, M.-B. & Was, I. (1972). Nitrogen fixation in Swedish soils by blue-green algae. In *Taxonomy and Biology of Blue-green Algae, Madras 1970* (ed. Desikachery, T. V.), pp. 265–9.

Hitch, C. J. B. & Stewart, W. D. P. (1973). Nitrogen fixation by lichens in Scotland. *New Phytol.*, **72**, 509–24.

Jurgensen, M. F. & Davey, C. B. (1968). Nitrogen-fixing blue-green algae in acid forest and nursery soils. *Can. J. Microbiol.*, **14**, 1179–83.

Kallio, P., Suhonen, S. & Kallio, H. (1972). The ecology of nitrogen fixation in *Nephroma arcticum* and *Solorina crocea*. *Rep. Kevo Sub. Res. Stat.*, no. **9**, 7–14.

Mayland, H. F., McIntosh, T. H. & Fuller, W. H. (1966). Fixation of isotopic nitrogen on a semi-arid soil by algal crust organisms. *Soil Sci. Soc. Am. Proc.*, **30**, 59–60.

Schell, D. M. & Alexander, V. (1973). Nitrogen fixation in arctic coastal tundra in relation to vegetation and micro-relief. *Arctic*, **26** (2), 130–7.

Shields, L. M. & Durrell, L. W. (1964). Algae in relation to soil fertility. *Bot. Rev.*, **30**, 92–128.

Shtina, E. A. (1969). Some regularities in the distribution of blue-green algae in soils. *Biologija sine-zelenikh vodoroslěj*, **2**, 21–45.

Shtina, E. A. & Roizin, M. B. (1966). Algae in podzolic soils of Chibin region. *Bot. Zh.*, **51**, 509–19.

Stewart, W. D. P. (1965). Nitrogen turnover in marine and brackish habitats. I. Nitrogen fixation. *Ann. Bot.*, **20**, 229–39.

(1966). *Nitrogen Fixation in Plants*. Athlone Press: London, 168 pp.

(1967*a*). Nitrogen turnover in marine and brackish habitats. II. Use of ^{15}N in measuring nitrogen fixation in the field. *Ann. Bot., N.S.*, **31**, 385–407.

(1967*b*). Transfer of biologically fixed nitrogen in a sand dune slack region. *Nature, Lond.*, **214**, 603–4.

(1969). Biological and ecological aspects of nitrogen fixation by free-living micro-organisms. *Proc. R. Soc. Lond. B*, **172**, 367–88.

(1970). Algal fixation of atmospheric nitrogen. *Pl. Soil*, **32**, 555–88.

Svensson, B. (1973). Methane production in a subarctic mire. In *Progress Report 1972, IBP Swedish Tundra Project, Tech. Rep.* (ed. Sonesson, M.), **14**, 159–66.

13. A comparison of nitrogen fixation by algae of temperate and tropical soils

ELISABET HENRIKSSON, L. E. HENRIKSSON
& E. J. DASILVA

Compared with investigations on algal nitrogen fixation in paddy fields and its importance to their nitrogen economy, only a few studies have been carried out to determine the importance of algal nitrogen fixation in other cultivated soils of tropical and temperate regions. The present study was carried out therefore to investigate the comparative nitrogen-fixing capacity under aerobic conditions of arable soils from temperate and tropical regions. For that reason soils were sampled in wheat fields in the surroundings of Uppsala (Sweden), in rice-fields near Bombay (India) and in a wheat field near New Delhi (India).

In addition, attention was paid to some of the factors likely to affect nitrogen fixation rates there. In a particular soil, the amount of nitrogen fixed depends on (*a*) the microbial nitrogen fixing potential of the soil and (*b*) on the environmental conditions. Environmental conditions found in the field usually result in suboptimum rates of nitrogen fixation since, in nature, the conditions of light, temperature and water are optimal only for short periods of time. In cultivated soils the effects of mechanical treatments such as ploughing, harrowing, sowing, and planting, which break up the algal layer in the soil surface may also affect the rates of nitrogen fixation. The importance of these factors is discussed.

Materials and methods

Soils

All samples were collected in November–December 1972. At the sampling sites, the plants of the autumn-sown wheat in the Swedish wheat fields were 5–10 cm high, soil temperature being approximately 0 °C. Table 13.1 presents the chemical data on the soils at the start of the experiments. These analyses were carried out by the National Swedish Laboratory for Agricultural Chemistry, Uppsala. Relevant information

Table 13.1. *Description of the soils studied, including sampling sites and chemical analysis*

Nos. 1–6 Swedish soils, Nos. 7–13 Indian soils. Soil types: Nos. 1, 2, 4–6 and 8–12 loamy. Nos. 3 and 7 sandy.

Number and locality	Cereal sown	pH	% H_2O	Specific soil con-ductivity	mg/100 g air-dried soil						
					P-AL	P-HCl	K-AL	K-HCl	Mg-AL	NH_4-N	NO_3-N
1 Börje	Wheat	6.9	37.9	1.0	9.2	62	19.0	345	39.1	10	2.9
2 Håga	Wheat	6.5	26.4	0.6	2.7	50	20.5	265	26.8	5	1.7
3 Flottsund	Wheat	6.2	12.8	0.5	9.5	44	24.0	55	6.8	7	1.7
4 Lilla Ultuna	Wheat	7.2	28.4	1.8	9.7	78	31.0	480	58.2	6	2.5
5 Linnés Hammarby	Wheat	7.5	21.6	1.1	22.6	100	38.0	535	20.0	5	1.4
6 Bärby	Wheat	7.4	26.7	1.9	14.2	80	33.0	645	26.7	3	5.7
7 Delhi	Wheat	8.8	0.6	1.1	1.5	34	9.0	130	17.3	< 1	0.8
8 Bassein	Rice	5.9	25.1	1.1	0.4	31	21.5	215	94.0	< 1	< 0.5
9 Bassein	Rice	6.2	22.5	0.7	0.2	32	18.5	80	94.0	9	0.6
10 Bassein	Rice	7.3	16.6	5.5	2.0	47	27.0	100	160.0	< 1	2.6
11 Bassein	Rice	7.6	10.1	3.0	42.0	126	34.0	150	153.0	< 1	2.4
12 Bassein	Rice	6.2	7.9	5.8	0.4	62	13.0	65	148.0	< 1	2.2
13 Bassein	Rice	6.3	8.6	82.0	4.0	50	100.0	255	277.0	< 1	17.5

The readily soluble constituents, marked AL in the table, were extracted in standard AL-solution (0.10 M NH_4-lactate, 0.40 M HAc). Values from 2.0 M HCl extractions, marked -HCl in the table, also include bound constituents. The analytical standard methods used are those decreed by Royal Proclamation and adopted by the National Swedish Board of Agriculture (1965).

on sampling sites is also given. The Swedish soils are fertilized but there is no evidence that fertilizer had ever been added to the Indian soils.

Blue-green algae

The heterocystous blue-green algae in the experimental treatments were identified after cultivation in a nitrogen-free medium (Henriksson, Enckell & Henriksson, 1972). The identification of the algal species was carried out according to Geitler (1932) and Desikachary (1959). The algae were as follows: *Anabaena fertilissima* Rao, C.B. (9), *Anabaena variabilis* Kütz. (4, 5, 6, 8, 10, 13), *Cylindrospermum licheniforme* Kütz. (1, 4, 6), *Nostoc commune* Vaucher (7), *Nostoc microscopicum* Carm. (1, 2, 3, 5, 6), *Nostoc muscorum* Ag. (11, 12), *Microchaete tenera* Thuret (8, 9, 11), *Scytonema* sp. (10). The figures in parentheses refer to the sampling locations given in Tables 13.1 and 13.2.

Acetylene-reduction assay

For the determination of the nitrogen-fixing capacity of the soils, the modified acetylene-reduction technique described by Henriksson, Enckell & Henriksson (1972) was used. This method involves measuring the maximal rate of the nitrogen fixation obtained when 1 g of soil is treated with 0.5 ml distilled water. The values obtained are then calculated as μg N fixed/g dry weight of soil. Special consideration was given to the occurrence of ethylene-producing micro-organisms in the soils (DaSilva, Henriksson & Henriksson, 1974). The experiments were carried out over an 18 week period and the measurements of the nitrogen fixation were made about every ten days. In addition, nitrogenase activity was measured in the dark in samples which had been kept in the dark for at least 24 hours prior to assay (Henriksson, Englund, Héden & Was, 1972).

Incubation

Each soil sample was incubated at 20 °C and also at 25 °C. The light was intermittent (16 hours of light and 8 hours of darkness) with the intensity during the light period being 3000 lx (Philips TL/33). Measurements of nitrogen fixation in light were made after a 2 h preincubation period in the light.

Table 13.2. *Capacity of algal nitrogen fixation in soil samples from cultivated fields in Sweden and India*

The values were determined by estimating the maximal nitrogen fixation in 1 g soil/h at 20 °C and 25 °C respectively. The maximum values were recorded 53–56 days after the start.

Swedish soils				Indian soils			
	μg N_2 fixed/g/h		Number of		μg N_2 fixed/g/h		Number of
Number and locality	20 °C	25 °C	N_2-fixing algal spp.	Number and locality	20 °C	25 °C	N_2-fixing algal spp.
1 Börje	3.70	1.23	2	7 Delhi	0.33	0.78	1
2 Håga	0.96	1.28	1	8 Bassein	0.10	1.36	2
3 Flottsund	1.35	0.19	1	9 Bassein	0.03	0.62	2
4 Lilla Ultuna	2.25	1.28	2	10 Bassein	0.02	0.13	2
5 Linnés Hammarby	3.48	1.34	2	11 Bassein	0.04	0.14	2
6 Bärby	0.92	0.03	3	12 Bassein	0.01	0.01	1
				13 Bassein	< 0.01	< 0.01	1
Average of nos. 1–6	2.11 ± 0.51	0.89 ± 0.24		Average of nos. 8–12	0.04 ± 0.00	0.45 ± 0.03	

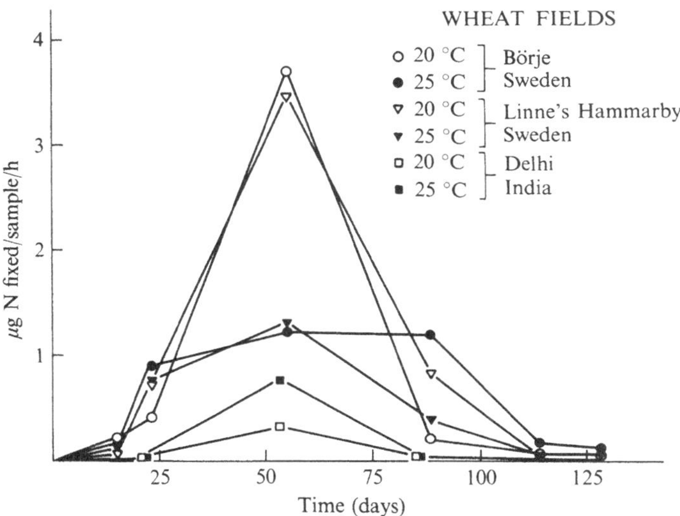

Fig. 13.1. Nitrogen fixation in light under aerobic conditions in soil samples from two Swedish and one Indian wheat field during an incubation time of 18 weeks. Nitrogen fixation was measured at two incubation temperatures, 20 °C and 25 °C respectively.

Results

The maximal rates of nitrogen fixation in the soils studied are presented in Table 13.2. Nitrogen fixation rates by one Indian and two Swedish soil samples from wheat-fields over the experimental period are given in Fig. 13.1. The data suggest that Swedish soils have a much higher nitrogen-fixing capacity than Indian soils. Not unexpectedly maximal nitrogen fixation values occurred at 20 °C for the Swedish soils and at 25 °C for the Indian soils, due presumably to the algae having adapted to the different climatic conditions of Sweden and India.

The time required to reach maximum nitrogen fixation in these cultivated soils is twice as long as for a virgin Swedish soil studied by Henriksson, Enckell & Henriksson (1972). The prolonged lag-phase may be because nitrogen is not fixed in these until the algae have depleted them of combined nitrogen.

The differences in the nitrogen-fixing activities of the soils studied can hardly be explained on the basis of differences in the nutrient content of the soils (see Table 13.1). Furthermore, the different pH values of the soils studied do not seem to be of importance. The saline, soil 13, which has a very high specific soil conductivity and a high level of nitrates, shows very low nitrogen-fixing activity.

Nitrogen fixation rates in the dark were very low (0.01 μg N/g dry soil/h), indicating that heterotrophic bacteria are not of appreciable importance as nitrogen-fixing agents in these soils under aerobic conditions.

Discussion

The nitrogen-fixing capacity of the wheat fields in the surroundings of Uppsala are of the same order as has been measured for a virgin soil at Bergsbrunna near Uppsala (Henriksson, Enckell & Henriksson, 1972). This value is approximately 1.5–5.1 g N/m^2/yr (Henriksson, 1971; Henriksson, Englund, Héden & Was, 1972). If the effect of the mechanical treatment of the soils is ignored, the biological aerobic nitrogen fixation of the wheat fields near Uppsala is about the same order, that is 15–51 kg N/ha/yr. These findings contrast with the conclusions of Jahnke (1967) who studied the occurrence of potential nitrogen-fixing algae in a corn-field in Mecklenburg (GDR) where 30 % of the soil surface could be covered by these algae. She concluded from laboratory experiments and analysis that the algae were of little importance in the nitrogen economy of that soil, because combined nitrogen in the soil inhibited the nitrogen fixation. She estimated that the maximum levels of nitrogen fixation in the corn field were equivalent to 2 kg N/ha/yr and on average were probably much lower (0.5 kg N/ha/yr). Paul, Myers & Rice (1971) and Vlassak, Paul & Harris (1973), likewise studied non-symbiotic nitrogen fixation in some cultivated and virgin soils in Saskatchewan (Canada) and their data agreed with the results of Jahnke in that nitrogen fertilization of the soils inhibited nitrogen fixation. In Australian rice fields, Bunt (1961) showed that potential nitrogen-fixing algae were a minor part of the algal flora, possibly because of the heavy dressings of ammonium sulphate applied to the soils. On the contrary, Bortels (1940) found that nitrogen-fixing algae increased with the fertility of the soil, and Shtina (1969*a*, *b*) studying cultivated soils in the Soviet Union, showed that an increase in soil fertility by addition of nitrogen alone, or in combination with phosphate and potassium, resulted in a development of these algae.

It is possible that on the surface of our soils the high rates result because the available nitrogen is rapidly removed from the soil by rain and assimilation by micro-organisms including blue-green algae. That is, the situation may be similar to that reported by Overrein (1969) who found that 12 weeks after adding a fertilizer treatment of 250 kg N/ha to forest soils 21.5 % and 91.8 % of the nitrogen added in the form of ammonium and nitrate respectively, was leached out of the soil. Irrespective of whether this is the case in our soils or not the differences in soil nutrient contents are obviously not reflected in their nitrogen-fixing capacity, and it is probable that the magnitude of the algal nitrogen fixation varies largely in different soils and in different parts of the world.

The better environmental conditions in the Indian rice fields compensate for the lower nitrogen-fixing capacity of these soils compared to the Swedish ones in this study. Prasad (1949) estimated the algal nitrogen fixation in rice fields in Bihar (India) to be about 14 kg N/ha during the plant-growing season. De & Mandal (1956) studying algal nitrogen fixation in rice fields in West Bengal (Inaia) estimated a fixation rate of 15–49 kg N/ha over a six week period. If the soils were fertilized with phosphorus, however, the amount of nitrogen fixed was calculated as 18–69 kg N/ha during the same period. Watanabe, Nishigaki & Konishi (1951) reported nitrogen fixation by *Tolypothrix tenuis* inoculated in a rice field in Japan to be about 22 kg N/ha. Similarly Singh (1961) estimated the nitrogen fixation by *Cylindrospermum licheniforme* in fields of sugarcanes and maize to be approximately 88 kg N/ha in about 75 days. These contributions are not insignificant and biofertilization with blue-green algae offers promise as a means of increasing rice paddy field yield (Venkataraman, 1972; Chapter 14).

The authors are indebted to Professor Nils Fries for the facilities provided. One of us (E. J. DaSilva) gratefully acknowledges the award of a UNESCO Fellowship.

References

Bortels, H. (1940). Über die Bedeutung des Molybdäns für stickstoffbindendende Nostocaceen. *Arch. Mikrobiol.*, **11**, 155–86.

Bunt, J. S. (1961). Nitrogen-fixing blue-green algae in Australian rice soils. *Nature, Lond.*, **192**, 479–80.

DaSilva, E., Henriksson, E. & Henriksson, L. E. (1974). Ethylene production by fungi. *Pl. Sci. Lett.*, **2**, 63–6.

De, P. K. & Mandal, L. N. (1956). Fixation of nitrogen by algae in rice soils. *Soil Sci.*, **81**, 453–8.

Desikachary, T. V. (1959). *Cyanophyta.* ICAR: New Delhi.

Geitler, L. (1932). Cyanophyceae von Europa unter Berücksichtigung der anderen Kontinente. *Rabenhorsts Kryptogamenflora*, **14**. Leipzig.

Henriksson, E. (1971). Algal nitrogen fixation in temperate regions. *Plant and Soil, Special Volume*, pp. 415–19.

Henriksson, L. E., Enckell, P. H. & Henriksson, E. (1972). Determination of the nitrogen fixing capacity of algae in soil. *Oikos*, **23**, 420–3.

Henriksson, E., Englund, B., Héden, M.-B. & Was, I. (1972). Nitrogen fixation in Swedish soils by blue-green algae. In *Taxonomy and Biology of Blue-green Algae* (ed. Desikachary, T. V.), pp. 269–73. University of Madras: Madras.

Jahnke, E. (1967). Die Rolle stickstoffbindender Blaualgen in mecklenburgischen Böden. *Zentr. Bakt. ParasitKde Abt. II*, **121**, 636–42.

National Swedish Board of Agriculture (1965). Royal Proclamation regarding the stipulations for soil analyses and methods. *Proclam. no. 1 & suppls.* (in Swedish).

Overrein, L. N. (1969). Lysimeter studies on tracer nitrogen in forest soil. II: Comparative losses of nitrogen through leaching and volatilization after addition of urea-, ammonium-, and nitrate-N^{15}. *Soil Sci.*, **107**, 149–59.

Paul, E. A., Myers, R. J. K. & Rice, W. A. (1971). Nitrogen fixation in grassland and associated cultivated ecosystems. *Plant and Soil, Special Volume*, pp. 495–507.

Prasad, S. (1949). In: Singh, R. N. (1961). *Role of blue-green algae in nitrogen economy of Indian agriculture.* ICAR: New Delhi.

Shtina, E. A. (1969*a*). Über die Verbreitung und ökologische Bedeutung der Algen in Ackerböden. *Pedobiologia*, **9**, 226–42.

Shtina, E. A. (1969*b*). Some regularities in the distribution of blue-green algae in soils. *Biology of the Cyanophyta*, **2**, 21–45. (In Russian with English summary.)

Singh, R. N. (1961). *Role of blue-green algae in nitrogen economy of Indian agriculture.* ICAR: New Delhi.

Venkataraman, G. S. (1972). *Algal Biofertilizers and Rice Cultivation.* Today & Tomorrow's Printers & Publishers: New Delhi.

Vlassak, K., Paul, E. A. & Harris, R. E. (1973). Assessment of biological nitrogen fixation in grassland and associated sites. *Pl. Soil*, **38**, 637–49.

Watanabe, A., Nishigaki, S. & Konishi, C. (1951). Effect of nitrogen fixing blue-green algae on the growth of rice plants. *Nature, Lond.*, **168**, 748–9.

14. The role of blue-green algae in tropical rice cultivation

G. S. VENKATARAMAN

The efficacy of biological inputs in improving soil fertility and crop productivity has been demonstrated many times. The classic example is nitrogen input from the legume–*Rhizobium* association, which has now assumed industrial dimensions in many countries. Certain non-symbiotic, autotrophic, blue-green algae, because of their ability to carry out both photosynthesis and nitrogen fixation, are ecologically and agriculturally important, particularly in tropical rice fields (De, 1939; Singh, 1961). A great deal of information is now available from laboratory experiments about the species of blue-green algae that fix nitrogen and also about the effect of various factors on the rate of nitrogen fixation. No extensive review of literature is called for here. What is proposed is to examine the possibility of harnessing these natural resources in agricultural production programmes.

Rice is the major food crop in South and South East Asia and covers an area of about 36 million hectares in India. Until the mid-1960s rice yields in India were generally low and in the face of an increasing population pressure, the future was looked upon with concern. However, during the last decade, scientific developments have shown that tropical rice yields could be raised several-fold and within a short period some rice growing countries have converted chronic deficits into self-sufficiency by the application of new production technology. In areas of intensive agriculture, it is now economically possible or at least feasible to provide most of a crop's nitrogen needs in the form of chemically produced fertilizer inputs. The question, therefore, is: what is the relevance of blue-green algae as a biological input under the modern methods of rice cultivation, which involve high yielding, photoinsensitive, fertilizer-responsive dwarf rice genotypes, efficient pest control measures with powerful agricultural chemicals, and other improved management practices?

The present review is based mainly on the work done by the author and his research group at the Indian Agricultural Research Institute, in this direction.

Table 14.1. *Percentage distribution of Indian rice-field soils harbouring nitrogen-fixing, blue-green algae*

States	Total no. of samples	No. of samples showing N_2-fixing forms	Percentage	Dominant form
Assam	17	3	17.6	*Nostoc*
Andhra Pradesh	273	64	23.07	*Anabaena*
Gujrat	14	3	21.5	*Mastigocladus*
Haryana	32	8	21.2	*Nostoc*
Himachal Pradesh	12	2	16.6	*Anabaena*
Kashmir	14	1	7.14	*Anabaena*
Kerala	352	32	9.09	*Nostoc*
Madhya Pradesh	72	9	12.5	*Anabaena*
Maharashtra	84	14	16.6	*Westiella*
Mysore	192	32	11.4	*Cylindrospermum*
Orissa	178	87	43.2	*Aulosira*
Panjab	57	12	21.5	*Calothrix*
Tamil Nadu	401	40	9.9	*Nostoc*
Uttar Pradesh	450	392	86.9	*Aulosira*
West Bengal	65	39	60.0	*Nostoc*
Total	2213	738	32.89	

Distribution of nitrogen-fixing blue-green algae in rice-field soils

Contrary to the general belief, nitrogen-fixing, blue-green algae are not invariably present in tropical rice soils, although recurrent combinations of algal species occur, particularly in comparable habitats. The distribution of soils harbouring nitrogen-fixing, blue-green algae varies from as low as 4 % to as high as 80 % (Aiyer, 1965; Materassi & Balloni, 1965; Watanabe, 1959; Singh, 1961; Pandey, 1965). A recent all-India survey showed that out of 2213 soil samples from rice fields, only about 33 % were found to harbour nitrogen-fixing forms (Table 14.1). These algae were particularly rich in the soils of Uttar Pradesh (87 %), West Bengal (60 %) and Orissa (43 %) and the dominant forms were *Aulosira fertilissima* and species of *Nostoc*. Soils of Tamil Nadu (10 %), Kashmir (7 %) and Kerala (9 %) were comparatively poor in these algae. Interestingly, regional differences could be seen in the distribution of the dominant forms. While Uttar Pradesh soils were rich in *Aulosira*, *Mastigocladus* was dominant in Gujarat, *Westiella* in Maharashtra, *Cylindrospermum* in Mysore and *Calothrix* in Panjab. Species of *Nostoc* and *Anabaena* were ubiquitous in all regions in varying proportions and formed the major constituents of the algal biotypes in 9 out of

15 States. These two genera have been found even in peaty soils of Kuttanad in Kerala (Amma, Aiyer & Subramoney, 1966).

Algal inoculation and soil properties

The physical properties of the soil seem to be little affected by algal inoculation, although the latter may improve soil aggregation (Sankaram, 1971). The percentage increase in soil aggregation varied from 35.7–42.8 in garden soil and from 57.8–78 in sandy soil mixtures (Marathe, 1972). Using ^{15}N as tracer, Nishigaki, Shibuya & Kodaira (1951; 1953) showed that algae fixed about 22.5 kg N/ha and the crop absorbed about 4.8 kg N. An in-situ fixation of about 80 kg N/ha was attributed by Singh (1961) to the algal activity in maize fields of Uttar Pradesh. In terms of nitrogen uptake by rice plants, algal fixation seems to contribute about 20 kg N/ha (Sahay, 1966; Sankaram, 1967). However, in a long-term field trial, no appreciable increase in the amount of the residual nitrogen in the soil after harvest could be detected (Aiyer, Salahudeen & Venkataraman, 1972). Algal inoculation results in the maintenance of both total and organic nitrogen levels beyond the tillering stage of the crop and the amounts of added organic nitrogen due to algae were not depressed even in presence of chemical nitrogen (Chopra & Dube, 1971).

Of interest is the reduction in the contents of oxidizable matter, total sulphides and ferrous iron in the soil due to algal inoculation, particularly in areas where iron and sulphide toxicity is a major problem (Aiyer *et al.*, 1972). The decrease in these reduced compounds is probably due to an increased oxygen tension in the rice field water, resulting from algal photosynthesis.

No substantial increase in the organic matter content of the soil could be observed as a result of algal activity. This is understandable since organic matter is readily broken down and lost in warm humid tropical regions.

In-situ nitrogen fixation by blue-green algae

Although considerable information is available on the nitrogen-fixing abilities of different blue-green algal strains, information on the actual amount of nitrogen fixed by these organisms in nature under field conditions and the effect of different agro-climatic factors on nitrogen turnover is meagre. The observed high correlation between the number of heterocysts and the ability to fix nitrogen in any particular species and the suggestion that the heterocyst is the site of nitrogen fixation offers a rough method for a semi-quantitative index of nitrogen-fixing capacity. Determination of rates of fixation by acetylene-reduction technique supplemented by heterocyst counts could give us information about the distribution and cycle of activity of nitrogen-fixing algae in rice fields. In terms of nitrogen-fixing ability, considerable variations exist even under laboratory conditions, not only between different genera but also within the species (Venkataraman, 1972).

Induction of mutations to enhance nitrogen-fixing power does not seem to be a profitable approach to improve strain efficiency. In fact, most of the mutants seem to have lost their nitrogen-fixing ability (Ahmad & Venkataraman, 1971). Since nitrogen fixation is determined by growth, the problem seems to be one of selecting rapidly growing strains and ensuring their growth requirements in the field. Laboratory conditions are obviously different from the situations occurring in the field, so that the best strain in a simple laboratory test is not necessarily the one which will be of practical use. However, production of strains with enhanced synthesis of growth-promoting substances and tolerance to pesticides may be useful.

Algalization and pesticide application

The introduction of chemical pest-control measures has brought into use many powerful pesticides, some of which are added even to irrigation water. Although some strains of algae have been used to assay soil-applied herbicides (Atkins & Tchan, 1967; Pillay & Tchan, 1969; 1970), very little information is available on the effect of pesticides on survival and growth of blue-green algae (Shtina, 1957; 1969; Balesina, 1967; Hamdi, El-Nawawy & Tewfic, 1970). γ-BHC application to rice fields has been reported to stimulate the algal population by inhibiting the predators which feed on algae (Anonymous, 1965). Both dichlorophenoxyacetic acid (2,4-D) and 4-chloro-2-methylphenoxyacetic acid (MCPA) affect nitrogen fixation by *Nostoc muscorum*, *N. punctiforme* and *Cylindrospermum* sp. at concentrations recommended for field application, but stimulate nitrogen fixation at 10^{-4} to 10^{-5} M concentration (Inger, 1970). Recently it has been shown that many of the nitrogen-fixing blue-green algal strains used for soil inoculation can grow in the presence of high levels of some of these chemicals, although there is a wide range of tolerance limits (Venkataraman & Rajyalakshmi, 1971;

1972; Ahmad & Venkataraman, 1973). This indicates not only the possibility of using these algae along with our pest control measures, but also the importance of the choice of strains with particular pesticides. In fact, in field trials, an algal effect has been observed when pesticides were used routinely to control the pests.

Algalization and crop yield

Crop yield provides an indirect measure of nitrogen fixation in a particular soil. Experiments in which rice plants have been shown to respond to algal inoculation provide presumptive evidence of this, although as Venkataraman & Neelakantan (1967) pointed out, the production of growth substances and vitamins by algae seems to be largely responsible for the higher plant-yield. Definite proof for a nitrogen transfer to higher plants has been obtained, using ^{15}N as a tracer (Mayland & McIntosh, 1966; Stewart, 1967).

Based on long-term field trials, a progressive increase in the yield of rice was obtained by Watanabe (1962; 1965) and the effect of algal inoculation was found to be comparable to the application of ammonium sulphate at the rate of 71.74 kg/ha (Hosoda & Takata, 1955). In the USSR, an increase in the crop yield of 13–20 % has been attributed to the application of *Amorphonostoc punctiforme* (Shtina, 1965; see also Kuksa & Orleanskii, 1965). *Anabaena azotica* was reported to increase the rice yield by about 24 % under field conditions in China (Ley, 1959).

In India, the beneficial effect of algal inoculation has been demonstrated in a number of localities in terms of the grain yield of many varieties (T9, Ptb 10, CRT 141, NP 130, Co 25, ASD 5, Co 29, 498-2A Aman, S 1092, BAM 3, SLO 16, Hx 35, NC 1281) (Venkataraman, 1972). Subrahmanyan and his co-workers (Subrahmanyan, Relwani & Manna, 1964*a*, *b*; Subrahmanyan, Manna & Patnaik, 1965; Subrahmanyan, Relwani & Manna, 1965) suggested the addition of 100 kg superphosphate and 0.25 kg sodium molybdate per hectare without the addition of any nitrogenous fertilizer to improve soil fertility. Partial soil sterilization, combined with the above treatment was also beneficial (Relwani & Subrahmanyan, 1963; Subrahmanyan, 1972). In the absence of any added nitrogen, an increase of about 15 % in the grain yield of ASD 5 rice variety was recorded by Venkataraman & Goyal (1968) in the field. Similar effects have been observed in many localities (Jha, Ali, Singh & Battacharya, 1965; Sankaram, 1971; El-

Nawawy, 1969; El-Nawawy, Lotfi & Fahmy, 1958; Eid, Hamissa & Shoukry, 1962; Aboul-Fadl *et al.*, 1967) and algal inoculation is effective under different agro-climatic conditions and soil types (Subrahmanyan, Manna & Patnaik 1965; Sankaram, Mudholker & Sahay, 1967; Goyal & Venkataraman, 1971).

Recent investigations show that irrespective of the varietal differences among the high-yielding, fertilizer-responsive dwarf rice varieties, all respond to algal inoculation (10–15 % increase in the grain yield) even in the presence of high levels of nitrogen fertilization (100–150 kg N/ha). This results in an increased output per unit input of added fertilizer (3–5 kg grain/kg N) with a consequently enhanced economic return (Rs. 400–600/ha) (Table 14.2) (Venkataraman & Goyal, 1968; 1969*a*, *b*; Goyal & Venkataraman, 1960; Aiyer *et al.*, 1972; Venkataraman, 1972). Biological nitrogen fixation is known to be suppressed by exogenous nitrogen. The observed algal effect in presence of high levels of nitrogen fertilizers suggests that it is largely due to the supply of growth-promoting substances to the crop, as well as to the nitrogen fixed by these organisms (Venkataraman & Neelakantan, 1967; Gupta & Lata, 1964; Gupta & Shukla, 1964; 1967). Excretion of considerable amounts of ascorbic acid by blue-green algae has been reported, which might play a dual role: (*a*) it can accelerate the growth and development of the rice plants directly and (*b*) it may in some way participate in the processes of nitrogen fixation and nitrate reduction, ultimately influencing rice cultivation although there is no direct evidence of this (Vaidya, Patel & Jani, 1970). Thus, these algae serve not as a substitute to mineral fertilization, but as a supplement to it. Of the different methods of inoculation, soil application either one week before or after transplantation seems to be effective. Soil application of algae, combined with half the dose of urea foliar spray may be as effective as the application of the full dose of urea to the soil. This is particularly relevant when conservation of available fertilizers is contemplated. Delaying the application of ammonium sulphate by three to four weeks after transplantation and applying superphosphate either two weeks before or four weeks after transplantation have been reported to be effective for the growth and activity of *Tolypothrix tenuis* in the Egyptian rice fields (El-Nawawy, 1969).

The percentage nitrogen, protein and free amino acid contents in the rice grain have been reported to increase as a result of algal inoculation (Sundara Rao, Goyal & Venkataraman, 1963; Gupta & Shukla, 1967). An increase in soluble plant nitrogen attributed to rhizosphere

Table 14.2. *Response of rice crops to algal inoculation under field conditions (grain yield kg/ha; figures in parentheses represent percentage increase). (From Venkataraman, 1972)*

	Lowland						Upland		
	Trivandrum	Moradabad	Burari	Kadayanallur	Tirunelveli	Kanpur	Kanpur		
	IR 8	IR 8	Jaya	IR 20	IR 8	Jaya	IET 1410	Pusa 2–21	IR 20
60 kg N/ha	—	—	—	—	—	—	4983	5083	4567
60 kg N/ha + algae	—	—	—	—	—	—	5900	5817	5033
							(18.4)	(14.4)	(10.2)
100 kg N/ha	—	3567	—	—	—	—	—	—	—
100 kg N/ha + algae	—	4007	—	—	—	—	—	—	—
		(12.34)							
120 kg N/ha	4504	—	—	4822	5895	5699	5967	6067	5817
120 kg N/ha + algae	5317	—	—	5852	6453	6332	6330	6300	5867
	(18.03)			(21.1)	(10.13)	(11.1)	(11.1)	(3.84)	(0.85)
150 kg N/ha	—	—	6562	—	—	—	—	—	—
150 kg N/ha + algae	—	—	7907	—	—	—	—	—	—
			(20.4)						
Increase/kg nitrogen	6.77	3.66	8.9	8.5	4.11	5.2	15.2*	12.2*	7.7*
							3.0**	1.5**	4.0*

* At 60 kg N/ha level. ** At 120 kg N/ha level.

micro-organisms has previously been noted (Aseeva & Kirillova, 1960; Krasil'nikov, 1961; Rempe & Kaltagova, 1965). The explanation for these microbial influences awaits further experimentation.

Establishment of the inoculum

The successful introduction of a particular strain in an area, or the strengthening of certain forms to the exclusion of others depends largely on their ability to survive and establish in the habitat. In soil, there exists a biocoenotic association of complex interacting factors. It may be possible to use the technique of immune diffusion in agar to identify the establishment of the inoculum, although we have at present no evidence for the stability of the serological characters of blue-green algae in soil. In many of our experiments, the inoculated fields showed abundant algal growth and the introduced species formed the dominant constituents. Equally very little is known about the possible biotic changes brought about by algalization. Indications that the quantitative incidence of other groups of micro-organisms is affected suggest the possibility of biotic engineering of rice field soils (Venkataraman, 1966).

The crop plant also may exert an influence on the algae occurring in the rhizosphere. Gonzalves & Yalvigi (1960) found that the rhizosphere effect varied with the crop and different species of algae were associated with different crop plants. This aspect has so far received meagre attention.

Algal establishment in the field may fail due to contact with acid fertilizers, lack of competitive ability, restricted tolerance to pesticides, presence of predators and/or phages, or nutrient deficiencies. The minimum moisture level that can support the growth and activity of these algae needs critical examination. This will be particularly relevant in areas without assured water supply.

Production of the algal inoculum

For a large-scale application of algae in rice cultivation, the 'open-air soil culture' method developed by the author has the advantage of being run continuously in the tropics and adopted by farmers in the form of algal nursery beds in their own fields. The other possibility of growing these algae for agricultural purposes is to grow them in sewage. This will be economically attractive, because of the involved reclamation of the sewage water that can be subsequently used for irrigation purposes. However, the presence of cyanophages in sewage requires the use of phage-resistant strains. Another promising line of approach is the introduction of *Azolla* into the rice fields. Promising claims have been made in Viet Nam with this practice (Khien, 1957; Le wan Kan & Sobachkin, 1963; Mishustin, 1964; Norris, 1962).

We have a vast reservoir of biological resources in the blue-green algae and a unique opportunity to harness them for maximizing ecological benefits and minimizing the ecological hazards. In this way we may make a positive impact on agrarian economy without undermining the long-term productivity of the soil.

References

Aboul-Fadl, M., Taha, E. M., Hamissa, M. R., El-Nawawy, A. S. & Shoukry, A. (1967). The effect of the nitrogen fixing blue-green alga. *Tolypothrix tenuis* on the yield of paddy. *J. Microbiol., UAR*, **2**, 241–9.

Ahmad, M. H. & Venkataraman, G. S. (1971). Note on a strain of *Nostoc linckia* unable to use molecular nitrogen. *Curr. Sci.*, **40**, 67–8.

Ahmad, M. H. & Venkataraman, G. S. (1973). Tolerance of *Aulosira fertilissima* to pesticides. *Curr. Sci.*, **42**, 108.

Aiyer, R. S. (1965). Comparative algological studies in rice fields in Kerala State. *Agri. Res. J., Kerala*, **3**, 100–4.

Aiyer, R. S., Salahudeen, S. & Venkataraman, G. S. (1972). On a long term algalization field trial with high yielding rice varieties: yield and economics. *Indian J. agri. Sci.*, **42**, 380–3.

Amma, P. S., Aiyer, R. S. & Subramoney, N. (1966). Occurrence of blue-green algae in acid soils of Kerala. *Agri. Res. J., Kerala*, **4**, 141–3.

Anonymous (1965). *Annual Report of the International Rice Research Institute*, pp. 176–9. IRRI: Manila, Philippines.

Aseeva, I. V. & Kirillova, N. F. (1960). Actions of soil dwelling bacteria on free amino acid content of leguminosae. *Nauch Dokl. vyseh. Shkoly giol. Nauk*, *1960*, 139–44.

Atkins, C. A. & Tchan, Y. T. (1967). Study of soil algae. II, Bioassay of atrazine and the prediction of its toxicity in soil using algal growth method. *Pl. Soil*, **17**, 432–42.

Balesina, L. S. (1967). Einfluss einiger Herbidine auf die Entwicklung der Bodenalgen. *Mikrobiologiya*, **36**, 1963–7.

Chopra, T. S. & Dube, J. N. (1971). Changes of nitrogen content of a rice soil inoculated with *Tolypothrix tenuis. Pl. Soil*, **35**, 453–62.

De, P. K. (1939). The role of blue-green algae in nitrogen fixation in rice fields. *Proc. R. Soc. Lond. B*, **127**, 121–39.

Eid, M. T., Hamissa, M. R. & Shoukry, A. (1962). Paddy fertilization trials in nursery and field. *Agri. Res. Rev. Cairo*, **40**, 136–46.

El-Nawawy, A. S. (1969). Research programme on nitrogen fixing blue-green algae in Agricultural Microbiology Research Division, Ministry of Agriculture, A.R.E. *Agri. Res. Rev. Cairo*, **50**, 117–28.

El-Nawawy, A. S., Lotfi, M. & Fahmy, M. (1958). Studies on the ability of some blue-green algae to fix atmospheric nitrogen and their effect on growth and yield of paddy. *Agri. Res. Rev. Cairo*, **36**, 308–19.

Gonzalves, E. A. & Yalvigi, V. S. (1960). Algae in the rhizosphere. *Proc. Symp. Algology, New Delhi*, pp. 335–44.

Goyal, S. K. & Venkataraman, G. S. (1970). Response of high yielding rice varieties to algalization. I. Effect on the yield of four rice varietes. *Phykos*, **9**, 137–8.

Goyal, S. K. & Venkataraman, G. S. (1971). Response of high yielding rice varieties to algalization. II. Interaction of soil types to algal inoculation. *Phykos*, **10**, 32–3.

Gupta, A. B. & Lata, K. (1964). Effects of algal growth hormones on the germination of paddy seeds. *Hydrobiologiya*, **24**, 430.

Gupta, A. B. & Shukla, A. C. (1964). The effect of algal hormones on the growth and development of rice seedlings. *Labdav J. Sci. Technol. Kanpur*, **2**, 204.

Gupta, A. B. & Shukla, A. C. (1967). Studies on the nature of algal growth substances and their influence on growth, yield and protein contents of rice plants. *Labdav J. Sci. Technol. Kanpur*, **5**, 162–3.

Hamdi, Y. A., El-Nawawy, A. S. & Tewfic, M. S. (1970). Effect of herbicides on growth and nitrogen fixation by the alga, *Tolypothrix tenuis*. *Acta Microbiol., Polonica*, **2**, 53–6.

Hosoda, K. & Takata, H. (1955). *Trans. Tottari Soc. agri. Sci.*, **10**, 1–10 (in Japanese, cited by Watanabe, 1965).

Inger, L. 1970. Effect of two herbicides on nitrogen fixation by blue-green algae. *Sv. Bot. Tidskr.*, **64**, 460–1.

Jha, K. K., Ali, M. A., Singh, R. & Battacharya, P. B. (1965). Increasing rice production through the introduction of *Tolypothrix tenuis*, a nitrogen fixing blue-green alga. *J. Indian Soc. Soil Sci.*, **13**, 161–6.

Khien, Z. K. (1957). The use of *Azolla* as fertilizer for rice cultivation in Democratic Republic of Vietnam. In *Conference on the Problems of Fertilizers*. Pub. Acad. Sci.: Moscow, USSR.

Krasil'nikov, N. (1961). On the role of soil bacteria in plant nutrition. *J. gen. appl. Microbiol.*, **7**, 128–44.

Kuksa, I. N. & Orleanskii, V. (1965). Development of scientific research on nitrogen fixing blue-green algae and their practical use in agriculture. *Mikrobiologiya*, **34**, 743–7.

Le wan Kan & Sobachkin, A. A. (1963). Problems of utilizing *Azolla* as fertilizers in the Democratic Republic of Vietnam. *Report in TSHA*, **94**, 93.

Ley, S. H. (1959). The effect of nitrogen fixing blue-green algae on the yields of rice plant. *Acta hydrobiol, Sinica*, **4**, 440–4.

Marathe, K. V. (1972). Role of some blue-green algae in soil aggregation. In *Taxonomy and Biology of Blue-green Algae* (ed. Desikachary, T. V.), pp. 328–31. University of Madras: Madras.

Materassi, R. & Balloni, W. (1965). Quelques observations sur la présence de microorganisme autotrophes fixateurs d'azote dans les rivières. *Ann. Inst. Pasteur, Paris*, **3**, *Suppl.*, 218–23.

Mayland, H. F. & McIntosh, T. H. (1966). Availability of biologically fixed atmospheric nitrogen-15 to higher plants. *Nature, Lond.*, **209**, 421–2.

Mishustin, E. H. (1964). Biological nitrogen fixation in agriculture and prospects of utilizing nitrogen fixing blue-green algae in agriculture. In *Proc. Sci. Adv. Comm. for Physiology and Biochemistry of Microorganisms*, pp. 1–47. Acad. Sci.: USSR.

Nishigaki, S., Shibuya, M. & Kodaira, K. (1951). *J. Sci. Soil and Manures*, **22**, 69. (Cited from Watanabe, 1965: in Japanese.)

Nishigaki, S., Shibuya, M. & Kodaira, K. (1953). *J. Sci. Soil and Manures*, **23**, 150. (Cited from Watanabe, 1965: in Japanese.)

Norris, D. O. (1962). *A Review of Nitrogen in the Tropics with particular reference to Pastures*, pp. 113–29. Alden Press: Oxford.

Pandey, D. C. (1965). A study of the algae from paddy soils of Ballia and Ghazipur districts of Uttar Pradesh, India. I. Culture and ecological conditions. *Nova Hedwigia*, **9**, 299–334.

Pillay, A. R. & Tchan, Y. T. (1969). Paper disc method for rapid bioassay of *S*-phenylurea herbicides. *Soil Biol.*, **10**, 23.

Pillay, A. R. & Tchan, Y. T. (1970). The use of algae for the bioassay and study of tri-allate (herbicides). *Soil Biol.*, **12**, 20–2.

Relwani, L. L. & Subrahmanyan, R. (1963). Role of blue-green algae, chemical nutrients and partial soil sterilization on paddy yield. *Curr. Sci.*, **32**, 441–3.

Rempe, J. K. & Kaltagova, O. G. (1964). Influence of root microflora on the increased development and activity of physiological processes in plants. In *Plant Microbes Relationships* (eds. Macura, J. and Vancura, V.), pp. 178–85. Czechosl. Acad. Sci.: Prague.

Sahay, M. N. (1966). Nitrogen uptake by the rice crop in the experiments with blue-green algae. *Proc. Indian acad. Sci.* **638**, 223–33.

Sankaram, A. (1967). Blue-green algae: Role in rice culture. *Farmer and Parliament*, **11**, 9–10.

Sankaram, A. (1971). *Work done on blue-green algae in relation to agriculture.* Indian Coun. Agri. Research: New Delhi, 28 pp.

Sankaram, A., Mudholkar, N. J. & Sahay, M. N. (1967). Inoculation of blue-green algae on the yield of rice under field conditions. *Indian J. Microbiol.*, **7**, 57–62.

Shtina, E. A. (1957). Die Entwicklung des Herbizids 2,4-D auf die Bodenalgen. *Tr. Kirovsk. sel'choz. Inst.*, **12**, 29–34.

Shtina, E. A. (1965). Fixation of free nitrogen in blue-green algae. In *The Ecology and Physiology of Blue-green Algae* (ed. Federov, V. D. and Tellichenko, M. M.), pp. 66–79. Moscow University Press: USSR. (In Russian.)

Shtina, E. A. (1969). Über der verbreitung und Okologische Bedeutung der algen in Ackerboden. *Pedobiologia*, **9**, 226–42.

Singh, R. N. (1961). *The role of blue-green algae in nitrogen economy of Indian agriculture.* Indian Council for Agricultural Research: New Delhi, 175 pp.

Stewart, W. D. P. (1967). Transfer of biologically fixed nitrogen in a sand dune slack region. *Nature, Lond.*, **214**, 603–4.

Subrahmanyan, R. (1972). Some observations on the utilization of blue-green algal mixtures in rice cultivation in India. In *Taxonomy and Biology of Blue-green Algae* (ed. Desikachary, T. V.), pp. 281–93. University of Madras: Madras.

Subrahmanyan, R., Manna, G. B. & Patnaik, S. (1965). Preliminary observations on the interaction of different rice soil types to inoculation of blue-green algae in relation to rice culture. *Proc. Indian Acad. Sci.*, **62B**, 171–5.

Subrahmanyan, R., Relwani, L. L. & Manna, G. B. (1964*a*). Observations on the role of blue-green algae on rice yield compared with that of conventional fertilizers. *Curr. Sci.*, **33**, 485–6.

Subrahmanyan, R., Relwani, L. L. & Manna, G. B. (1964*b*). Role of blue-green algae and different methods of partial soil sterilization on rice yield. *Proc. Indian Acad. Sci.*, **60B**, 293–7.

Subrahmanyan, R., Relwani, L. L. & Manna, G. B. (1965). Fertility build-up of rice field soils by blue-green algae. *Proc. Indian Acad. Sci.*, **62B**, 252–72.

Sundara Rao, W. V. B., Goyal, S. K. & Venkataraman, G. S. (1963). Effect of inoculation of *Aulosira fertilissima* on rice plants. *Curr. Sci.*, **32**, 366–7.

Vaidya, B. S., Patel, I. M. & Jani, V. M. (1970). Secretion of a highly reducing substance by algae in media and its possible role in crop physiology. *Sci. Culture*, **37**, 383–4.

Venkataraman, G. S. (1966). Algalization. *Phykos*, **5**, 164–74.

Venkataraman, G. S. (1972). *Algal Biofertilizers and Rice Cultivation*, 75 pp. Today & Tomorrow's Printers and Publishers: New Delhi.

Venkataraman, G. S. & Goyal, S. K. (1968). Influence of blue-green algal inoculation on crop yields of rice plants. *Soil Sci. Pl. Nutr.*, **14**, 249–51.

Venkataraman, G. S. & Goyal, S. K. (1969*a*). Influence of blue-green algae on the high yielding paddy variety IR-8. *Sci. Culture*, **35**, 58–9.

Venkataraman, G. S. & Goyal, S. K. (1969*b*). Some recent observations on the effect of nitrogen fixing blue-green algae on crop plants. *Mikrobiologiya*, **38**, 709–12.

Venkataraman, G. S. & Neelakantan, S. (1967). Effect of the cellular constituents of the nitrogen fixing blue-green alga, *Cylindrospermum muscicola* on the root growth of rice seedlings. *J. gen. appl. Microbiol.*, **13**, 53–61.

Venkataraman, G. S. & Rajyalakshmi, B. (1971). Tolerance of blue-green algae to pesticides. *Curr. Sci.*, **40**, 143–4.

Venkataraman, G. S. & Rajyalakshmi, B. (1972). Relative tolerance of blue-green algae to pesticides. *Indian J. agric. Sci.*, **42**, 119–21.

Watanabe, A. (1959). Distribution of nitrogen fixing blue-green algae in various areas of South and East Asia. *J. gen. appl. Microbiol.*, **5**, 21–9.

Watanabe, A. (1962). Effect of nitrogen fixing blue-green alga, *Tolypothrix tenuis* on the nitrogenous fertility of paddy soils and on the crop yield of rice plant. *J. gen. Microbiol.*, **8**, 85–91.

Watanabe, A. (1965). Studies on the blue-green algae as green manure in Japan. *Proc. natn. Acad. Sci. India*, **35**A, 361–9.

15. Research on blue-green algae in Egypt, 1958–1972

A. S. EL-NAWAWY & Y. A. HAMDI

The aim of this chapter is to review research work on blue-green algae carried out by Egyptian workers both at home and abroad during the period 1958–1972. It is not intended to review all aspects of research on blue-green algae, other such reviews already being available (Fogg & Stewart, 1965; Laporte & Pourroit, 1967; Stewart, 1970; 1973; Venkataraman, 1972; Fogg, Stewart, Fay & Walsby, 1973). Particular attention is paid in this review to four main aspects: (1) the Egyptian flora of blue-green algae; (2) the physiology of certain of those species; (3) pot and field experiments on the inoculation of paddy soils with blue-green algae; (4) future investigations.

The flora of blue-green algae in soil and water resources

The blue-green algal flora of Egypt has scarcely been studied but the algae which have been recorded are considered here. Only in a few cases have these algae been tested for nitrogen fixation (Taha & El-Refai, 1963*a*). El-Nawawy, El-Nawawy, Abou-el-Fadl & Nada (1962) found that the genera *Calothrix* and *Anabaena* are dominant in Egyptian soils and shortly thereafter Taha & El-Refai (1963*a*) reported on ten species of blue-green algae. Taha (1963*a*) isolated *Hapalosiphon fontinalis*, *Anabaena variabilis*, and *Calothrix elenkinii* from Egyptian soils. El-Ayouty (1966) isolated *Anabaena variabilis*, *A. torulosa*, *Cylindrospermum muscicola*, *Calothrix parietina*, *Nostoc muscorum*, *N. paludosum*, *Nostoc* spp. and *Lyngbya aeruginosa-coerula* from Giessen soils. Kampur soils contained *Calothrix braunii*, *Scytonema javanicum*, *Nostoc* spp., *Phormidium molle*, *Symploca muscorum* and *Microcoleus sociatus*. Shakeeb (1970) isolated *Nostoc muscorum* and *Phormidium fragile* while El-Borollosy (1972) isolated *Nostoc linckia* var. *arvense*, *N. piscinale*, *N. paludosum* and *Cylindrospermum* sp. from soils of the Delta. Ramadan, Shehata & Roushdy (1973) identified species of *Microcystis*, *Anabaena*, *Oscillatoria*, *Lyngbya* and *Nodularia* in Nile water before erection of the High Dam in 1965–1966. Shalaan (unpublished data, 1973) isolated species of *Nostoc*, *Anabaena* and *Cylindro-*

Table 15.1. *A summary of Egyptian physiological studies on the nutrition of blue-green algae*

Topic	Reference
Culture media	Taha (1963*b*)
	El-Nawawy, Lotfi & Fahmy (1958)
pH	Taha (1963*c*)
Molybdenum	Taha & El-Refai (1963*b*)
	El-Nawawy, Hamdi, El-Sayed & Shaalan (1973)
Cobalt	Taha & El-Refai (1963*b*)
	El-Nawawy, Hamdi, El-Sayed & Shaalan (1973)
Combined nitrogen	Taha (1964)
	El-Nawawy, Ibrahim & Abou-el-Fadl (1968)
	Shakeeb (1970)
	El-Borollosy (1972)
	El-Sayed (1973)
Sulphur	Shakeeb (1970)
Phosphorus	El-Nawawy, Kamal & Abou-el-Fadl (1970)
NaCl	El-Nawawy *et al.* (1968)
	El-Borollosy (1972)
Sulphide	El-Nawawy *et al.* (1968)

spermum. The conclusion is that a wide variety of blue-green algae, including many nitrogen-fixing species, occur in Egyptian habitats.

Physiological studies on blue-green algae

Nutrition

Physiological studies on growth and nutrition of the blue-green algae, have been investigated by several workers and references to published work are summarised in Table 15.1. The findings do not differ significantly from those obtained by workers in other parts of the world. For further details the original papers should be consulted.

Effects of pesticides

The algicidal properties of the sodium salt and six *S*-alkylisothiouronium salts of arylmercaptoacetic acid were studied in a glucose salts medium (El-Nawawy *et al.*, 1962). The inoculum was a mixed algal culture from Egyptian rice fields. The *S*-alkylisothiouronium compounds at a concentration of 0.5 g anion/l destroyed all algae and some were effective at lower concentrations. The sodium salt was less toxic to algae than other salts tested. Re-inoculation after 15 days revealed that all or some of the algicidal properties persisted. All the compounds tested at lower

Table 15.2. *Effect of herbicides on dry weight, nitrogen, and chlorophyll content of* T. tenuis (Hamdi *et al.*, 1970)

	Herbicides added					
	After inoculation			On 10-day-old cultures		
Herbicides (ppm)	Dry wt (g)	Nitrogen (mg)	Chloro-phyll (mg)	Dry wt (g)	Nitrogen (mg)	Chloro-phyll (mg)
Ordram						
0.25	280	8.3	3.2	395	10.6	3.2
2.50	290	7.8	2.0	371	6.7	3.2
25.0	101	6.6	1.0	340	8.2	2.9
Trifluralin						
0.25	307	8.5	3.3	221	10.3	2.7
2.50	243	7.7	2.1	289	9.2	3.5
25.0	96	8.6	0.6	309	7.9	2.5
2,4-D						
0.045	316	10.4	8.6	276	9.5	2.8
0.45	240	9.5	2.7	280	8.3	2.9
4.50	375	9.2	3.1	270	9.1	2.5
Stam						
0.18	308	9.2	3.3	407	10.4	3.8
1.8	231	4.6	1.0	427	6.4	2.8
18.0		No growth		199	1.8	1.1
Control	448	13.0	2.0	448	13.0	2.0

concentrations showed a selective action, being less effective against unicellular green algae than against filamentous forms.

Hamdi, El-Nawawy & Tewfik (1970) evaluated the effect of the herbicides Ordram (*S*-ethylhexahydro-1-M-azepine-1-carbothionate), Stam (2,4-dichloropropion-anilide), Trifluralin (α,α,α-trifluoro-2,6-dinitro-*N*,*N*-dipropyl-*p*-toluidine and 2,4-D (dichlorophenoxyacetic acid) on growth and nitrogen fixation by *Tolypothrix tenuis*. The herbicides were added to algal cultures (*a*) just after inoculation, or (*b*) 10 days after the growth of the algae, at concentrations equal to 0.1, 1.0 and 10 times the recommended dose in the field. Dry weights and the nitrogen-contents of the algae were reduced in the presence of the Ordram, Trifluralin, 2,4-D and Stam added either at the beginning, or 10 days after the start of growth (Table 15.2). Chlorophyll synthesis was stimulated by low levels of Ordram, Trifluralin and 2,4-D in both treatments, but Stam inhibited chlorophyll synthesis.

In another study, Ibrahim (1970) reported the effect of the herbicides

Table 15.3. *Effect of certain herbicides on algal growth and nitrogen fixation* (Ibrahim, 1970)

	T. tenuis		*Calothrix brevissima*	
Herbicides (ppm)	Dry wt (g/l)	Total-N (mg)	Dry wt (g/l)	Total-N (mg)
Eptam 6-E				
0.01	2.760	197.06	2.028	88.42
0.10	1.492	66.54	1.043	44.64
1.0	1.425	58.71	0.094	26.13
10.0	—	—	—	—
100.0	—	—	—	—
Stam F-34				
0.01	2.962	211.49	2.013	86.16
0.10	2.038	145.51	1.817	78.49
1.0	0.916	49.65	1.642	70.93
10.0	—	—	—	—
100.0	—	—	—	—
Ordram				
0.01	2.285	174.57	1.423	65.74
0.1	2.137	153.01	1.291	51.64
1.0	2.153	152.43	0.214	4.41
10.0	1.292	79.33	—	—
100.0	0.148	6.10	—	—
Trifluralin				
0.01	2.137	163.27	1.654	76.41
0.10	2.051	116.44	1.872	80.12
1.0	0.680	16.86	0.681	14.57
10.0	—	—	0.442	10.96
100.0	—	—	—	—

Eptam 6-D, Stam F-34, Ordram and Trifluralin (0.01, 0.1, 1.0, 10 and 100 ppm on the basis of active ingredients) on nitrogen fixation by *T. tenuis* and *Calothrix brevissima*. Eptam, Stam and Trifluralin were most toxic to *T. tenuis* with no growth being recorded in the presence of 10 and 100 ppm (Table 15.3). Similar results were obtained with *C. brevissima* which, however, is more sensitive to Ordram and less sensitive to Trifluralin.

Extracellular nitrogen production

The nature of the nitrogenous products formed in the cells and culture solutions of *Nostoc commune* was investigated by Taha & El-Refai (1962) who found that 13% of the cellular nitrogen was present in soluble form with aspartic acid, glutamic acid and alanine as the only free amino acids. The extracellular nitrogen amounted to 30 % of the total nitrogen fixed. Glutamic acid and aspartic acid predominated in the filtrates.

Pot and field experiments on the inoculation of paddy soils

Studies have been carried out to evaluate the effect of inoculating paddy soils with blue-green algae in the greenhouse and in the field. El-Nawawy *et al.* (1958) inoculated one gram (dry weight) of *T. tenuis* into lysimeters (20 × 2 m) planted with rice to give a final concentration of 200 g/acre. This was found to be sufficient to satisfy the nitrogen requirements of the rice. Abou-el-Fadl *et al.* (1964) studied the response of rice to inoculation with *T. tenuis* together with ammonium sulphate, compost, straw, and superphosphate. In general, inoculation with *T. tenuis* significantly increased soil nitrogen but had no effect on rice yield. The response differed with the type of fertilizer applied. While ammonium sulphate was inhibitory, the addition of organic matter (compost or straw) stimulated nitrogen fixation. The addition of superphosphate together with ammonium sulphate, or compost, markedly inhibited nitrogen fixation.

Ibrahim, Kamel & El-Sherbeny (1971) studied the effect on rice yield and soil nitrogen of adding *T. tenuis*, phosphorus and nitrogen fertilizer. Addition of the alga increased grain, straw and nitrogen yield, as shown in Table 15.4. Algal inoculation, without fertilizer, increased the yield of rice grain and straw by 4.2 and 19.3 % respectively. In the presence of phosphorus, the increase was 7.0 and 56.6 % with maximum yield being obtained with alga plus 0.5 g phosphorus/pot.

El-Borollosy (1972) evaluated the response of rice to inoculation with the following blue-green algae: *N. linckia* var. *arvense*, *Cylindrospermum* sp., *N. paludosum*, *N. piscinale*, *T. tenuis*, and *N. commune*. Five rice seeds were planted in washed sand in pot cultures. These pots were divided into two series, the first was irrigated with nitrogen-free nutrient solution, while the second was treated in the same manner but supplemented with nitrogen. Highest total nitrogen levels in the plants and in sand cultures, as well as highest fresh and dry weight of rice plants after 115 days were observed in pots inoculated with *T. tenuis* followed by the local isolate *N. linckia* var. *arvense* then in decreasing order: *N. piscinale*, *Cylindrospermum* sp., *N. commune* and *N. paludosum* (Table 15.5).

Field experiments on the inoculation of rice paddy by *T. tenuis* were

Table 15.4 *Effect of inoculation with alga* T. tenuis *on the yield of rice (g/pot)* (Ibrahim *et al.*, 1971)

Treatment	Grain			Straw		
	−Alga	+Alga	% Increase	−Alga	+Alga	% Increase
Control	10.45	10.89	4.2	20.42	24.37	19.3
0 g N+P	12.37	13.23	7.0	21.06	32.99	56.6
0.25 g N+P	20.15	21.61	7.2	28.45	47.52	67.0
0.5 g N+P	24.14	28.03	16.1	32.35	52.96	63.7
1.0 g N+P	36.44	39.20	7.5	41.98	62.77	39.5
2.0 g N+P	39.15	41.48	5.9	45.68	63.63	39.3
L.S.D. 5 %		1.97			14.56	

Table 15.5. *Dry weight of rice plants inoculated with different species of algae* (El-Borollosy, 1972)

Algal inoculation	N-deficient			N-supplemented		
	Roots	Shoots	Total	Roots	Shoots	Total
Control	0.25	0.24	0.49	1.01	1.05	2.06
N. linckia	2.65	2.87	5.52	4.26	3.25	7.51
Cylindrospermum sp.	2.05	1.29	3.84	4.00	3.21	7.21
N. paludosum	1.15	1.31	2.46	2.72	2.41	5.13
N. piscinale	2.54	2.08	4.62	4.20	3.05	7.25
T. tenuis	3.71	3.21	6.92	4.99	3.38	8.37
N. commune	1.99	1.92	3.91	3.86	3.00	6.86

carried out over two successive years (Abou-el-Fadl *et al.*, 1967). In the first year, rice followed bean crops; in the second year, it followed wheat. Two levels of algal inoculum (100 and 200 g/feddan) and ammonium sulphate (10 and 20 kg N/feddan) were added. These treatments were repeated once with and once without the addition of calcium superphosphate (15 kg P_2O_5/feddan). The results (Table 15.6) showed that the average rice yield following beans was much higher than that planted after wheat. When no phosphate was added, the application of alga, or nitrogen, at the two levels gave approximately similar results. Yields were increased by 16.56, 19.56, 16.68 and 16.62 % by the application of 10 kg N, 100 g alga, 20 kg N and 200 g alga, respectively. With calcium superphosphate, the effect of 100 g alga was generally less than that caused by 10 kg N, but the difference was in-

Table 15.6. *Effect of nitrogen and algal inoculation in the absence or presence of phosphate on rice yield* (Abou-el-Fadl *et al.*, 1967)

	Preceding crop			
	Horse bean		Wheat	
Treatment	Yield (ardeb/feddan)	Increase (%)	Yield (ardeb/feddan)	Increase (%)
	Without Ca-superphosphate			
10 kg N	22.73	9.28	16.83	28.18
20 kg N	22.43	7.84	17.16	30.69
100 g alga	23.09	11.01	17.49	33.21
200 g alga	21.58	3.75	17.99	37.01
	With Ca-superphosphate (15 kg P_2O_5/feddan)			
10 kg N	24.75	18.99	19.50	48.51
20 kg N	25.30	21.64	18.41	40.21
100 g alga	23.72	14.04	17.03	29.70
200 g alga	19.99	—	16.68	25.50

significant when rice followed bean. Moreover, the effect of 200 g alga in the presence of phosphorus was much smaller than in its absence.

Abou-el-Fadl, Hamissa, El-Nawawy & Abd-el-Aziz (1970) studied the response of rice to algal inoculation as affected by rotation and the time of application of nitrogen and phosphorus fertilizers. The data obtained (Table 15.7) indicate that: (1) In the first four experiments the best rice yield occurs on fertilization with 20 kg N and 15 kg P_2O_5. Algal inoculation gave a yield similar to fertilization with 10–20 kg N. The algae + 15 kg P_2O_5 did not give a better yield than that without P_2O_5. (2) With the application of superphosphate two weeks before and ammonium sulphate four weeks after transplanting, algal inoculation caused an increase of 23.1 % in yield (experiment 5). (3) In experiment 6, where 30 kg N/feddan was added four weeks after transplanting, algal inoculation increased rice yield by 22.3 %. (4) When the amount of ammonium sulphate was reduced to two thirds the recommended dose (experiments 7 and 8), inoculation with alga gave a greater yield than when maximum levels of ammonium sulphate alone were used. In general, the greenhouse and field experiments indicated that blue-green algae served as a useful nitrogen source for rice growth.

Table 15.7. *Effect of certain fertilizers and the preceding crops on rice yield* (Abou-el-Fadl *et al.*, 1970)

Experimental applications	Time (weeks)		Preceding crops*** (ardeb/feddan)							
Constituents	N	P	Horse bean	Barley	Clover	Flax	Wheat	Bean	Clover	Wheat
(1) No fertilizer	—	—	100	100	100	100	—	—	—	—
(2) 10 kg N	2*	—	109.3	128.2	—	—	—	—	—	—
(3) 20 kg N	2	—	107.3	130.2	119	113.4	—	—	—	—
(4) 20 kg N + 15 kg P_2O_5	2	2*	121.6	140.2	136.0	129.6	—	—	—	—
(5) as (4)	2	2**	—	—	—	—	100	—	—	—
(6) 100 g alga	—	—	11.1	133.2	120.5	115.3	—	—	—	—
(7) 100 g alga + 15 kg P_2O_5	—	2	114.1	129.7	—	—	—	—	—	—
(8) 100 g alga + 20 kg N	2	—	—	—	—	—	—	107.2	—	—
(9) 100 g alga + 20 kg N	2	—	—	—	—	—	—	—	—	—
15 kg P_2O_5	—	2	—	—	—	127.8	—	—	—	—
(10) as (9)	4	2**	—	—	—	—	123.1	—	—	—
(11) 100 g alga + 30 kg N	4	—	—	—	—	—	—	122.3	—	—
(12) 30 kg N	2	—	—	—	—	—	—	100	100	—
(13) as (11) + 15 kg P_2O_5	—	—	—	—	—	—	—	—	—	100
(14) 100 g alga + 20 kg N	—	—	—	—	—	—	—	—	—	—
+ 15 kg P_2O_5	2	2**	—	—	—	—	—	—	—	107.2

* After transplanting. ** Before transplanting.
*** Relative yield compared with untreated series.

Future investigations

Work on blue-green algae in Egypt has only begun recently and many problems remain. A detailed survey of the flora of blue-green algae and their contribution to nitrogen economy is needed. In addition a critical evaluation of nitrogen fixation *in situ* by the acetylene-reduction technique is required. The use of blue-green algae in rice cultivation requires the selection of suitable strains, the mass production of the alga, the development of a carrier, determination of the time and the amount of fertilizers to be applied and modern management approaches. Studies on many of these aspects are under way in Egypt.

References

Abou-el-Fadl, M., Eid, M. T., Hamissa, M. R., El-Nawawy, A. S. & Shoukry, A. (1967). The effect of nitrogen-fixing blue-green alga, *Tolypothrix tenuis* on the yield of paddy in U.A.R. *J. Microbiol., UAR*, **2**, 241–9.

Abou-el-Fadl, M., El-Nawawy, A. S., El-Mofty, M., El-Nady, M. & Farag, F. A. (1964). Nitrogen fixation by the blue-green alga, *Tolypothrix tenuis*, as influenced by ammonium sulphate, compost, straw and superphosphate with special reference to their effect on rice yield. *J. Soil Sci., UAR*, **4**, 91–104.

Abou-el-Fadl, M., Hamissa, M. R., El-Nawawy, A. S. & Abd-el-Aziz, M. S. (1970). Evaluation of the blue-green alga *Tolypothrix tenuis* as a nitrogen source for rice plants. *The Agricultural Conference of Rice, Cairo, 1970.*

El-Ayouty, E. Y. M. (1966). Systematik und stickstoffbindung einiger Blaualgen in Lehmboden aus einem humiden und einem semiariden Gebiet. Ph.D. Thesis, Faculty of Justus Liebig-University of Giessen, Germany.

El-Borollosy, M. A. (1972). Studies on nitrogen-fixing blue-green algae in A.R.E. M.Sc. Thesis, Faculty of Agriculutre, Ain Shams University, Cairo, ARE.

El-Nawawy, A. S., El-Nawawy, A. S., Abou-el-Fadl, M. & Nada, M. M. (1962). Economical studies on Algae in Egypt (Part I). Effect of new isothiouronium derivatives of arylmercaptoalkane carboxylic acids on the paddy soil flora of algae in Egypt. *Soil Sci., UAR*, **2**, 3–14.

El-Nawawy, A. S., Hamdi, Y., El-Sayed, M. & Shaalan, S. N. (1973). Growth of the blue-green alga *Tolypothrix tenuis* as affected by cobalt and molybdenum. *Zbl. Bakt. Abt.* II, 182, S. 452–6.

El-Nawawy, A. S., Ibrahim, A. N. & Abou-el-Fadl, M. (1968). Nitrogen fixation by *Calothrix* sp. as influenced by certain sodium salts and nitrogenous compounds. *Acta Agron.*, **17**, 323–7.

El-Nawawy, A. S., Kamal, R. M. & Abou-el-Fadl, M. (1970). Growth and nitrogen fixation by blue-green alga *Tolypothrix tenuis* as affected by phosphorus content of media. *4th Conference of Soil Science*, Cairo.

El-Nawawy, A. S., Lotfi, M. & Fahmy, M. (1958). Studies on the ability of some blue-green algae to fix atmospheric nitrogen and their effect on growth and yield of paddy. *Agric. Res. Rev.*, Ministry of Agriculture, UAR, **36**, 308–20.

El-Sayed, M. (1973). Some factors affecting propagation of certain N_2-fixing blue-green algae. M.Sc. Thesis, Faculty of Agriculture, Cairo University, Egypt.

Fogg, G. E. & Stewart, W. D. P. (1965). Nitrogen fixation in blue-green algae. *Sci. Prog.*, **53**, 181–201.

Fogg, G. E., Stewart, W. D. P., Fay, P. & Walsby, A. E. (1973). *The Blue-green Algae.* Academic Press: London, New York.

Hamdi, Y. A., El-Nawawy, A. S. & Tewfik, M. S. (1970). Effect of herbicides on growth and nitrogen-fixation of alga *Tolypothrix tenuis. Acta Microbiol. Polon. ser. B*, **2** (19), no. 1, 53–6.

Ibrahim, A. N. (1970). Effect of certain herbicides on growth of nitrogen-fixing algae and rice plants. *Symposium on soil microbiology*, pp. 15–20. Hungarian Acad. Sci., Budapest.

Ibrahim, A. N., Kamel, M. & El-Sherbeny, M. (1971). Effect of inoculation with alga *Tolypothrix tenuis* on the yield of rice and soil nitrogen balance. *Agrok. Talajtan*, **20**, 389–400.

Laporte, G. & Pourroit, R. (1967). Fixation de l'azote atmosphérique par les algues Cyanophycées. *Rev. Ecol. Biol. Sol*, **4**, 81–112.

Ramadan, F. M., Shehata, S. A. & Roushdy, M. (1973). Studies on the Nile water plankton prior to High Dam. *J. Wasser* and *Abwasser Forschung* (in press).

Shakeeb, M. A. (1970). Nitrogen fixation in some Cyanophycean algal members. M.Sc. Thesis, Faculty of Sciences, University of Cairo, A.R.E.

Stewart, W. D. P. (1970). Algal fixation of atmospheric nitrogen. *Pl. Soil*, **32**, 555–88.

Stewart, W. D. P. (1973). Nitrogen fixation. In *The Biology of Blue-Green Algae* (eds. Carr, N. G. and Whitton, B. A.), pp. 260–78. Blackwells: Oxford.

Taha, M. S. (1963*a*). Isolation of some nitrogen-fixing blue-green algae from the rice fields of Egypt, in pure cultures. *Mikrobiologiya*, USSR, **32**, 493–7.

Taha, M. S. (1963*b*). The effect of concentration of different components of the medium on growth and nitrogen-fixation by blue-green algae. *Mikrobiologiya*, USSR, **32**, 582–9.

Taha, M. S. (1963*c*). The influence of the hydrogen ion concentration in the medium, and the temperature on the growth and nitrogen fixation by blue-green algae. *Mikrobiologiya*, USSR, **32**, 968–72.

Taha, M. S. (1964). The effect of nitrogen compounds upon growth of blue-green algae and fixation of molecular oxygen by them. *Mikrobiologiya*, USSR, **33**, 397–403.

Taha, E. E. M. & El-Refai, A. E. H. (1962). Physiological and biochemical studies on nitrogen fixing blue-green algae. I. On the nature of cellular and extracellular nitrogenous substances formed by *Nostoc commune*. *Arch. Mikrobiol.*, **41**, 307–12.

Taha, E. E. M. & El-Refai, A. E. H. (1963*a*). On the nitrogen fixation by Egyptian blue-green algae. *Z. Allg-Mikrobiol.*, **3**, 382–8.

Taha, E. E. M. & El-Refai, A. E. H. (1963*b*). Physiological and biochemical studies on nitrogen fixation by blue-green algae. III. The growth and nitrogen-fixation of *Nostoc commune* as influenced by culture conditions. *Arch. Mikrobiol.*, **44**, 356–65.

Venkataraman, G. S. (1972). *Algal Biofertilizers and Rice Cultivation*. Today & Tomorrow's Printers & Publishers: New Delhi.

16. Nitrogen-fixing algae in Morocco

J. RENAUT, A. SASSON, H. W. PEARSON
& W. D. P. STEWART

A good deal of information has existed for some time on the ecological distribution of blue-green algae in Africa (see e.g. Steyn, 1945; Stephens, 1949; Talling, 1957; Prowse & Talling, 1958; Talling & Rzoska, 1967; Carmouse *et al.*, 1972) but until recently such studies were not concerned with nitrogen fixation by these algae. The earliest studies on nitrogen fixation were those of Taha and co-workers (Taha, 1963*a–c*; 1964*a*, *b*; Taha & El-Refai, 1962; 1963*a*, *b*) on Egyptian species, and information on algae from Egypt is given in Chapter 15. Recently data for Lake George (Uganda), Senegal, Mali and Morocco have also become available.

Data for Lake George obtained by Horne & Viner (1971) suggest that nitrogen fixation by algae may account for approximately 30 % of the total particulate nitrogen in this lake, where the dominant blue-green algae (mainly *Microcystis aeruginosa*, *M. flos-aquae* and *Anabaenopsis* spp.) comprise about 80 % of the total phytoplankton (Burgis *et al.*, 1973).

In Senegal, the distribution of blue-green algae in several paddy soils has been studied, and nitrogen fixation estimated in a variety of soil types including sandy paddy soils, sugar cane fields and other neutral and acid soils. It was also shown that there was antagonism between blue-green algae and sulphate-reducing bacteria in these soils (Roger, 1973; Roger & Jacq, 1972) and it has been suggested that it may be possible to prevent sulphate reduction there by introducing blue-green algae, possibly by pelleting the rice seeds with algae.

In Mali, K. Traore (1973) and T. Traore (1973) carried out ecological and taxonomic studies on nitrogen fixation in streams near Bamako and found that of about 60 species recorded, many were heterocystous forms and potentially capable of fixing nitrogen. These workers also carried out acetylene-reduction tests on two paddy soils in Mali and recorded that around noon during the wet season (August 1973) a maximum rate of acetylene reduction of 983 nmoles $C_2H_4/m^2/h$ was obtained.

Nitrogen fixation by free-living, blue-green algae

In Morocco, we have carried out a survey of nitrogen fixation by blue-green algae from a variety of habitats and some of the preliminary results are presented here.

Materials and methods

Sampling sites

A variety of diverse habitats within 250 km of Rabat were examined. These ranged from arid desert in the east, to coastal plains and rocky shores on the west, as well as paddy soils, freshwater lakes and a river estuary. Some of the characteristics of the areas studied in most detail are presented in Table 16.1.

Table 16.1. *Main locations surveyed for blue-green algae in Morocco*

Sampling site	Habitat	pH Soil	pH Water	Organic matter	Soil data (g/kg)* C	Soil data (g/kg)* N
Bou-Regreg	River estuary	7.7	7.5–8.0	24.0	13.7	1.71
Sidi-Bou Rhaba	Coastal freshwater lake	7.8–8.8	9.1–9.7	23.0	13.1	1.18
Dayet er Roumi	Inland freshwater lake	9.0	8.0–9.0	—	—	—
Moghrane	Commercial paddy fields	6.5–7.7	8.8	18.0	10.7	1.10
Sidi-Amira	Temporary shallow forest pond	7.45	7.2	25.0	14.3	1.24
El-Haroura	Rocky shore	—	8.2	—	—	—

* The soil data are for the soil surrounding the water mass except for the paddy field where the actual paddy soil has been sampled.

Acetylene-reduction assays

Acetylene-reduction assays were carried out using the technique of Stewart, Fitzgerald & Burris (1967) except that acetylene was added without removing the natural gas phase (Stewart, Mague, Fitzgerald & Burris, 1971). Experiments were terminated by removing a sample of the gas phase into an evacuated container and analysing this gas phase subsequently for acetylene and ethylene. Exposures to acetylene were carried out in 7.0 ml, 25 ml, or 250 ml bottles as required. With the soil samples, the soils were sampled from the top 1.0 cm of soil either on a unit area basis using a cork borer, or on a fresh weight basis. The usual period of incubation was 30 min.

Isolation of algae

The algae were isolated in unialgal or axenic culture by repeated subculturing from natural populations on to the nitrogen-free medium of Allen & Arnon (1955), solidified with 1.5 % agar, and incubated at 25 °C under a light intensity of 2000–4000 lx.

Table 16.2. *Nitrogen-fixing blue-green algae and lichens from various Moroccan habitats*

Species	Sampling site	Sample
Anabaena anomala	Moghrane	Rice paddy soil
A. variabilis	Sidi-Bou Rhaba	Lake, water column
	Sidi-Amira	Lake, water column
	Dayet er Roumi	Lake, water column
Anabaenopsis milleri	Sidi-Bou Rhaba	Lake, water column
Calothrix braunii	Sidi-Bou Rhaba	Soil core
	Dayet er Roumi	Freshwater epilithic sample
C. crustacea	El Haroura	Marine epilithic sample
C. membranacea	Sidi-Bou Rhaba	Soil core, axenic culture
Collema sp.	Sidi-Bou Rhaba	Terrestrial epilithic sample
Cylindrospermum alatosporum	Sidi-Bou Rhaba	Soil core
C. licheniforme	Sidi-Amira	Soil core
Gloeotrichia sp.	Moghrane	Rice paddy water
Lichina pygmaea	El-Haroura	Marine epilithic sample
Lyngbya sp. (red pigmented)	El-Haroura	Marine sand
Lyngbya sp.	Bou-Regreg	Unialgal culture
Nodularia spumigena	Sidi-Bou Rhaba	Unialgal culture
Nostoc calcicola	Sidi-Bou Rhaba	Axenic culture
Nostoc carneum	Sidi-Bou Rhaba	Soil core
N. commune	Sidi-Bou Rhaba	Unialgal culture
N. ellipsosporum	Sidi-Amira	Soil core
N. microscopicum	Sidi-Amira	Soil core
N. muscorum	Sidi-Amira	Axenic culture
N. punctiforme	Sidi-Amira	Axenic culture
Nostoc sp.	Moghrane	Unialgal culture
Nostoc sp.	Sidi-Bou Rhaba	Unialgal culture
Tolypothrix tenuis	Sidi-Bou Rhaba	Unialgal culture
	Moghrane	Axenic culture

Results

Tests on a variety of blue-green algae from Morocco for a capacity to reduce acetylene to ethylene

Tests were carried out using algae isolated in culture in the laboratory, and using field populations where blue-green algae and lichens containing blue-green algae occurred. Table 16.2 summarises those algal species which showed light-dependent acetylene reduction under aerobic conditions, and the habitats from which they were obtained. All the habitats were of neutral or slightly alkaline pH, and blue-green algae in general were rare in acidic habitats. It may be noted that all the species of blue-green algae in this Table, with the exception of a red pigmented *Lyngbya* species, are heterocystous forms. Acetylene reduction by this *Lyngbya* species is considered in more detail below. No positive data were obtained for acetylene reduction by any other non-heterocystous algae although microaerobic tests were not carried out on algae apart from the *Lyngbya*, because of the unavailability of suitable anaerobic gas mixtures in Morocco.

Nitrogenase activity in terrestrial habitats

Over 500 soil samples from various terrestrial habitats in Morocco were tested for acetylene reduction and three main soil types were distinguished. In arid and semi-arid soils of central and west-central Morocco no evidence of nitrogen-fixing algae was found except in a few habitats after the rainy season, particularly round transient pools and puddles, where growths of algae occurred under stones and translucent pebbles. The occurrence of blue-green algae in such situations is similar to that found not only in other parts of Africa (Lund, 1967), but also in other desert regions of the world (Shields & Durrell, 1964). The rates of nitrogenase activity in these habitats were very low (< 1.5 nmoles $C_2H_4/cm^2/h$). During the dry season there was no evidence of nitrogenase activity in these habitats.

A second type of habitat where nitrogen-fixing algae were found was damp angiosperm-colonised ground along the coastal plain. There, blue-green algae occurred in low numbers in neutral and alkaline habitats, the dominant species being *Oscillatoria*, *Spirulina*, *Lyngbya* and *Microcoleus* and nitrogenase activity, which again was low (< 0.95 nmoles $C_2H_4/cm^2/h$), occurred in those samples which also contained species of *Nostoc*, *Anabaena* or *Nodularia*. Unlike most other habitats (see below) moisture did not seem to be the main factor limiting the distribution of algae in these situations because even on the river banks which were permanently moist, blue-green algae were rare. This may have been due to the polluted nature of many of these rivers, which contained Chlorophyceae as the dominant algae.

The other type of habitat found frequently along the coastal plain was semi-bare areas which were protected by shading from excessive sunlight. An example of this type is the coastal dune systems (e.g. between

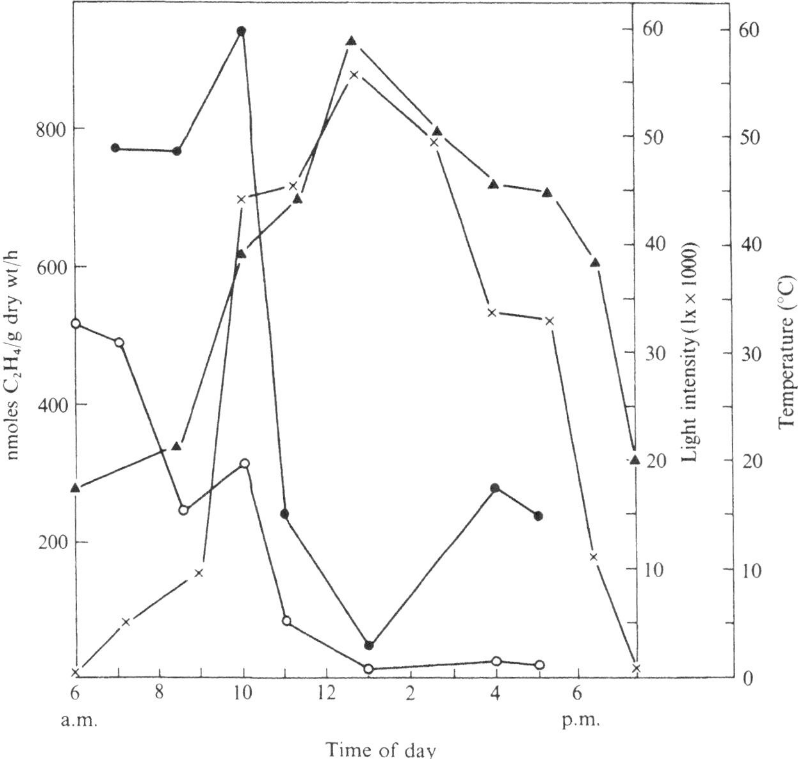

Fig. 16.1. Acetylene reduction *in situ* by untreated (○) and moistened (●) terrestrial *Nostoc* samples from Sidi-Bou Rhaba throughout the day. ×, Light intensity; ▲, temperature. Each value is the mean of triplicate determinations.

Sidi-Bou Rhaba and the sea) where good growths of free-living *Nostoc* species and the lichen *Collema* are to be found. These algae develop profusely during the rainy season, but dry up in the dry season and do not usually fix nitrogen then. During the early part of the dry season, however, nitrogenase activity occurs in the morning when the algae are wet with dew, the temperature is below 38 °C and the light intensity is below 40000 lx (Fig. 16.1). In untreated samples, activity then decreases in the morning and is negligible in the afternoon and early evening. In artificially wetted (distilled water) samples the rates of acetylene reduction are higher than in the untreated samples indicating that desiccation is an important regulatory factor. However as these samples also show a decrease in activity around mid-day and in the afternoon, this suggests that high temperatures (up to 56 °C) and/or

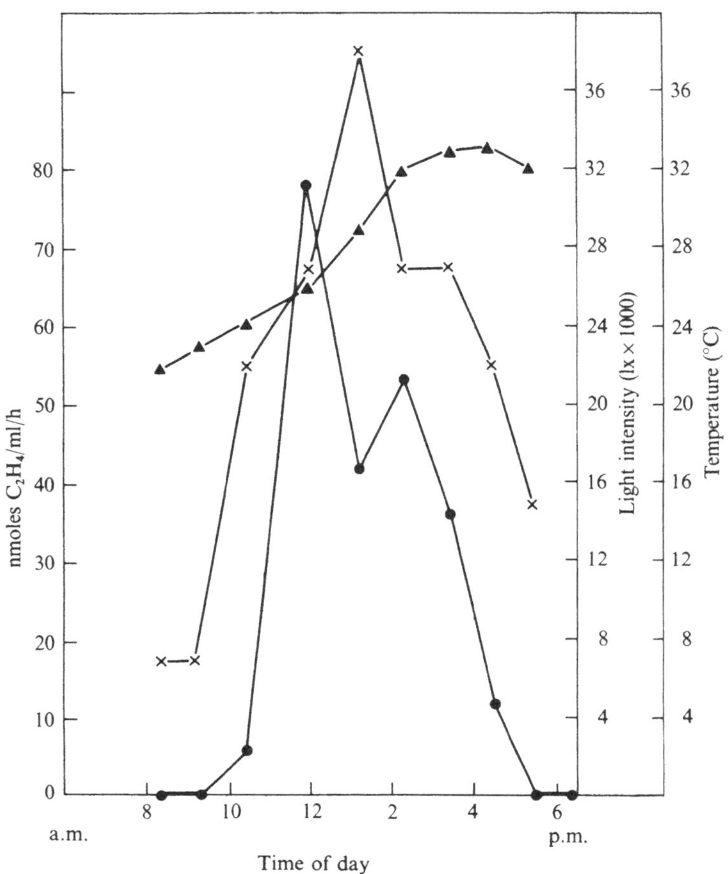

Fig. 16.2. Acetylene reduction (●) *in situ* by a mixed planktonic population of *Anabaena* sp. and *Merismopedia* sp. in Sidi Amira. The algae were taken from the 0–0.25 m layer of the water column. ×, Light intensity; ▲, temperature. Each value is the mean of triplicate determinations.

light intensity (up to 58000 lx) may also be inhibitory to nitrogenase activity.

Nitrogenase activity in freshwater lakes

Studies on nitrogenase activity in fresh water were carried out on three lakes. At Sidi Amira, a shallow forest pond, north-east of Rabat, a dense bloom of a *Merismopedia* sp. and an *Anabaena* sp. was present throughout much of June–July 1971 and was distributed evenly throughout the water column. Data obtained on acetylene reduction during the day are presented in Fig. 16.2. It is seen that activity was barely measurable at daybreak but then increased rapidly during the morning and reached a maximum just before noon. It then decreased sharply, increased slightly once more and then fell at a steady rate until at 6 p.m. activity had ceased completely. Throughout the day there was no detectable change in the chlorophyll content of the water column (1.0 m). As Fig. 16.2 shows there was a direct correlation between light intensity and nitrogenase activity except that at the highest light intensity (38000 lx) there was a drop-off in acetylene reduction, and that in the late afternoon nitrogenase activity ceased while the light intensity was as high as 15000 lux, an intensity which permitted nitrogenase activity earlier in the day. The reason for the latter cessation of nitrogenase activity is unknown.

At the second lake, Dayet er Roumi, no evidence of planktonic blue-green algae or acetylene reduction was obtained in the water column throughout the summers of 1968 and 1971, or spring, 1973. However, high rates of acetylene reduction were associated, during these periods, with an epilithic growth of *Calothrix* which was present as a mat on boulders which occurred extensively on the bottom of this lake. The growth of *Calothrix* was particularly profuse on boulders near the shallow lake edge where light penetration to the algal mats was high. These algal populations showed light-dependent acetylene reduction and the pattern of activity during the day was very similar to that obtained in the water column of Sidi Amira. As Fig. 16.3 shows, nitrogenase activity in the light increased in the submerged algal mat from dawn until around 10.30 a.m. and then decreased throughout the afternoon, apart for a short period around 4.30 p.m. when it increased briefly again. As at Dayet er Roumi there was a general direct correlation between light intensity and nitrogenase activity, apart from the period of highest light intensity where there may have been some photoinhibition of metabolism. There was negligible nitrogenase activity in the dark and the temperature of the water during the experiment did not appear to be a critical factor.

Comparative tests were also carried out on *Calothrix* present on boulders which were exposed to air above the water, and in this study a somewhat different pattern of acetylene reduction was noted during the day (Fig. 16.4). Activity paralleled that in submerged samples up to 11 a.m. although it was only 50 % as high on a unit area basis. It then decreased very rapidly between 10.30 and 12.30 probably due mainly to increasing desiccation, and then decreased at a much slower

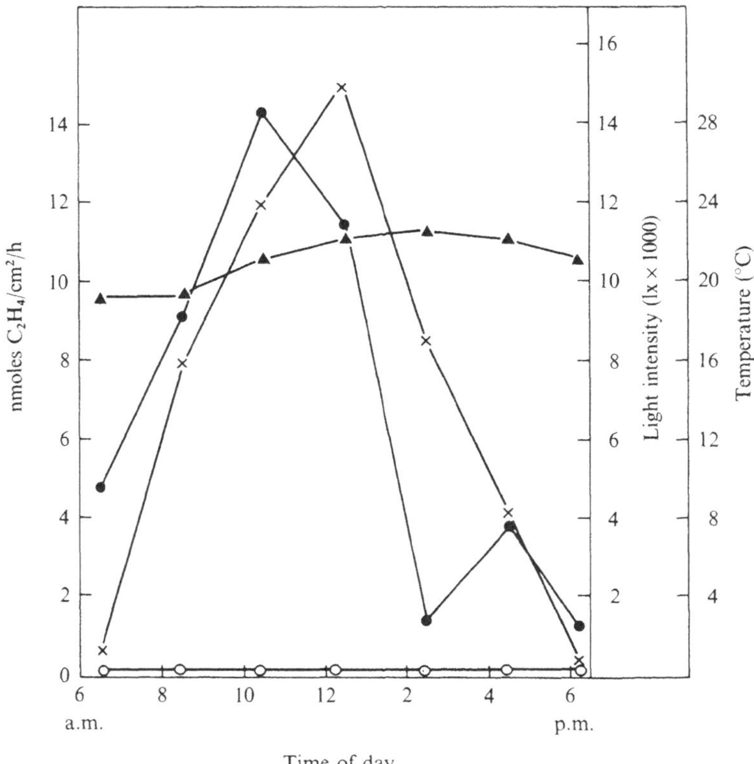

Fig. 16.3. Acetylene reduction *in situ* in the light (●) and in the dark (○) by epilithic submerged *Calothrix* at Dayet er Roumi. ×, Light intensity; ▲, temperature. Each value is the mean of triplicate determinations.

rate until 6 p.m. when activity had almost ceased. Again there was negligible nitrogenase activity in samples incubated in the dark.

The third lake studied, Sidi-Bou Rhaba, was quite different from either of the two lakes discussed above in its algal flora and in its pattern of acetylene reduction. Here low rates of acetylene reduction occurred on occasions in the water column when *Anabaena* was present, but usually *Microcystis* sp. was the dominant blue-green alga together with some simple green algae such as *Scenedesmus*. In addition, this lake had a mud bottom and there was no extensive boulder cover dominated by *Calothrix*, as occurred in Dayet er Roumi. However, a marshy area extended round the north edge of the lake and here a variety of filamentous blue-green algae occurred, with acetylene reduction being noted in eight of the nine sampling areas tested. All sample

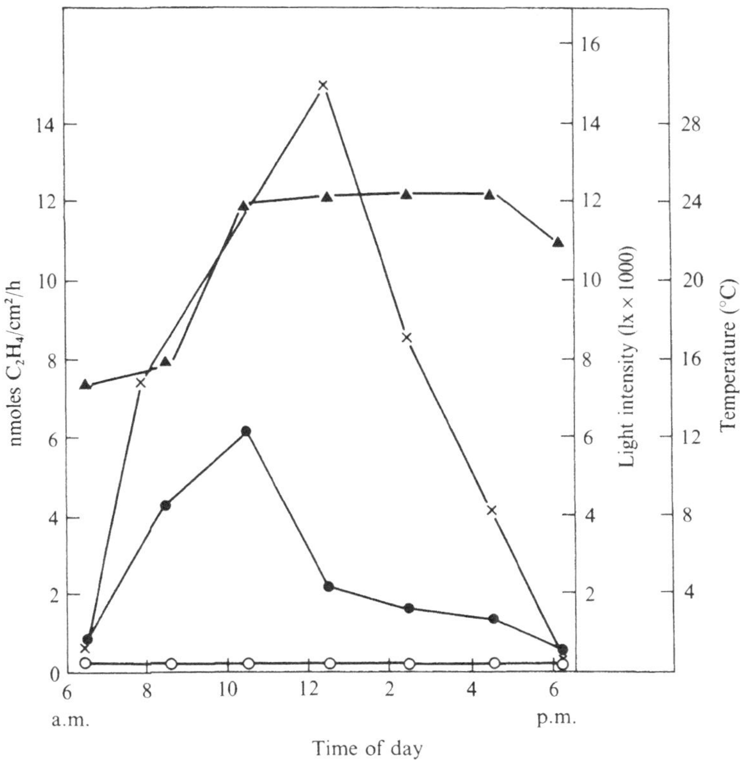

Fig. 16.4. Acetylene reduction *in situ* in the light (●) and in the dark (○) by epilithic *Calothrix* exposed to air at Dayet er Roumi. ×, Light intensity; ▲, temperature. Each value is the mean of triplicate determinations.

sites had an alkaline pH (8.1–9.5) and showed redox (Eh) values which ranged from 0.0 to −2.60 V. Acetylene reduction, which was largely light dependent, was highest in samples which contained species of *Anabaena*, *Nodularia* and *Nostoc*. Rates were negligible in samples rich in unicellular algae, and were low but very variable in samples where filamentous, non-heterocystous species predominated (Table 16.3).

In sum, therefore, we found nitrogen-fixing species of blue-green algae in the three lakes examined, but their distribution varied in each body of water. In Sidi Amira the main nitrogen-fixing species were planktonic; in Dayet er Roumi they were epilithic on boulders on the lake bottom, and in Sidi-Bou Rhaba they were associated

Table 16.3. *Acetylene reduction and nitrogen fixation by various samples of supralittoral blue-green algae*

Sampling site	Treatment	nmoles C_2H_4/g fresh weight soil/h	Genera*
1	Light	0.0	*Chroococcus turgidus*, *Spirulina*, *Arthrospira*,
	Dark	0.0	*Scenedesmus*, numerous diatoms
2	Light	0.2	*Lyngbya*, *Phormidium*, *Spirulina*, *Euglena*,
	Dark	0.0	diatoms
3	Light	25.5	*Spirulina*, *Lyngbya*, ***Anabaena***, *Phormidium*
	Dark	9.4	***Nodularia***, *Euglena*
4	Light	0.6	*Chroococcus*, *Arthrospira*, *Lyngbya*, *Phormidium*
5	Light	28.7	*Spirulina*, *Lyngbya*, *Phormidium*, ***Anabaena***, ***Nodularia***
6	Light	3.3	*Lyngbya*, *Phormidium*, *Oscillatoria*, *Scenedesmus*, diatoms
7	Light	2.9	*Phormidium*, *Oscillatoria*, ***Nodularia***, diatoms, *Euglena*, *Scenedesmus*
8	Light	3.2	*Arthrospira*, *Lyngbya*, *Oscillatoria*, *Phormidium*, *Euglena*
9	Light	0.2	*Chroococcus*, *Arthrospira*, *Lyngbya*, *Phormidium*

* Genera which fix nitrogen in air are in bold type.

particularly with mud round the edge of the lake although planktonic nitrogen-fixing *Anabaena* also occurred in the water column on occasion.

Nitrogenase activity in rice paddy soils

Rice has been grown in Morocco for many years although it is not a widespread crop. Tests for acetylene reduction were carried out on rice paddy soils in the Moghrane area. In these paddy soils, blue-green algae occurred in the soil but the dominant forms present were not heterocystous species, but species of *Phormidium*, *Lyngbya* and *Oscillatoria* and only low rates of acetylene reduction (< 1.5 nmoles C_2H_4/gm fresh weight soil/h) were obtained. On incubating such soil samples in culture medium a variety of blue-green algae were isolated including *Chroococcus turgidus*, *Lyngbya aerugineo-coerulea*, *L. aestuarii*, *L. limnetica*, *Oscillatoria formosa*, *O. pseudogeminata*, *Phormidium molle*, *P. autumnale* ,*P. papyraceus*, *Anabaena anomala*, *Gloeotrichia* sp. and *Tolypothrix tenuis*. Only the latter three species were able to reduce acetylene aerobically. Despite the low rates of acetylene reduction by the soil algae, very much higher rates (44 nmoles C_2H_4/ml/h) were obtained in the water layer above the mud where extensive floating colonies of *Gloeotrichia* species developed.

In rice paddy soil situations, it has always been assumed that the blue-green algae benefit the rice plant by supplying combined nitrogen (see Singh, 1961) and possibly growth substances (see Venkataraman, 1972) but there has been no direct demonstration using ^{15}N tracer techniques of the transfer of nitrogen from the alga to the rice. There is evidence of this, however, in other situations (Mayland & McIntosh, 1966; Stewart, 1967). This release of nitrogen by the alga is facilitated by intermittent periods of wetting and drying such as are common in paddy soils, as well as by cell autolysis. We have carried out tests using ^{15}N as tracer in the laboratory using *Westiellopsis prolifica*, a common blue-green alga from Indian paddy soils, and as the data in Fig. 16.5 show clearly at least some of the nitrogen fixed and liberated by the algae is assimilated by the rice plant. Detailed data on the uptake of ^{15}N-labelled, algal, extracellular nitrogen by the rice plant are presented elsewhere (Stewart, 1976).

Nitrogenase activity in intertidal marine habitats

Along much of the rocky west coast of Morocco there is a rich algal flora (see Gayral, 1958) and in view of the finding of nitrogen-fixing algae in intertidal marine habitats in temperate regions (see Stewart, 1972) a study was made, for comparative purposes, of nitrogen fixation by algae along the rocky coastline south-west of Rabat at El Haroura. On this shore, which contains both sheltered and exposed habitats, several nitrogen-fixing algae, either free-living or in symbiotic association, were abundant. On the softer sand-stone rocks, particularly in sheltered habitats, a rich growth of species of *Calothrix* (mainly *C. crustacea*) developed, while on the more exposed and harder rocks, there was a good growth of the black, nitrogen-fixing lichen, *Lichina pygmaea*. Nitrogenase activity was detected readily in samples of both these genera, and in addition, high but variable rates of acetylene reduction were also found to be associated consistently with a red pigmented (phycoerythrin-rich) species of the non-heterocystous alga *Lyngbya* (up to 60 nmoles/g fresh weight/min). Surprisingly, these natural algal populations reduced acetylene both under aerobic and microaerobic conditions and reminded us of the position in the red-pigmented

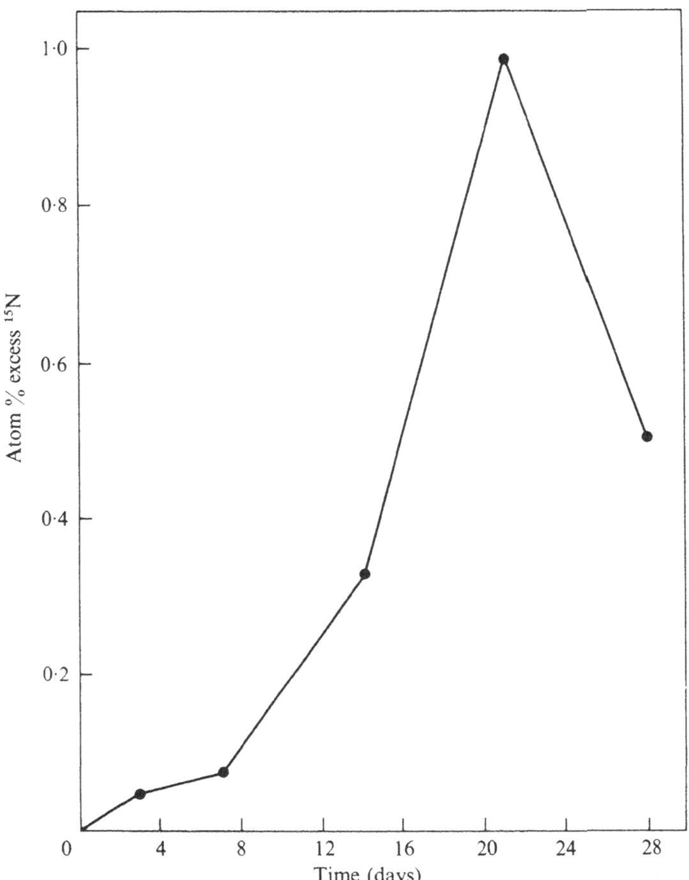

Fig. 16.5. ^{15}N labelling of *Oryzae sativa* seedlings grown in 100 cc of paddy soil to which a *Westiellopsis* culture (10 ml containing alga equivalent to 10 mg dry weight) was added at the soil surface. The ^{15}N labelling of the alga at the start of the experiment (time 0) was 25.137 atom % ^{15}N.

planktonic *Trichodesmium*, natural populations of which reduce acetylene (Dugdale, Goering & Ryther, 1962; Taylor *et al.*, 1972). Further studies on this *Lyngbya* species are required. There was no evidence of acetylene reduction by other genera of blue-green algae (*Gloeocapsa crepidum* and *Entophysalis granulosa*) or by any eukaryotic algae (Table 16.4).

The rates of acetylene reduction by *Calothrix* and *Lichina* varied markedly throughout the day, the over-riding regulatory factor by far being desiccation, with the algae and lichen drying out on the hot rock

Table 16.4. *Rates of acetylene reduction* in situ *by various algae at El Haroura*

Species	nmoles C_2H_4/g sample/ min*
Calothrix crustacea	13
Codium tomentosum	0
Enteromorpha intestinalis	0
Entophysalis granulosa	0
Gloeocapsa crepidum	0
Gracilaria sp.	0
Jania rubens	0
Lichina pygmaea	155
Lyngbya sp.	59
Pleurocapsa sp.	0
Ulva lactuca	0

* Each sample is of 1.0 g fresh weight, except for *Calothrix crustacea* where the data are expressed per square cm of *Calothrix* mat.

surface within about 1 h of uncoverage by the sea and activity restarting when the plants are submerged rapidly in the sea. This is exemplified by the results for *Lichina pygmaea* presented in Fig. 16.6. This shows that, when *Lichina* is uncovered by the sea, acetylene reduction occurs in the morning when the atmosphere is damp and the plants are fairly moist. However, with increase in temperature and light intensity during the morning, acetylene reduction drops off. This is a desiccation effect in that activity restarts within 30 min of the lichen being resubmerged, and reduces acetylene while submerged at a rate which is three times higher than the highest rate found earlier in the day. On subsequent uncoverage by the tide in the afternoon, when the desiccating effects of the sun's rays are severe, acetylene reduction rapidly falls off and activity remains negligible throughout the remainder of the uncoverage period.

Nitrogenase activity by cycad root nodules

Cycads occur in a variety of habitats in Morocco, mainly as ornamental plants (they are not native to Morocco but have been introduced). Tests showed that these plants all bore abundant root nodules at depths down to 25 cm below the soil surface, but particularly at a depth of 1–10 cm. All nodules examined contained blue-green algae and acetyl-

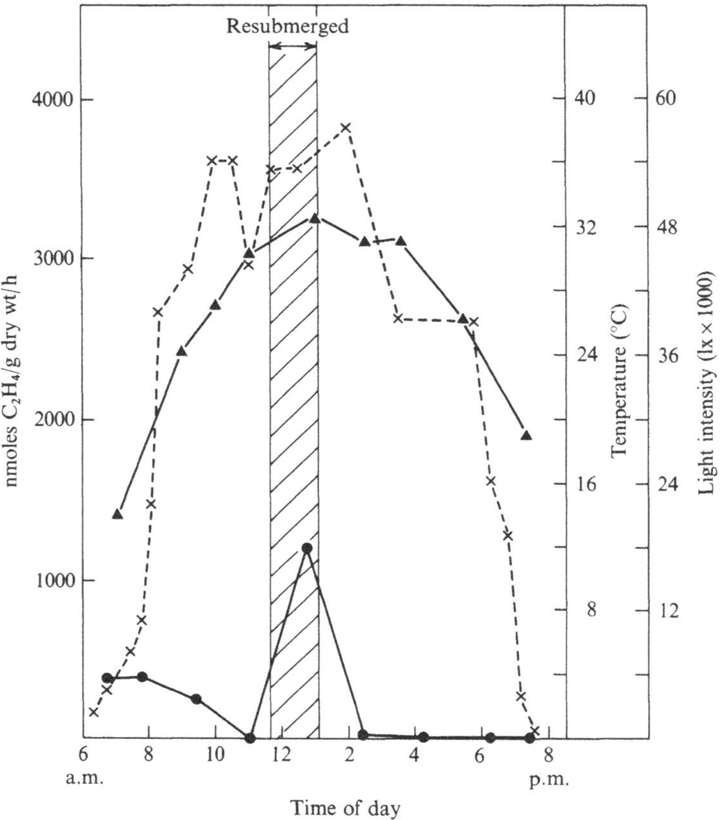

Fig. 16.6. Acetylene reduction in the field by *Lichina pygmaea* samples when uncovered and covered by the sea (●). ▲, Temperature; ×, light intensity. The plants were first uncovered by the sea at 6.0 a.m., were resubmerged by the incoming tide at 11.45 a.m. and uncovered again by the outgoing tide at 1.15 p.m. Each value is the mean of triplicate determinations.

ene-reduction tests on *Cycas revoluta* plants showed that at 25 °C in the light (3600 lx) intact nodules reduced acetylene at an average rate of 22.5 nmoles C_2H_4/gm dry weight/h. Tests using ^{15}N as tracer and carried out on *Cycas revoluta* in the laboratory have confirmed the data of Bergersen, Kennedy & Wittmann (1965) that the nitrogen fixed by these root nodules is transferred rapidly from the root nodules to other parts of the plant (Table 16.5). It is likely that the particularly healthy growth of ornamental cycads in Morocco is due in part to the available nitrogen which they receive from the symbiotic blue-green alga.

Table 16.5. *Distribution and* ^{15}N *label in various parts of* Cycas *after exposure of the root nodules to* $^{15}N_2$

Plant material	Sample no.	Atom % ^{15}N	Atom % excess ^{15}N
Nodules	1	1.206	0.840
	2	1.654 (1.301)	1.288 (0.934)
	3	1.042	0.676
Roots	1	0.649	0.282
	2	0.422 (0.518)	0.055 (0.151)
	3	0.483	0.116
Leaf bases	1	0.485	0.119
	2	0.397 (0.436)	0.031 (0.070)
	3	0.424	0.059
Leaf lamina	1	0.374	0.007
	2	0.382 (0.379)	0.015 (0.12)
	3	0.380	0.013

The initial gas phase was $A/N_2/O_2/CO_2$: 59.96/20.00/20.00/0.04, v/v; the ^{15}N label of the N_2 was 94 atom % ^{15}N; the exposure period was 43 h. Unexposed samples served as control.

Discussion

The above results provide preliminary information on the distribution, and nitrogenase activity, of blue-green algae in certain habitats in Morocco. These show that blue-green algae are common and locally abundant in a variety of habitats, particularly in areas of western Morocco where moisture is regularly available, for example, along the rocky coast-line, in lakes, along the edges of lakes, etc. In other areas they develop only during the wet season. Of the nitrogen-fixing algae which we found, the most widespread and efficient in fixing nitrogen were species of *Nostoc*. These occur in terrestrial habitats, either free-living, or in symbiotic association in *Collema* or cycad root nodules. Two other common nitrogen-fixing algae belonged to the family Rivulariaceae. Thus *Calothrix* species occurred free-living as an epilithic mat, in the marine intertidal region, as well as in symbiotic association in the lichen *Lichina*, and *Gloeotrichia* species, which occur in rice paddy fields, were also abundant. The occurrence of *Gloeotrichia* in such habitats is unusual, although it is characteristic of certain eutrophic waters in temperate regions, e.g. Lake Mendota, where it also fixes nitrogen (Stewart *et al.*, 1967).

Despite the similarity of the various algal species and the general types of habitats where they are found, to those of temperate regions, there is a striking difference in the diurnal pattern of nitrogenase activity. In temperate regions activity is usually low in the morning, increasing to a maximum around mid-day and in the early afternoon, and then decreasing during late afternoon (see e.g. Hitch & Stewart, 1973). In Morocco, on the other hand, the diurnal pattern is governed largely by desiccation and possibly by inhibitory temperatures and light intensities around mid-day. Thus high rates of nitrogenase activity occur in the early morning, possibly during the night, and to a lesser extent in the evening when the algae are moist with dew, and not around mid-day and in the afternoon as in temperate regions. We have noticed this in terrestrial habitats (Fig. 16.1), in lakes (Figs. 16.2–16.4) and in marine habitats when the algae are not covered by the sea (Fig. 16.6).

Because of the extreme environments, and desiccation in particular, it appears that apart from during the wet season, and possibly along coastal regions, nitrogen fixation by blue-green algae is unlikely to be a major source of nitrogen input to terrestrial habitats at present. Nevertheless, the algae are present, they do fix nitrogen when moist, and it is probable that as more efficient use of water is made in Morocco, through irrigation, an additional benefit will be the increased input of newly-fixed nitrogen which will result from algal nitrogen fixation. However, whether this input of nitrogen will match the increased crop yields which will also result from irrigation is questionable.

This study was made possible by British Council support, which enabled Professor W. D. P. Stewart and Dr H. W. Pearson to visit Morocco to initiate acetylene-reduction studies there.

References

Allen, M. B. & Arnon, D. I. (1955). Studies on nitrogen-fixing blue-green algae. I. Growth and nitrogen fixation by *Anabaena cylindrica* Lemm. *Pl. Physiol.*, **30**, 366–72.

Bergersen, F. J., Kennedy, G. S. & Wittmann, W. (1965). Nitrogen fixation in the coralloid roots of *Macrozamia communis*. *Aust. J. biol. Sci.*, **18**, 1135–42.

Burgis, M. J., Darlington, J. P. E. C., Dunn, I. G., Ganf, G. G., Gwahaba, J. J. & McGowan, L. M. (1973). The biomass and distribution of organisms in Lake George, Uganda. *Proc. R. Soc. Lond. B*, **184**, 271–98.

Carmouze, J.-P., Dejoux, C., Durand, J. R., Gras, R., Iltis, A., Lauzanne, L., Lemoalle, L., Leveque, C., Loubens, G. & Saint-Jean, L. (1972). Grandes zones écologiques du lac Tchad. *Cah. off. Rech. Sci. Tech. Outre-Mer, ser. Hydrobiol.*, **4**, 103–69.

Dugdale, R. C., Goering, J. J. & Ryther, J. H. (1964). High nitrogen fixation rates in the Sargasso Sea and the Arabian Sea. *Limnol. Oceanogr.*, **9**, 507–10.

Hitch, C. J. B. & Stewart, W. D. P. (1973). Nitrogen fixation by lichens in Scotland. *New Phytol.*, **72**, 509–24.

Horne, A. J. & Viner, A. B. (1971). Nitrogen fixation and its significance in tropical Lake George, Uganda. *Nature, Lond.*, **232**, 417–18.

Lund, J. W. G. (1967). Soil algae. In *Soil Biology* (ed. Burges, A. and Raw, F.), pp. 129–48. Academic Press: London, New York.

Mayland, H. F. & McIntosh, T. H. (1966). Availability of biologically fixed nitrogen-15 to higher plants. *Nature, Lond.*, **209**, 421–2.

Prowse, G. A. & Talling, J. F. (1958). The seasonal growth and succession of plankton algae in the White Nile. *Limnol. Oceanogr.*, **3**, 222–38.

Roger, P. (1973). *Recherches préliminaires sur les Cyanophycées des sols de rizière du Sénégal. Rapport de stage 1971–72*, Orstom, Dakar, publ. roneo., diff. limitée.

Roger, P. & Jacq, V. (1972). *Rapport préliminaire concernant un antagonisme entre des Bactéries sulfato-réductrices et une souche de Cyanophycée dans la rhizosphère et la spermosphère du riz.* Orstom, Dakar, pub. roneo., diff. limitée.

Shields, L. M. & Durrell, L. W. (1964). Algae in relation to soil fertility. *Bot. Rev.*, **30**, 92–128.

Singh, R. N. (1961). *Role of Blue-Green Algae in Nitrogen Economy of Indian Agriculture.* Indian Council of Agricultural Research: New Delhi.

Stephens, E. L. (1949). *Microcystis toxica* sp.nov.: a poisonous alga from the Transvaal and Orange Free State. *Tr. R. Soc. South Africa*, **32**, 105.

Stewart, W. D. P. (1967). Transfer of biologically fixed nitrogen in a sand-dune-slack region. *Nature, Lond.*, **214**, 603–4.

Stewart, W. D. P. (1972). Nitrogen fixation in the sea. In *Fertility of the Sea* (ed. Costlow, J. D.), pp. 537–64. Gordon and Breach Science Publishers Ltd: London, England.

Stewart, W. D. P. (1976). Nitrogen fixation by free-living micro-organisms of tropical and sub-tropical soils. I: Interactions between the nitrogen-fixing alga *Westiellopsis prolifica* Janet and the rice plant *Oryzae sativa* L. (in preparation).

Stewart, W. D. P., Fitzgerald, G. P. & Burris, R. H. (1967). *In situ* studies on N_2 fixation using the acetylene reduction technique. *Proc. natn. Acad. Sci. USA*, **58**, 2071–8.

Stewart, W. D. P., Mague, T., Fitzgerald, G. P. & Burris, R. H. (1971). Nitrogenase activity in Wisconsin lakes of differing degrees of eutrophication. *New Phytol.*, **70**, 497–509.

Steyn, D. G. (1945). Poisoning of animals and human beings by algae. *S. Afr. J. Sci.*, **41**, 243.

Taha, M. S. (1963*a*). Isolation of some nitrogen-fixing blue-green algae from the rice fields of Egypt in pure cultures. *Mikrobiologiya*, **32**, 493–7.

(1963*b*). The effect of concentration of different components of the medium on growth and nitrogen-fixation by blue-green algae. *Mikrobiologiya*, **32**, 582–9.

(1963*c*). The influence of the hydrogen ion concentration in the medium, and the temperature on the growth and nitrogen fixation by blue-green algae. *Mikrobiologiya*, **32**, 968–72.

(1964*a*). The effect of nitrogen compounds upon growth of blue-green algae and fixation of molecular oxygen by them. *Mikrobiologiya*, **33**, 397–403.

(1964*b*). Effect of light on growth and nitrogen-fixation by blue-green algae. *Pl. Physiol.*, USSR, **11**, 424–31.

Taha, E. E. & El-Refai, A. H. (1962). Physiological and biochemical studies on nitrogen-fixing blue-green algae. I: On the nature of cellular and extracellular nitrogenous substances formed by *Nostoc commune*. *Arch. Mikrobiol.*, **41**, 307–12.

(1963*a*). On the nitrogen fixation by Egyptian blue-green algae. *Z. Allg. Mikrobiol.*, **3**, 382–8.

(1963*b*). Physiological and biochemical studies on nitrogen fixation by blue-green algae. III. The growth and nitrogen-fixation of *Nostoc commune* as influenced by culture conditions. *Arch. Mikrobiol.*, **44**, 356–65.

Talling, J. F. (1957). Diurnal changes of stratification and photosynthesis in some tropical African waters. *Proc. R. Soc. Lond. B*, **147**, 56–83.

Talling, J. F. & Rzoskà, J. (1967). The development of plankton in relation to hydrological regime in the Blue Nile. *J. Ecol.*, **55**, 637–62.

Traore, K. (1973). Première contribution à l'étude floristique des Cyanophytes du Mali – Thèse de Doctorat de spécialité, Centre pédagogique supérieur, Bamako, Mali.

Traore, T. (1973). Recherches préliminaires sur la fixation de l'azote atmosphérique dans quelques sols maliens. Thèse de Doctorat de spécialité, Centre pédagogique supérieur, Bamako, Mali.

Venkataraman, G. S. (1972). *Algal Biofertilizers and Rice Cultivation*. Today and Tomorrow's Printers & Publishers: New Delhi, 75 pp.

PART III

The acetylene-reduction technique

17. The acetylene-reduction technique

R. H. BURRIS

The acetylene-reduction technique has been used extensively in the last five years as an index of biological nitrogen fixation. In evaluating its present application it may be useful to look back to the origin of the method.

Wilson & Umbriet (1937) reported that hydrogen inhibited biological nitrogen fixation, and this was the first indication that a gas other than N_2 could bind to nitrogenase. Subsequently carbon monoxide and nitrous oxide also were shown to be gaseous inhibitors of nitrogenase, but it was not until 1954 that Mozen & Burris (1954) reported the reduction of nitrous oxide as the first example of reduction of a substrate other than N_2 by nitrogenase. Hoch, Schneider & Burris (1960) expanded on these observations and Lockshin & Burris (1965) demonstrated that cell-free extracts from *Clostridium pasteurianum* could reduce nitrous oxide.

During 1965 and early 1966, R. Schöllhorn was working in my laboratory and observed that azide and acetylene also were reduced by extracts from *C. pasteurianum*. These observations were reported informally at a conference at the Sagehen Field Station of the University of California on 1 October 1965, and at the Federation of American Societies for Experimental Biology (FASEB) in April 1966 (Schöllhorn & Burris, 1966).

Late in December of 1965 we mailed our abstract of Schöllhorn's FASEB paper after having received a letter early in December from M. J. Dilworth which he had written 30 November; Dilworth had spent the previous year in my laboratory. In this letter he stated:

> Things are moving well here now, although in a quite unexpected direction. I found about two months ago that acetylene is a very potent inhibitor of nitrogen fixation with clostridial extracts using either pyruvate or H_2 as the substrate. It is about as effective as CO – 100 % inhibition at about 1 % C_2H_2 and about 50 % with 0.35 % C_2H_2. It is reversible, at least up to 30 min preincubation before changing the gas phase. Inhibition is apparently non-competitive with respect to N_2 for either the H_2 or pyruvate systems, but there are some problems here.

To my great surprise, I found that clostridial extracts can reduce C_2H_2 with H_2 – the product is ethylene (confirmed by a mass peak at 28 and by its infrared spectrum). Ethylene itself is inert – it is not reduced and inhibits neither N_2 nor C_2H_2 reduction at levels up to 0.4 atm.

The exciting part is that all the requirements for C_2H_2 reduction are the same as for N_2 reduction – absolute requirements for enzyme, H_2, acetyl-P, ATP, Mg^{++} and ferredoxin. I should add that the nitrogen-fixing system is responsible for C_2H_2 reduction because (a) the cofactor needs are the same, (b) extracts of ammonia-grown cells do not reduce acetylene, and (c) CO, which is not itself reduced, inhibits C_2H_2 reduction at the same pressures as it inhibits N_2 fixation. The ATP consumed per H_2 molecule activated is only about half that for N_2 reduction. So at last we have an intermediate which DOES dissociate from the nitrogenase.

The non-competitive inhibition is surprising, and I'm still working on this aspect, but I think the results are correct. There is a slight but definite depression of H_2-mediated ATP hydrolysis in the presence of acetylene, but this is very small compared to the effects on fixation of N_2. Maybe this is obscuring the picture – I don't know yet.

I am about to have a go at studying the actual reduction mechanism – using D_2O and H_2, it should be possible to establish the structure of the resulting ethylene by infrared methods. I have also to attempt a K_m determination for C_2H_2 – difficult because it appears to be about 0.01 atm and I have no way to measure reduction except manometry and this isn't exactly suited to measuring uptakes at those pressures. Anyway, I hope to tidy up these aspects and write this up before Christmas, but the schedule may be a bit tight. I wished I had got onto this earlier in the year – might have made the N_2 fixation meeting in California somehow.

On 17 December I wrote to M. Dilworth to indicate as follows that we also had observed acetylene reduction:

It was most interesting to find that you had observed the inhibition by acetylene. We ran across the same observation six weeks or so ago. [Inhibition was first seen by Schöllhorn on 9 August 1965.] Dr. Schöllhorn from Bonn, Germany is working in my lab this year. He is a hydrazine chemist and is interested in intermediates. This fall we obtained a tank of acetylene because he thought it should serve as an inhibitor and in fact he did find inhibition with it. However, the acetylene disappeared suggestive that it was being reduced. He had not characterized the reduction product but suspected it was ethylene and ordered a special chromatographic column to achieve a separation from other gases. So far the column has not arrived and as a result he has not checked the product.

You indicate that the inhibition is noncompetitive. This also is Schöllhorn's impression although he has found the inhibition pattern to be a bit confusing. I suppose this is partly the result of the loss of the inhibitor during the course of the reaction since it is present in relatively low partial pressures.

Schöllhorn has not yet checked ethylene although we also had a tank of this material on order. It was very interesting to hear that you find that it is not inhibitory. Schöllhorn has not followed up the observations as completely as you have at this stage. We will be particularly curious to carefully check the nature of the inhibition as one *a priori* would guess that it should be competitive.

I will keep you informed of any further developments on the acetylene frontier.

We have no intention of rushing anything into print immediately on acetylene, so rest assured on that score. In any case we should keep each other informed so that we can be mutually helpful.

Mike Dilworth replied on 21 December with more details:

I must admit to being very surprised to find that you had also been working with acetylene – I thought I was the only one crazy enough to try it! However, when you mentioned that Dr. Schöllhorn was a hydrazine chemist it started to make pretty good sense.

Thank you for the information concerning the inhibition pattern – I fear it means more work for both of us. I must say I was very surprised to see non-competitive inhibition as, like you, I expected competition and simple kinetics would predict it provided the electron supply is not limited. However, none of my runs has given a hint of competition.

To give you a few more details on mine – it was all done in 30 min assays in serum bottles in the usual way, with $pC_2H_2 = 0.0035$ atm, $pH_2 = 0.1$ atm and pN_2 from 0.04 to 0.2 atm. The inhibition was between 35 and 40 % and the control K_m was 0.1 atm N_2. I haven't gone as high as 7 mm C_2H_2, mainly because the inhibition at this level has been too high for convenience. However, I do know that the acetylene I have been using has a small amount of an unknown contaminant with a mass peak of 43 (almost certainly $CH_3.CO^+$ from the acetone in the acetylene), but unless this is far more potent than acetylene itself I can't see this producing the difference. Nevertheless I will remove this and see what that brings.

As I told you, I was interested to know how the reduction took place, so I ran the reaction in D_2O under C_2H_2/H_2 and measured the infra-red spectrum. Fortunately, it turns out to be a pure *cis*-$C_2H_2D_2$ which is produced, for which the bands are very easy to distinguish. Although the main bands for the *trans*-isomer would be obscured by acetylene, one of them would be visible and this was certainly not produced.

I have also done the K_m determinations for C_2H_2 and H_2 for acetylene reduction. As I thought the K_m for acetylene is very low – between 0.007 and 0.01 atm, which has made precise determination very difficult. This makes it very likely that acetylene is consumed when it is used as an inhibitor of N_2 fixation. The K_m for H_2 is about 0.06 atm, which checks fairly well with the value for N_2 reduction (about 0.1 atm).

As to publication – I will hold off submitting my paper for the time being, and do some more work on the inhibition pattern. If you and Dr. Schöllhorn can sort out the inhibition to your own satisfaction, we can compare notes again and see if we get agreement. I can easily drop the inhibition stuff out of my paper, and it could then be submitted from Madison at the same time I submit mine.

On 31 December 1965 I sent M. Dilworth a copy of our FASEB abstract and wrote:

I also am enclosing an abstract of a paper which Schöllhorn will give at the Federation meetings in the middle of April this year. You will notice that he is confident enough to stick his neck out about the nature of the inhibition.

I have been thinking a good bit about the proper publication of this material. What do you think of the possibility of submitting a short Dilworth, Schöllhorn, Burris paper to PNAS, such a paper to be concerned only with the inhibition by acetylene? You then could submit all of your other work independently and the material that we both have chanced upon would be published very promptly. The thing that it involves is attaining a meeting of the minds as to the nature of the inhibition.

The things that have occurred to us regarding the discrepancies in the nature of the inhibition are as follows:

First, Schöllhorn has always used the acetylphosphate H_2 system for supporting nitrogen fixation. If you have been using the pyruvate system, this conceivably could introduce some differences. Both of you have been using about 30 minutes exposure so there should be no substantial difference there. One thing that may well be bothersome is the disappearance of some of the acetylene during the course of the reaction. It may be necessary to go to somewhat shorter times to control this factor. The levels of acetylene you have been using are very close to those in which Schöllhorn finds competitive inhibition. You will remember from our earlier letter that when he operated at 7 mm that the data were rather erratic and certainly not sufficient to establish the competitive nature of the inhibition. 2–3 mm is better.

Schöllhorn has operated in serum bottles in the same fashion as you have. We will be interested to hear whether your purified acetylene gives any different reaction. We purchased the purest material obtainable from a commercial supplier. Schöllhorn's initial tests were done with welding acetylene and he had a very strong inhibition but did not pursue the matter at that point. As a matter of fact, these first experiments had such a high level of acetylene that inhibition was almost complete.

We were very interested to hear about your experiments with D_2O. It is indeed significant that the deuterium is added in the *cis* position suggesting that the acetylene is rigidly held during the reduction process. This is particularly significant since a comparable type of test cannot be done with a comparable nitrogenous compound.

Please let me have your reaction to the suggestion of a joint publication in PNAS. If we could check the data during January, we could have the material published before the time of the Federation meetings this spring. I would like to be perfectly fair to both you and Schöllhorn on this matter and a joint publication on the inhibition observations might serve the trick.

The correspondence was continued with a letter from Dilworth on 17 January 1966. With this letter he sent a complete protocol of an experiment and a figure which was very convincing for the non-competitive nature of C_2H_2 inhibition of N_2 reduction.

Many thanks for your letter and the abstract of the paper for the Federation meetings, not to mention the whereabouts of the Yellow Springs symposium.

Your suggestion about the publication of the acetylene material seems to me very generous indeed, and is accepted with pleasure. However, there seems to be quite a lot to do before then.

To clear up one or two points first – I too have mostly used the H_2/acetyl phosphate system for this work – in fact I have only done two experiments with the pyruvate system, one of which was no good, and the other showed non-competitive inhibition.

I have just done another couple of experiments on the inhibition. The acetylene used was a welding grade, but guaranteed by the manufacturers as at least 99.5 % C_2H_2, with the major impurity CO_2. This I passed through alkaline pyrogallol to remove CO_2, some acetone and H_2S (though I have never detected any in the mass spectrometer) and then through conc. H_2SO_4 to remove acetone and any ammonia. Our mass spectrometer is shut down for the moment, so I haven't any analytical data on this preparation, but it should be pretty clean especially as I ran 2 litres to waste before collection.

This made no difference whatever to the inhibition – I have enclosed my complete notes and a copy of the Lineweaver–Burk plot for this run. As you will see, I used a high pC_2H_2 this time (0.01 atm) to give inhibition of at least 65 %. The K_m for N_2 is rather higher than usual (0.25 atm) but otherwise I can see nothing wrong with the experiment. Both lines were computed by the least squares method and are significant to better than 0.1 %.

One thing I have noticed, and which I think I mentioned before – the presence of acetylene, even at low pressures, does slightly depress acetyl phosphate consumption by the extracts. The effect is small, about 10 %, but in the experiment attached it was statistically significant to a 't' test to better than 0.1 %.

As I see it, there are two possible sources of difference: (1) differences in extract – in enzyme balance which might give different kinetic interpretations, (b) differences in gases – which I now find hard to buy.

How about sending me some of your clostridial material to test under my conditions, so that at least we can be sure just what the variable factor is? I would do the reciprocal cross, except that I am very low on cells at the moment. As I am still mystified as to how this situation arises when I know that the acetylene is reduced, this seems to be the most appropriate thing to try. Perhaps we should both try short term experiments (15 min) and see if this gives anything different.

I was intrigued about the azide reduction you and Schöllhorn have found – is this being followed up at the moment? I checked the stoichiometry using manometric assays, and the H_2 uptake and NH_3 production certainly agree remarkably well with a complete reduction, although I didn't measure azide.

My reply on 26 January was as follows:

The discrepancy in results gets more and more puzzling. This is particularly true since Schöllhorn now has run experiments with the *Azotobacter* preparation and has found competitive inhibition with it. I will enclose a copy of his data on the *Azotobacter* run.

Your data certainly look very good and obviously display noncompetitive inhibition. We will send you some cells so that you may try this as a source of the difference. My own reaction is that the difference may lie in the gases rather than in the cells that are being used. Not only will we send you some cells, but we also will send you a sample of our acetylene gas enclosed in glass. This is a high purity Matheson product. Unfortunately, acetylene is not bottled in a conveniently small lecture bottle or we would send you one. The special packaging required for acetylene limits their cylinder sizes to a couple of somewhat larger units for the high purity gas. If necessary you could order such a cylinder from Matheson but the small amount we will send you should be sufficient for quite a number of tests.

Dr. Schöllhorn wonders about the possibility of impurities in the form of phosphine or diphosphine P_2H_4. Apparently these are generated from calcium carbide in the production of acetylene. He thinks they might be removed from gas by scrubbing through a strong solution of potassium permanganate.

Schöllhorn has been studying the azide reduction and as you indicate there is a nearly stoichiometric production of ammonia from it. This is quite different from the usual chemical reaction of azide in which N_2 is produced.

To cover our embarrassment at a delay in shipping the promised C_2H_2, I wrote 18 February and mentioned our intention to try C_2H_2 reduction for detecting N_2 fixation.

I suppose you have been waiting for that shipment of cells and gas. Just after I wrote you, Schöllhorn had to make a quick trip to New York to arrange his next stop in postdoctoral training at Columbia University. When he returned and was ready to fill the flask for you, it turned out that the tank of acetylene was empty. We immediately shipped it to have a refill but have not received the new tank as yet. As soon as it arrives we will air mail some of the gas to you.

I am sorry for the delay but we will try to get the materials to you as promptly as possible now.

We have some ^{14}C-labelled acetylene on order for use in studying the mechanism. We also intend to try this as a means for detecting nitrogen fixation. If one can convert the ^{14}C-labelled acetylene to ethylene and then recover the ethylene by gas chromatography, it should make a sensitive means of detecting fixation. Obviously one has to carefully establish that the same enzymes are involved but in some cases this might be a useful additional tool for detecting weak fixation.

There were further letters from Mike Dilworth on 23 February, 16 March and 17 June and from me on 8 and 22 March, 19 May and

8 July. They are interesting to me and perhaps to Mike but I will not burden you with them. They involved further speculation as to the basis for discrepancies in our observations, and a decision on Mike's part to submit his paper to *Biochimica et Biophysica Acta* (Dilworth, 1966) but to omit discussion of inhibition with the thought that data could be published jointly later. The cells and C_2H_2 we sent to Dilworth did nothing to clarify the issue. We did not resolve the discrepancy on inhibition as Dr Schöllhorn left our lab in early March of 1966. Subsequently several others in my lab checked the inhibition and found that Mike had been correct, and we had been wrong (Schöllhorn & Burris, 1967); the inhibition is clearly noncompetitive.

In any case, each group observed inhibition by acetylene independently. Although Mike Dilworth had spent the previous year in my lab., there was absolutely no hint that anything said or done there triggered his concept of acetylene reduction. He and Schöllhorn each clearly conceived the logical possibility of the inhibition by himself.

While all this was occurring, R. W. F. Hardy and his colleagues had been investigating reduction of cyanide by nitrogenase (Hardy & Knight, 1967) and I mentioned in my 8 March 1966, letter to Dilworth, 'I had a chat with Hardy on the phone the other day and he indicates that they also are working on reduction processes. They find that the enzyme system is relatively non-specific and will reduce cyanide. Acetylene, cyanide, azide are reduced and I don't know what other ones they may be working on.'

Hardy discussed with H. Evans and W. Silver the possibility of following cyanide reduction as a measure of nitrogenase activity. (The correspondence would be an interesting addition to that of Dilworth and Burris.) Methane produced by reduction of cyanide could be detected by gas chromatography. The reason cyanide has never been accepted for this use is that it is reduced much more slowly than N_2 because it inhibits nitrogenase. When acetylene was recognized as a substrate for nitrogenase, its superiority over cyanide was apparent and attention was shifted to it. As a flame ionization detector was the standard and highly sensitive detector for hydrocarbons in gas chromatographic effluents, Hardy suggested to Evans and Silver that they try gas chromatographic apparatus with a flame ionization detector to measure ethylene formed from acetylene. Before Silver had assembled the apparatus to test cyanide reduction, Hardy suggested to him that acetylene probably would be a superior substrate to try.

Evans, likewise concentrated his efforts on acetylene rather than cyanide.

The first recorded application of the acetylene-reduction technique for measuring N_2 fixation was the paper by Koch & Evans (1966). Shortly thereafter Sloger & Silver (1967) and Stewart, Fitzgerald & Burris (1967) used the method, and Hardy & Knight (1967) in their paper discussing reduction of azide and cyanide by nitrogenase stated: 'Utilization of the reduction of HCN to CH_4, of $H^{14}CN$ to $^{14}CH_3NH_2$ or of C_2H_2 to C_2H_4, and detection of CH_4 and C_2H_4 by hydrogen flame ionization after gas chromatography or detection of $^{14}CH_3NH_2$ may provide a sensitive new assay for detection of the N_2-fixing system.' The Hardy & Knight manuscript was received on 24 October 1966 by *Biochimica et Biophysica Acta.* The paper by Hardy, Holsten, Jackson & Burns (1968) thoroughly discussed the acetylene-reduction method.

From these rather modest beginnings we have seen the adoption of the acetylene-reduction method as a major tool for study of N_2 fixation both in the field and in the laboratory. It has its limitations but its wide acceptance attests to its usefulness.

References

Dilworth, M. J. (1966). Acetylene reduction by nitrogen-fixing preparations from *Clostridium pasteurianum. Biochim. biophys. Acta*, **127**, 285–94.

Hardy, R. W. F., Holsten, R. D., Jackson, E. K. & Burns, R. C. (1968). The acetylene–ethylene assay for N_2 fixation; laboratory and field evaluation. *Pl. Physiol.*, **43**, 1185–1207.

Hardy, R. W. F. & Knight, E., Jr (1967). ATP-dependent reduction of azide and HCN by N_2-fixing enzymes of *Azotobacter vinelandii* and *Clostridium pasteurianum. Biochim. biophys. Acta*, **139**, 69–90.

Hoch, G. E., Schneider, K. C. & Burris, R. H. (1960). Hydrogen evolution and exchange, and conversion of N_2O to N_2 by soybean root nodules. *Biochim. biophys. Acta*, **37**, 273–9.

Koch, B. & Evans, H. J. (1966). Reduction of acetylene to ethylene by soybean root nodules. *Pl. Physiol.*, **41**, 1748–50.

Lockshin, A. & Burris, R. H. (1965). Inhibitors of nitrogen fixation in extracts from *Clostridium pasteurianum. Biochim. biophys. Acta*, **111**, 1–10.

Mozen, M. M. & Burris, R. H. (1954). The incorporation of ^{15}N-labelled nitrous oxide by nitrogen fixing agents. *Biochim. biophys. Acta*, **14**, 577–8.

Schöllhorn, R. & Burris, R. H. (1966). Study of intermediates in nitrogen fixation. *Fedn Proc.*, **25**, 710.

Schöllhorn, R. & Burris, R. H. (1967). Acetylene as a competitive inhibitor of N_2 fixation. *Proc. natn. Acad. Sci. USA*, **58**, 213–16.

Sloger, C. & Silver, W. S. (1967). Biological reductions catalyzed by symbiotic nitrogen-fixing tissues. *Bacteriol. Proc.*, p. 112.
Stewart, W. D. P., Fitzgerald, G. P. & Burris, R. H. (1967). *In situ* studies on N_2 fixation, using the acetylene reduction technique. *Proc. natn. Acad. Sci. USA*, **58**, 2071–8.
Wilson, P. W. & Umbriet, W. W. (1937). Mechanism of symbiotic nitrogen fixation: III. Hydrogen as a specific inhibitor. *Arch. Mikrobiol.*, **8**, 440–57.

Comment by Dr Michael J. Dilworth

Michael J. Dilworth paid tribute to Dr Malcolm Winfield, who had suggested the possibility of using acetylene as an inhibitor of nitrogen fixation several years prior to Dilworth's experiments. Dilworth added that a series of unsuccessful experiments had been the stimulus necessary to make him try it out.

18. Recent studies using the acetylene-reduction technique as an assay for field nitrogen fixation levels

E. A. PAUL

Advances in science are as often limited by the availability of simple, reliable techniques to test hypotheses as by the lack of new concepts. This statement is well documented by the surge of information resulting from the application of the acetylene assay to nitrogen fixation studies. The ease and applicability of this assay have stimulated field studies in a wide array of disciplines, and the different environments now under investigation have required an adaptability in sampling techniques and canopy design at least equal to the versatility shown by the enzymes involved in the process. The application of the acetylene–ethylene assay and the significance of asymbiotic nitrogen fixation have been reviewed in detail by Hardy, Burns & Holsten (1973), Hardy & Holsten (1972), and Knowles (personal communication). Their data show that most habitats contain some sites for nitrogen fixation. The Russian literature also contains an increasing number of publications on both the methodology and results obtained with the acetylene–ethylene assay (Skalon, 1971; Fedorova, Milekhina, Ilyukhina & Brazhnikov, 1973; Maltseva, 1973; Trepachev, Atrashkova & Khabarova, 1971).

Recent studies on terrestrial sites range from Devon Island in the Arctic (Stutz & Bliss, 1973) to the Orkney Islands in the Antarctic (Horne, 1972). The extent of nitrogen fixation and its significance in the ecosystem are being studied in: new volcanic soil (Henriksson, Henriksson & Pejler, 1972), subarctic mire (Granhall & Selander, 1973), peat (Waughman & Bellamy, 1972), the epiphytes of sea grasses (Goering & Parker, 1972), the Sargasso Sea (Carpenter, 1972), lakes (Granhall & Lundgren, 1971; Horne, Dillard, Fujita & Goldman, 1972; Rusness & Burris, 1970; Vanderhoef, Dana, Emerich & Burris, 1972; Mague & Burris, 1973), fish culture ponds (Sugahara, Swada & Kawai, 1971), estuarine environments (Brooks, Brezonik, Putnam & Keirn, 1971) and coral reefs (Wiebe, personal communication).

The acetylene-reduction technique

The nature of fixation dictates that activity is concentrated in microsites in nearly all ecosystems, and that it varies greatly on a seasonal or even on a daily basis. Determination of the site of fixation is relatively simple with the acetylene–ethylene assay. However, extrapolation of the data to an area or ecosystem basis is difficult. The following examples of fixation studies with special reference to asymbiotic fixation in terrestrial soils or sediments illustrate some of the factors involved.

Fixation by algae

The blue-green algae are probably the most ubiquitous fixers in aerobic environments. The surface of the new volcanic island, Surtsey, supports blue-green algae growing in association with each other and a moss *Funaria hygrometrica.* Fixation as measured by ethylene production ranged from 2–54 ng N/g sample/h in the light, and 0–9 ng in the dark. This was compared to 133 to 540 ng in the light and 4 to 56 ng in the dark on the south of Iceland (Henriksson, Henriksson & Pejler, 1972). The lack of water holding power and the heavy metal and salt concentrations in the substrate were cited as factors making extrapolation of the data to a seasonal basis very difficult. The use of small (1 g) samples of algal soil (Henriksson, Enckell & Henriksson, 1972) resulted in the measurement of fixation ranging from 0 to 2.7 μg N/g/h in calcareous soils or loess of Sweden and Germany.

Moisture variations, even in permafrost areas, appear to be one of the major factors controlling the level of fixation. Schell & Alexander (1973) found ethylene production of 10.5 μmoles/m^2/h on damp inter-polygonal troughs where *Nostoc* mats were abundant. A value of 2.8 μmoles was found on dry, high-centered polygons where the lichens *Peltigera* and *Stereocaulon* were most active. This study utilized large cores (16 cm^2) and found a Q_{10} of 3.7 over the incubation temperature of 0–18 °C under the continuous light of the arctic summer.

Nostoc species are responsible for much of the fixation on dry prairie and desert sites. These organisms occur on the surface either in crusts (MacGregor & Johnson, 1971) or associated with lichens and mosses. Expression of results to either a square meter or hectare basis is usually based on the percent distribution of the algae and an estimate of the length of time the organisms are moist. *Nostoc* crusts have been found to respond to different light levels (Henriksson & Simu, 1971) and acetylene reduction can occur at very low moisture levels (45 % H_2O on a dry weight basis) with increasing water resulting in higher fixation.

However, excess water can cause a decrease in activity (Paul, Myers & Rice, 1971).

Fixation in grasslands

The acetylene–ethylene assay has confirmed earlier results with ^{15}N (Delwiche & Wijler, 1956; Porter & Grable, 1969; Fehr, Pang, Hedlin & Cho, 1972) that the total annual fixation by the blue-green algae, clostridial systems and the native legumes of native grasslands seldom exceeds a few kilograms per hectare or 0.1 to 0.3 g/m^2 (Vlassak, Paul & Harris, 1973; Steyn & Delwiche, 1970). Copley & Reuss (1972) found an acetylene-reduction rate equivalent to less than 0.1 mg N/m^2/day under field moisture levels at the USA Pawnee Grassland site. Saturation of the soil without further amendment resulted in acetylene reduction equivalent to 1 mg N/m^2/day. Consideration of the occurrence of native legumes plus the asymbiotic fixation for a 100 day period of activity resulted in an overall maximum estimate of 0.1 g N/m^2/year. Reuss & Cole (1973) indicated an input of 0.4 g N/m^2 by precipitation at the Pawnee site. They also noted an annual flow of 3.5 g N/m^2/year from the plant roots to the tops. Volatilization of nitrogen from animal wastes and losses during litter decomposition are the two most likely areas of nitrogen loss from a grassland ecosystem. Copley & Ruess (1972) concluded that if losses are less than the sum of nitrogen input (0.5 g N/m^2/year, which is 14 % of the annual flow of nitrogen to the plant tops), the grassland ecosystem will be self-sustaining with regard to nitrogen.

Investigation of a similar grassland, in Canada, showed that an average of 4.55 g N/m^2/year were utilized in above-ground production and 2 g/m^2/year were utilized in root growth (Coupland, personal communication). There was a precipitation input of 0.3 g N/m^2/year. The majority of this nitrogen was organic, with 0.13 g of ammonium-nitrogen and nitrate-nitrogen being added by the rainfall. Detailed sampling and exposure of 6.9 cm cores of native grassland to acetylene showed a 1971 growing season cumulative rate of 0.2 g N/m^2. Fixation by native legumes and algae comprised a portion of the 0.2 g N/m^2 overall estimate (Vlassak *et al.*, 1973). These data are three times as high as fixation levels found on the same site in the summer of 1970 (Paul *et al.*, 1971). Comparison of these data with the above shows that nitrogen fixation provided an average of 2 % of the nitrogen annually utilized by primary production, and nitrogen added in precipitation a total of 5 % of the flux. Moisture and temperature were the major factors affecting fixation. The response to temperature was positive in spring, to at least 15 °C, but the overall correlation with temperature was negative because the soils warmed up drastically as they dried in the summer with a corresponding decrease in nitrogen fixation. Regression analyses showed that soil moisture accounted for 64 % of the variability, but the effects of temperature could not be delineated by regression analysis of the field data.

The continued high agricultural production of former grassland sites in the absence of nitrogen fertilization has been attributed to asymbiotic nitrogen fixation (Mishustin & Shilnikova, 1971). The Canadian prairie soils have very low fixation levels under cultivation (Vlassak *et al.*, 1973) and the continued crop growth has not been sustained by fixation but by a 40 % drop in the level of total organic nitrogen, since the initiation of intensive cultivation 50 to 70 years ago. The organic nitrogen level of most ecosystems is very high in comparison to one year's plant productivity. Even minor changes in the turnover rate of this large reservoir will greatly influence the available mineral nitrogen. This affects the level of fixation and complicates the interpretation of nitrogen fixation measurements, and means that true estimates will have to be based on the data from a number of years which will hopefully include the extremes in abiotic conditions encountered.

The problems of extrapolation of nitrogen fixation levels within and between ecosystems are also pointed out for marine angiosperms (Patriquin & Knowles, 1972). Nitrogen fixation of a typical stand was estimated to be 100–500 kg N/ha/year. This was said to equal the nitrogen requirements of the plants in this system. These high levels are the opposite of those ascertained for grassland, where input was very low and the dependence on turnover of organic matter high. In an ecosystem study, the fate of the nitrogen fixed in the sediment must be taken into consideration. Continued high levels of fixation should indicate complete export of all the primary productivity to other ecosystems. Alternatively, organic deposits could rapidly be formed from a portion of the net fixed nitrogen.

Nitrogen fixation in forests

Assays of forest ecosystems have shown a diversity of microsites capable of fixing nitrogen. These include termites (Breznak, Brill, Mertins & Coppel, 1973; Benemann, 1973), decaying portions of living pine

(Seidler, Aho, Raju & Evans, 1972) and chestnut logs (Evans, personal communication, 1973). The root–soil association of conifers has been investigated both with ^{15}N (Richards, 1973) and the acetylene–ethylene assay (Silvester & Bennett, 1973). Activity was present only if it was found also in the surrounding soils, and rates of fixation recorded in the *Fh* layer were ten times the maximum root rates. Silvester's & Bennett's data when converted to nitrogen fixed gave a maximum *Fh* rate of 6.7 μg N/g/day, while the soil–root rate could be as high as 0.1 μg N/g/day. Fixation within termites and during the microbial degradation of the low-nitrogen wood has been considered to provide nitrogen for the organisms carrying out the decay process in the generally low-nitrogen substrate. Harris & Dart (1973) in attempting to ascertain the sites of increased nitrogen content in Broadbalk wilderness, found nitrogenase activity in the rhizosphere of *Stachys sylvatica* and other dicotyledonous plants. Phylloplane associations have been found in tropical rain forests (Edmisten, 1970) and in a range of untreated soils from Nigeria (Spiff & Odu, 1972). The generally low levels of nitrogen fixation reported for forests by the above workers and others (Knowles; Todd, personal communications) make it necessary to examine the total nitrogen inputs relative to the growth requirements and losses of the system.

Jorgensen & Wells (1971) found that although the average fixation of a Carolina forest amounted to only 25 g N/ha/year in the 0–4 cm depth of soil, plots in which the understory had been burnt fixed 949 grams. Long-term studies indicated a large increase in Kjeldahl nitrogen in the burned plots. The 949 grams represented about 4 % of the actual increase in Kjeldahl nitrogen in the 0–10 cm layer. Rhizosphere effects are probably included in the overall estimates for fixation if representative, adequate-sized soils cores are utilized, and the above data should indicate the nitrogen fixation potential of the system. Modelling of nitrogen flow in such forest ecosystems will have to take into account the potential for large amounts of nitrogen incorporation into standing biomass or litter build-up in addition to losses. This can involve a lot of nitrogen. To date, most nitrogen fixation studies in forests have not indicated the source of this nitrogen.

Expression of nitrogen fixation data

The conversion of nitrogen fixation data to an area basis can be calculated on a plant or algal crust distribution basis, or on a volume basis if undisturbed cores are utilized. This involves either an adequate number of randomized cores or sampling on a distribution pattern taking into account the different plant types or microbial habitats in the area under investigation.

Conversion of literature data based on unit weight of the soil to an area basis, either g/m^2 per unit time or kg/ha requires measurement of the bulk density. Jorgensen & Wells (1971) used a density of 0.2 g/cc in the organic layer of forest soil, 0.8 in the 0–1 cm layer and 1.4 for the 3–4 cm layer. The cultivated and grassland prairie soils in our own investigations in Canada had a bulk density of 1.0 in the 0–10 cm layer and 1.3 at 20–30 cm. Utilization of the common conversion factor of 2×10^6 lb/acre-6″ or 2.24 kg/ha-15 cm assumes a bulk density of 1.49. This is probably higher than encountered in many soils, and a greater depth must be taken into account if this conversion is utilized.

Measurement of undisturbed soil or sediment systems in adequately designed canopies such that field conditions are approximated as closely as possible over a number of years is probably the surest way of obtaining reliable estimates of nitrogen fixation actually occurring in an ecosystem. Fairly large undisturbed cores (6.9 cm diameter) give higher fixation rates than small (2 cm diameter) cores (Vlassak *et al.*, 1973). The field techniques employing internal standards appear very useful for in-situ assays (Balandreau & Dommergues, 1973). However, extended measurements are impractical for many sites.

Extrapolation of limited data requires measurements of abiotic factors and estimates of their effects: oxygen tension (Knowles, Brouzes & O'Toole, 1973; Dobereiner, Day & Dart, 1973), soil moisture tension (Vlassak *et al.*, 1973; Stutz & Bliss, 1973), light (Okafor & MacRae, 1973), pH (Granhall, 1970), salinity (Hauke-Pacewiczowa, 1970), and available energy especially in the rhizosphere where diurnal effects may be noted (Balandreau, Millier & Dommergues, 1974; Dommergues, Balandreau, Rinaudo & Weinhard, 1973). Measurement of temperature response with either acetylene or other growth parameters has indicated a biphasic curve with higher activity below 25 °C than above (Brouzes & Knowles, 1973). The high Q_{10}, approaching 4 in the lower temperature ranges, does not take into account the greatly reduced fixation at temperatures of 5 °C and below. There are many portions of the earth in which nitrogen fixation is being studied at these low temperatures. A response surface was suggested by Russell (1975) in which ammonification and nitrification temperature response data were fitted to the normal curve. It appears that the use of a normal curve as a response surface for modifying rate contents would apply equally well to nitrogen fixation parameters, and overcome the poor fit now obtained with the very low temperatures.

The high levels of acetylene reduction in rice rhizosphere (Yoshida & Ancajas, 1973) and aquatic sediments (Bristow, 1973) have raised the question of N_2 diffusion to the site of fixation in a waterlogged system.

Studies of the ratio of N_2:C_2H_2 reduction continue to show a diversity of results. However, the very wide ratios previously reported have not been found to occur if acetylene assay conditions, N_2 partial pressure, and N_2 diffusion problems in waterlogged soils were overcome (Rice & Paul, 1971; Brouzes & Knowles, 1973). It is usually suggested that ^{15}N be used to corroborate the acetylene–ethylene assay. Discrimination against ^{15}N and the apparently false positives obtained with pure cultures tested for fixation with ^{13}N and ^{15}N in the past lead one to also recommend that ^{15}N studies be corroborated with the acetylene assay.

The contribution of nitrogen fixation to plant nutrition

Interpretation of the role of nitrogen fixation in the functioning of an ecosystem or in agricultural production requires an assessment of the relative efficiency of uptake of soil mineral nitrogen and applied fertilizer nitrogen and an accurate estimate of the nitrogen supplying power of the soil or sediment systems, in addition to measurements of the nitrogen fixation potential of the system. Indirect estimates of some of these parameters can be obtained by utilizing non-nitrogen fixing plants as controls and measurements of the nitrate-nitrogen and ammonium-nitrogen status of fields utilized for fixation studies.

Long-term crop production data in England have indicated that fertilizer nitrogen has only been used with a general efficiency of 33 %. Cooke (1971) has calculated that in 1888 in all of England no more than 30 tons of fertilizer nitrogen were applied to a crop containing 270000 tons of nitrogen. In 1970, 400000 tons of nitrogen were applied and 340000 tons of nitrogen were harvested. The total acreage was similar in the two years. On the assumption that the total soil nitrogen was similar in 1888 and 1970, one must come to the conclusion that a great deal of the fertilizer nitrogen was wasted.

Bartholomew (1971) quotes a general efficiency of uptake of ^{15}N fertilizer of 50 to 60 % and states that native soil nitrogen coming from the mineralization of plant residues and soil organic matter is not used more efficiently by crops than is added fertilizer nitrogen.

The acetylene-reduction technique

It is difficult in modelling nitrogen flow through an ecosystem to evaluate the role that the various nitrogen inputs have because of the many possibilities for escape of nitrogen from the system, and the relative inaccuracy of the Kjeldahl technique for total nitrogen, especially in the light of the poor utilization of other nitrogen reserves in cultivated crops. The use of labelled organic matter as a tracer material to measure the native nitrogen-supplying power and its effect on nitrogen fixation has been suggested on the assumption that ^{15}N released is representative of the nitrogen supplied by the soil (Hauck, 1973). In such a study, growth of the nitrogen-fixing systems in an ordinary atmosphere, with appropriate acetylene–ethylene tests, would be utilized to estimate the contribution of nitrogen fixation, but the measurement of the nitrogen-supplying power of the soil by this technique can be questioned.

The nitrogen fixation data whether obtained by ^{15}N, the acetylene assay or measurements of nitrogen accumulation over long periods of time, by Kjeldahl analyses, is best interpreted if the investigation is a portion of an overall ecosystem or soil fertility study. In such a study, many of the parameters affecting this process probably are being measured. Problems in equipment, size of sample to be utilized, length of exposure to acetylene and the application of a theoretical ratio of acetylene/N_2 are probably more easily overcome than is the problem of obtaining representative samples characteristic of all parts of the ecosystem under study over a long enough period to make meaningful estimates possible.

The nitrogen fixation data presently being collected are being well utilized in the models of ecosystems and the nitrogen balances that are being constructed for a number of research stations in the different biomes in the International Biological Programme. Further modelling no doubt will point to areas where the present measurements of nitrogen fixation have been inadequate. However, the data obtained have been of great use in the modelling conducted to date.

Published as Canadian Committee for the IBP No. 246 and Saskatchewan Institute of Pedology No. 128.

References

Balandreau, J. P. & Dommergues, Y. (1973). Assaying nitrogenase C_2H_2 activity in the field. *Bull. Ecol. Res. Comm. Stockholm*, **17**, 247–54.

Balandreau, J. P., Millier, C. R. & Dommergues, Y. R. (1974). Diurnal variations in non-symbiotic nitrogen fixing systems. *Appl. Microbiol.*, **27**, 662–80.

Bartholomew, W. V. (1971). ^{15}N in research on the availability and crop use of nitrogen. In *Nitrogen in Soil–Plant Studies*, pp. 1–20. I.A.E.A.: Vienna.

Benemann, J. R. (1973). Nitrogen fixation in termites. *Science*, **181**, 164–5.

Breznak, J. A., Brill, W. J., Mertins, J. W. & Coppel, H. C. (1973). Nitrogen fixation in termites. *Abst. Am. Soc. Microbiol. Annual Meeting*, p. 164.

Bristow, J. M. (1973). Nitrogen fixation in rhizosphere of aquatic angiosperms. *Pl. Physiol.*, **51**, 34.

Brooks, R. H., Jr, Brezonik, P. L., Putnam, H. D. & Keirn, M. A. (1971). Nitrogen fixation in an estuarine environment: The Waccasassa on the Florida Gulf Coast. *Limnol. Oceanogr.*, **16**, 701–10.

Brouzes, R. & Knowles, R. (1973). Kinetics of nitrogen fixation in a glucose-amended, anaerobically incubated soil. *Soil Biol. Biochem.*, **5**, 223–9.

Carpenter, E. J. (1972). Nitrogen fixation by a blue-green epiphyte on pelagic *Sargassum*. *Science*, **178**, 1207–9.

Cooke, G. W. (1971). Fertilizers and society. *Proc. Fert. Soc. Lond.*, **121**, 1–48.

Copley, P. W. & Reuss, J. O. (1972). Evaluation of biological N_2 fixation in a grassland ecosystem. *Technical Report* no. 152. US IBP Grassland Biome: Fort Collins, Colo.

Delwiche, C. C. & Wijler, J. (1956). Non-symbiotic nitrogen fixation in soil. *Pl. Soil*, **7**, 113–29.

Dobereiner, J., Day, J. M. & Dart, P. J. (1973). Rhizosphere associations between grasses and nitrogen-fixing bacteria: Effect of O_2 on nitrogenase activity in the rhizosphere of *Paspalum notatum*. *Soil Biol. Biochem.*, **5**, 157–9.

Dommergues, Y., Balandreau, J. P., Rinaudo, G. & Weinhard, P. (1973). Non-symbiotic nitrogen fixation in the rhizospheres of rice, maize and different tropical grasses. *Soil Biol. Biochem.*, **5**, 83–9.

Edmisten, J. (1970). Preliminary studies of the nitrogen budget of a tropical rain forest. In *A Tropical Rain Forest* (ed. Odum, H. T. and Pigeon, R. F.), pp. 211–15. US Atomic Energy Commission.

Fedorova, R. I., Milekhina, E. I., Ilyukhina, N. I. & Brazhnikov, V. V. (1973). Question of application of acetylene method for determination of nitrogen fixing ability of soils. *Izv. Akad. Nauk USSR*, **1**, 106–14.

Fehr, P. I., Pang, P. C., Hedlin, R. A. & Cho, C. M. (1972). Some factors affecting asymbiotic nitrogen fixation in soils as measured by ^{15}N enrichment. *Agron. J.*, **64**, 251–4.

Goering, J. J. & Parker, P. L. (1972). Nitrogen fixation by epiphytes on sea grasses. *Limnol. Oceanogr.*, **17**, 320–3.

Granhall, U. (1970). Acetylene reduction by blue-green algae isolated from Swedish soils. *Oikos*, **21**, 330–2.

Granhall, U. & Lundgren, A. (1971). Nitrogen fixation in Lake Erken. *Limnol. Oceanogr.*, **16**, 711–19.

Granhall, U. & Selander, H. (1973). Nitrogen fixation in a subarctic mire. *Oikos*, **24**, 8–15.

Hardy, R. W. F., Burns, R. C. & Holsten, R. D. (1973). Applications of the acetylene–ethylene assay for measurement of nitrogen fixation. *Soil Biol. Biochem.*, **5**, 47–81.

Hardy, R. W. F. & Holsten, R. D. (1972). Global nitrogen cycling: Pools, evolution, transformations, transfers, quantitation and research needs, pp. 87–134. *Proc. Environ. Protection Agency, Washington, D.C.*

Harris, D. & Dart, P. J. (1973). Nitrogenase activity in the rhizosphere of *Stachys sylvatica* and some dicotyledenous plants. *Soil Biol. Biochem.*, **5**, 277–9.

Hauck, R. D. (1973). Nitrogen tracers in nitrogen cycle studies – past use and future needs. *J. environ. Qual.*, **2**, 317–27.

Hauke-Pacewiczowa, T. (1970). Fixation microbienne de l'azote dans un sol salin tunisien. *Soil Biol. Biochem.*, **2**, 47–53.

Henriksson, L. E., Enckell, P. H. & Henriksson, E. (1972). Determination of the nitrogen-fixing capacity of algae in soil. *Oikos*, **23**, 420–3.

Henriksson, E., Henriksson, L. E. & Pejler, B. (1972). Nitrogen fixation by blue-green algae on the Island of Surtsey, Iceland. *Surtsey Research Progress Report VI*, pp. 66–8.

Henriksson, E. & Simu, B. (1971). Nitrogen fixation by lichens. *Oikos*, **22**, 119–21.

Horne, A. J. (1972). Ecology of nitrogen fixation on Signy Island, South Orkney Islands. *Br. Antarctic Surv. Bull.*, **27**, 1–18.

Horne, A. J., Dillard, J. E., Fujita, D. K. & Goldman, C. R. (1972). Nitrogen fixation in Clear Lake, California. II. Synoptic studies on the autumn *Anabaena* bloom. *Limnol. Oceanogr.*, **17**, 693–703.

Jorgensen, J. R. & Wells, C. G. (1971). Apparent nitrogen fixation in soil influenced by prescribed burning. *Soil Sci. Soc. Am. Proc.*, **35**, 806–10.

Knowles, R., Brouzes, R. & O'Toole, P. (1973). Kinetics of nitrogen fixation and acetylene reduction, and effects of oxygen and of acetylene on these processes in a soil system. *Bull. Ecol. Res. Comm. Stockholm*, **17**, 255–62.

MacGregor, A. N. & Johnson, D. E. (1961). Capacity of desert algal crusts to fix atmospheric nitrogen. *Soil Sci. Soc. Am. Proc.*, **35**, 843–4.

Mague, T. H. & Burris, R. H. (1973). Biological nitrogen fixation in the Great Lakes. *Bioscience*, **23**, 236–39.

Maltseva, N. N. (1973). Nitrogen-fixing ability of the main soil types in the Ukrainian SSR. *Mikrobiol. Zh.*, Kiev, **35**, 143–8.

Mishustin, E. N. & Shilnikova, U. K. (1971). *Biological fixation of atmospheric nitrogen* (translation by P. Corzy). Macmillan Press, Toronto.

Okafor, N. & MacRae, I. C. (1973). The influence of moisture level, light, aeration and glucose upon acetylene reduction by a black earth soil. *Soil Biol. Biochem.*, **5**, 181–6.

Patriquin, D. & Knowles, R. (1972). Nitrogen fixation in the rhizosphere of marine angiosperms. *Marine Biol.*, **16**, 49–58.

Paul, E. A., Myers, R. J. K. & Rice, W. A. (1971). Nitrogen fixation in grassland and associated cultivated ecosystems. *Plant and Soil, Special Volume*, 495–507.

Porter, L. K. & Grable, A. R. (1969). Fixation of atmospheric nitrogen by non-legumes in wet mountain meadows. *Agron. J.*, **61**, 521–3.
Reuss, J. O. & Cole, C. V. (1973). *Simulation of nitrogen flow in a grassland ecosystem.* Proc. US IBP Grassland Biome: Fort Collins, Colo.
Rice, W. A. & Paul, E. A. (1971). The acetylene reduction assay for measuring nitrogen fixation in waterlogged soil. *Can. J. Microbiol.*, **17**, 1049–56.
Richards, B. N. (1973). Nitrogen fixation in the rhizosphere of conifers. *Soil Biol. Biochem.*, **5**, 149–52.
Rusness, D. & Burris, R. H. (1970). Acetylene reduction (nitrogen fixation) in Wisconsin lakes. *Limnol. Oceanogr.*, **15**, 808–13.
Russell, J. S. (1975). Systems analysis of soil ecosystems. In *Soil Biology and Biochemistry*, vol. **3** (eds. Paul, E. A. and McLaren, A. D.), pp. 38–82. Marcel Dekker, Inc.: New York.
Schell, D. A. & Alexander, V. (1973). Nitrogen fixation in Arctic coastal tundra in relation to vegetation and micro-relief. *Arctic*, **26**, 130–7.
Seidler, R. J., Aho, P. E., Raju, P. N. & Evans, H. J. (1972). Nitrogen fixation by bacterial isolates from decay in living white fir (*Abies concolor* (Gord, and Glend.) Lindl.). *J. gen. Microbiol.*, **73**, 414–16.
Silvester, W. B. & Bennett, K. J. (1973). Acetylene reduction by roots and associated soil of New Zealand conifers. *Soil Biol. Biochem.*, **5**, 171–9.
Skalon, I. S. (1971). Nitrogen-fixing microorgansisms and biological activity of soils of steppe and desert communities in connection with a study of the biological productivity of plants. *Biol. Prod. Krugovorot Khim. Elem. Rast. Soobshchestvak*, *1971*, pp. 176–81.
Spiff, E. D. & Odu, C. T. I. (1972). An assessment of non-symbiotic nitrogen fixation in some Nigerian soils by the acetylene reduction technique. *Soil Biol. Biochem.*, **4**, 71–7.
Steyn, P. L. & Delwiche, C. C. (1970). Nitrogen fixation by non-symbiotic microorganisms in some California soils. *Environ. Sci. Technol.*, **4**, 1122–8.
Stutz, R. C. & Bliss, L. C. (1973). Acetylene reduction assay for nitrogen fixation under field conditions in remote areas. *Pl. Soil*, **38**, 209–13.
Sugahara, I., Sawada, T. & Kawai, A. (1971). Microbiological studies on nitrogen fixation in aquatic environments. VI. On the *in situ* nitrogen fixation in water regions. *Bull. Jap. Soc. Sci. Fish.*, **37**, 1093–9.
Trepachev, E. P., Atrashkova, N. A. & Khabarova, A. I. (1971). Experimental verification of the methods of studying the rates of nitrogen fixation by leguminous plants. *Nov. Izuch. Biol. Fiksatsii Azota, 1971*, pp. 145–53.
Vanderhoef, J. N., Dana, B., Emerich, D. & Burris, R. H. (1972). Acetylene reduction in relation to levels of phosphate and fixed nitrogen in Green Bay. *New Phytol.*, **71**, 1097–105.
Vlassak, K., Paul, E. A. & Harris, R. (1973). Assessment of biological nitrogen fixation in grassland and associated sites. *Pl. Soil*, **38**, 637–49.
Waughman, G. J. & Bellamy, D. J. (1972). Acetylene reduction in surface peat. *Oikos*, **23**, 353–8.
Yoshida, T. & Ancajas, R. (1973). The fixation of atmospheric nitrogen in the rice rhizosphere. *Soil Biol. Biochem.*, **5**, 153–5.

19. Seasonal and diurnal variations in $N_2(C_2H_2)$-fixing activity in field soybeans

C. SLOGER, D. BEZDICEK, R. MILBERG & N. BOONKERD

Our knowledge about the important symbiosis between soybean and *Rhizobium japonicum* has advanced in recent years. Many aspects of symbiotic nitrogen (N_2) fixation have been studied using the acetylene-reduction method which is a simple, rapid, but an indirect measure of N_2 fixation (Hardy, Burns & Holsten, 1973). Although symbiotic N_2 fixation by nodulated soybeans is detected within several weeks after planting, the major portion of N_2 is fixed between flowering and approaching green bean stages of development (Hardy, Holsten, Jackson & Burns, 1968; Weber, Caldwell, Sloger & Vest, 1971). Symbiotic N_2 fixation is an important system in supplying nitrogen during reproductive growth while nitrate utilization, via the enzyme nitrate reductase, is an important system in supplying nitrogen during vegetative growth (Harper & Hageman, 1972). Symbiotic N_2 fixation by soybeans is enhanced by efficient soybean-*Rhizobium* strain associations (Sloger, 1969; Weber *et al.* 1971; Mague & Burris, 1972), carbon dioxide enrichment (Hardy & Havelka, 1973), proper amounts of water (Sprent, 1971, 1972*a*, *b*) and sunlight (Mague & Burris, 1972).

We report here on the seasonal variation in $N_2(C_2H_2)$-fixing activity by nodules located on the tap and lateral roots and its correlation with certain environmental parameters.

Materials and methods

Plants and field locations

Soybeans (*Glycine max* (L.) Merr.) variety York (maturity group V) were planted on 24 May 1972, at Marlboro, Maryland, in fields which were essentially void of nodulating *Rhizobium japonicum* (Weber *et al.*, 1971). Seeds were inoculated with a conventional peat inoculum containing *R. japonicum* strain 6 from the USDA Beltsville *Rhizobium* Collection. The varieties York and Kent (maturity group IV) were planted on 13 June 1972, at Beltsville in the USDA South Farm fields which have a diverse population of nodulating *R. japonicum* strains (10^5 to 10^6 organisms per gram of soil).

Soybeans were planted in three row plots with four replicates. All rows were 6.09 m long and 1.016 m and 0.914 m wide at Beltsville and Marlboro, respectively. In the diurnal study at Beltsville, Kent soybeans were grown in 15 row plots with four replicates. The average plant count at four weeks after planting was six plants per 0.31 m. The fields received an unusual 20 cm of rain during 21 to 22 June 1972, from tropical storm Agnes.

Acetylene-reduction method

For the nodule study, a 0.31 m row of York soybeans was dug carefully and their roots rinsed quickly in water to remove soil. Nodulated roots were detached from the stem at the cotyledonary node and were placed in pint jars (474 ml) or, after flowering in wide-mouth quart jars (989 ml). For the diurnal study, 0.62 m rows of Kent soybeans were sampled until the eighth week after planting and 0.31 m rows were sampled thereafter. The nodulated roots were placed in quart jars, and screw cap lids equipped with two serum stoppers were tightly fastened. A gas mixture of 75 % Ar and 25 % O_2 was purged through all jars for three minutes. The assay was started by addition of 60 or 120 ml C_2H_2 from a large plastic syringe into each pint or quart jar, respectively. The jars were quickly shaken and vented to 1 atm leaving about 11 % C_2H_2, 21 % O_2 and remainder argon. The jars in the diurnal study were incubated in the soil where plants were dug. All other jars were incubated at 27 ± 2 °C in a water bath in an insulated container. Incubation time was half an hour in the diurnal study and one hour for all other assays.

The $N_2(C_2H_2)$-fixing activity of tap and lateral root nodules was determined by gas chromatography (g.c.) according to the method of Sloger (1969). $N_2(C_2H_2)$-fixing activity and carbon dioxide production in the diurnal study was determined by thermal conductivity gas chromatography. Separation of gases in a 1 ml sample was obtained in a 0.6×184 cm teflon column containing a stationary phase of Porapak N* 100 to 120 mesh (column temperature 50 °C, helium

* Mention of a trademark or proprietary product does not constitute a guarantee or warranty of the product by the US Department of Agriculture, and does not imply its approval to the exclusion of other products that may also be suitable.

carrier gas flow rate 20 ml per minute, injection port temperature 75 °C, and detector current was 150 mA).

Relative turgidity of soybean leaves was determined by the method of Barrs & Weatherley (1962). Leaf disks (20 per replicate) were punched with a cork borer (19 mm diameter), weighed, and floated on 25 °C water. After four hours, leaf disks were blotted and turgid fresh weight determined.

The carbon dioxide production attributed to nodules was estimated by multiplying the ratio of nodule fresh weight to root fresh weight times the total carbon dioxide production.

Air temperature at time of sampling was determined by a shielded thermometer placed in the middle of the soybean row. Soil temperature was determined 8 cm deep in the soybean row. The solar radiation data by hour intervals were obtained from the Department of Meterology at the University of Maryland, College Park Campus, which was three miles from the fields at Beltsville.

Results and discussion

Tap and lateral root nodules

Nodule fresh weight increase on York soybean roots from 30 to 90 days after planting at Marlboro and Beltsville fields is shown in Fig. 19.1. Because nodule development at Marlboro was slowed due to cool soil temperature, the planting date was adjusted to the Beltsville planting date for these data. The composite data showed a linear increase in nodule development. Initial nodule development occurred first on the tap root.

The nodulation pattern at the two locations was different. Nodules on the inoculated plants at Marlboro were centered near the tap root and extended out about 8 cm along the lateral roots, while nodules were found throughout the root system at Beltsville, Also, nodules on plants at Marlboro were fewer and larger than those at Beltsville.

The tap root nodules were the first to senesce at the end of flowering, about 65 days after planting. These observations support the report by Bergersen (1958) that nodule senescence on greenhouse grown soybeans began by 67 days after planting. Interestingly, at Marlboro in the 1973 tests where specific *R. japonicum* strains were applied to Kent soybeans, tap root nodule senescence was delayed until mid-bean development, more than 90 days after planting. These observations were confirmed by Dr Deane Weber and Dr R. J. Roughley. Adequate soil moisture

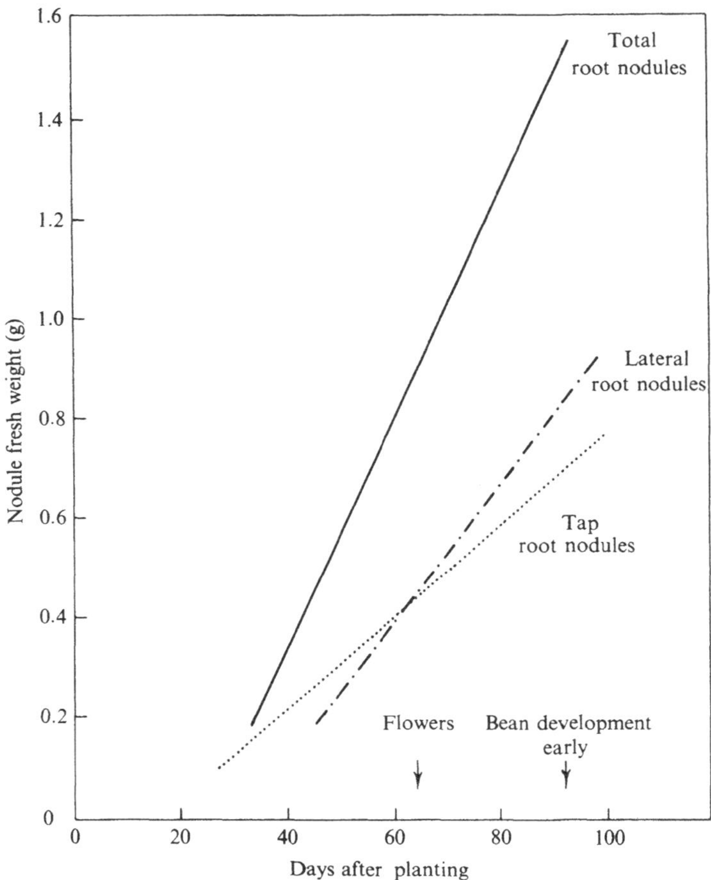

Fig. 19.1. Comparison of tap and lateral root nodule development.

during the growing season was probably the reason for the delay in nodule senescence.

The seasonal profile of $N_2(C_2H_2)$-fixing activity of tap and lateral root nodules at both locations is indicated in Fig. 19.2. Initially, the activity of tap root nodules was greater than the lateral root nodules, however, the lateral root nodules ultimately exhibited higher rates. The York × *R. japonicum* strain 6 association showed a greater peak rate of $N_2(C_2H_2)$-fixing activity than the York grown in Beltsville fields containing native nodulating *Rhizobium*. The difference in seasonal profile of $N_2(C_2H_2)$-fixing activity may be partly due to the difference in *R. japonicum* strains associated with the plant, since nodule fresh weights for tap and lateral root nodules were similar at both locations.

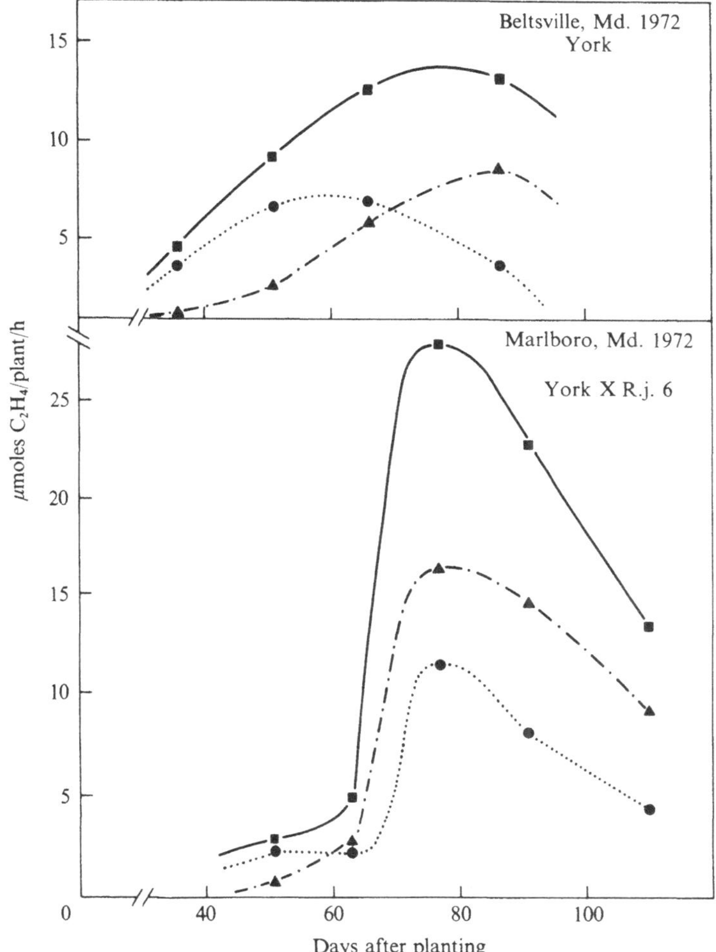

Fig. 19.2. Seasonal profile of $N_2(C_2H_2)$-fixing activity for nodulated York soybeans grown at two field locations. ■, Total root nodules; ▲, lateral root nodules; ●, tap root nodules.

Strain 6 associations with three local soybean varieties grown at Marlboro for the past three years produced superior yields in comparison with other strains (Weber, personal communication). In addition, different soil characteristics may account for part of the difference in the seasonal profiles.

The $N_2(C_2H_2)$-fixing activity by the tap and lateral root nodules changed during the season. At first the tap root nodules supported a high percentage of the activity per plant, but as the season progressed

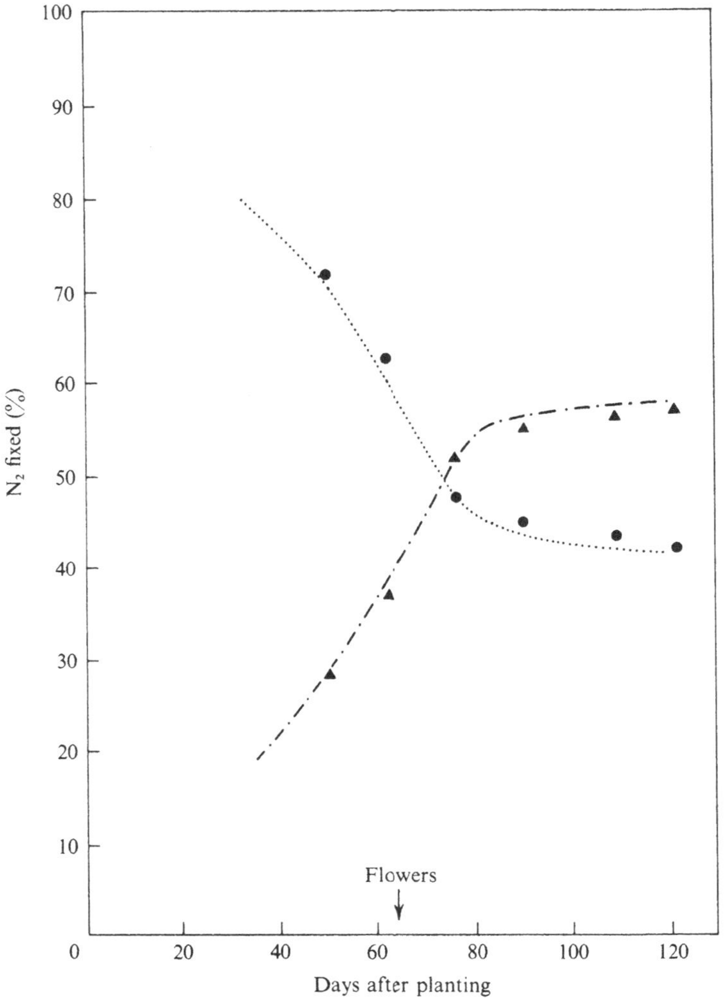

Fig. 19.3. Comparison of per cent N_2 fixed by tap (●) and lateral (▲) root nodules. Data from York soybeans grown at two locations was combined.

they supported a decreasing percentage of activity. The lateral root nodules supported an increasing percentage of the activity per plant during the season and exceeded the tap root nodules after flowering. Using these acetylene reduction data the N_2 fixed was estimated by the method of Hardy *et al.* (1968). The percentage of N_2 fixed by tap and lateral root nodules during the season is shown in Fig. 19.3. Lateral root nodules fixed about 60 % of N_2 while tap root nodules fixed about

40 %. The seasonal profiles of N_2 fixed and $N_2(C_2H_2)$-fixing activity by tap and lateral root nodules were similar.

The extent of nodule loss from roots when plants were dug at the Beltsville field was determined. Plants were sampled as usual from 0.31 m of the row at 70 days after planting. Nodules were collected from roots and weighed. The soil in the row where plants were dug was collected to a depth of 0.31 m. The area was 1.01 m wide so that half a row width was included on each side of the sample row. The soil was screened, and the nodules were collected and weighed. The average of two replicate samples indicated that 28 % of the nodules remained in the soil after digging. The majority of the nodules were from the lateral roots. This loss of nodules introduces a negative error in the estimation of $N_2(C_2H_2)$-fixing activity.

The initial infection and nodule development occurs on the soybean tap root. As these nodules develop they are the major sites of symbiotic N_2 fixation during vegetative growth. The continued importance of tap root nodules in the nutritional status of the plant is curtailed by senescence, and lack of continued infections on the tap root. The lateral root nodule development is abundant and they fixed most of the N_2 during peak fixation after flowering. Weber *et al.* (1971), showed that populations of native *Rhizobium* strains in the soil at Beltsville shifted during the season as the soil temperature increases resulting in redistribution of *Rhizobium* serogroups within the nodules on the roots of soybeans. Clearly, it is desirable to understand these shifts in order to maintain highly efficient soybean–*R. japonicum* strain associations in the tap and especially the lateral root nodules.

Diurnal study

The diurnal variations of $N_2(C_2H_2)$-fixing activity, soil and air temperature, solar radiation, relative turgidity of leaves, and nodule respiration are shown in Figs. 19.4 and 19.5. The diurnal variations of $N_2(C_2H_2)$-fixing activity and nodule respiration were similar. The diurnal variations in relative turgidity of leaves for ten weekly samplings were also similar. Soil moisture data verified that soil moisture changed only slightly and remained adequate for the season. The $N_2(C_2H_2)$-fixing activity of the root nodules was dependent upon solar radiation. On cloudy mornings, such as in the sixth and ninth weeks after planting, the activity remained unchanged until the sky cleared and solar radiation increased. A cloudy period accompanied with a light rain caused

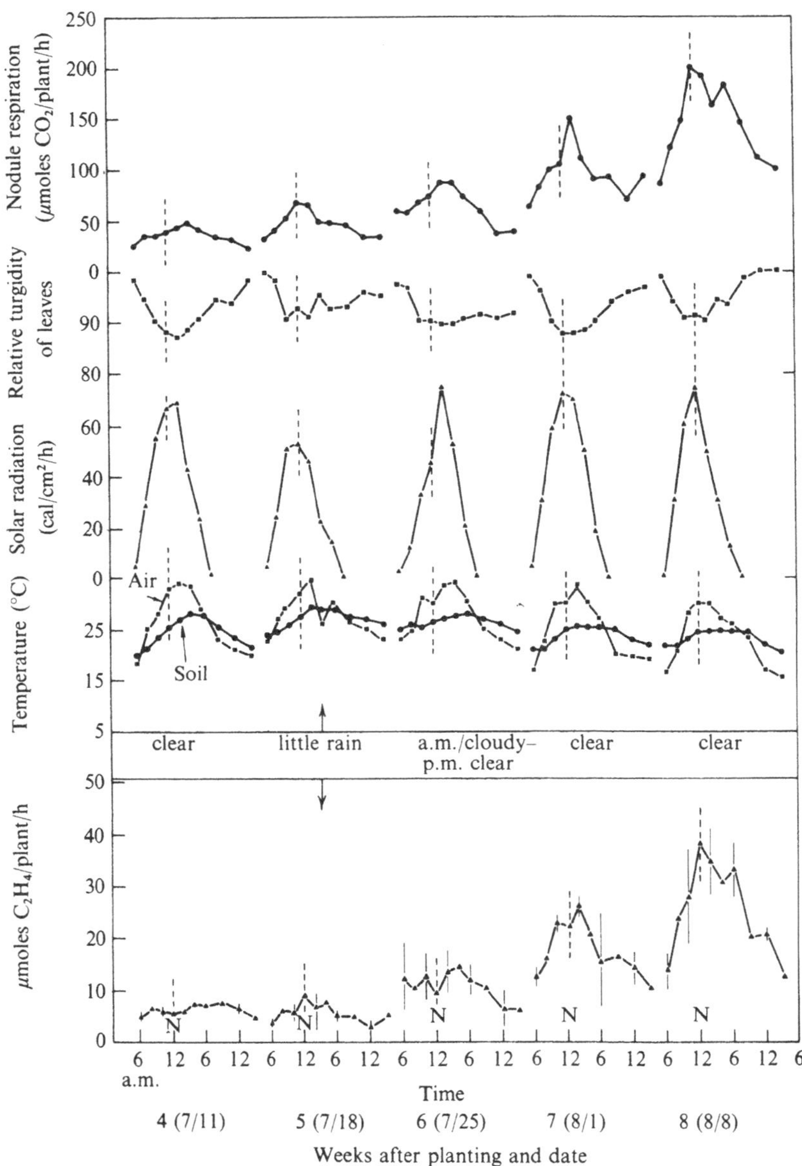

Fig. 19.4. Comparison of diurnal variation in $N_2(C_2H_2)$-fixing activity, soil and air temperature, solar radiation, relative turgidity of leaves and nodule respiration. Samples were taken starting at 6 a.m., 8, 10, 12, 2, 4, 6 p.m., 9, 12 and 3 a.m. The solid and dash vertical lines show the standard deviations and 12 noon, respectively. Each point is the average of four replicates.

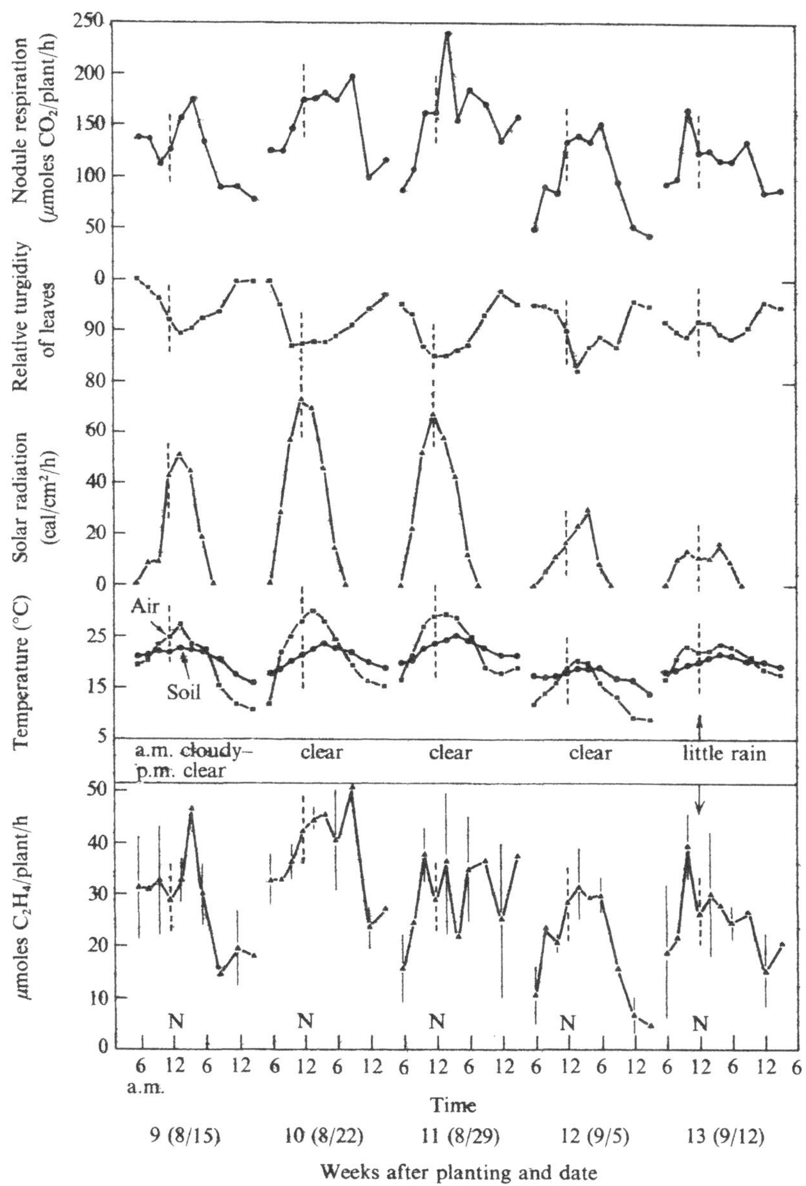

Fig. 19.5. Comparison of diurnal variation in $N_2(C_2H_2)$-fixing activity, soil and air temperature, solar radiation, relative turgidity of leaves and nodule respiration.

a decrease in activity. On clear days, such as in the seventh and eighth weeks after planting, the diurnal variation of $N_2(C_2H_2)$-fixing activity and air temperature were similar. The diurnal variation in $N_2(C_2H_2)$-fixing activity and air temperature during the last five sample weeks on clear days had some unexplained differences. In two cases, nine and 12 weeks after planting, the low $N_2(C_2H_2)$-fixing activity at night was associated with low air temperature. As expected, the diurnal variation in soil temperature was much less than variation in air temperature and $N_2(C_2H_2)$-fixing activity.

The diurnal variations for all dates were analyzed. The diurnal variation in $N_2(C_2H_2)$-fixing activity per gram fresh weight of nodules per hour (specific activity) was similar to the diurnal variation in activity per plant per hour. The average specific activity for all sample times was highest in the afternoon from 2 to 4 p.m. The average specific activity for each date was significantly correlated (r = 0.859**)† with the cumulative solar radiation for each date but not significantly correlated (r = 0.560) with the cumulative solar radiation for the day previous to the sample date. The average specific activity of nodules for each date was significantly correlated (r = 0.666*)† with average air temperature for each date but not significantly correlated (r = 0.578) with average soil temperature for each date. The average air temperature for each date was significantly correlated to the cumulative solar radiation.

The comparison of the average $N_2(C_2H_2)$-fixing activity per plant for each date with the product of the average nodule fresh weight times the average air temperature for each date is shown in Fig. 19.6. As much as 59 % of the variation in $N_2(C_2H_2)$-fixing activity per plant for the 10 weeks is explained by this variable.

The comparison of the estimated N_2 fixed and nodule fresh weight during the season is shown in Fig. 19.7. The accumulation of fixed N paralleled the increase in nodule fresh weight for the season. From the summation of acetylene-reduction data the N_2 fixed for the season was estimated to be 84 kg N/ha. This is about $\frac{1}{3}$ the plant and seed nitrogen in a 2219 kg/ha (33 bu/a) yield. Accounting for $N_2(C_2H_2)$-fixing activity loss due to nodule loss, the estimated fixed nitrogen rose to 103 kg N/ha, about 40 % of the plant and seed nitrogen. These estimates are similar to calculated values reported in a review by Hardy *et al.* (1973).

For calculation of the total N_2 fixed for the season the average daily

† *, ** Significant at the 0.05 and 0.01 levels, respectively.

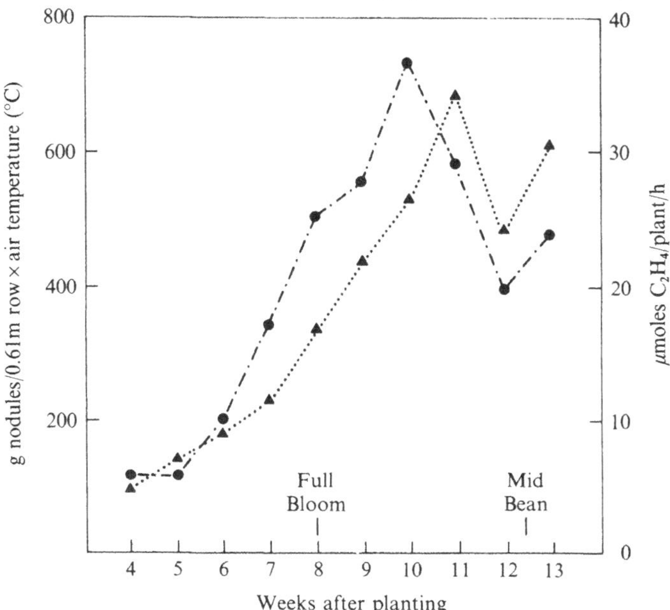

Fig. 19.6. Relationship between $N_2(C_2H_2)$-fixing activity per plant (●) with the variable nodule fresh weight × air temperature (▲). The variable represents the product of average nodule fresh weight times the average air temperature for each date. The correlation coefficient is $r = 0.769$ ($r^2 = 0.59$).

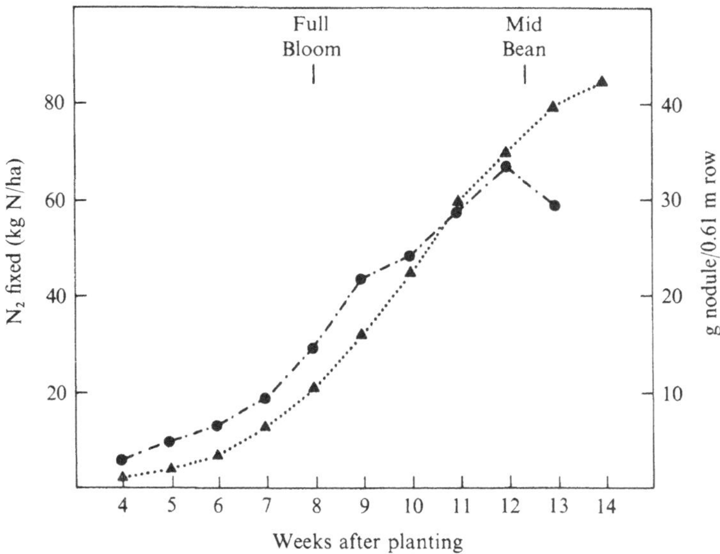

Fig. 19.7. Relationship between estimated N_2 fixed (▲) and nodule development (●) for field grown soybeans. Each value is the average of four replicates.

Table 19.1. *The r^2 values for relationships between average daily $N_2(C_2H_2)$-fixing activity computed from ten sample times per date for ten dates and average daily $N_2(C_2H_2)$-fixing activity computed from only one to four selected sample times per date for ten dates*

Selected sample times (a.m.)	r^2
6, 8, 10, 12	0.98*
6, 9, 12**	0.98*
6, 8, 10	0.96*
6, 12	0.94*
8, 10	0.96*
7, 9, 11**	0.92*
8, 10, 12	0.94
10	0.88

* Student's *t* values for paired differences test do not differ significantly at the 5 % probability level from average daily $N_2(C_2H_2)$-fixing activity computed from 10 sample times for 10 sample dates.

** Activity at 7, 9, 11 are calculated values. $N_2(C_2H_2)$-fixing activity was expressed as μmoles C_2H_4/plant/h.

$N_2(C_2H_2)$-fixing rate has to be determined (Hardy *et al.*, 1971). Ideally, we would prefer to sample plants a minimum number of times and still accurately determine the average daily rate. An analysis of our diurnal variation data shown in Table 19.1 indicated that the average rate for samples at 6, 8, 10 and 12 noon is significantly correlated to the average daily rate computed from 10 sample times. Sampling two and three times in the morning also showed significant correlations. Determining the average daily rate from one sampling (10 a.m.) was undesirable because of variation.

In a nearby field, soybeans growing in a specific gravel area (an old stream bed) showed signs of water stress, i.e. wilting leaves during the day. The diurnal variation in $N_2(C_2H_2)$-fixing activity and relative turgidity of leaves of water-stressed plants as compared to unstressed plants in the same field are shown in Fig. 19.8. The $N_2(C_2H_2)$-fixing activity of the plants under water stress supported only a low rate of activity during the day. These plants supported an increased $N_2(C_2H_2)$-fixing activity from 6 to 8 a.m. but then declined because of water stress. The relative turgidity of leaves of the water-stressed plants declined faster and to a lower value than the unstressed plants.

These data confirm the importance of solar radiation, air temperature, and soil moisture to symbiotic N_2 fixation. Using equations derived

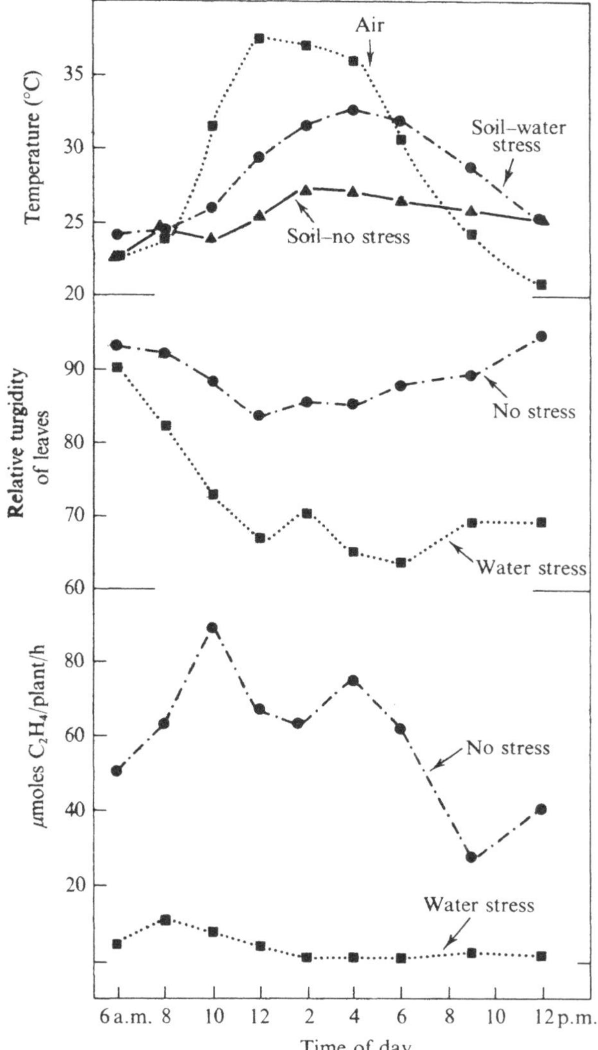

Fig. 19.8. Comparison of diurnal variation of $N_2(C_2H_2)$-fixing activity, relative turgidity of leaves and temperature for plants under water stress and no stress. Kent soybeans were sampled 25 August 1972, 54 days after planting. Each value is the average of four replicates. The weather was clear.

from long-term yield studies, Runge & Carmer (1963) reported that average maximum air temperature and average rainfall during an 88 day interval in the growing season accounted for 68 % of the variation in soybean yields in Illinois for a period covering 49 years.

References

Barrs, H. D. & Weatherley, P. E. (1962). A re-examination of the relative turgidity technique for estimating water deficits in leaves. *Austr. J. biol. Sci., Melbourne*, **15**, 415–28.

Bergersen, F. J. (1958). The bacterial component of soybean root nodules; changes in respiratory activity, cell dry weight and nucleic acid content with increasing nodule age. *J. gen. Microbiol.*, **19**, 312–23.

Hardy, R. W. F., Holsten, R. D., Jackson, E. K. & Burns, R. C. (1968). The acetylene–ethylene assay for N_2 fixation: laboratory and field evaluation. *Pl. Physiol., Baltimore*, **43**, 1185–1207.

Hardy, R. W. F., Burns, R. C., Hebert, R. R., Holsten, R. D. & Jackson, E. K. (1971). Biological nitrogen fixation: a key to world protein. *Plant and Soil, Special Volume*, 561–90.

Hardy, R. W. F., Burns, R. C. & Holsten, R. D. (1973). Applications of the acetylene–ethylene assay for measurement of nitrogen fixation. *Soil Biol. Biochem.*, **5**, 57–81.

Hardy, R. W. F. & Havelka, U. D. (1973). Symbiotic N_2 fixation: multifold enhancement by CO_2-enrichment of field-grown soybeans. *Pl. Physiol., Baltimore*, **51**., *Suppl.*, 35.

Hardy, R. W. F., Holsten, R. D., Jackson, E. K. & Burns, R. C. (1968). The acetylene–ethylene assay for N_2 fixation: laboratory and field evaluation. *Pl. Physiol., Baltimore*, **43**, 1185–207.

Harper, J. E. & Hageman, R. H. (1972). Canopy and seasonal profiles of nitrate reductase in soybeans (*Glycine max* (L.) Merr.). *Pl. Physiol., Baltimore*, **49**, 146–54.

Mague, T. H. & Burris, R. H. (1972). Reduction of acetylene and nitrogen by field-grown soybeans. *New Phytol.*, **71**, 275–86.

Runge, E. C. A. & Carmer, S. G. (1973). How weather affects corn and soybean yields. *Illinois Research, Urbana*, **15**, 6–7.

Sloger, C. (1969). Symbiotic effectiveness and N_2 fixation in nodulated soybean. *Pl. Physiol., Baltimore*, **44**, 1666–8.

Sprent, J. I. (1971). The effect of water stress on nitrogen-fixing root nodules. I. Effects on the physiology of detached soybean nodules. *New Phytol.*, **70**, 9–17.

(1972*a*). The effect of water stress on nitrogen-fixing root nodules. II. Effects on the fine structure of detached soybean nodules. *New Phytol.*, **71**, 443–50.

(1972*b*). The effects of water stress on nitrogen-fixing root nodules. IV. Effects on whole plants of *Vicia faba* and *Glycine max*. *New Phytol.*, **71**, 603–11.

Weber, D. F., Caldwell, B. E., Sloger, C. & Vest, G. H. (1971). Some USDA studies on the soybean-*Rhizobium* symbiosis. *Plant and Soil, Special Volume*, 293–304.

20. Acetylene-reduction assay at ambient P_{O_2} of field and forest soils: laboratory and field core studies*

R. KNOWLES & P. O'TOOLE

Since oxygen affects aerobic nitrogen fixers (Drozd & Postgate, 1970) as well as anaerobic and facultative nitrogen fixers, the problem of matching oxygen supply rates during assay (Bergersen, 1970) is a perplexing one. This is particularly true in the assay of unamended soils which, because of their low nitrogenase activity (see Knowles, 1975), require the use of long assay times whether $^{15}N_2$ or acetylene is used. On the other hand the oxygen consumption rates of unamended soils are relatively low compared to those of carbohydrate-amended systems.

Long-term assays are thus open to criticism because of the possible changes in gas phase composition during the course of the assay. Moreover acetylene may exert side effects through its competition with N_2 (Brouzes & Knowles, 1973), or its other inhibitory effects on nitrogen fixers (Brouzes & Knowles, 1971) and possibly on other organisms (Knowles, Brouzes & O'Toole, 1973).

The present paper reports a laboratory study in which two glucose-amended soil model systems incubated at different P_{O_2} values were subjected to three different types of assay regimes. This paper also presents data obtained from long-term acetylene-reduction assays of cores from a pasture soil, a maple forest soil and an aspen forest soil.

Material and methods

Soil sampling sites were (*a*) a pasture plough layer in a St Bernard sandy loam, (*b*) mull *L* and *A1* horizons (*F* and *H* layers were absent) from the same soil series in an adjacent maple (*Acer saccharum*) stand in the Morgan Arboretum of McGill University and (*c*) the forest floor in an aspen poplar (*Populus grandidentata* and *P. tremuloides*) stand near Chalk River, Ontario. This profile consisted of the litter (*L*) layer, of

* Canadian IBP Contribution No. 276.

fresh leaf material; the *F* layer, of leaves in early stages of decomposition; the *H* layer, of black amorphous organic matter with occasional charcoal present; the *A2*, of greyish red sand; and the *B* horizon, of reddish brown sand. It was shown earlier (Brouzes, Lasik & Knowles, 1969) that whereas soils (*a*), (*b*) and (*c*) all contained *Clostridium*, only soil (*a*) contained *Azotobacter*.

For the laboratory model system study the soils were sampled, processed and stored as described by Brouzes, Mayfield & Knowles (1971). Seven-gram aliquots of moist soil in 10 ml side-arm test tubes were amended with glucose solution to give a final concentration of 2 % w/w and were incubated attached to gas-manifolds through which flowed appropriate mixtures of N_2 and air (Brouzes *et al.*, 1971). Acetylene-reduction assays were performed repeatedly on the same samples (O'Toole & Knowles, 1973) and were of 1 h duration at 30 °C in the dark using 0.05 atm $P_{C_2H_2}$. Aerobic assays were performed by evacuating and filling the tubes with an argon/oxygen mixture (4:1) three times to one atmosphere and then injecting acetylene. Anaerobic assays employed the same procedure using argon alone. In so-called ambient assays, acetylene was injected without removal of the atmosphere in the tube and the assay thus began at close to the ambient P_{O_2} of the incubated soil (O'Toole & Knowles, 1973). Results of this part of the study are expressed per gram of oven-dry soil and are the means of triplicates.

Because of the large variation encountered in soil core experiments and the difficulty of evaluating precisely the various assay procedures (see Results), the methods used in the core study were selected somewhat arbitrarily on the basis of the preliminary experiments described. Soil cores were obtained using a steel core sampler of 48 mm internal diameter and were placed for assay in PVC cylinders (51 mm i.d. by 305 mm long). The cylinders were closed with rubber stoppers fitted with glass tubes and serum stoppers for venting and sampling. Samples from sites (*a*) pasture field and (*b*) maple forest were (unless otherwise stated) evacuated and filled three times with argon/oxygen mixture (4:1), then partially evacuated and 0.1 atm $P_{C_2H_2}$ introduced (aerobic assay). Gas phase samples were analyzed after 1, 2 and 3 days of assay in the dark at 30 °C. Core samples from site (*c*) aspen forest were merely partially evacuated and 0.1 atm $P_{C_2H_2}$ introduced (ambient assay). Gas phase samples were analyzed after 2, 12, 24, 36 and 48 h of assay in the dark at 30 °C. Results of the soil core studies are expressed per gram of fresh soil or per unit area and are the means of the numbers of replicates indicated in each case.

Analyses for acetylene, ethylene, oxygen, nitrogen and carbon dioxide were performed as described by Brouzes *et al.* (1971) Analysis of samples from site (*c*) aspen forest were made using a Carle 9000 gas chromatograph (hydrogen flame ionization detector) fitted with a 1830 × 3.2 mm column of Porapak R (60–80 mesh) at 50 °C using N_2 carrier at 25 ml per min.

Results

Data from an experiment in which glucose-amended soils were incubated on manifolds at various P_{O_2} values and were then subjected to three kinds of acetylene-reduction assay are shown in Fig. 20.1. In the field soil, negligible activity was observed at 0.5 atm P_{O_2} but in the maple forest soil activity was suppressed down to 0.09 atm P_{O_2}. Activity was observed in aerobic assays of field soil previously incubated between 0.19 and 0.04 atm P_{O_2} but not at any P_{O_2} values in the maple forest soil. Activity in anaerobic assay was seen in the field soil preincubated at 0.15 atm P_{O_2} or less and in the maple soil when preincubated at 0.05 atm P_{O_2} or less. The activity in assay carried out at the ambient P_{O_2} of preincubation was equivalent to the aerobic assay activity in soil preincubated at higher P_{O_2} values and to the anaerobic activity in soil preincubated at lower P_{O_2} values, whereas at intermediate oxygen concentrations the ambient activity was greater than either the aerobic or the anaerobic assay activities.

As an approach to soil core studies, homogenized pasture field soil was moistened with glucose solution and 500 g quantities assayed in PVC cylinders. Acetylene-reduction assays of the aerobic and of the ambient type were performed. The data in Table 20.1 show that because of the high coefficients of variation (24 and 97 %) the difference between aerobic and ambient assays was not significant.

Table 20.1. *Production of ethylene from acetylene (0.1 atm) by sandy loam field soil in an Ar/O_2 (4:1) mixture and in air*

Gas phase	nmoles C_2H_4/g/h	CV %*
Ar:O_2 (aerobic)	1.94 ± 0.94	97
Air (ambient)	1.43 ± 0.17	24

Hand-mixed soil (500 g) was moistened with a 1 % glucose solution.
Values are means of 4 replicates ± S.E.
Rates were derived from the slope of the line of best fit through points in cumulative assays at 1, 2 and 3 days.
* CV = coefficient of variation.

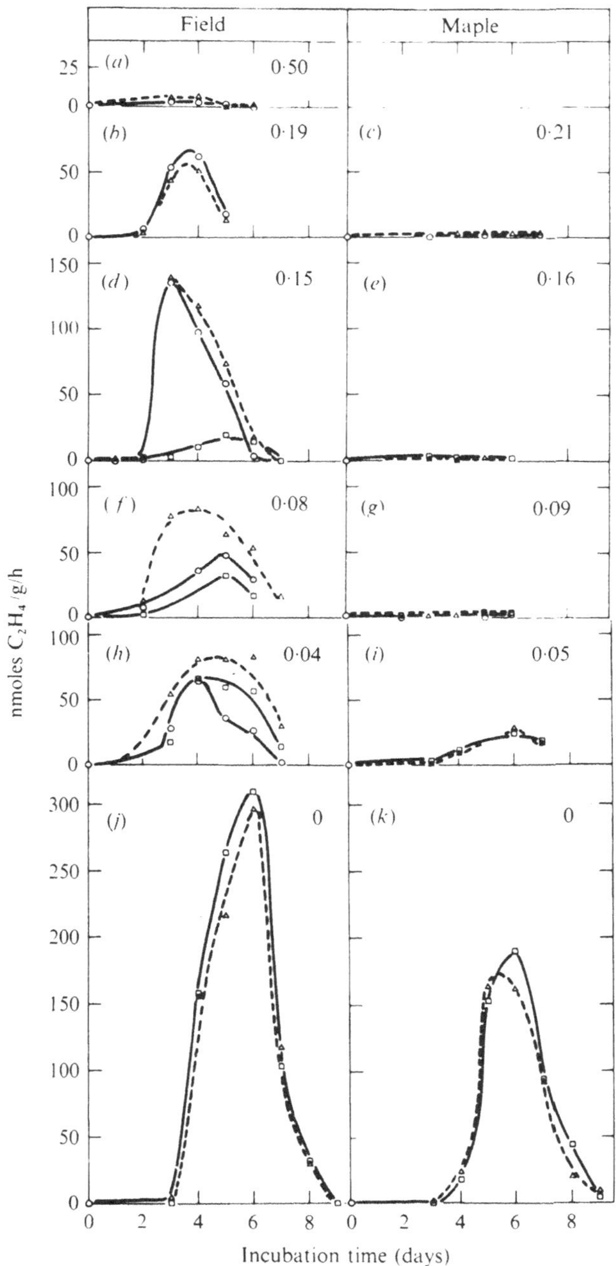

Fig. 20.1. Rates of production of ethylene from acetylene by pasture field (*a*, *b*, *d*, *f*, *h*, *j*) and maple forest (*c*, *e*, *g*, *i*, *k*) soils amended at zero time with 2 % (w/w) glucose and incubated at the P_{O_2} values shown. Acetylene-reduction assays were aerobic (○), anaerobic (□) and ambient P_{O_2} (△).

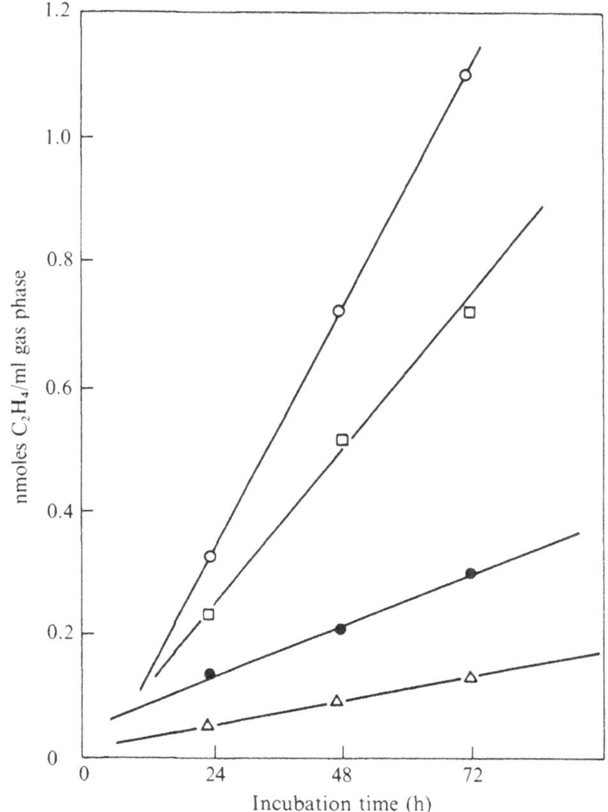

Fig. 20.2. Production of ethylene from acetylene by cores of pasture field and maple forest soils. Representative time course curves for individual cores.

Figure 20.2 shows some representative data from pasture field and maple forest soil cores subjected to aerobic (argon/oxygen) assays. Relative linearity and absence of appreciable lag was observed in most of the time course curves and therefore rates were derived from the slope of lines fitted as shown in Fig. 20.2.

The results of assays conducted between May and November on cores from the pasture field and maple forest sites are shown in Fig. 20.3 along with certain environmental data. Marked fluctuation in the mean activities of maple forest soil cores occurred but this was less pronounced in the pasture cores. The pasture core activities indeed showed a relation to the moisture contents but not to temperature (Fig. 20.3*b*, *c*). Since all assays were done at 30 °C no relation to the seasonal air temperatures shown in Fig. 20.3*c* was to be expected.

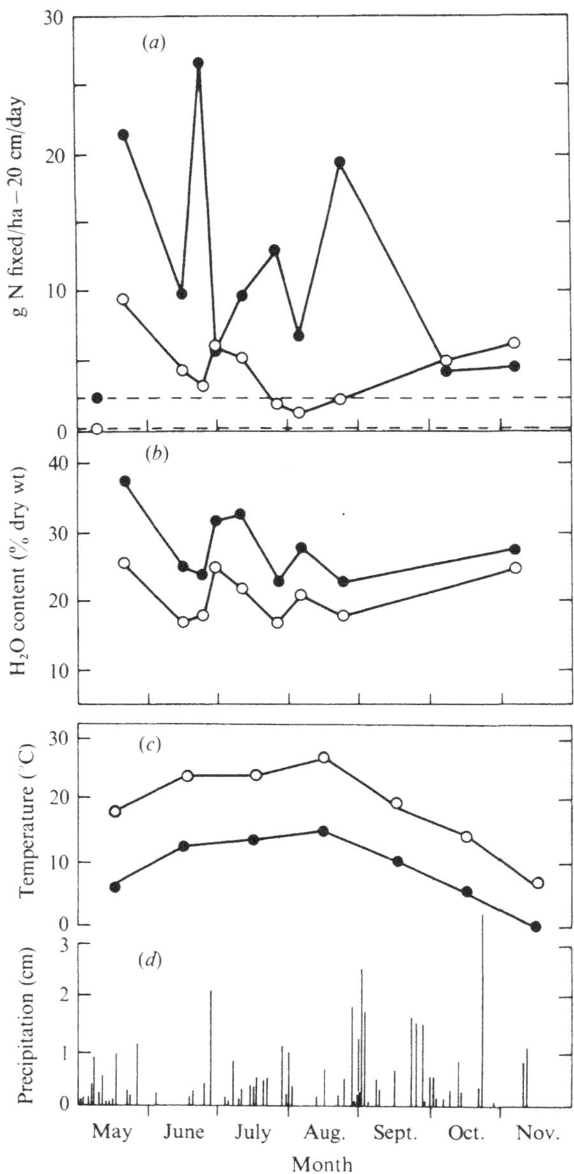

Fig. 20.3. Seasonal data from pasture field (○) and maple forest sites (●), Morgan Arboretum of McGill University. (*a*) Activities calculated as rate of $N_2(C_2H_2)$ fixation assuming an C_2H_4/N_2 conversion factor of 3.0. Each point represents the mean of 8 to 12 replicates; (*b*) soil moisture content (% oven-dry weight). Each point represents the mean of triplicates; (*c*) monthly mean maximum and minimum temperatures; (*d*) daily precipitation.

Table 20.2. *Mean rates of $N_2(C_2H_2)$-fixation by cores of soil from a pasture field and from the forest floor of a stand of Maple* (Acer saccharum). *Values are in g N/ha-20 cm/day*

	Control	Assay	Assay minus control
Field	0.43 ± 1.63 (1970)	4.46 ± 5.77 (398)	4.03 ± 5.99
Maple	2.48 ± 5.77 (1250)	12.6 ± 15.3 (274)	10.2 ± 16.3

Values are the means (± S.E.) of all observations made during the season of study. Figures in parentheses are the mean percentage coefficients of variation.

A consistent feature of the acetylene-reduction data shown in Fig. 20.3*a* was the large sample variation. Table 20.2 shows that the mean of the coefficients of variation observed for control cores of field and maple soils was 1970 % and 1250 %, respectively, and for assay cores was 398 % and 274 %, respectively. The Table also shows the means and standard errors for all seasonal data combined. After correction for ethylene in controls, the calculated mean activities for the pasture field and maple forest soils were 4 and 10 g N/ha-20 cm/day, respectively.

Fig. 20.4 shows representative time-course curves of the production of ethylene by cores of aspen forest soil incubated without acetylene as controls (Fig. 20.4*a*) and with 0.1 atm $P_{C_2H_2}$ in ambient assay (Fig. 20.4*b*). In many core assays (Fig. 20.4*b*) there was a change in the rate of ethylene production at about 24 h. In such cases rates were therefore calculated from the 0 to 24 h period.

The aspen forest soil core studies were performed during the periods 9 to 12 July and 31 July to 3 August. However, analysis of variance of the data showed that results of the two different times could be pooled. All data are therefore summarized in Table 20.3 which shows the results of assays of various components of the organic and mineral parts of the soil profile. Coefficients of variation ranged from 45 to 91 % and were thus much smaller than those observed in the pasture and maple forest study. The data show that on a unit weight basis the highest activity (13.5 nmoles C_2H_4/g/day) was located in the *LF* layer material. The calculated values for N_2 fixation suggest an overall activity for the whole profile of about 6 grams N_2 fixed per hectare per day.

The data shown in Tables 20.2 and 20.3 indicate that endogenous production of ethylene was sufficient, especially in the two forest soils

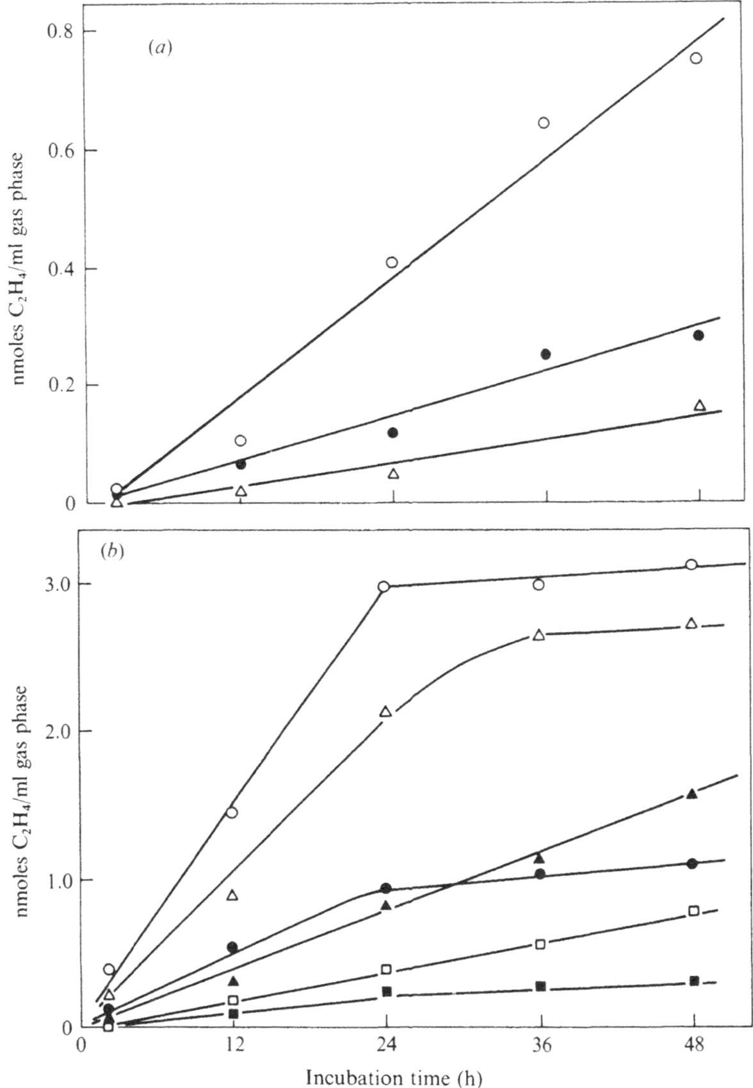

Fig. 20.4. Production of ethylene by cores of aspen forest soil. Representative time course curves for individual cores. (*a*) Control cores incubated without acetylene; (*b*) ambient assay of cores incubated with 0.1 atm $P_{C_2H_2}$.

studied, to warrant the establishment of equal numbers of control and assay cores.

Table 20.3. *Production of ethylene by cores from the forest floor in a stand of aspen* (Populus tremuloides). *Values are in nmoles $C_2H_4/g/day$*

Horizons sampled	Control	Assay	Assay minus control	Calculated** g N/ha/d
LFH	0.873±0.161 (74)	8.44±1.27 (60)	7.57±1.28	3.53
LF	1.86±0.366 (79)	15.4±3.49 (91)	13.5±3.51	4.15
HA2	0.336±0.064 (76)	1.06±0.167 (63)	0.724±0.179	2.43
L-A2/B	0.164±0.026* (45)	0.498±0.093 (75)	0.334±0.096	5.53

Values are means of 16 observations ± S.E. (8 observations for value marked *).
Figures in parentheses are the mean percentage coefficients of variation.
** Calculated assuming a C_2H_4/N_2 conversion factor of 3.0 and using unpublished estimates of horizon weight per unit area.

Discussion

The data from the glucose-amended pasture and maple forest soils are consistent with the known occurrence of *Azotobacter* in only the pasture soil and of *Clostridium* (and possibly facultatively anaerobic N_2 fixers) in both soils (Brouzes *et al.*, 1969). The suppression of development of nitrogenase activity at 0.5 atm P_{O_2} is in agreement with the known sensitivity of *Azotobacter* to oxygen (e.g. Drozd & Postgate, 1970). O'Toole & Knowles (1973) also observed in the same pasture field soil that compared to activity in ambient assay, the exposure to oxygen during aerobic assay caused inhibition of nitrogenase activity and that the degree of inhibition increased as the P_{O_2} of incubation decreased and as the concentration of added glucose decreased. The data therefore suggest that, at least in amended soils, ambient assay without replacement of the soil atmosphere may be preferable.

The large variation associated with the replicate soil core assays in the present study made it difficult not only to obtain precise estimates of activities in the soils examined but to design experiments to establish the most appropriate conditions for assay. The coefficients of variation in the aspen forest soil samples of the present study are within the range of values (17 to 101 %) obtained from data of Stutz & Bliss (1973) in in-situ assays of arctic soils.

The rates of endogenous production of ethylene observed necessitated correction of the acetylene-reduction data. Although comparison is difficult, the ethylene production here appeared to be of the same order of magnitude as that observed by Smith & Restall (1971) in an anaerobic soil.

The acetylene-reduction technique

The mean rates of $N_2(C_2H_2)$ fixation observed in the pasture field, the maple forest and the aspen forest soils were 4, 10 and 6 grams nitrogen per hectare per day, respectively, measured at 30 °C. These rates are similar to some of the lowest of the rates observed in equivalent samples by Brouzes *et al.* (1969) using ^{15}N procedures. Extrapolated to an annual basis, and making rough temperature corrections, they suggest fixed nitrogen contributions of the order of 0.5 to 1.0 kg N/ha/yr.

The authors acknowledge the support of the National Research Council of Canada and of the Canadian Committee for the IBP. Grateful thanks are due to Dr C. I. Mayfield for help in initiating the core study and to Mr D. Denike for able technical assistance. Thanks are also due to Mr D. W. Maclean, Petawawa Forest Experiment Station, Chalk River, Ontario, for providing laboratory space

References

Bergersen, F. J. (1970). The quantitative relationship between nitrogen fixation and the acetylene reduction assay. *Aust. J. Biol. Sci.*, **23**, 1015–25.

Brouzes, R. & Knowles, R. (1971). Inhibition of growth of *Clostridium pasteurianum* by acetylene: implication for nitrogen fixation assay. *Can. J. Microbiol.*, **17**, 1483–9.

Brouzes, R. & Knowles, R. (1973). Kinetics of nitrogen fixation in a glucose-amended, anaerobically incubated soil. *Soil Biol. Biochem.*, **5**, 223–9.

Brouzes, R., Lasik, J. & Knowles, R. (1969). The effect of organic amendment, water content, and oxygen on the incorporation of $^{15}N_2$ by some agricultural and forest soils. *Can. J. Microbiol.*, **15**, 899–905.

Brouzes, R., Mayfield, C. I. & Knowles, R. (1971). Effect of oxygen partial pressure on nitrogen fixation and acetylene reduction in a sandy loam soil amended with glucose. *Plant and Soil, Special Volume*, 481–94.

Drozd, J. & Postgate, J. R. (1970). Interference by oxygen in the acetylene reduction test for aerobic nitrogen-fixing bacteria. *J. gen. Microbiol.*, **60**, 427–9.

Knowles, R. (1975). The significance of asymbiotic dinitrogen fixation by bacteria. In *Dinitrogen* (N_2) *Fixation* (ed. Hardy, R. W. F.). Wiley-Interscience, New York (in press).

Knowles, R., Brouzes, R. & O'Toole, P. (1973). Kinetics of nitrogen fixation and acetylene reduction, and effects of oxygen and acetylene on these processes, in a soil system. In *Modern Methods in the Study of Microbial Ecology* (ed. Rosswall, T.), *Bulletin of the Ecological Research Committee, Stockholm*, **17**, 255–61.

O'Toole, P. & Knowles, R. (1973). Oxygen inhibition of acetylene reduction (nitrogen fixation) in soil: effect of glucose and oxygen concentrations. *Soil Biol. Biochem.*, **5**, 783–8.

Smith, K. A. & Restall, S. W. F. (1971). The occurrence of ethylene in anaerobic soil. *J. Soil Sci.*, **22**, 430–43.

Stutz, R. C. & Bliss, L. C. (1973). Acetylene reduction assay for nitrogen fixation under field conditions in remote areas. *Pl. Soil*, **38**, 209–13.

PART IV

The biochemistry of nitrogen fixation

21. The biochemistry of nitrogen fixation – an introduction

F. J. BERGERSEN

There were good reasons for the inclusion of a theme on the biochemistry of nitrogen fixation in the over-all PP–N programme, while omitting an operational plan. Extensive research on nitrogen-fixation biochemistry was already in progress in several countries when the preparative phases of the IBP began in 1966. Little was therefore to be gained by attempting to superimpose an international programme of research. However, the PP–N subcommittee considered that the results of biochemical work would probably influence what was done in the more biological themes and also that the IBP could provide opportunities for communication between biochemical workers and between them and workers in other PP–N themes. Both of these considerations proved to have been correct. There is no better example of the first than the widespread use of the acetylene-reduction technique in biological and ecological studies within the IBP. The publication of the report of the basic biochemistry of this technique and its first applications coincided with the beginning of the operational phase. Further, the understanding of the nature of nitrogenase and the conditions necessary for the maintenance of its activity, which has developed in the past five years, is an important part of the background upon which much of the work in the area of nitrogen fixation is now based. Many reviews and books published recently or about to be published, have contributed to this (e.g. Bergersen, 1971; 1973; Chatt & Fogg, 1969; Dalton & Mortenson, 1972; Evans, 1969; Fottrell, 1968; Hardy & Burns, 1968; 1972; Hardy, Burns & Parshall, 1971; 1972; Hardy, 1975; Murray & Smith, 1968; Postgate, 1970; 1971*a*, *b*; Quispel, 1974). In addition, conferences such as those held under IBP auspices in 1967 in Addis Ababa, at Prague and Wageningen in 1970 and in Canberra in 1971, have provided opportunities for the results of research in biochemical and physiological aspects of the subject to be passed on to workers concerned with other aspects of nitrogen fixation. These opportunities are not usually presented at the more restricted specialist conferences which have become the usual form of international meetings.

It is against this background that the main contributors to theme 10 have prepared their papers. The collation of reported progress in the biochemistry of nitrogen fixation has been divided into several aspects, each of which is of importance to the more applied study of this process.

The serology of *Rhizobium*

Theme 10 also contains reference to the serology of *Rhizobium* spp. This topic will not be specifically dealt with by any other collator so I have included a brief review in this introductory paper. Its inclusion in theme 10 is somewhat incongruous and its connection with nitrogen fixation is indirect. However, serological methods have been widely used in the study of the legume symbiosis and have played a part in IBP experiments. The following is not an exhaustive treatment of the subject, but examples of the various types of research which have been done during the period 1967–72 inclusive, will be cited.

Methods

The Ouchterlony technique of immune precipitation in agar gels has been extensively developed for work with rhizobia, supplementing the agglutination methods which have been used in the past (Dudman, 1964; Skrdleta, 1967; 1969*a*; Vincent, 1970). Bacterial cells are used in the antigen wells, from which specific soluble antigens diffuse in sufficient quantity for reaction with the antibodies diffusing from the antiserum wells. *Rhizobium japonicum* antigens are released slowly and poor results ensue unless the cells are first sonicated (Dudman, 1971) or heated (Skrdleta, 1969*b*; Dudman, 1971). Strain specific antigens of this species are heat-stable (Means & Johnson, 1968; Skrdleta, 1969*b*) while heat-labile antigens are generally non-specific. Ultra-sonic treatment may cause loss of antigenicity with *R. meliloti* (Gibbins, 1967).

The use of crushed, fresh soybean nodules as a direct source of rhizobial antigens in agglutination reactions was developed earlier (Means, Johnson & Date, 1964; Skrdleta, 1967) and has been extended successfully to the more specific immune-diffusion reaction (Skrdleta, 1969*c*). However, the use of dried nodules led to reduced numbers of identifications (Skrdleta & Mareckova, 1971), presumably because of occlusion or inactivation of antigenic material during drying. A similar technique has been applied also to the smaller nodules of clover plants (Parker & Grove, 1970).

Fluorescent antibody techniques have proved to be particularly useful in the detection of sero-types of free-living rhizobia in soil (Bohlool, Gray & Schmidt, 1968; Schmidt, Bankole & Bohlool, 1968; Bohlool & Schmidt, 1970). The method has also been applied to the identification of rhizobial strains in crushed preparations from small nodules which are not suitable for direct agglutination or immune diffusion (Trinick, 1969). In this way, the step of isolating and culturing nodule bacteria from small nodules prior to serological typing, is avoided. Jones & Russell (1972) have also used fluorescent antibody methods in the study of the early stages of the infection of white clover roots and for the identification of rhizobia in nodule tissue.

The use of serology in strain identification

Serological methods have been widely used in the evaluation of legume inoculation. The ability of inoculant strains to persist in soils has generally been found to be limited to one to three years after which the recovery from nodules of bacteria which are serologically identical with the inoculant strains becomes increasingly difficult (Dudman & Brockwell, 1968; Brockwell, Bryant & Gault, 1972; DeEscuder, 1972). The ability to live saprophytically in the soil was considered to be an important character for inoculant strains by Chatel, Greenwood & Parker (1968) who used serological methods to evaluate their results. The persistence of the serological characters in strains introduced into soils should be investigated further before these characters alone are used as criteria for strain persistence. Genetic modification may be widespread in soils, perhaps mediated by wide-range, temperate bacteriophages (Bergersen, Brockwell, Gibson & Schwinghamer, 1971) which may be potent agents of transduction. The use of multiple, unlinked marker characters, including serological, antibiotic resistance (Schwinghamer & Dudman, 1973) and symbiotic (Brockwell, Hely & Neale-Smith, 1966) characters, may be the only way by which the true persistence of unaltered inoculant rhizobia may be measured.

Serological methods have been extensively used to investigate the ability of strains of rhizobia to produce nodules in competition with other strains inoculated simultaneously or subsequently or with rhizobia already present in the soil (Brockwell & Dudman, 1968; Caldwell, 1969; Ham, Cardwell & Johnson, 1971; Skrdleta, 1968; 1970; 1971; Skrdleta & Karimova, 1969). The generalized findings were as follows: (*a*) There are strong competitive differences between strains of the same species (Gibson, 1968; Brockwell & Dudman, 1968; Caldwell, 1969) but differences also occur between greenhouse and field results (Skrdleta, 1968). (*b*) Strains applied early have an advantage in nodule formation (Skrdleta, 1970). (*c*) With strains of equal competitiveness, the proportion of nodules formed by one of them is usually related to the proportion present in the inoculum (Skrdleta, 1971; Skrdleta & Karimova, 1969). (*d*) Nodules containing two serotypes of rhizobia occurred, especially when both strains were applied simultaneously (Skrdleta, 1970). (*e*) Effective strains of *R. trifolii* had a competitive advantage over ineffective strains under conditions of restricted rhizosphere multiplication and serological methods were used to show that the discrimination occurred early in the infection process (Robinson, 1969) but this has not been confirmed in other work (Jones & Russell, 1972).

In the USA there has been considerable research on the occurrence of various sero-groups of *R. japonicum* in the nodules of soybean crops. The identification of a sero group is usually made on the basis of the pattern of agglutination of bacteroids in crushed nodules using antisera to certain standard laboratory strains. In Iowa soils, 97 % of nodules reacted with one or more of seven antisera (Ham, Frederick & Anderson, 1967; 1971). Sero-group 123 was found in 40–63 % of nodules on some soils while on some alkaline soils, group 135 was dominant; other sero-groups occurred at much lower frequencies and the frequency of all sero-types was affected by soil type (Damirgi, Frederick & Anderson, 1971). Later work on other soils has shown effects of soil type on sero-group frequency but no consistent correlations with identifiable soil properties have been obtained (Caldwell & Hartwig, 1970; Bezdicek, 1972). Host variety (Caldwell & Vest, 1968) and planting date (Caldwell & Weber, 1970) also influenced the sero-group distribution. Vest, Caldwell & Petersen (1971) studied sampling procedures and found considerable variability in sero-group distribution, which originated mainly in 'replicate in area' and 'samples in replicate' effects. Gibson *et al.* (1971) studied individual cultures isolated from nodules containing serotypes identified by agglutination and found considerable serological diversity within groups by the use of immune diffusion reactions, although one common antigen was usually present.

Serological characters have been included in genetic studies. For example, Mareckova (1970) used serological typing to check against the presence of surviving donor cells in transformation experiments and Macgregor & Alexander (1972) found minor serological differences between an *R. trifolii* strain and a non-nodulating mutant derived from it.

Barnet & Vincent (1970) found a simultaneous change in the somatic antigen of a strain of *R. trifolii* when it was lysogenized, coupled with a loss of ability to adsorb the same or related phages. However, later work revealed no relationships between the susceptibility of strains to a particular phage and the sharing of a somatic antigen (Barnet, 1972).

The investigation of the taxonomic position of the rhizobia has employed serological techniques. For example, Chen, Hsu, Stevenson & Gainer (1967) found cross-reactions between *Agrobacterium tumefaciens* and *R. meliloti* and Graham & O'Brien (1968) found similarities in the chemical compositions of lipopolysaccharides isolated from nine strains of *Agrobacterium* spp. and 16 strains of *Rhizobium* spp. Later work with antisera prepared against three *A. radiobacter* and six *A. tumefaciens* strains, showed the presence of three common and five different lipopolysaccharide antigens. These antisera cross-reacted strongly with 37 strains of fast-growing *Rhizobium* spp. but only one out of seven *R. lupini* strains (representing slow-growing rhizobia) produced any reaction (Graham, 1971). Similar results were obtained by Vincent & Humphrey (1970) using immune-diffusion reactions of antisera prepared against *R. trifolii* and *R. meliloti* and homogenates of rhizobia and agrobacteria. Relationships between fast-growing rhizobia and agrobacteria were attributed to related internal antigens released from within the cells.

Immunochemical studies

These studies have concerned the chemical nature of the somatic antigens isolated from rhizobia and the study of their extracellular polysaccharides.

Vincent & Humphrey (1968) studied the properties of calcium-deficient cells of *R. trifolii* which appeared from electron microscope studies to have defective cell walls. Quantitative antibody absorption studies using antisera prepared against whole cells and somatic agglutination reactions suggested that calcium in normal cell walls obscured or modified an antigenic group. Treatment of normal cells with EDTA or growth in calcium-deficient media released this antigen. The calcium-content of the cell walls did not influence electrophoretic mobility of the cells (Humphrey, Marshall & Vincent, 1968) and the calcium was therefore not apparently concerned with the surface charge of the bacteria. However, there was a correlation between the surface charge-density of whole cells and the electrophoretic mobility of isolated somatic antigen (Humphrey & Vincent, 1969*b*).

The somatic antigens of two strains of *R. trifolii* were found to resemble somatic antigens of the Enterobacteriaceae in containing firmly bound lipid, 2-*keto*-3-deoxyoctonate (KDO), glucose, mannose and fucose. However, the *R. trifolii* antigens differed from the enterobacterial antigens in containing glucuronic acid but little phosphorus (Humphrey & Vincent 1969*a*). The presence of KDO in *R. trifolii* somatic antigens was confirmed by Ellwood (1970) and Lorkiewicz & Russa (1971). The latter authors also concluded that the antigens of smooth cultures were more complex than those of rough cultures.

Exopolysaccharides of rhizobia have been studied with regard to their composition and structure and their antigenic properties. Keele, Elkan & Wheat (1967; 1968) reported that protein contaminated preparations made by ethanol precipitation of culture supernatants; purer preparations resulted when the fluids were dialyzed and treated with phenol before precipitation. These observations emphasize a deficiency in many of the reported studies, i.e. a lack of criteria of homogeneity in descriptions of the composition of rhizobial polysaccharides. Dudman (1964) used the Sevag procedure to remove protein, followed by dialysis to purify *R. meliloti* polysaccharides and also showed that several different physical types of polysaccharides were found in 23 representative strains of rhizobia (Dudman, 1968). Electrophoresis in gels using borate buffer with detection by means of dyes or DEAE-dextran (for acidic polysaccharides) can be used for the analysis of homogeneity (Dudman & Bishop, 1968; Dudman, 1972) but generally, preparative purification of rhizobial polysaccharides prior to analysis has been minimal. Stevenson, Clapp & Molina (1970) found traces (0.1–0.7 %) of glucosamine, galactosamine, muramic acid and possibly fucosamine in preparations from all of seven strains representing *R. leguminosarum*, *R. trifolii*, *R. meliloti* and *R. lupini*, suggesting that contamination with cell wall materials is widespread in such preparations. This is not surprising in view of the diffusible nature of somatic antigens shown from immune diffusion reactions with whole cells. Contamination of culture fluids with these materials is therefore inevitable, although representing a small proportion of the extracellular materials. Reports of the presence of small amounts of components (say 1 % or less) should be viewed critically against data about the homogeneity of the preparation.

A number of wide-ranging surveys of polysaccharide composition have been reported. The polysaccharides of ten strains of *R. meliloti* were consistently glucose (83–87 %), galactose (13–17 %) and glucuronic acid (0.4–1.2 %) (Amarger, Obaton & Blachère, 1967). In an examination of 10 strains of *R. trifolii*, chosen to include four effective, four ineffective and two avirulent representatives, no significant difference was found in the composition of any of their polysaccharides (Hepper, 1972). Nine strains produced polysaccharides with constituents in the same approximate proportions: glucose (25–43 %), galactose (6–11 %), glucuronic acid (15–20 %), pyruvic acid (6–11 %) and acetyl groups (5–6 %); the glucose : galactose ratio in all was close to 4:1. The polysaccharide of one effective strain differed by containing 47 % glucose and only 1 % galactose. It was concluded that there was no correlation between the composition of the polysaccharides and the ability of the strain to nodulate, or, if nodulation occurs, the ability to fix nitrogen. It is possible that although similar in composition these polysaccharides may differ in structural detail.

Zevenhuizen (1971) studied the polysaccharides of 21 strains, representing five taxonomic groups of rhizobia. All had glucose as the main monosaccharide (*R. leguminosarum*, *R. phaseoli*, *R. trifolii*, 58–61 %, *R. meliloti* 76–83 %) with galactose and sometimes mannose. The only uronic acid to be identified was glucuronic acid, detected as the free acid and the lactone; aldobiuronic acids, products of incomplete hydrolysis were present and could not be completely hydrolyzed even under more drastic conditions. No 4-*O*-methylglucuronic acid was found. Bailey, Greenwood & Craig (1971) found that differences in sugar compositions appeared to be correlated with the division into acid-producing and non-acid-producing types in a study of the exopolysaccharides of lotus rhizobia. Some reports showed deficits of 10–20 % between dry weight and sugar composition (e.g. Dudman, 1964). Recent studies (Dudman & Heidelberger, 1969; Heidelberger, Dudman & Nimmich, 1970; Zevenhuizen, 1971) have shown that this was due to the presence of acyl substituents in rhizobial polysaccharides. Pyruvate has been found to the extent of 3–15 % and acetyl to 2–13 %.

Chaudhari, Bishop & Dudman (1973) have proposed a structure for the repeating unit of the polysaccharide produced by *R. trifolii* strain TA1 (which is shown on the next page) but the location of the 4 % acetyl which occurs in this material (Dudman & Heidelberger, 1969) has not been determined.

Bjorndahl *et al.* (1971) studied the extracellular polysaccharide of *R. meliloti* and in agreement with Dudman & Heidelberger (1969),

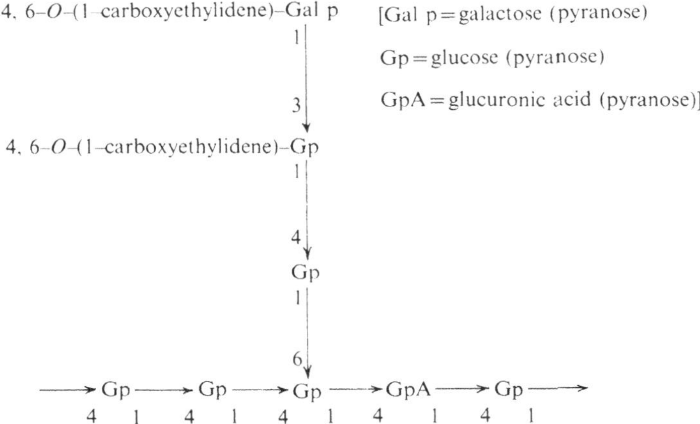

found that it contained glucose, galactose, pyruvate and *O*-acetyl in the ratio of 7:1:1:1. All the sugars were pyranosidic and *β*-linked. The pyruvate was present as terminal 4,6-*O*-(1-carboxyethylidene)-D-glucose and the acetyl groups were attached at the 6-position of 3- and 4-substituted glucose residues. However, the order of the linkages of the sugar residues was not determined.

The antigenic properties of the extracellular polysaccharides of the rhizobia have been studied by several workers. In general, these materials are poorly antigenic when injected into rabbits, but antisera prepared against whole cells contain antibodies which react with them. Amarger *et al.* (1967) identified a polysaccharide precipitin band in immune diffusion reactions with whole cells of 8 out of 10 strains of *R. meliloti* and a common polysaccharide band with some strains of *R. trifolii*, while Failly & Blachère (1968) found a common polysaccharide band in reactions with all 10 strains of *R. meliloti* which they studied. Dudman & Heidelberger (1969) found that although the polysaccharides of five strains of *R. trifolii* contained very similar proportions of glucose, galactose and glucuronic acid, they displayed little serological cross-reactivity, while reacting strongly with their homologous antisera. They all cross-reacted with type 27 anti-pneumococcal antisera. These properties were attributed to the presence of pyruvyl- and acetyl-substituents. Removal of pyruvate from TA1 polysaccharide by mild acid treatment removed the activity with TA1 antiserum and with anti-pneumococcus type 27. Removal of the acetyl group with dilute alkali also removed the activity with the homologous antiserum but the cross-reactivity with pneumococcus type 27 remained and was attributed to the presence of pyruvate groups. When depyruvilated, TA1 polysaccharide cross-reacted with antisera to pneumococcus types 6–10 and 14 which contain no pyruvate. These studies have contributed significantly to the establishment of the importance of acyl-substituents in the immunology of bacterial polysaccharides (Heidelberger *et al.*, 1970) and have established beyond doubt that the serological activity of rhizobial polysaccharides is not due to the presence of traces of somatic antigens in the preparations. The presence of specific chemical configurations in these polysaccharides, which represent the outermost surfaces of the rhizobia, again raises the possibility that they may be concerned in specific processes connected with the infection of legume roots, as suggested by Bjorndahl *et al.* (1971).

References

Amarger, N., Obaton, M. & Blachère, H. (1967). Polysaccharides extracellulaires de *Rhizobium meliloti*. *Can. J. Microbiol.*, **13**, 99–105.

Bailey, R. W., Greenwood, R. M. & Craig, A. (1971). Extracellular polysaccharides of *Rhizobium* strains associated with *Lotus* species. *J. gen. Microbiol.*, **65**, 315–24.

Barnet, Y. M. (1972). Bacteriophages of *Rhizobium trifolii*, 1. Morphology and host range. *J. gen. Virol.*, **15**, 1–15.

Barnet, Y. M. & Vincent, J. M. (1970). Lysogenic conversion of *Rhizobium trifolii*. *J. gen. Microbiol.*, **61**, 319–25.

Bergersen, F. J. (1971). Biochemistry of symbiotic nitrogen fixation in legumes. *Ann. Rev. Pl. Physiol.*, **22**, 121–40.

Bergersen, F. J. (1973). Symbiotic nitrogen fixation by legumes. In *Chemistry and Biochemistry of Herbage* (ed. Butler, G. W. and Bailey, R. W.), vol. 2, pp. 189–226. Academic Press: London, New York.

Bergersen, F. J., Brockwell, J., Gibson, A. H. & Schwinghamer, E. A. (1971). Studies of natural populations and mutants of *Rhizobium* in the improvement of legume inoculants. *Plant and Soil, Special Volume*, 3–16.

Bezdicek, D. F. (1972). Effect of soil factors on the distribution of *Rhizobium japonicum* serogroups. *Soil Sci. Soc. Amer. Proc.*, **36**, 305–7.

Bjorndahl, H., Erbing, C., Lindberg, B., Fahreus, G. & Ljunggren, H. (1971). Studies on an extracellular polysaccharide from *Rhizobium meliloti*. *Acta chem. scand.*, **25**, 1281–6.

Bohlool, B. B., Gray, T. A. G. & Schmidt, E. L. (1968). Immunofluorescence observations on *Rhizobium japonicum* in soils and nodules. *Bacteriol. Proc.*, p. 4.

Bohlool, B. B. & Schmidt, E. L. (1970). Immunofluorescent detection of *Rhizobium japonicum* in soils. *Soil Sci.*, **110**, 229–36.

Brockwell, J., Bryant, W. G. & Gault, R. R. (1972). Ecological studies of root-nodule bacteria introduced into field environments, 3. Persistence of *Rhizobium trifolii* in association with white clover at high elevations. *Aust. J. exp. Agric. Anim. Husb.*, **12**, 407–13.

Brockwell, J. & Dudman, W. F. (1968). Ecological studies of root-nodule bacteria introduced into field environments, 2. Initial competition between seed inocula in the nodulation of *Trifolium subterraneum* L. seedlings. *Aust. J. agric. Res.*, **19**, 749–57.

Brockwell, J., Hely, F. W. & Neal-Smith, C. (1966). Some symbiotic characteristics of rhizobia responsible for spontaneous, effective field nodulation of *Lotus hispidus*. *Aust. J. exp. Agric. Anim. Husb.*, **6**, 365–70.

Caldwell, B. E. (1969). Initial competition of root nodule bacteria on soybean in a field environment. *Agron. J.*, **61**, 813–15.

Caldwell, B. E. & Hartwig, E. E. (1970). Serological distribution of soybean root nodule bacteria in soils of south eastern U.S.A. *Agron. J.*, **62**, 621–2.

Caldwell, B. E. & Vest, G. (1968). Nodulation interactions between soybean (*Glycine max*) genotypes and serogroups of *Rhizobium japonicum*. *Crop Sci.*, **8**, 680–2.

Caldwell, B. E. & Weber, D. F. (1970). Distribution of *Rhizobium japonicum* serogroups in soybean nodules as affected by planting dates. *Agron. J.*, **62**, 12–14.

Chatel, D. L., Greenwood, R. M. & Parker, C. A. (1968). Saprophytic competence as an important character in the selection of *Rhizobium* for inoculation. *IX Int. congr. Soil Sci. Trans.* II, 65–73.

Chatt, J. & Fogg, G. E. (1969). A discussion on nitrogen fixation. *Proc. R. Soc. Lond. B*, **172**, 317–437.

Chaudhari, A. S., Bishop, C. T. & Dudman, W. F. (1973). Structural studies on the specific capsular polysaccharide from *Rhizobium trifolii* TA1. *Carbohydrate Res.*, **28**, 221–31.

Chen, P. K., Hsu, J. C., Stevenson, T. & Gainor, C. (1967). Serological studies with Agrobacteria. *Bacteriol. Proc.*, p. 11.

Dalton, H. & Mortenson, L. E. (1972). Dinitrogen (N_2) fixation (with a biochemical emphasis). *Bacteriol. Rev.*, **36**, 231–60.

Damirgi, S. M., Frederick, L. R. & Anderson, I. C. (1967). Serogroups of *Rhizobium japonicum* in soybean nodules as affected by soil types. *Agron. J.*, **59**, 10–12.

DeEscuder, A. M. Q. (1972). A survey of rhizobia in farm soils at Wye College, Kent. *J. appl. Bacteriol.*, **35**, 109–18.

Dudman, W. F. (1964). Immune diffusion analysis of the extracellular soluble antigens of *Rhizobium meliloti*. *J. Bacteriol.*, **88**, 782–94.

(1968). Capsulation in *Rhizobium* species. *J. Bacteriol.*, **95**, 1200–1.

(1971). Antigenic analysis of *Rhizobium japonicum* by immunodiffusion. *Appl. Microbiol.*, **21**, 973–85.

(1972). Detection of acidic polysaccharides in gels by DEAE dextran. *Analyt. Biochem.*, **46**, 668–73.

Dudman, W. F. & Bishop, C. T. (1968). Electrophoresis of dyed polysaccharides on cellulose acetate. *Can. J. Chem.*, **46**, 3097–84.

Dudman, W. F. & Brockwell, J. (1968). Ecological studies of root-nodule bacteria introduced into field environments. 1. A survey of field performance of clover inoculants by gel immune diffusion serology. *Aust. J. agric. Res.*, **19**, 739–47.

Dudman, W. F. & Heidelberger, M. (1969). Immunochemistry of newly found substituents of polysaccharides of *Rhizobium* species. *Science*, **164**, 954–5.

Ellwood, D. C. (1970). The distribution of 2-keto-3-deoxyoctonic acid in bacterial walls. *J. gen. Microbiol.*, **60**, 373–80.

Evans, H. J. (1969). How legumes fix nitrogen. In *How Crops Grow* (ed. Horsfall, J. G.), Bull. 708. Conn. Exp. Sta.: New Haven, USA.

Failly, C. & Blachère, H., (1968). Identité immunochemique de certains polysaccharides extracellulaires chez diverses espêces de *Rhizobium*, 1. Polysaccharides extracellulaires solubles de *Rhizobium meliloti. Can. J. Microbiol.*, **14**, 247–51.

Fottrell, P. F. (1968). Recent advances in biological nitrogen fixation. *Sci. Progr. Oxford*, **56**, 541–55.

Gibbons, L. N. (1967). The preparation of antigens of *Rhizobium meliloti* by ultrasonic disruption: an anomoly. *Can. J. Microbiol.*, **13**, 1375–8.

Gibson, A. H. (1968). Nodulation failure in *Trifolium subterraneum* L. cv. Woogenelup (syn. Marrar). *Aust. J. agric. Res.*, **19**, 907–18.

Gibson, A. H., Dudman, W. F., Weaver, R. W., Horton, J. C. & Anderson, I. C. (1971). Variations within serogroup 123 of *Rhizobium japonicum*. *Pl. Soil*, *Special Volume*, 33–7.

Graham, P. H. (1971). Serological studies with *Agrobacterium radiobacter*, *A. tumefaciens*, and *Rhizobium* strains. *Arch. Mikrobiol.*, **78**, 70–5.

Graham, P. H. & O'Brien, M. A. (1968). Composition of lipopolysaccharides from *Rhizobium* and *Agrobacterium*. *Antonie van Leeuwenhoek*, **34**, 326–30.

Ham, G. E., Cardwell, V. B. & Johnson, H. W. (1971). Evaluation of *Rhizobium japonicum* inoculants in soils containing natural populations of rhizobia. *Agron. J.*, **63**, 301–3.

Ham, G. E., Frederick, L. R. & Anderson, I. C. (1967). Serogroups of *Rhizobium japonicum* in relationship to soil properties and soybean varieties. *Bacteriol Proc.*, p. 4.

Ham, G. E., Frederick, L. R. & Anderson, I. C. (1971). Serogroups of *Rhizobium japonicum* in soybean nodules. *Agron. J.*, **63**, 69–72.

Hardy, R. W. F. (ed.) (1975). *Dinitrogen* (N_2) *Fixation*. Wiley-Interscience: New York (in press).

Hardy, R. W. F. & Burns, R. C. (1968). Biological nitrogen fixation. *Ann. Rev. Biochem.*, **37**, 331–58.

Hardy, R. W. F. & Burns, R. C. (1972). Comparative biochemistry of iron sulfur proteins and dinitrogen fixation. In *Iron Sulfur Proteins* (ed. Lovenberg, W.), pp. 65–110. Academic Press: New York, London.

Hardy, R. W. F., Burns, R. C. & Parshall, G. W. (1971). The biochemistry of N_2 fixation. *Adv. Chem. Series*, **100**, 219–47.

Hardy, R. W. F., Burns, R. C. & Parshall, G. W. (1972). Bioinorganic chemistry of dinitrogen fixation. In *Inorganic Biochemistry* (ed. Eichorn, G.). Elsevier: Amsterdam (in press).

Heidelberger, M., Dudman, W. F. & Nimmich, W. (1970). Immunochemical relationships of certain capsular polysaccharides of Klebsiella, Pneumococci and Rhizobia. *J. Immunol.*, **104**, 1321–8.

Hepper, C. M. (1972). Composition of extracellular polysaccharides of *Rhizobium trifolii. Antonie van Leeuwenhoek*, **38**, 437–45.

Humphrey, B. A., Marshall, K. C. & Vincent, J. M. (1968). Electrophoretic mobility of calcium adequate and calcium deprived *Rhizobium trifolii. J. Bacteriol.*, **95**, 721.

Humphrey, B. A. & Vincent, J. M. (1969*a*). The somatic antigens of two strains of *Rhizobium trifolii. J. gen. Microbiol.*, **59**, 411–25.

Humphrey, B. A. & Vincent, J. M. (1969*b*). Correlation between the surface charge densities of whole cells and electrophoretic movement of isolated somatic antigen of *Rhizobium trifolii. J. Bacteriol.*, **98**, 845–6.

Jones, D. G. & Russell, P. E. (1972). The application of immunofluorescence techniques to host plant/nodule bacteria selectivity experiments using *Trifolium repens. Soil Biol. Biochem.*, **4**, 277–82.

Keele, B. B., Elkan, G. H. & Wheat, R. W. (1967). Composition of the extracellular material of four strains of *Rhizobium japonicum. Bacteriol. Proc.*, p. 31.

Keele, B. B., Elkan, G. H. & Wheat, R. W. (1968). Alcohol precipitation as a method of collecting extracellular material from *Rhizobium japonicum. Appl. Microbiol.*, **16**, 155–6.

Lorkiewicz, Z. & Russa, R. (1971). Immunochemical studies of *Rhizobium* mutants. *Plant and Soil, Special Volume*, 105–9.

Mareckova, H. (1970). Transformation of infectivity in *Rhizobium japonicum. Zentbl. Bakt. ParasitKde Abt. II*, **125**, 594–6.

Macgregor, A. N. & Alexander, M. (1972). Comparison of nodulating and non-nodulating strains of *Rhizobium trifolii. Pl. Soil*, **36**, 129–39.

Means, U. M. & Johnson, H. W. (1968). Thermostability of antigens associated with serotypes of *Rhizobium japonicum. Appl. Microbiol.*, **16**, 203–6.

Means, U. M., Johnson, H. W. & Date, R. A. (1964). Quick serological method of classifying strains of *Rhizobium japonicum* in nodules. *J. Bacteriol.*, **87**, 547–53.

Murray, R. & Smith, D. C. (1968). The activation of molecular nitrogen. *Coordin. Chem. Rev.*, **3**, 429–70.

Parker, C. A. & Grove, P. L. (1970). The rapid serological identification of rhizobia in small nodules. *J. appl. Bacteriol.*, **33**, 248–52.

Postgate, J. R. (1970). Biological nitrogen fixation. *Nature, Lond.*, **226**, 25–7.

(1971*a*). *The Chemistry and Biochemistry of Nitrogen Fixation.* Plenum Press: London, New York, 326 pp.

(1971*b*). Relevant aspects of the physiological chemistry of nitrogen fixation. In *Microbes and Biological Productivity, Symp. Soc. Gen. Microbiol.*, vol. XXI, pp. 287–307.

Quispel, A. (1974). *Biological Nitrogen Fixation.* North-Holland: Amsterdam. 769 pp.

Robinson, A. C. (1969). Competition between effective and ineffective strains of *Rhizobium trifolii* in the nodulation of *Trifolium subterraneum*. *Aust. J. agric. Res.*, **20**, 827–41.

Schmidt, E. L., Bankole, R. O. & Bohlool, B. B. (1968). Fluorescent antibody approach to the study of rhizobia in soil. *J. Bacteriol.*, **95**, 1987–92.

Schwinghamer, E. A. & Dudman, W. F. (1973). Evaluation of spectinomycin resistance as a marker for ecological work with rhizobium. *J. appl. Bacteriol.*, **36**, 263–72.

Skrdleta, V. (1967). The serological identity of agar cultures of *Rhizobium japonicum* and bacterial component of soybean root nodules (in Czech.). *Rostl. Vyrob.*, **13**, 671–8.

(1968). Inoculation of soybean with mixtures of different somatic serogroup strains (in Czech.). *Rostl. Vyrob.*, **14**, 969–78.

(1969*a*). Determination of soybean root nodule origin by the Ouchterlony method. *Zentbl. Bakt. ParasitKde Abt. II*, **123**, 111–15.

(1969*b*). Serological analysis of eleven strains of *Rhizobium japonicum*. *Antonie van Leeuwenhoek*, **35**, 77–83.

(1969*c*). Application of immunoprecipitation in agar gel for the serological typing of soybean root nodules. *Folia microbiol.*, **14**, 32–5.

(1970). Competition for nodule sites between two inoculum strains of *Rhizobium japonicum*. *Soil Biol. Biochem.*, **2**, 167–71.

(1971). Competition between inoculum strains of *Rhizobium japonicum* in nodulation of soybean during three planting periods. *Folia Microbiol.*, **16**, 515.

Skrdleta, V. & Karimova, J. (1969). Competition between two somatic serotypes of *Rhizobium japonicum* used as double strain inocula in varying proportions. *Arch. Mikrobiol.*, **66**, 25–8.

Skrdleta, V. & Mareckova, H. (1971). Serotyping of fresh and dried soybean root nodules. *Zentbl. Bakt. ParasitKde Abt. II*, **126**, 656–8.

Stevenson, F. J., Clapp, C. E. & Molina, J. A. E. (1970). Occurrence of amino sugars in extracellular polysaccharides of the genus *Rhizobium* and some observations regarding their distribution in somatic components. *Soil Sci. Soc. Am. Proc.*, **34**, 759–62.

Trinick, M. J. (1969). Identification of legume root nodule bacteroids by fluorescent antibody reactions. *J. appl. Bacteriol.*, **32**, 181–6.

Vest, G., Caldwell, B. E. & Petersen, H. D. (1971). Variability associated with sampling *Rhizobium japonicum* populations in soybeans. *Crop Sci.*, **11**, 780–2.

Vincent, J. M. (1970). A manual for the practical study of root nodule bacteria. *IBP Handbook*, no. 15, pp. 34–40. Blackwell: Oxford and Edinburgh.

Vincent, J. M. & Humphrey, B. A. (1968). Modification of the antigenic surface of *Rhizobium trifolii* by a deficiency of calcium. *J. gen. Microbiol.*, **54**, 397–405.

Vincent, J. M. & Humphrey, B. A. (1970). Taxonomically significant group antigens in Rhizobium. *J. gen. Microbiol.*, **63**, 379–82.

Zevenhuizen, L. P. T. M. (1971). Chemical composition of polysaccharides of *Rhizobium* and *Agrobacterium*. *J. gen. Microbiol.*, **68**, 239–43.

22. Some aspects of leghaemoglobin biosynthesis

C. A. GODFREY, D. R. COVENTRY & M. J. DILWORTH

Leghaemoglobins (Lb) are a unique group of haemoproteins, identified only in the *Rhizobium*-containing cells of nitrogen-fixing legume root nodules. The recently reported association between *Rhizobium* and the non-legume *Trema* (Trinick, 1973) is currently being examined for the occurrence of haemoglobins. The properties of Lb suggest that it functions in transporting oxygen to nodule bacteroids, a proposal supported by recent experiments with soybean nodules (Tjepkema, 1971; Bergersen, Turner & Appleby, 1973). The localization of Lb within the membrane envelopes surrounding the bacteroids in nodules (Dilworth & Kidby, 1968; Truchet, 1972; Bergersen & Goodchild, 1973*b*), would be compatible with this function.

The unique occurrence of Lb in a symbiotic bacteria–plant association raises questions of their evolutionary and biochemical relationships to other haemoglobins (Dilworth & Parker, 1969). The internal organization of Lb biosynthesis in the nodule will be considered in three areas; the synthesis of the haem moiety, the synthesis of the apoprotein, and the combination of the two.

Haem synthesis

Protohaem IX for Lb is not the only product of tetrapyrrole biosynthesis in the nodule, since mitochondrial haemoproteins, free porphyrins, bacteroid corrins and bacteroid haemoproteins must be considered.

Mitochondria are seen in the cytoplasm of infected plant cells (Dart & Mercer, 1963; Kidby & Goodchild, 1966; Bergersen & Goodchild, 1973*a*) and have been isolated in active form (Muecke & Wiskich, 1969). Although mitochondrial numbers per cell are not accurately known, they are small compared to bacteroid numbers, and their haem an insignificant part of the nodule total. They may, nevertheless, participate in haem synthesis as in other eukaryotic cells.

Free porphyrins in the non-bacteroid fractions of soybean nodules amount to less than 0.01 μmole/g fresh weight (Klüver 1948; Falk,

Appleby & Porra, 1959). These porphyrins are probably released by the bacteroids, since laboratory cultures of *R. meliloti* excrete copious amounts of coproporphyrin under certain conditions (Hendry & Jordan, 1969).

The corrin content of bacteroids may reach 0.05 μmoles/g fresh weight, but is not significantly greater than in laboratory cultures (Kliewer & Evans, 1963). The early steps in corrin and haem biosynthesis appear to be the same in other micro-organisms.

Bacteroid haem per g bacterial protein is double that of laboratory-grown *R. japonicum* (Tuzimura & Watanabe, 1964; Appleby, 1969*a*, *b*) and probably amounts to about 0.01 μmole/g fresh weight of soybean nodules.

When these figures are compared with a maximum soybean nodule content of about 0.3 μmole Lb haem/g fresh weight of nodule (Bergersen & Goodchild, 1973*b*), it is obvious that haem synthesis for Lb involves a marked increase in haem biosynthetic activity.

In soybean nodule breis, label is incorporated into haem from radioactive glycine and acetate (Richmond & Salomon, 1955), and from succinate, propionate, 2-oxoglutarate, lactate, propionate and 5-aminolaevulinic acid (ALA) (Jackson & Evans, 1966; Cutting & Schulman, 1969). Laboratory-grown *R. japonicum* and nodule breis convert porphobilinogen (PBG) into porphyrins, but bacteroids do so poorly (Falk *et al.*, 1959), suggesting defective porphyrin synthesis in isolated bacteroids.

Haem synthesis in nodules for Lb has been attributed to the bacteroids (Cutting & Schulman, 1969). Experiments in which intact bacteroids incubated in the presence of Lb were compared with the cytoplasmic (plant) fraction of soybean nodules for ability to incorporate [^{14}C]ALA into haem showed that the bacteroid was the major contributor. Later work (Cutting & Schulman, 1972) on haem inhibition of ALA incorporation into extracellular haem was interpreted in terms of haem synthesis being regulated by haem by negative feedback control in bacteroids.

Several questions arise from this work. If the bacteroids synthesize the haem for Lb, besides making normal amounts of corrins and haem for cytochromes, their haem synthetic activity should increase by a factor of about 20 in relation to the laboratory form. This question has been studied with laboratory-grown *Rhizobium lupini*, strain WU8, and bacteroids of predominantly the same strain from the nodules of lupin (*Lupinus luteus* L.) and serradella (*Ornithopus sativus* Brot.). The ability of the two forms to incorporate [^{14}C]ALA into extracellular haem, and the activities of selected enzymes in the pathway(s) of haem synthesis, have been compared.

Secondly, disruption of membrane structures during isolation of bacteroids may seriously affect haem synthetic activity in the plant fraction, introducing a bias into the incorporation results in favour of the intact bacteroids. Plant fraction has therefore been examined for the presence of selected enzyme activities possibly reflecting potential levels of haem synthesis.

Thirdly, ALA formation has not been described in legume root nodules or in rhizobia. Since it is commonly the rate-limiting step in bacterial haem synthesis (Tait, 1968), ALA incorporation into haem may not give a true picture of the distribution of haem biosynthetic activity. ALA synthesis has, therefore, been measured in both laboratory-grown and nodule forms of *R. lupini*, and in the plant extract.

Lastly, addition of plant fraction to soybean bacteroids (Cutting & Schulman, 1969) causes a synergistic effect on ALA incorporation into exogenous haem. We have investigated this phenomenon with lupin and serradella bacteroids using laboratory-grown rhizobia as a control.

Globin synthesis

While Lb patterns are genetically determined by the plant in the systems so far studied (Dilworth, 1969; Peive, Yagodin, Zhiznevskaya & Borodenko, 1970; Broughton & Dilworth, 1971; Cutting & Schulman, 1971), globin synthesis could occur on either plant or bacteroid ribosomes. Experiments with lupin nodules incorporating photosynthetically-fixed ^{14}C into Lb in the presence and absence of D(−)-chloramphenicol indicate that the plant cytoplasmic ribosomes are responsible for globin synthesis.

Leghaemoglobin assembly

Cutting & Schulman (1972) propose that the apoLb is formed in the plant cytoplasm and haem in the bacteroid, implying that haem must be excreted into the medium surrounding the bacteroid. They also showed that label from [^{14}C]ALA appeared in haem attached to apoLb. The Lb probably forms wherever haem and apoLb meet, since conjugation occurs with ease *in vitro* under mild conditions (Ellfolk & Sievers, 1964) without a specific enzyme. If this model is correct, Lb becomes a true product of the symbiosis. The supporting evidence is limited to soybeans, and has the drawbacks noted above.

Methods

Plants and bacteria

Serradella and yellow lupin were inoculated with *R. lupini*, strain WU8, and grown as described by Dilworth (1969). Nodules were harvested, washed with distilled water, dried on paper towelling, and used immediately. *R. lupini* was grown in liquid culture, and counted as described by Godfrey & Dilworth (1971).

[^{14}C]ALA incorporation

Nodule fractions were prepared and incubated using the methods and Honda-type medium of Cutting & Schulman (1969). The incubation was started by simultaneously adding ALA and 4-[^{14}C]ALA (53 mCi/mmole) to bring the concentration to 1 mM ALA and the specific activity to 0.2 μCi/μmole in most cases. Five or ten millilitre suspensions were incubated in 125 ml cotton-plugged flasks on a shaking water bath at 28 °C for the required period, usually 6 h. Haem was also isolated using the methods of Cutting & Schulman (1969).

Enzyme assays

Nodules were ground with a mortar and pestle in two volumes of cold 0.1 M potassium phosphate buffer (pH 7.0) containing one-third tissue weight of solid polyvinylpyrrolidone. The homogenate was squeezed through nylon mesh, and the filtrate centrifuged at 5000 × **g** for 15 min. The supernatant or plant fraction, was used directly for enzyme assays. The bacteroid pellet was washed with 1 tissue volume of buffer, suspended in four times its weight of buffer and treated for 20 min in a Raytheon 9 kHz sonicator at 0 °C. Laboratory-grown bacteria were treated similarly. All enzyme experiments were completed in a single day.

3-Hydroxybutyrate dehydrogenase (D-3-hydroxybutyrate : NAD oxidoreductase, EC 1.1.1.20) was assayed according to Delafield & Doudoroff (1969), succinyl-CoA synthetase (succinate:CoA ligase (ADP), EC 6.2.1.5), ALA synthetase (succinyl-CoA-glycine succinyltransferase), and ALA dehydratase [5-aminolaevulinate hydrolyase (adding 5-aminolaevulinate and cyclizing), EC 4.2.1.24] by the methods of Burnham & Lascelles (1963), or slight modifications thereof, and L-alanine:4,5-dioxovalerate transaminase according to Gassman, Plusec & Bogorad (1968).

ALA dehydratase from lupin nodules was salted out between 25 and 50 % saturation with ammonium sulphate, thus removing most of the Lb. The pellet was dissolved in and dialysed against 10 mM Tris-chloride, pH 7.8, and used for pH dependence experiments. Fractionation of the resultant solution up to 35 % saturation with ammonium sulphate gave a ten-fold purification.

Lupin Lb was purified according to Dilworth (1969).

Analytical methods

Soluble protein was determined by the Lowry method (Lowry, Rosebrough, Farr & Randall, 1951) with lysozyme as standard. ALA and porphobilinogen (PBG) were measured by the methods of Moore & Labbe (1964). Radioactive samples were counted according to Godfrey & Dilworth (1971); haem was chromatographed according to Broughton, Dilworth & Godfrey (1972). Carbon dioxide was collected on paper discs saturated in KOH suspended in closed incubation flasks. Porphyrins were isolated and fractionated according to Doss (1969).

Haemin solutions were freshly prepared in 50 mM NH_4OH, diluted and centrifuged before adding a 50 μl aliquot to the incubation mixture. All concentrations in text or tables are final concentrations.

Labelling of Lb in nodules

Field-grown yellow lupin plants were transferred from soil to aerated nutrient solution (Broughton & Dilworth, 1971) 15 h before exposure to $^{14}CO_2$. The plant tops were sealed into a 55 l plastic chamber, and the roots dipped into nutrient solution with or without 1 mM D(−)-threo-chloramphenicol 2 h prior to liberation of $^{14}CO_2$ (200 μCi) from $Ba^{14}CO_3$ to give a final carbon dioxide concentration of 1 % by volume. At 1.5, 3, 5 and 7 h, root nodules were harvested and ground in a mortar in 0.1 M barbiturate buffer (pH 8.6) at 4 °C. The homogenate was filtered through Mira-cloth, and bacteroids sedimented by centrifuging at 12000 ×g for 20 min.

The supernatant was fractionated on a 90 cm × 1.8 cm diameter column of Sephadex G75 in 0.1 M barbiturate, pH 8.6. A large molecular weight fraction containing soluble plant protein, and a lower molecular weight fraction containing Lb were collected. Electrophoresis in polyacrylamide gel showed that the Lb fraction was at least 95 % pure. Lb concentration was measured by radial immunodiffusion (Mancini, Carbonara & Heremans, 1965).

Soluble plant protein was precipitated with an equal volume of 5 % trichloracetic acid (TCA) and the precipitate washed several times with 5 % TCA and finally with absolute ethanol before being dissolved in 1 M hyamine hydroxide. Haem was removed from Lb by adding to a salt-free solution at pH 2 and 0 °C an equal volume of cold methylethyl ketone and shaking (Teale, 1959). The radioactivity of globin in the aqueous phase was then measured (Patterson & Greene, 1965).

Bacteroids were washed and suspended several times in the barbiturate buffer, and then disrupted by sonication for 15 min at 9 kHz at 0 °C. After centrifugation for 20 min at 12000 × *g*, an equal volume of 5 % TCA was added to the supernatant, and the resulting precipitate washed several times with 5 % TCA. After incubation at 90 °C for 15 min in 5 % TCA, the precipitate was finally washed with absolute ethanol before being dissolved in 1 M hyamine hydroxide.

Results

[^{14}C]*ALA metabolism*

ALA was metabolized very rapidly by aerobic rhizobia (Fig. 22.1). In both laboratory-grown and nodule rhizobia an incubation density of about 2×10^{10} cells was found optimal; most incubations therefore used

Table 22.1. *The incorporation of* [^{14}C]*ALA into extracellular haem by laboratory-grown rhizobia and serradella and lupin nodule fractions*

The range of results for lupin was obtained from three experiments using nodules from plants of age 50–84 days after sowing. The remaining results are the means and standard deviations of 8 replicate experiments. The serradella nodules were taken from plants between 63–105 days after sowing. 1 nmole of haem produced is equivalent to approximately 4400 dpm of ^{14}C incorporated

Extract	Incubation density (bacteria/ml)	Extracellular haem (dpm/6 h/10^{10} bacteria)
Lab-grown *R. lupini*, WU8	$2.7 \pm 0.7 \times 10^{10}$	325 ± 88
Serradella WU8 bacteroids	$2.2 \pm 0.8 \times 10^{10}$	100 ± 65
Lupin WU8 bacteroids	$1.0–1.4 \times 10^{10}$	15–76
Serradella nodule breis	$2.2 \pm 0.8 \times 10^{10}$	516 ± 180
Lupin nodule breis	$1.0–1.4 \times 10^{10}$	30–77

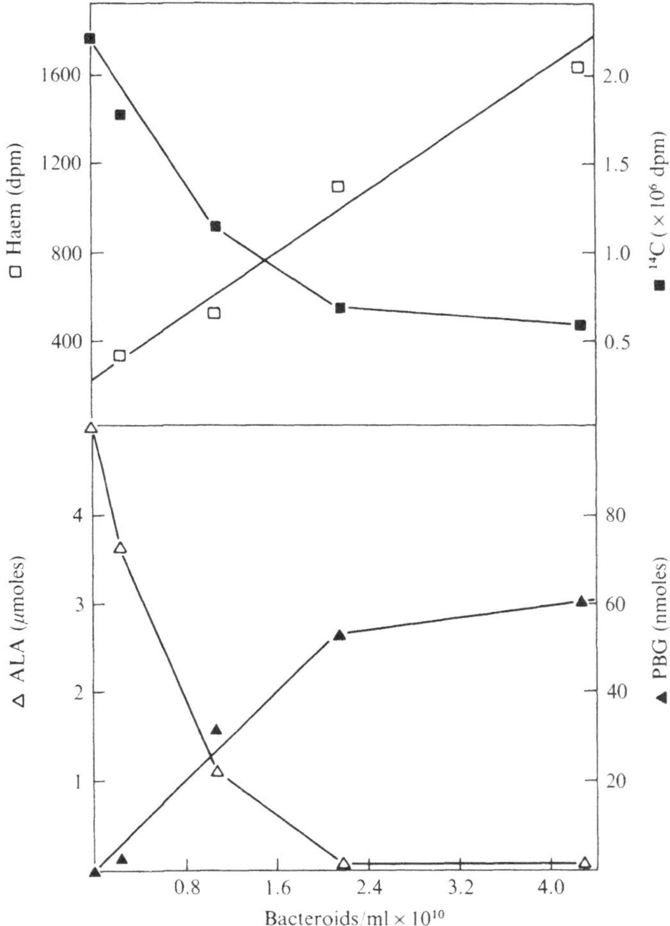

Fig. 22.1. [^{14}C]ALA metabolism by increasing densities of serradella bacteroids incubated for 6 hours.

this density (Table 22.1). At higher density, ALA quickly disappeared and the rates of PBG and subsequently extracellular haem production decreased as shown in Fig. 22.1. The pattern was alike for both nodule and laboratory rhizobia; production of extracellular haem was linear during the 6 h incubation, whereas the synthesis of PBG declined as the ALA was consumed (Fig. 22.2).

Strain identity of bacteroids was tested serologically on three occasions. Serradella nodules contained 50–70 % strain WU8, and lupin nodules over 90 %; there was no apparent effect of WU8 content on

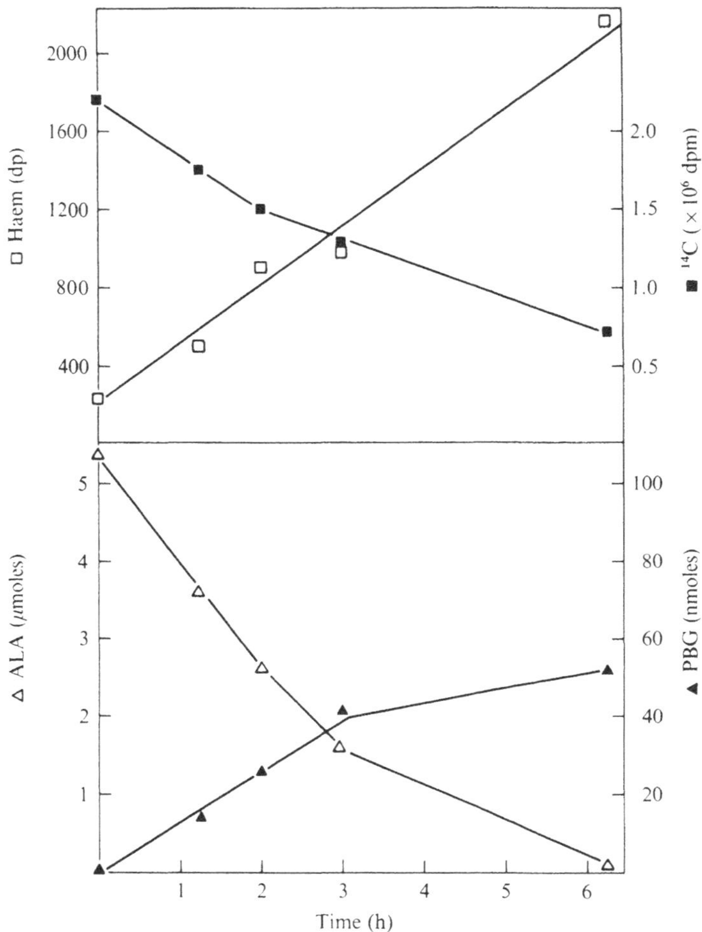

Fig. 22.2. [^{14}C]ALA metabolism to haem and PBG as a function of time during incubation with 2×10^{10} serradella bacteroids/ml.

haem formation. Activity of serradella bacteroids fell from 200 dpm/6 h/10^{10} bacteroids from young 63-day-old plants (58 mg nodules/plant) to 33 dpm/6 h/10^{10} bacteroids from greening nodules on 105-day-old plants (164 mg nodules/plant). The most active bacteroid preparation was barely as active as the least active laboratory culture, whether they are compared on the basis of bacterial wet weight, Lowry protein content, or numbers (Table 22.2). Lupin bacteroids were only weakly active and their activity did not increase in nodule breis (Table 22.1). Significant differences exist in morphology of lupin and serradella

Table 22.2. *The comparison of the incorporation of* [^{14}C]*ALA into haem by preparations of laboratory-grown rhizobia and the particular fraction of serradella nodules*

Rhizobium	Extracellular haem (dpm/6 h) (per g wet bacteria)	(per mg protein)	(per 10^{10} bacteria)
Lab-grown *R. lupini*, WU8	13000	810	410
Serradella WU8 bacteroids	6700	300	200

bacteroids (Kidby & Goodchild, 1966), and these results indicate a difference in function also.

Although ALA was rapidly metabolized by rhizobia, less than 0.1 % appeared as extracellular haem. In both serradella nodule breis and laboratory rhizobia, 20 % of the total radioactivity appeared as carbon dioxide during incubation. Less than 1 % was isolated as porphyrins (mainly coproporphyrin), while total haem, porphyrins, and PBG together, accounted for less than 5 % of total radioactivity. The rhizobia obviously catabolize ALA readily, and could provide ready information on pathways of bacterial ALA degradation. ALA consumption decreased markedly under anaerobic conditions.

Plant fraction metabolized only 2–4 % of added ALA during incubation, mostly to PBG; of this, contamination with bacteroids accounted for about 10 % of the activity. In relation to total nodule brei activity (100 %), bacteroids (17 ± 6 %) were around three times more active than plant fraction (5 ± 3 %).

Stimulation and inhibition of rhizobial [^{14}C]*ALA incorporation*

A synergistic effect of four- to five-fold was observed in nodule breis or when plant and bacteroid fractions of serradella nodules were recombined. A similar but much smaller synergism was reported with soybean nodules (Cutting & Schulman, 1969).

Serradella bacteroids were very sensitive to added plant fraction: about 20 % of the equivalent plant fraction gave a maximal response, and increasing inhibition then occurred (Fig. 22.3). Lupin plant fraction also stimulated haem synthesis from ALA by serradella bacteroids (Table 22.3), indicating that low activity in lupin nodule breis was not due to inhibition by the plant fraction. No such stimulation by plant

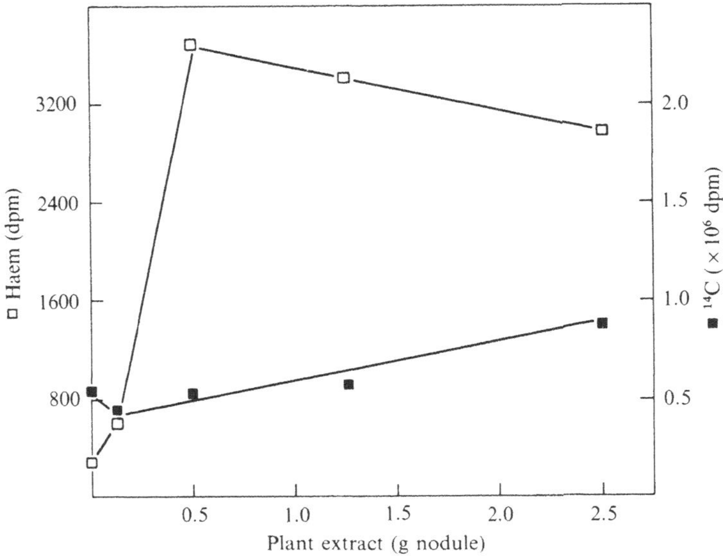

Fig. 22.3. The effect of plant fraction on the [^{14}C]ALA metabolism of serradella bacteroids. Bacteroids from 2.5 g of nodule were incubated with varying amounts of plant fraction for 6 hours.

fractions occurred with laboratory-grown rhizobia, a significant difference in behaviour.

The stimulating factor(s) in the plant fraction was heat-labile, suggesting that it was a protein. Iron is not limiting in the system as 1 mM ferrous sulphate was without effect. A range of proteins such as partially purified lupin ALA dehydratase, lupin Lb or bovine serum albumin all resulted in marked stimulation of activity (Table 22.3). It seems probable that the effect is a general stimulation of haem release by inclusion of protein. A similar effect has been found when cell sap or proteins are added to liver mitochondria incubated in sucrose-salts medium (Yoda & Israels, 1972).

For soybean bacteroids incubated in Lb solution 0.1 mM haemin inhibited incorporation of [^{14}C]ALA into haem by 50 % (Cutting & Schulman, 1972). With *R. lupini*, haemin inhibited ALA incorporation in nodule breis, but had no effect on laboratory-grown cells. Exogenous haemin did not affect the rate of PBG production or of ALA degradation; only the appearance of extracellular haem was reduced.

Addition of 10 mM laevulinic acid, a competitive inhibitor of ALA dehydratase (Nandi & Shemin, 1968; Beale, 1971; Yamasaki & Mori-

Table 22.3. *Effects of selected compounds and treatments on* [^{14}C]*ALA incorporation into extracellular haem by laboratory-grown* R. lupini, *serradella bacteroids and serradella nodule bries*

Treatment	Incorporation of [^{14}C]ALA into extracellular haem as percentage of control values		
	Lab.-grown rhizobia	Serradella bacteroids	Serradella nodule breis
Control	100 %	100 %	100 %
Serradella plant fraction (21 mg protein)	88–92	910	—
Serradella plant fraction (100 °C for 5 min)	—	109	—
Lupin plant fraction (29 mg protein)	—	760	—
Lupin ALA dehydratase (1 mg protein)	—	330	—
Lupin Lb (5 mg protein)	—	1130	—
Bovine serum albumin (25 mg)	189	400	—
10^{-5} M haemin	91	—	77
10^{-4} M haemin	92	—	72
10^{-3} M haemin	100	—	42
10 mM laevulinic acid	14–30	33–44	19–26

yama, 1971) inhibited incorporation in each system tested, and reduced the rate of ALA consumption.

Enzyme assays

The main aims behind the enzyme assays were to indicate the likely sites for ALA synthesis and conversion to PBG in the nodule and to indicate if any of them was significantly increased or decreased by transformation from laboratory rhizobia to nodule bacteroids. PBG formation and consumption were also followed to confirm or contradict the conclusions from the ALA incorporation experiments that the bacteroid was the probable site of total haem synthesis for Lb.

Since the conclusions depended on a satisfactory fractionation of plant from bacteroid enzymes, an indicator enzyme for the bacteroids (3-hydroxybutyrate dehydrogenase) was assayed in the plant fraction as a control on leakage from the bacteroids. Although soybean bacteroids apparently contained no malate dehydrogenase, allowing it to be used as an indicator of plant contamination of bacteroid extracts (Cutting & Schulman, 1969), the high activities in *R. lupini* bacteroids render it unsuitable, and no satisfactory marker was found.

Leakage of enzyme occurred from the bacteroids to the extent of a maximum of about 20 % of total activity in the case of lupin nodules,

Table 22.4. *Distribution of 3-hydroxybutyrate dehydrogenase activity within lupin and serradella nodules and activity of laboratory-grown* Rhizobium lupini, *strain WU8*

Duplicate determinations of enzyme activity were made in each case and the range of values was established from three experiments. There was no measurable reduction of NAD by endogenous substrate

Organism	Fraction	Specific activity (μmoles/min/ mg protein)	Total activity/ g nodule	Nodule activity (%)
R. lupini	Soluble	0.03–0.04	—	—
Lupinus luteus	Plant	0.03–0.04	0.31–0.42	20–24
	Bacteroid	0.30–0.40	1.17–1.32	80–76
Ornithopus	Plant	0.00–0.01	0.05–0.10	9–13
sativus	Bacteroid	0.09–0.12	0.55–0.78	91–87

and of 10 % for serradella (Table 22.4). At least for serradella, those enzymes found predominantly in the plant fraction are not artefacts of the extraction procedure.

Succinyl-CoA synthetase

Laboratory-grown *R. lupini* extracts, bacteroids from all plants, and the plant extracts themselves, all had levels of activity of succinyl-CoA synthetase sufficiently high to allow for haem synthesis at high rates (Table 22.5).

Table 22.5. *Succinyl CoA synthetase activity in extracts of laboratory-grown* Rhizobium lupini (*WU8*) *and of legume nodule fractions*

The ranges of results below were collected from duplicate determinations of enzyme activity using from 0.5 to 4 mg of enzyme extract in at least two experiments

Organism	Fraction	Specific activity (μmoles/min/ mg protein)	Total activity/g nodule	Nodule activity (%)
R. lupini	Soluble	0.025–0.034	—	—
Lupinus luteus	Plant	0.004–0.006	0.052–0.072	30
	Bacteroid	0.019–0.026	0.120–0.160	70
Ornithopus	Plant	0.005–0.007	0.060–0.075	30
sativus	Bacteroid	0.016–0.024	0.140–0.195	70

Table 22.6. *ALA synthetase activity in extracts of laboratory-grown* Rhizobium lupini, *and in serradella and lupin nodule fractions. The results are taken from three different experiments*

Organism	Fraction	Specific activity (nmoles/min/ mg protein)	Total activity/ g nodule	Nodule activity (%)
R. lupini	Sonicate	0.08–0.19	—	—
L. luteus	Plant	0.00–0.02	0.00–0.32	0–30
	Bacteroid	0.05–0.20	0.25–0.81	100–70
O. sativus	Plant	0.00–0.15	0.00–1.3	0–30
	Bacteroid	0.15–0.40	0.9–2.8	100–70

ALA synthetase

ALA synthetase activity was low in each case, but definitely present in bacterial extracts (Table 22.6). No significant amino acetone formation was detected using the method of Granick (1966). Absorbance differences in the colorimetric determination of ALA synthetase in bacteroid extracts were always greater than 0.03 (555 nm), and assays were linear with both time and protein concentration. Variations in specific activity were, however, greater than for the other enzymes measured.

Measurements in the plant fraction were complicated by the greater dilution of the plant fraction, the higher background absorbance, and the small differences between blanks and samples (less than 0.02 at 555 nm). Consequently the apparently positive values for the plant fractions may well have resulted from errors in measurement. Since bacteroid tissue is only about 8–10 % of the nodule, this error is compounded when making comparisons of total activity per unit weight of nodule.

ALA dehydratase

ALA dehydratase occurred predominantly in the plant fraction, and bacteroids had lower specific activities than laboratory-grown bacteria (Table 20.7). These activities are so low as to preclude purification on a practical scale, or even separation on polyacrylamide gels to demonstrate the identity or otherwise of plant and bacteroid enzymes.

When an oxidizing agent present in the plant material was removed, all extracts were similar in their requirements for cysteine or reducing

Table 22.7. *ALA dehydratase activity in extracts of laboratory-grown* Rhizobium lupini *and of lupin and serradella nodule fractions. The range of values presented is from at least three different experiments*

Organism	Fraction	Specific activity (nmoles/min/ mg protein)	Total activity/ g nodule	Nodule activity (%)
R. lupini	Soluble	0.05–0.11	—	—
L. luteus	Plant	0.03–0.05	0.40–0.57	92–75
	Bacteroid	0.02–0.03	0.07–0.20	8–25
O. sativus	Plant	0.04–0.08	0.38–0.70	80–70
	Bacteroid	0.02–0.03	0.14–0.20	20–30

agent, substrate, and Mg^{2+} for activity. The pH optima of the dehydratases from plant tissue, bacteroid extracts, and extracts from laboratory cultures were very similar, with sharp optima at pH 7.5 in Tris buffers for the bacterial extracts, and at 7.4 for the plant enzyme. Inhibition by EDTA and laevulinic acid was similar for both bacterial and plant enzymes.

PBG consumption

In soybean nodule breis, added PBG has been found to disappear rapidly, but soybean bacteroid preparations did not cause PBG to disappear (Falk *et al.*, 1959).

Extracts of both bacteroids and laboratory-grown rhizobia consumed PBG during a 2 h incubation (Fig. 22.4), but little PBG disappeared in plant extracts or controls over this period. Addition of plant extract to bacteroids to reconstitute a brei slightly reduced the rate of PBG consumption, while treatment for 3 min at 100 °C prevented it completely.

Disappearance of PBG might have led to underestimation of the ALA dehydratase activity of bacteroids or laboratory-grown rhizobia, so an attempt was made to eliminate this possibility by decreasing PBG utilization without affecting ALA dehydratase activity. In bacteroid extracts stored at −20 °C, ALA dehydratase activity remained constant while the ability to metabolize PBG dropped to one-sixth (Table 22.8). Coupled with the findings that enzyme assays were linear, that plant and bacterial extract activities were additive, and that stepwise ammonium sulphate fractionation did not produce any significant

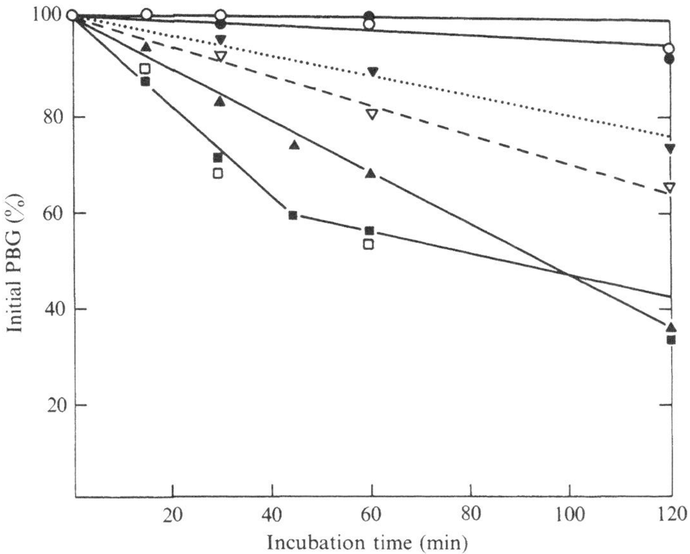

Fig. 22.4. The utilization of PBG by serradella and lupin nodule extracts, and by extracts of laboratory-grown *R. lupini* WU8. Extract from the stated fresh weight of nodules was incubated in the ALA dehydratase assay system minus ALA, but with approx. 100 nmoles of PBG. ○, 0.1 g Serradella nodule plant fraction; ▽, 0.4 g serradella nodule bacteroid fraction; ▼, serradella nodule plant and bacteroid fractions combined; ●, 0.2 g lupin nodule plant fraction; ■, 0.8 g lupin nodule bacteroid fraction; □, 0.7 g lupin nodule bacteroid fraction; ▲, laboratory-grown *R. lupini* extract.

increase in apparent enzyme activity, the work with the frozen extracts indicates that PBG consumption did not seriously affect assays for ALA dehydratase.

Alternative pathways for ALA synthesis

The first step specific to tetrapyrrole biosynthesis, the formation of ALA via ALA synthetase, has only been demonstrated in a few plant tissues (De Xifra, Batlle & Tigier, 1971; Ramaswamy & Nair, 1973), while evidence from incorporation of ^{14}C-labelled precursors suggests that another pathway for ALA synthesis operates in some plants (Beale & Castelfranco, 1973). In view of our failure to demonstrate definite ALA synthetase activity in the plant fractions of nodules, alternative routes for ALA synthesis were examined.

L-Alanine/4,5-dioxovalerate transaminase activity was found in extracts of both nodule and laboratory forms of *R. lupini* (Table 22.9); activity in the plant fraction was low and often non-linear with time or enzyme concentration. Attempts to demonstrate formation of 4,5-

Table 22.8. *Effects of freezing and thawing upon PBG formation and disappearance in bacteroid extracts*

Bacteroid extracts were freshly prepared and aliquots were incubated with 30 nmoles per ml of PBG or with ALA in the assay for ALA dehydratase. The remaining extract was then frozen at −20 °C and stored 1 week in the case of serradella and 7 weeks in the case of lupin before reassay

	Fresh extract		Frozen extract	
	PBG consumed/ 60 min (nmoles)	PBG formed/ 60 min (nmoles)	PBG consumed/ 60 min (nmoles)	PBG formed/ 60 min (nmoles)
Serradella bacteroid	10	7.0	1.6	6.0
Lupin bacteroid	17	3.6	3.0	3.8

Table 22.9. *L-Alanine/4,5-dioxovalerate transaminase activity in extracts of laboratory-grown* Rhizobium lupini *and of legume nodule fractions. The results shown are from either one or two experiments as indicated*

Organism	Fraction	Specific activity (nmoles ALA/ min/mg protein)	Total activity/ g nodule	Nodule activity (%)
R. lupini	Soluble	8.7–11.1	—	—
Lupinus luteus	Plant	0.05–0.42	0.8–6.3	2–15
	Bacteroid	6.7–6.4	41–33	98–85
Ornithopus sativus	Plant	0.18	1.4	2
	Bacteroid	7.9	63.4	98

dioxovalerate from 2-oxoglutarate using the transaminase to convert it to ALA were unsuccessful. Since selected organic acids at substrate concentrations inhibited the transamination (precluding assay in the reverse direction), this result is inconclusive.

5-Hydroxylaevulinic acid, whose formation via the 2-oxoglutarate/glyoxylate carboligase reaction could be demonstrated in rhizobial extracts, was inert in all extracts tested even when coenzymes and cofactors were added.

No conclusive evidence was obtained for an alternative pathway of ALA biosynthesis in the legume nodule. The function of the high level of L-alanine/4,5-dioxovalerate transaminase in the rhizobia is not known, but it is likely to be involved in the rapid dissimilation of ALA by extracts from them.

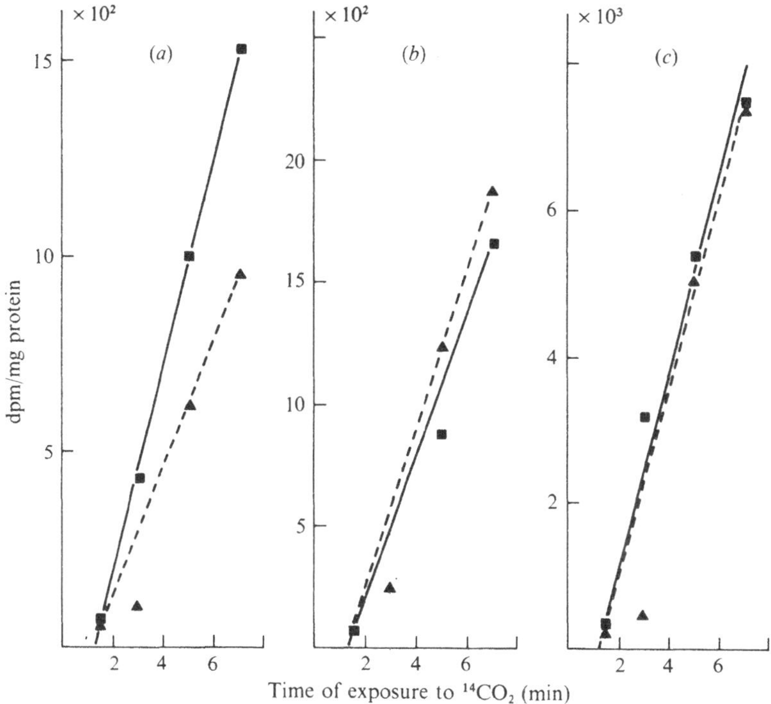

Fig. 22.5. The incorporation of ^{14}C from $^{14}CO_2$ into soluble plant protein (*b*), soluble bacteroid protein (*a*) and Lb (*c*) in nodules of yellow lupin, in the presence and absence of D(−)-chloramphenicol. --▲-- with 1 mM D(−)-CAP; —■— without CAP.

Globin synthesis

In these experiments, the selective inhibition of bacterial ribosomes by chloramphenicol in the intact nodule was attempted. In preliminary experiments it was found that laboratory cultures of *R. lupini* WU8 are insensitive to chloramphenicol, lincomycin and erythromycin. Incorporation of [^{3}H]leucine into soluble protein by bacteroid suspensions, however, was found to be sensitive to D(−)-chloramphenicol, though not to erythromycin or lincomycin. Incorporation of ^{14}C from photosynthetically fixed $^{14}CO_2$ into Lb is linear for some hours (Fig. 22.5) after translocation of [^{14}C]sugars to the nodules commences. Similarly, linear incorporation occurs into general plant protein, and into protein released from the bacteroids by sonication. In the presence of 1 mM D(−)-chloramphenicol, ^{14}C incorporation into general plant protein and into Lb is unaffected, while incorporation into bacteroid protein is inhibited by over 40 %. While it remains to be shown that synthesis of all bacteroid proteins is inhibited by chloramphenicol to the same extent, these results are consistent with globin being synthesized on plant cytoplasmic ribosomes.

The induction of globin synthesis in nodulating yellow lupins has been followed by immunological assay of Lb. At temperatures ranging from 15° to 25 °C, globin was detected several days before nitrogenase activity, although the acetylene reduction assay for nitrogenase was still more sensitive than the immunological one for Lb.

Discussion

Certain general conclusions may be drawn in relation to haem biosynthesis in legume root nodules.

With the demonstration of ALA synthetase activity in bacteroids, the complete pathway for haem synthesis has been shown to occur in bacteroids.

Generation of succinyl-CoA in lupin and serradella nodules cannot be rate-limiting in either plant or bacteroid fractions (Table 22.5). However, with the failure to determine reliably either the route or the potential rate of ALA synthesis in the plant fraction, and the occurrence of an inhibitor for ALA synthetase in it, ALA synthesis for Lb haem by the plant fraction cannot be excluded. Bacteroid ALA synthetase activity is adequate to explain nodule haem synthesis, even though no definite increase in enzyme activity appears to occur during transition from laboratory culture to nodule bacteroids. No evidence has been found for a substantial derepression of haem biosynthetic enzymes in bacteroids, despite the large increase in haem requirements in the nodule, possibly suggesting that turnover of Lb haem is very slight.

Although our results on lupin and serradella reinforce those of Cutting & Schulman (1969) that ALA incorporation into haem is mainly bacteroid property, reliable measurements of ALA synthetase in the plant fraction are needed before this can be taken as established.

ALA dehydratase activity appears to be significantly lower in the bacteroids, but high in the plant fraction. Whether the enzyme in the plant fractions is made by the plant or possibly secreted from the rhizobia remains to be established, by immunological or electrophoretic techniques. At the present time, synthesis of intermediates for haem biosynthesis up to and including PBG could occur in either plant or bacteroid tissues.

[^{14}C]ALA is poorly incorporated into haem by the plant fraction of soybean nodules (Cutting & Schulman, 1969) or of lupin or serradella nodules. Enzyme assays suggest that it is not ALA dehydratase which is limiting in the latter cases (Table 22.7), but further metabolism of PBG (Fig. 22.4).

The synergism of plant and bacteroid fractions in haem synthesis appears to occur with soybean and serradella but not with lupin. The effects seen with serradella bacteroids (Table 22.3) did not occur with lupin bacteroids; the effects with soybean bacteroids (Cutting & Schulman, 1969) were smaller, but could well have been larger in the absence of Lb since with serradella bacteroids this protein stimulates incorporation of ALA into extracellular haem (Table 22.3). Laboratory-grown rhizobia show no effect from adding plant extracts, suggesting a physiological importance for the effect. It cannot be said whether the synergism results from the plant extract's ALA dehydratase activity or simply from its protein content, although the latter appears the more probable.

It is suggested that both the haemin inhibition and the protein stimulation effects observed with soybean (Cutting & Schulman, 1969; 1972) and serradella bacteroids (neither of which occur with laboratory cultures) reflect a regulation on the rate of haem release from the bacteroid membrane, and that a separate internal negative-feedback control probably functions to regulate haem synthesis within the bacteroid. The rate of globin production would then regulate the rate of haem release from the bacteroid, which would in turn regulate the rate of internal synthesis.

Certain parallels between bacteroids in nodule cells and mitochondria in animal cells can be drawn, particularly in relation to cytochrome *c* synthesis. In the latter, the apoprotein is a product of cytoplasmic ribosomes (as Lb appears to be in nodules), and haem is produced and released from the mitochondria (or the bacteroids). Haem release in both appears to require external proteins (Yoda & Israels, 1972; Table 22.3). Compartmentation of enzymes may be similar also – ALA synthetase being mitochondrial in location (or possibly exclusively bacteroid) and ALA dehydratase being largely cytoplasmic in both cases.

References

Appleby, C. A. (1969*a*). Electron transport systems of *Rhizobium japonicum*. I. Haemoprotein P-450, other CO-reactive pigments, cytochromes and oxidases in bacteroids from N_2-fixing root nodules. *Biochim. biophys. Acta*, **172**, 71–87.

Appleby, C. A. (1969*b*). Electron transport systems of *Rhizobium japonicum*. II. *Rhizobium* haemoglobin, cytochromes and oxidases in free-living (cultured) cells. *Biochim. biophys. Acta*, **172**, 88–105.

Beale, S. I. (1971). Studies on the biosynthesis and metabolism of δ-aminolevulinic acid in *Chlorella*. *Pl. Physiol*, **48**, 316–19.

Beale, S. I. & Castelfranco, P. A. (1973). ^{14}C incorporation from exogenous compounds into δ-aminolevulinic acid by greening cucumber cotyledons. *Biochem. biophys. Res. Commun.*, **52**, 143–9.

Bergersen, F. J. & Goodchild, D. J. (1973*a*). Aeration pathways in soybean root nodules. *Aust. J. biol. Sci.*, **26**, 729–40.

Bergersen, F. J. & Goodchild, D. J. (1973*b*). Cellular location and concentration of leghaemoglobin in soybean root nodules. *Aust. J. biol. Sci.*, **26**, 741–56.

Bergersen, F. J., Turner, G. L. & Appleby, C. A. (1973). Studies on the physiological role of leghaemoglobin in soybean root nodules. *Biochim. biophys. Acta*, **292**, 271–82.

Broughton, W. J. & Dilworth, M. J. (1971). Control of leghaemoglobin synthesis in snake beans. *Biochem. J.*, **125**, 1075–81.

Broughton, W. J., Dilworth, M. J. & Godfrey, C. A. (1972). Molecular properties of lupin and serradella leghaemoglobins. *Biochem. J.*, **127**, 309–14.

Burnham, B. & Lascelles, J. (1963). Control of porphyrin biosynthesis through a negative-feedback mechanism. *Biochem. J.*, **87**, 462–72.

Cutting, J. A. & Schulman, H. M. (1969). The site of heme synthesis in soybean nodules. *Biochim. biophys. Acta*, **192**, 486–93.

(1971). The biogenesis of leghemoglobin. The determinant in the *Rhizobium*–legume symbiosis for leghemoglobin specificity. *Biochim. biophys. Acta*, **229**, 58–62.

(1972). The control of heme synthesis in soybean root nodules. *Biochim. biophys. Acta*, **261**, 321–27.

Dart, P. J. & Mercer, F. V. (1963). Development of the bacteroid in the root nodule of barrel medic (*Medicago tribuloides* Desr.) and subterraneum clover (*Trifolium subterraneum* L.). *Arch. Mikrobiol.*, **46**, 382–401.

Delafield, F. P. & Doudoroff, M. (1969). β-Hydroxybutyrate dehydrogenase from *Pseudomonas lemoignei*. *Methods in Enzymol.*, vol. XIV, pp. 227–31.

De Xifra, E. A. W., Batlle, A. M. del C. & Tigier, H. A. (1971). δ-Aminolaevulinate synthetase in extracts of cultured soybean cells. *Biochem. biophys. Acta*, **235**, 511–17.

Dilworth, M. J. (1969). The plant as the genetic determinant of leghaemoglobin production in the legume root nodule. *Biochim. biophys. Acta*, **184**, 432–41.

Dilworth, M. J. & Kidby, D. K. (1968). Localization of iron and leghaemoglobin in the legume root nodule by electron microscope autoradiography. *Exp. Cell Res.*, **49**, 148–59.

Dilworth, M. J. & Parker, C. A. (1969). Development of the nitrogen-fixing system in legumes. *J. theor. Biol.*, **25**, 208–18.

Doss, M. (1969). Trennung, Isolierung und Bestimmung von Proto-, Kopro-, Pentacarboxy-, Hexacarboxy-, Heptacarboxy- und Uroporphyrin. *Hoppe-Seyler's Z. Physiol. Chem.*, **350**, 499–502.

Ellfolk, N. & Sievers, G. (1965). Crystalline leghemoglobin. IX. Artifical leghemoglobins. *Acta chem. scand.*, **19**, 2409–19.

Falk, J. E., Appleby, C. A. & Porra, R. J. (1959). The nature, function and biosynthesis of the haem compounds and porphyrins of legume root nodules. *Sym. Soc. exp. Biol.*, **13**, 73–86.

Gassman, M. L., Plusec, J. & Bogorad, L. (1968). δ-Aminolevulinic acid transaminase in *Chlorella vulgaris*. *Pl. Physiol.*, **43**, 1411–14.

Godfrey, C. A. & Dilworth, M. J. (1971). Haem biosynthesis from ^{14}C-δ-aminolaevulinic acid in laboratory-grown and root nodule *Rhizobium lupini*. *J. gen. Microbiol.*, **69**, 385–90.

Granick, S. (1966). The induction *in vitro* of the synthesis of δ-aminolevulinic acid synthetase in chemical porphyria: A response to certain drugs, sex hormones and foreign chemicals. *J. biol. Chem*, **241**, 1359–75.

Hendry, G. S. & Jordan, D. C. (1969). Coproporphyrin excretion by *Rhizobium meliloti*. *Can. J. Microbiol.*, **15**, 242–4.

Jackson, E. K. & Evans, H. J. (1966). Propionate in heme biosynthesis in soybean nodules. *Pl. Physiol.*, **41**, 1673–80.

Kidby, D. K. & Goodchild, D. J. (1966). Host influence on ultrastructure in root nodules of *Lupinus luteus* and *Ornithopus sativus*. *J. gen. Microbiol.*, **45**, 147–52.

Kliewer, M. & Evans, H. J. (1963). Identification of cobamide coenzyme in nodules of symbionts and isolation of the B_{12} coenzyme from *Rhizobium meliloti*. *Pl. Physiol.*, **38**, 55–9.

Klüver, H. (1948). On a possible use of the root nodules of leguminous plants for research in neurology and psychiatry (preliminary report on a free porphyrin-hemoglobin system). *J. Psychol.*, **25**, 331–56.

Lowry, O. H., Rosebrough, N. J., Farr, A. L. & Randall, R. J. (1951). Protein measurement with the Folin phenol reagent. *J. biol. Chem.*, **193**, 265–75.

Mancini, G., Carbonara, A. O. & Heremans, J. F. (1956). Immunochemical quantitation of antigens by single radial immunodiffusion. *Immunochemistry*, **2**, 235–54.

Moore, D. J. & Labbe, R. F. (1964). A quantitative assay for urinary porphobilinogens. *Clin. Chem.*, **10**, 1105–11.

Muecke, P. S. & Wiskich, J. T. (1969). Respiratory activity of mitochondria from legume root nodules. *Nature, Lond.*, **221**, 674–5.

Nandi, D. L. & Shemin, D. (1968). δ-Aminolevulinic acid dehydratase of *Rhodopseudomonas spheroides*. III. Mechanism of porphobilinogen synthesis. *J. biol. Chem.*, **243**, 1236–42.

Patterson, M. S. & Greene, R. C. (1965). Measurement of low energy beta-emitters in aqueous solution by liquid scintillation counting of emulsions. *Analyt. Chem.*, **37**, 854–7.

Peive, Ya. V., Yagodin, B. A., Zhiznevskaya, G. Ya. & Borodenko, L. I. (1970). Polymorphism of hemoglobin crystals from nodules of various lupin species. *Fiziol. Rast.* (Transl.), **17**, 240–3.

Ramaswamy, N. K. & Nair, P. M. (1973). δ-Aminolevulinic acid synthetase from cold-stored potatoes. *Biochim. biophys. Acta*, **293**, 269–77.

Richmond, J. E. & Salomon, K. (1955). Studies on the biosynthesis of hemin in soybean nodules. *Biochim. biophys. Acta*, **17**, 48–55.

Tait, G. H. (1968). General aspects of haem synthesis. *Biochem. Soc. Symp.*, **28**, 19–34.

Teale, F. W. J. (1959). Cleavage of the hemoprotein link by acid methyl ethyl ketone. *Biochim. biophys. Acta*, **35**, 543.

Tjepkema, J. D. (1971). Oxygen transport in the soybean nodule and the function of leghaemoglobin. Ph.D. thesis, University of Michigan, U.S.A.

Trinick, M. T. (1973). Symbiosis between *Rhizobium* and the non-legume *Trema aspera*. *Nature, Lond.*, **244**, 459–60.

Truchet, G. (1972). Mise en évidence de l'activité peroxydasique dans les différentes zones des nodules radiculaires de pois (*Pisum sativum* L.). Localisation de la leghémoglobine. *C. r. Acad. Sci. Paris*, **274**, 1290–3.

Tuzimura, K. & Watanabe, I. (1964). Electron transport systems of *Rhizobium* grown in nodules and in laboratory medium. *Pl. Cell Physiol.*, **5**, 157–70.

Yamasaki, H. & Moriyama, T. (1971). δ-Aminolevulinic acid dehydratase of *Mycobacterium phlei*. *Biochim. biophys. Acta*, **227**, 698–705.

Yoda, B. & Israels, L. G. (1972). Transfer of heme from mitochondria in rat liver cells. *Can. J. Biochem.*, **50**, 633–7.

23. Preparation and properties of nitrogenase proteins

R. H. BURRIS

In recent years not only has it been possible to prepare N_2-fixing extracts from a variety of micro-organisms, but it also has been possible to purify the constituents of nitrogenase to essential homogeneity. Thus it is possible to study the properties of these individual proteins. We will discuss the methods of purification and the properties of the constituents of nitrogenase.

Preparation of cells

To obtain adequate nitrogenase proteins for study it is necessary to grow substantial quantities of micro-organisms. Usually the individual investigator is well acquainted with the growth characteristics of the organism he wishes to produce, and hence he can adapt the growth conditions by scaling up the usual laboratory culture for a large preparation. It is often possible to substitute relatively cheap nutrients for expensive nutrients used in small volume laboratory cultures. When growing large volumes of culture it is helpful to use a substantial-sized inoculum, perhaps 10 % of the volume of the final culture. This insures a rapid start from exponentially growing cells and a minimum lag period. If an anaerobic organism is being grown the use of a large inoculum will minimize the difficulties encountered in scrubbing the last traces of oxygen from the medium.

In growing an organism such as *Azotobacter* it is necessary to supply large volumes of air. The organism uses oxygen so rapidly that air must be bubbled into the culture at a high rate, and the medium must be stirred vigorously to insure that the level of dissolved oxygen is maintained. It is useful to have an oxygen probe in the fermentation tank to give a continuous indication of dissolved oxygen. Maintenance of pH poses no problem with *Azotobacter*.

In growing *Clostridium pasteurianum* attention must be paid to maintaining anaerobic conditions. This is easily done by attaching a tank of high purity N_2 to the fermentor and bubbling N_2 through the medium. This supplies nitrogen necessary for growth of the organisms and sweeps oxygen from the tank. The clostridia produce large quantities of acid so it is necessary that this be neutralized during the fermentation. A pH-stat should be attached to the fermentor so that a glass electrode senses a change in pH; the instrument will add sufficient sodium hydroxide to neutralize the acid produced. High levels of phosphate are detrimental to the extracts from *C. pasteurianum*, so the medium employed is not as highly buffered as the usual medium for acid-producing anaerobes. When they are growing vigorously clostridia, like *Azotobacter*, should have generation times between two and three hours. It is often convenient to grow the cultures in series by retaining 10 % of the culture as an inoculum for a subsequent batch. It may be advisable to grow organisms in continuous culture. For example, Munson & Burris (1969) were able to obtain much more consistent N_2 fixation from extracts of *Rhodospirillum rubrum* grown in continuous culture. The overflow from the continuous culture can be collected in a chilled container and kept for the time necessary (12–24 h) to accumulate enough culture to make recovery of the cells worthwhile.

Cells are normally recovered from the culture medium by centrifugation. A Sharples continuous centrifuge is suitable for medium-size batches, and for larger quantities a sludge separator is very helpful because of the large volume which it can handle. Centrifugation yields a rather densely packed mass of wet cells.

After recovering the cells, an investigator is usually faced with the necessity of preserving the cells for later preparations. With some organisms it may be necessary to rupture the organisms immediately to obtain active nitrogenase preparations. In most instances, however, it is possible to devise methods of preservation which will keep the cells in an active state for extended periods. The most generally applicable method is to freeze the cell paste rapidly in liquid nitrogen. With organisms such as *Azotobacter*, it then is possible to store the material at −20 °C for months without appreciable further inactivation, but with other organisms it may be necessary to keep them in liquid nitrogen to preserve activity.

Carnahan, Mortenson, Mower & Castle (1960) demonstrated that dried *C. pasteurianum* cells retained their activity. The cells were dried in a laboratory rotary vacuum evaporator with the bath at a temperature of about 45 °C. It is important that cells never freeze, as this inactivates them; hence they are dried at a relatively high temperature while being evacuated. The completely dried cells are scraped from the rotary evaporator flask and stored under an inert gas or in vacuum at freezer temperatures for months, with minimal loss of activity.

Disruption of cells

C. pasteurianum which has been dried as described does not need to be broken further to release its enzymes. The drying process alters the permeability of the cells, so that when they are suspended in buffer and shaken under an anaerobic atmosphere for 30 to 60 minutes they release the nitrogenase proteins into the medium. The nitrogenase is soluble in the sense that it cannot be sedimented by prolonged centrifugation at high speed. Much of the inert protein is sedimented and can be separated from the active fraction. The ease of preservation and extraction of nitrogenase from *C. pasteurianum* accounts partially for the popularity of this organism as a source of the N_2-fixing enzyme complex. Although there are literature reports of other active nitrogenase extracts prepared from cells dried in the same manner as *C. pasteurianum*, other methods of preparation appear to be preferable for most organisms.

The most widely used method for disrupting cells is by passing them through some type of press. Carnahan *et al.* (1960) described the preparation of active nitrogenase with the Hughes press at the same time they reported success with extraction from dried cells. A heavy paste of cells is placed in the pre-chilled block of the Hughes press, and a sudden blow from a fly press momentarily melts the cell mass and extrudes it through a fine slit in the press. The material quickly refreezes in the chilled block, and the extruded material is scraped out after it accumulates from a series of blows from the fly press. More commonly the French press is utilized for disrupting cells. In this press a heavy slurry of micro-organisms is placed in a cylinder, and a pressure of 10000 to 20000 pounds per square inch is applied with a plunger in a hydraulic press. The pressurized material is allowed to pass slowly through a needle valve, and the sudden decrease in pressure together with the shearing force of passing the orifice causes the disruption of the organisms. The method is simple, it requires apparatus of modest cost and it is applicable to many organisms. A modified, but more elaborate, instrument known as the Ribi fractionator operates on the same principle.

Osmotic shock provides another method for disrupting cells. Shah, Davis & Brill (1972) recently described the application of this method for making extracts from *Azotobacter vinelandii.* They placed the organisms in 4 M glycerol at pH 7.4 until they had equilibrated, sedimented the cells, decanted the glycerol, and lyzed the cells by shaking rapidly with four volumes of 0.025 M Tris HCl at pH 7.4; the osmotic shock produced by rapid dilution effects cell breakage. This method apparently liberates different proteins from those liberated by disruption with the French pressure cell, and purification procedures must be modified for the two types of extracts. Both of the nitrogenase proteins have been recovered in homogeneous state from osmotic extracts. The method is simple, requires minimal equipment and yields a product amenable to purification. The method also is effective for disrupting cells from *R. rubrum.*

Lysozyme has been used in preparing nitrogenase from several organisms. This appears to be the method of choice for preparations from *Bacillus polymyxa.* The organism is suspended in glycerol and, after incubation with the glycerol, lysozyme is added and incubation is continued. The protoplasts formed by this treatment then are disrupted by sedimenting them, decanting excess glycerol, and diluting them rapidly with buffer solution. The method combines protoplast production with osmotic shock.

A number of investigators have disrupted N_2-fixing organisms by sonic oscillation. This is the method most commonly used with preparations from *R. rubrum.* A suspension of the organisms is subjected to sonic oscillation, usually with a magnetostrictive device with a frequency of about 10000 cps. The method has the advantage that small amounts of material can be handled and the suspension can be cooled in a jacketed container during processing. High purity N_2 can be bubbled through the chamber to prevent oxygen inactivation during the treatment. The time of treatment is determined empirically.

Preliminary treatment of extracts

Extracts prepared as described will contain nucleic acids, which because of their high viscosity interfere with purification processes. Nucleic acids can be removed by treatment with some basic substance such as protamine sulfate or streptomycin sulfate. After addition of these materials the precipitate formed can be sedimented by centrifugation and removed. Care must be used, because excess protamine or streptomycin will bring down active protein. Each batch of protamine sulfate must be standardized for the preparation, because protamine sulfate is a variable product. In contrast, streptomycin sulfate is commercially available in high purity and gives reproducible results from batch to batch.

After removing nucleic acids, it is often possible to remove substantial amounts of contaminating proteins by relatively simple means. Many investigators have employed a heating step to precipitate inactive proteins. Characteristically, the preparation is heated at 60 to 65 °C for 5 to 10 minutes under inert gas to produce anaerobic conditions, and this is followed by centrifugation to sediment the inactivated protein. Up to 50 % of the protein may be removed from preparations in this way.

Treatment with polyethylene glycol (PEG) under standardized conditions furnishes a relatively simple means for removing inactive protein. Dry PEG is added to a vessel which is then connected to the vessel containing crude nitrogenase. The two vessels are evacuated in tandem, and then their contents are mixed to give a predetermined concentration of PEG. The material is centrifuged to sediment the precipitate. With *C. pasteurianum* we employ PEG 6000 at a concentration of 10 % for the initial precipitation, and after this precipitate is removed the PEG concentration is raised to 30 %; this precipitates the active nitrogenase proteins (Fig. 23.1). Not only does the PEG method remove many inactive proteins, but it also greatly decreases the concentration of contaminating hydrogenase. Hydrogenase is disruptive when it is added with nitrogenase to columns containing $Na_2S_2O_4$; the hydrogenase plus $Na_2S_2O_4$ generates hydrogen which disrupts the columns.

Purification of extracts

Not only can protamine sulfate be used for the removal of nucleic acids but its judicious addition can also effect a precipitation and purification of the nitrogenase components. This method, described in detail by Mortenson (1972), includes eight precipitation steps with protamine sulfate to recover purified Mo–Fe protein and Fe protein. Most investigators prefer purification by separation on DEAE-cellulose and Sephadex columns, and recently Zumft & Mortenson (1973) also have reduced their number of protamine sulfate precipitations and have incorporated DEAE and Sephadex columns in their purification scheme.

Purification on DEAE-cellulose columns usually is employed immediately following preliminary purifications. The concentrated enzyme extract is placed on a column of DEAE-cellulose and is eluted with salt

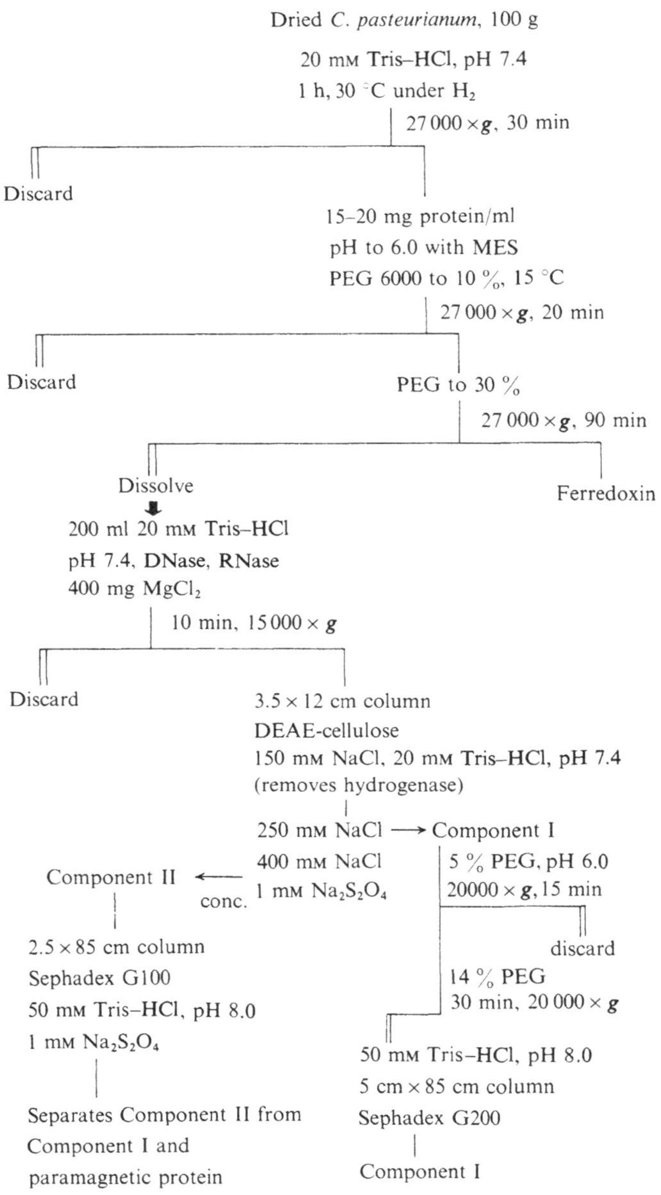

Fig. 23.1. Procedure for purifying the nitrogenase proteins from *C. pasteurianum* (Tso, Ljones & Burris, 1972).
MES = 2-(*N*-morpholino)ethanesulfonic acid; PEG = polyethylene glycol 6000
Tris = Tris(hydroxymethyl)aminomethane; DEAE = diethylaminoethyl
DNase = deoxyribonuclease; RNase = ribonuclease

solution. A concentration of 0.15 M NaCl will remove much of the unwanted protein, 0.25 M NaCl will displace the Mo–Fe protein, and 0.4 M NaCl will elute the Fe protein. It is also possible to use gradient elution, although most preparations have described stepwise increases in NaCl concentrations.

It should be stressed that all operations with nitrogenase proteins must be performed under strictly anaerobic conditions. This means that the anaerobic extract must be added to an anaerobic column, and each fraction must be collected without exposure to oxygen. Columns can be rendered anaerobic by passing a few displacement volumes of anaerobic buffer through them. Oxygen can be removed from the buffer by sparging high purity N_2, hydrogen or argon through the buffer for a few hours. Investigators often insure anaerobic conditions by adding about 0.001 M $Na_2S_2O_4$ to the buffer solution. The protein solution can be added to the anaerobic column by injecting it through rubber tubing connections with a hypodermic needle and syringe. The components of nitrogenase are colored and hence can be followed visually on the column of DEAE-cellulose. Samples can be collected in small serum bottles which have been evacuated and filled with argon or N_2. When the needle at the bottom of the column is inserted through the serum stopper of the bottle, another needle is inserted to relieve the pressure developed as liquid enters the bottle. When the sample is collected, the bottle is removed from the needles and left under the inert atmosphere.

To retain activity through a series of columns it is important that the constituents of nitrogenase be kept highly concentrated. The most effective way to achieve this is with an ultrafilter such as the Diaflo filters supplied by Amicon Corporation, Lexington, Mass. 02173, USA. It is particularly helpful to have a continuous, thin-layer concentrator which can be attached directly to the bottom of a column so that material is concentrated approximately five fold as it is eluted from the column. After such concentration the material is ready for addition to the next column without further concentration; this saves time and minimizes exposure of the labile proteins.

Following passage through DEAE-cellulose the proteins may be purified further on Sephadex columns. Sephadex G100 is useful for purifying Component II (the Fe protein), and Sephadex G200 is the material of choice for purification of Component I (the Mo–Fe protein). Concentrated proteins are added to Sephadex columns approximately one meter in length. As compaction of these columns, particularly the

G200 column, is troublesome, it is advantageous to develop the columns by upward flow. Passage through Sephadex is accompanied by a substantial dilution, so it is necessary again to concentrate the effluent by passage through an ultrafilter. We have found (Tso, Ljones & Burris, 1972; Fig. 23.1) that extracts from *C. pasteurianum* which have undergone preliminary treatment with PEG and have been passed through DEAE-cellulose columns can be purified on the respective Sephadex columns essentially to homogeneity. Operations should be rapid, and there should be particular emphasis on maintaining strictly anaerobic conditions. When these precautions are observed, proteins are consistently recovered which appear homogeneous by polyacrylamide gel electrophoresis. Specific activities we have obtained from such preparations are as high as any reported.

It has been possible to crystallize nitrogenase Component I from *A. vinelandii* (Burns, Holsten & Hardy, 1970). Crystallization is achieved by dilution of the protein-containing salt solution. We and others have repeated the crystallization by the technique described. Shah & Brill (1973) have obtained crystals which appear to be cleaner than those recovered by Burns *et al.* (1970); the earlier crystals were not freed from mother liquor, and optical absorption at 550 nm suggested contamination by cytochrome *c*. The preparations which Shah & Brill have recrystallized do not show an absorption band at 550 nm.

Various criteria have been cited to establish purity of the nitrogenase components. The shape of the peaks eluted from DEAE and Sephadex columns can be taken as an index of purity. Some protein preparations have been analyzed by ultracentrifugation and have met this criterion of purity. The most critical purity standard probably is posed by separation electrophoretically on anaerobic polyacrylamide gels. The components can be separated in a gel containing $Na_2S_2O_4$ in an Ortec pulsed-current electrophoretic apparatus; when heavy concentrations of our best preparations are separated electrophoretically and are stained they show only the faintest bands of any impurities.

The absorption spectra of the components are not definitive for purity, because their molar extinction coefficients are not high and the peaks are broad. However, an impurity such as cytochrome *c* shows up readily upon spectral examination. Absorption by the oxidized Component II near 375 nm is rather similar to that of certain bacterial ferredoxins.

Measurement of activity

Investigators should be urged to measure the reduction of N_2 to ammonia as their index of nitrogenase activity, because this is the reaction of nitrogenase which is of primary interest. Nitrogenase preparations normally have sufficient activity so that this reaction can be measured easily. The reaction is conveniently terminated by adding saturated potassium carbonate and immediately closing the vessel with a rubber stopper which carries an etched glass rod which has been dipped in weak sulfuric acid (Burris, 1972). If the bottles are rotated mechanically for an hour or so, the ammonia diffuses from the alkaline solution to the glass rod where it is captured. Ammonium sulfate on the tip of the rod can be transferred to a colorimetric reagent by using the glass rod to stir the reagent. Total ammonia then can be measured colorimetrically with Nessler's reagent or with the indophenol reagent which is several times more sensitive than Nessler's reagent. It is also possible to do a microtitration on the ammonia which is captured by microdiffusion (Mortenson, 1961), but the colorimetric methods are somewhat simpler.

Nitrogenase activity can be expressed in terms of hydrogen metabolism as indicated by Mortenson (1966). In this method both the uptake of N_2 and of hydrogen is measured and nitrogenase activity is calculated from the combined uptakes. One must compromise in selecting the amount of hydrogen to add to support the reaction, as hydrogen inhibits fixation of N_2 as well as supporting it as an electron donor.

One of the more popular methods for measuring nitrogenase is by determining its reduction of acetylene to ethylene. This is an extremely sensitive method and its simplicity recommends it. It should be kept in mind, however, that acetylene reduction is an indirect assay and acetylene reduction must be converted to equivalent N_2 reduction to make measurements meaningful. Because the acetylene/N_2 conversion factor is not constant under different conditions, it is less ambiguous and often virtually as easy to measure reduction of N_2 to ammonia as to measure ethylene production. In the field the use of the acetylene-reduction method has obvious advantages, but in the laboratory the use of N_2 reduction is so simple that it should be employed in most measurements.

A continuous assay method is particularly helpful for study of the kinetics of the nitrogenase reactions (Ljones & Burris, 1972). Sodium dithionite which has a strong absorption band which centers near

315 nm, can be used as the reductant in the nitrogenase reactions. Because the enzyme has a very low Michaelis constant for $Na_2S_2O_4$, the $Na_2S_2O_4$ is virtually exhausted before there is a change in the rate of reaction. One can continuously record the change in absorption at 315 nm caused by oxidation of $Na_2S_2O_4$ and thus continuously measure the nitrogenase activity of an extract.

Preservation

Component II is cold labile during at least some stages in its purification, hence, it cannot be stored at refrigerator temperatures. It is reasonably stable at 15 °C but is particularly susceptible to inactivation at 0 °C. It can be stored at liquid-nitrogen temperatures for months as demonstrated by Kelly, Klucas & Burris (1967). In contrast, the Component I can be stored successfully under anaerobic conditions at refrigerator temperatures. However, the method of choice for storage of both proteins is at liquid-nitrogen temperatures. Apparently all of the nitrogenase proteins from various organisms can be stored successfully at liquid-nitrogen temperatures. In any storage process, precautions must be taken to maintain anaerobic conditions at all times. Component II is particularly susceptible to oxygen inactivation, and this inactivation is irreversible.

Properties

Molecular weights of proteins and their subunits

A variety of methods has been used to determine the molecular weights of the proteins of nitrogenase. Both sedimentation velocity and sedimentation equilibrium measurements have been made. However, the most commonly used method has been by comparison of movement on Sephadex columns with movement of proteins of known molecular weight. The reported molecular weights of the Mo–Fe protein from *C. pasteurianum* have varied widely, hence people are still concerned with establishing the true molecular weight. Measurements from various laboratories are converging on molecular weight values near 220000 for the Mo–Fe protein from *C. pasteurianum*, and it is not anticipated that this value will change greatly (Table 23.1).

There has been more reluctance to report molecular weights of the highly labile Component II, but there is reasonable agreement that its molecular weight is near 55000. It apparently has two equal subunits,

Table 23.1 *Properties of the component proteins of nitrogenase from various sources*

	C. pasteurianum	*A. vinelandii*	*K. pneumoniae*	Soybean bacteroids
Properties of the Mo–Fe protein of nitrogenase				
Molecular weight	220000	220000	218000	200000
Subunits	4	4	4	4
	2 (50000)	4 (55000)	2 (50000)	4 (50000)
	2 (60000)	—	2 (60000)	—
Atoms of Mo	2	1.1	1	1.3
Atoms of Fe	22	24	17	28.8
Acid-labile S	22	20	17	26.2
Isoelectric pt	4.95	5.2	—	—
Crystals	–	+	–	–
Properties of the Fe protein of nitrogenase				
Molecular weight	55000	63000	66800	51000
Subunits	2 (27500)	2	2 (34600)	—
Atoms of Fe	4	4	4	—
Acid-labile S	4	4	4	—
Isoelectric pt	4.5	4.7	—	—

and the unit of 55000 carries four Fe and four acid-labile sulfur atoms (Dalton & Mortenson, 1972). Its absorption spectrum from *C. pasteurianum* is not unlike that of a bacterial ferredoxin, and it may well be that the iron and sulfur are clustered in the same manner as in ferredoxins. Table 23.1 also shows the molecular weights reported for the proteins of nitrogenase from other organisms. Although the reported value for Component I from *A. vinelandii* is considerably higher than that from *C. pasteurianum*, we find values very close to those reported for *C. pasteurianum*. It is apparent that there are points still to be resolved in establishing the true molecular weights of the nitrogenase components.

Other physical chemical properties

The nitrogenase proteins are quite acidic as one would conclude from the tenacity with which they are held on DEAE-cellulose. Measurements by isoelectric focusing indicate that the isoelectric point for the Component I protein is about 4.9 and for Component II protein about 4.6.

The EPR (electron paramagnetic resonance) spectra have been especially useful in interpreting the mechanism of nitrogenase activity (Fig. 23.2). Component II has an EPR spectrum with bands centering near $g = 2.0$; these bands are quite similar to those observed with the

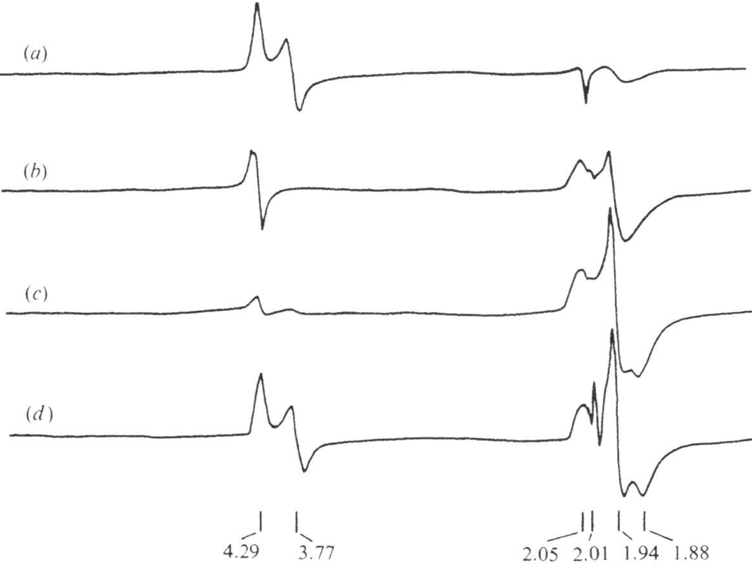

Fig. 23.2. Electron paramagnetic resonance spectra of nitrogenase components of *C. pasteurianum* (Orme-Johnson *et al.*, 1972), in the presence of Mg–ATP. Each sample contained an ATP-generating system to maintain the Mg–ATP concentration near 5 mM. (*a*) 10 mg Component I per ml, plus 5 mM $Na_2S_2O_4$; (*b*) 10 mg of Component II per ml, plus 5 mM $Na_2S_2O_4$; (*c*) the two proteins, 10 mg of each per ml (i.e., at a molar ratio of 1:4) $Na_2S_2O_4$, frozen 45 sec after mixing; (*d*) same as (*c*), except that 0.5 mM $Na_2S_2O_4$ was initially present, along with 40 μM methylviologen, and incubation after mixing and before freezing was for 90 sec, at which time the bluish cast of reduced viologen was no longer visible.

ferredoxins. Component I, on the other hand, has an unusual spectrum with bands near $g = 3.65$ and 4.3. Apparently iron is responsible for these signals in Component I. The EPR spectrum of Component II is changed when Mg–ATP is added and it is bound by the protein. The EPR spectra for Component I indicate that it is oxidized and reduced during its catalytic activation of N_2 fixation.

Composition of proteins

The amino acid compositions of the nitrogenase proteins have been reported, but apparently there is nothing unique about them. As mentioned, Component II carries four Fe atoms per molecule of 55000 molecular weight and it does not contain Mo. The Mo and Fe contents of Component I, on the other hand, are not as clearly established. There are reports of both one and two Mo per molecular weight unit of 220000, but the evidence appears to favor two Mo per unit. The Fe content has been reported variously from 18 to 40 atoms per molecule with an acid-labile sulfur for each one of the Fe atoms. Table 23.1 summarizes information on the composition of the nitrogenase proteins.

Role of Mg–ATP

ATP is an absolute requirement for biological N_2 fixation. ATP is hydrolyzed during N_2 fixation, and the enzyme complex acts like ATPase and yields ADP and orthophosphate. It is interesting that the Mg–ATP is bound specifically to Component II (Dalton & Mortenson, 1972). Addition of Mg–ATP is accompanied by a clear-cut shift in the EPR spectrum of this protein but does not influence the spectrum of Component I (Orme-Johnson *et al.*, 1972). Tso & Burris (1973) have demonstrated in another way that the binding of ATP is specific for the Fe protein. If ^{14}C-labeled Mg–ATP is mixed with a suspension of Sephadex G25 most of the low molecular weight Mg–ATP will be held by the Sephadex. However, if you add the Mg–ATP together with the Fe protein to which it binds, the high molecular weight complex of protein and Mg–ATP will be excluded from the Sephadex, and ^{14}C analysis will show that more ^{14}C-labeled ATP will appear in the supernatant above the Sephadex (Table 23.2). This experimental technique clearly shows that Component II protein binds Mg–ATP whereas Component I protein does not. Furthermore, Component I does not influence the binding of Mg–ATP by Component II.

Table 23.2. *Binding of ATP to the Component II protein of nitrogenase.* From Tso & Burris (1973)

Nitrogenase protein	[^{14}C]ATP outside the gel after equilibration (counts in 5 min)	% Increase in radioactivity due to binding
No protein as control	15594	0.0
20 μM Component II	18153	16.4
10 μM Component I	15447	−0.9
20 μM Component II + 10 μM Component I	18201	16.7

It has been assumed that ATP serves as an energy source in the reduction of N_2, but the mechanism by which it utilizes this energy has not been clear. Our tests with oxidation–reduction dyes, Mg–ATP and Component II indicate that the addition of Mg–ATP substantially lowers the potential of the Fe protein. Although our measurements are not precise at this point, they suggest that the Mg–ATP plus Component II has a potential near −490 millivolts. This is sufficiently negative to reduce Component I. It is interesting to note that the EPR signal of isolated Component I is not changed by the addition of $Na_2S_2O_4$. The only thing we know which will reduce Component I is the reduced Component II in combination with Mg–ATP. This reductant abolishes the characteristic EPR signal of the Component I protein, and the signal returns when reductant is exhausted and the Component I protein once more is oxidized. Further oxidation with ferricyanide produces a 'superoxidized' state of Component I devoid of the characteristic EPR signals near $g = 4.3$ and 3.65, but it is doubtful that the 'superoxidized' state has any physiological significance.

Inhibitors

Inhibitors have been very useful over the years in helping interpretation of the nature of the nitrogenase reaction. When a variety of substrates for nitrogenase which interact with each other became available it was useful to re-examine the inhibition patterns. Hwang, Chen & Burris (1973) reported the response of nitrogenase to a variety of inhibitors, and more recently Rivera-Ortiz (1973) has examined a few additional inhibitors, of the nitrogenase reaction. Table 23.3, which summarizes results from these studies, indicates a variety of types of inhibition. Hydrogen clearly is a competitive inhibitor, and other studies have indicated that nitrous oxide and probably nitric oxide are competitive inhibitors. Other substrates for nitrogenase are non-competitive inhibitors. These include cyanide, azide, methylisocyanide, and acetylene, and some of their analogs. Although they are non-competitive with N_2, the compounds cyanide, azide, and methylisocyanide are mutually competitive among themselves. Carbon monoxide is not reduced by nitrogenase and is a potent non-competitive inhibitor. One of the most interesting observations is that although acetylene is non-competitive with N_2, N_2 is competitive with acetylene.

Rivera-Ortiz (1973) has formulated the nitrogenase reaction in Fig. 23.3 in a way which seems to reconcile the observations. The interpretation depends upon the assumption that in the nitrogenase reaction it is necessary to store a substantial amount of reducing capacity before the N_2 molecule can be reduced to ammonia. During the process of energy storage for this six electron transfer, the reduc-

Table 23.3. *Inhibition of nitrogenase reactions in* Azotobacter vinelandii. From Rivera-Ortiz (1973)

Inhibitor ... Substrate	N_2	N_3^-	C_2H_2	HCN	N_2O	H_2	CO	CH_3NC
N_2	—	—	NC	NC*	C	C*	NC	—
N_3^-	—	—	NC*	C*	—	NI*	NC*	C*
C_2H_2	C	NC*	—	NC	NC	NI*	NC	—
HCN	I	C	E	—	E	E*	NC	—
H^+	I	I*	I	I	I†	NI*	NI*	I

* Hwang, Chen & Burris (1973).
† Bulen, W. A., LeComte, J. R., Burns, R. C. & Hinkson, J. (1965). In *Non-heme Iron Proteins: Role in Energy Conversion* (ed. San Pietro, A.), pp. 261–74. Antioch Press: Yellow Springs, Ohio.
Symbols: C, competitive inhibition; NC, non-competitive inhibition; NI, does not inhibit; I, inhibits; E, enhances.

tants can be tapped for their energy by two electron acceptors. Thus, compounds such as acetylene can utilize the partially reduced energy pool for reduction to ethylene, and this reaction establishes a non-competitive inhibition of N_2 fixation. In contrast, the reduction of N_2 must await the accumulation of the equivalent of six electrons in the reducing pool, and when these are utilized by nitrogenase for the reduction of N_2 their use is competitive with the reduction of acetylene. Carbon monoxide must block the overall reaction at a point before the electron transfer reaches the central reducing pool; hence, all reactions (except hydrogen formation) are non-competitively blocked by carbon monoxide. Carbon monoxide is unique because it is unable to block transfer of electrons to H^+ to form H_2. It also is interesting that H_2 blocks reduction of N_2 but does not inhibit the reduction of any of the other substrates.

These data with inhibitors give credence to the idea that in the reduction of N_2 it is necessary to accumulate a powerful reservoir of reducing capacity before the six electron transfer to N_2 is triggered. The other reductions are less demanding and hence can compete for the electron pool and block reduction of N_2.

My assignment was not to deal with the mechanism of N_2 fixation, but some of the studies of the properties give hints in that direction. I anticipate that the other speakers will develop the points in more detail and will indicate the current consensus on the mechanism of N_2 fixation.

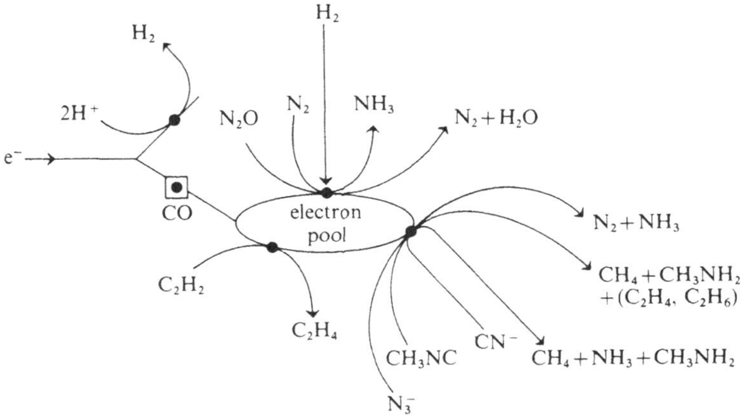

Fig. 23.3. Scheme proposed by Rivera-Ortiz (1973) to explain interaction of substrates and inhibitors of nitrogenase.

References

Burns, R. C., Holsten, R. D. & Hardy, R. W. F. (1970). Isolation by crystallization of the Mo–Fe protein of *Azotobacter* nitrogenase. *Biochem. biophys. Res. Commun.*, **39**, 90–9.

Burris, R. H. (1972). Nitrogen fixation – assay methods and techniques. *Meth. Enzym.*, **24***B*, 415–31.

Carnahan, J. E., Mortenson, L. E., Mower, H. F. & Castle, J. E. (1960). Nitrogen fixation in cell-free extracts of *Clostridium pasteurianum*. *Biochim. biophys. Acta*, **44**, 520–35.

Dalton, H. & Mortenson, L. E. (1972). Dinitrogen (N_2) fixation (with a biochemical emphasis). *Bacteriol. Rev.*, **36**, 231–60.

Hwang, J. C., Chen, C. H. & Burris, R. H. (1973). Inhibition of nitrogenase-catalyzed reductions. *Biochim. biophys. Acta*, **292**, 256–70.

Kelly, M., Klucas, R. V. & Burris, R. H. (1967). Fractionation and storage of nitrogenase from *Azotobacter vinelandii*. *Biochem. J.*, **105**, 3C–5C.

Ljones, T. & Burris, R. H. (1972). Continuous spectrophotometric assay for nitrogenase. *Analyt. Biochem.*, **45**, 448–52.

Mortenson, L. E. (1961). A simple method for measuring nitrogen fixation by cell-free enzyme preparations of *Clostridium pasteurianum*. *Analyt. Biochem.*, **2**, 216–20.

(1966). Components of cell-free extracts of *Clostridium pasteurianum* required for ATP-dependent H_2 evolution from dithionite and for N_2 fixation. *Biochim. biophys. Acta*, **127**, 18–25.

(1972). Purification of nitrogenase from *Clostridium pasteurianum*. *Meth. Enzym.*, **24***B*, 446–56.

Munson, T. O. & Burris, R. H. (1969). Nitrogen fixation by *Rhodospirillum rubrum* grown in nitrogen-limited continuous culture. *J. Bacteriol.*, **97**, 1093–8.

Orme-Johnson, W. H., Hamilton, W. D., Ljones, T., Tso, M.-Y. W., Burris, R. H., Shah, V. K. & Brill, W. J. (1972). Electron paramagnetic resonance of nitrogenase and nitrogenase components from *Clostridium pasteurianum* W5 and *Azotobacter vinelandii* OP. *Proc. natn. Acad. Sci. USA*, **69**, 3142–5.

Rivera-Ortiz, J. M. (1973). Interactions among alternative substrates and some inhibitors of nitrogenase from *Azotobacter vinelandii.* Thesis, Master of Science, University of Wisconsin, Madison.

Shah, V. K. & Brill, W. J. (1973). Nitrogenase. IV. Simple method of purification to homogeneity of nitrogenase components from *Azotobacter vinelandii. Biochim. biophys. Acta*, **305**, 445–54.

Shah, V. K., Davis, L. C. & Brill, W. J. (1972). Nitrogenase. I. Repression and derepression of the iron-molybdenum and iron proteins of nitrogenase in *Azotobacter vinelandii. Biochim. biophys. Acta*, **256**, 498–511.

Tso, M.-Y. W. & Burris, R. H. (1973). The binding of ATP and ADP by nitrogenase components from *Clostridium pasteurianum. Biochim. Biophys. Acta*, **309**, 263–70.

Tso, M.-Y. W., Ljones, T. & Burris, R. H. (1972). Purification of the nitrogenase proteins from *Clostridium pasteurianum. Biochim. biophys. Acta*, **267**, 600–4.

Zumft, W. G. & Mortenson, L. E. (1973). Evidence for a catalytic center heterogeneity of molybdoferredoxin from *Clostridium pasteurianum. Eur. J. Biochem.*, **35**, 401–9.

24. The nitrogenase reaction

I. Morphology and possible cellular localization of *Azotobacter* nitrogenase proteins

II. Established characteristics, relevant models, and possible modes of action of nitrogenase turnover and substrate reduction

III. Applications with emphasis on a novel abiological N_2 fixation system employing membranes

R. W. F. HARDY, R. C. BURNS, J. T. STASNY
& G. W. PARSHALL

This synthesis chapter emphasizes the nitrogenase reaction, the specific area of N_2 fixation biochemistry that has produced a major impact on biological N_2 fixation by providing the basis for the much used acetylene–ethylene assay. Logically, it follows the chapter (Chapter 23) summarizing the preparation and static properties which are now well established for two or three of the nitrogenase proteins, while it precedes the synthesis chapter (Chapter 25) and a paper (Chapter 28) discussing the kinetics of the reaction, an area which is not yet well defined.

Three aspects of the nitrogenase reaction are emphasized in this chapter: (1) the morphology of the nitrogenase proteins examined individually or as a N_2-fixing mixture as well as their possible localization within the cell; (2) tabulation of established characteristics, relevant models and possible modes of action of nitrogenase turnover and substrate reduction; and (3) possible and demonstrated applications based on the nitrogenase reaction with emphasis on an abiological N_2-fixing system employing membranes.

Morphology and possible cellular localization of *Azotobacter* nitrogenase proteins

We have utilized electron microscopy to determine the morphology of the component proteins of nitrogenase from *Azotobacter vinelandii* (Stasny, Burns, Korant & Hardy, 1974). Recombined nitrogenase proteins have also been examined in a search for a structure of the active catalyst. Ferritin-conjugated antibodies have been employed as an approach to locate nitrogenase within the cell.

Electron micrographs of nitrogenase Component I, the Mo–Fe protein

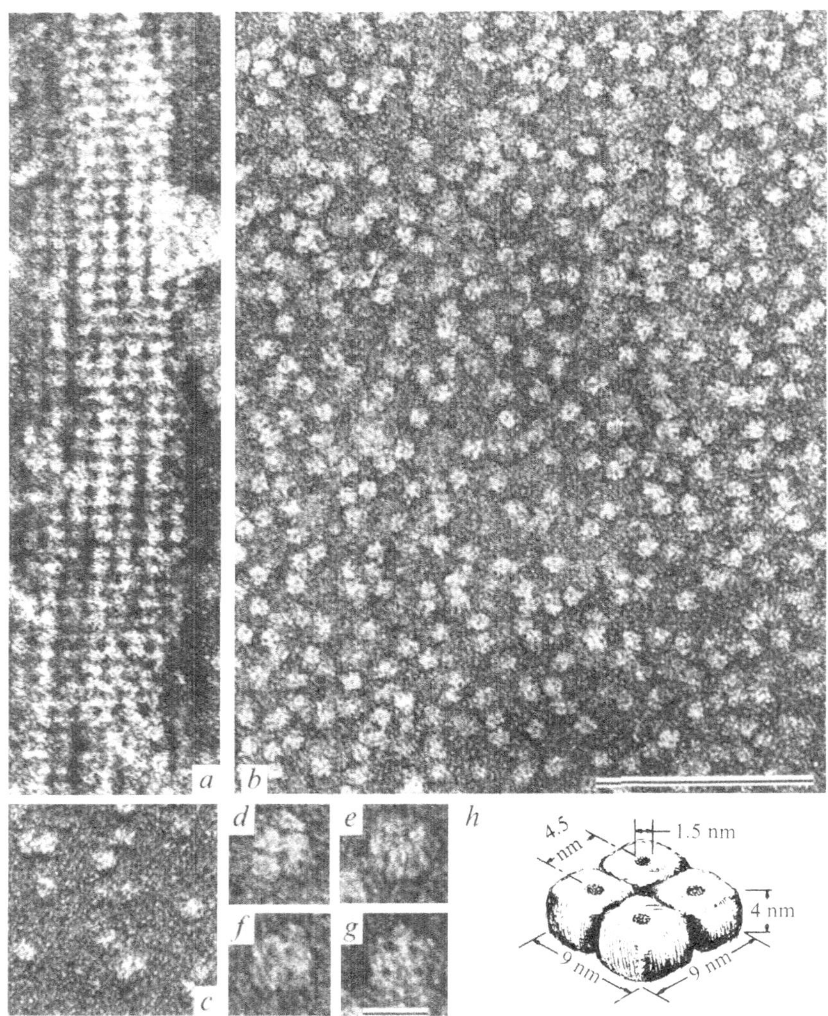

Plate 24.1. Electron micrographs and model of Component I (Mo–Fe protein) of *Azotobacter* nitrogenase, (*a*) 2x crystallized negatively stained with 1.5 % uranyl acetate; (*b*) resolubilized negatively stained (*c*) shadowed with fused platinum carbon; (*d*)–(*g*) enlargement of single particles as in (*b*), and (*h*) drawing of proposed structure of Component I. Marker bar for (*a*)–(*c*) is 1.0 μm and for (*d*)–(*g*) is 0.1 μm (Stasny, Burns, Korant & Hardy, 1974).

and a proposed model, are shown in Plate 24.1. Negatively-stained, ultracentrifugally homogeneous, crystalline (Plate 24.1*a*) or resolubilized Component I protein (Plate 24.1*b*) with a high specific activity of 1535 nmoles H_2/min/mg protein show only one type of particle with a square face, 9 × 9 nm, and a height of 4 nm based on shadow casting (Plate 24.1*c*). The estimated molecular weight for a rectangular parallelepiped, 9 × 9 × 4 nm, with an assumed $\bar{v}$ = 0.73 g/cc is 270000 daltons and is in close agreement with that of 270000–300000 daltons calculated for *Azotobacter* Component I protein from sedimentation data and Mo content (Burns & Hardy, 1972). The electron micrographs of the negatively stained Mo–Fe protein suggest four subunits (Plate 24.1*d–g*) of similar size, and electrophoresis of nitrogenase Component I in SDS-containing polyacrylamide gel produces a single band with an estimated molecular weight of 70000 daltons. More vigorous treatment has yielded smaller units of two types. Slepko, Uzenskaya, Linde & Levchenko (1971) also observed a tetramer of similar size for *Azotobacter* nitrogenase Component I, although their preparation was only 15 % as active as that used in our study. Our proposed model (Plate 24.1*h*) of the Mo–Fe protein based on both electrophoretic and electron microscopic data is a four subunit structure with each subunit 4 × 4.5 × 4.5 nm. It is not possible to indicate the distribution within or between subunits of the two Mo atoms and the approximately thirty Fe atoms. However, an electron dense region (~ 1.5 nm diameter) in one or more of the subunits may represent localization of Fe and/or Mo and/or an area of the protein which is reproducibly stained.

Electron micrographs of negatively stained purified *Azotobacter* nitrogenase Component II, the Fe protein, show an ellipsoidal monomer (Plate 24.2) of 3.5 × 4.0 nm as the predominant unit. This unit is usually observed as a multimer such as the tetramer and hexamer in Plate 24.2*b*, *c* as well as stacked hexamers. The calculated molecular weight for the ellipsoidal monomer is 24000 daltons.

Homogeneous Component I protein and highly purified Component II protein (the Fe protein) similar to those in the above studies, have been recombined under conditions used for N_2 fixation, i.e. with ATP, reductant and N_2 gas phase. Electron micrographs of these negatively stained and catalytically active recombinations revealed associations between the two proteins. Examples included two Component II monomers associated with one Component I tetramer (Plate 24.3*a*) and one Component II monomer with one Component I tetramer (Plate 24.3*b*). These ratios of Fe protein:Mo–Fe protein of one or two are consistent with those determined by activity measurement. The next chapter (Chapter 25) also suggests interactions between the two proteins based on sedimentation. However, it is not possible at this time to identify whether the associated structures observed in the electron micro-

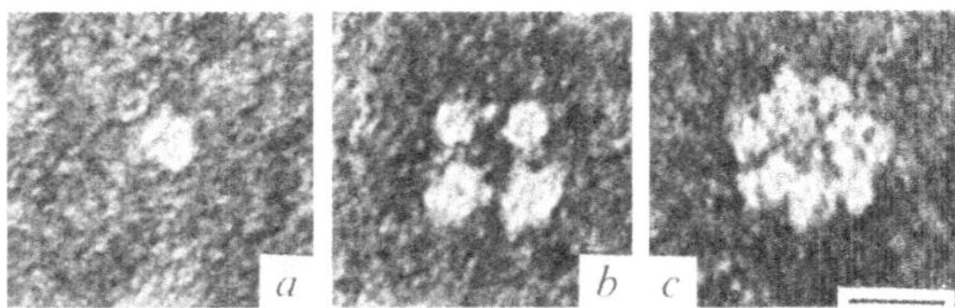

Plate 24.2. Electron micrographs of highly purified Component II (Fe protein) from *Azotobacter* nitrogenase showing (*a*) monomer, (*b*) tetramers and (*c*) hexamers stained as in Plate 24.1*a*. Scale bar for all three is 0.1 μm.

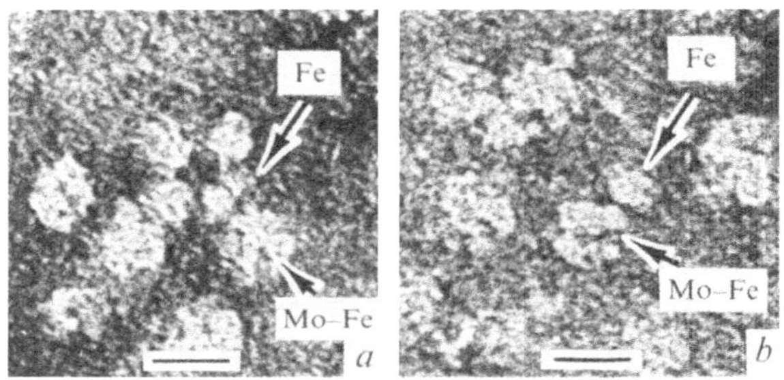

Plate 24.3. Electron micrographs (*a* and *b*) of Components I and II of *Azotobacter* nitrogenase recombined under N_2-fixing conditions and stained as in Plate 24.1*a*. Observed associations of Components I and II may represent nitrogenase structure.

graphs are nitrogenase or non-catalytic interactions of its component proteins.

The interaction between ferritin-conjugated (FC) antibodies and the nitrogenase proteins have been observed with the electron microscope and the electron-dense FC antibodies have been used to probe directly the cellular localization of nitrogenase (Stasny, Burns & Hardy, 1973). Previous indirect studies on localization were based on electron microscopic observations of the amount of internal membrane structure in N_2-grown versus NH_4^+-grown cells, and the results are variable (Oppenheim & Marcus, 1970; Hill, Drozd & Postgate, 1972; Pate, Shah & Brill, 1973). Ferritin-conjugated anti-Component I IgG attaches to the corners of Component I protein antigen where the antibody can be seen as a bridge joining the ferritin with its dense iron core to the antigen (Stasny *et al.*, 1974). Ferritin-conjugated antibody applied topically to thin sections of N_2-fixing *Azotobacter* cells is localized mainly around the cell periphery but also in an organized manner within the internal cytoplasm. Appropriate controls of ferritin or other FC antibodies show no interaction with *Azotobacter* cells. Further confirmation of the suggested localization is desirable, as for any study of this type, and might be provided by similar studies with FC antibody to Component II protein.

In summary, electron microscopic studies have been used to reveal the shape of the individual nitrogenase proteins directly, to demonstrate associations between these two proteins and to suggest but not establish a possible structure of nitrogenase and its localization within the cell. Additional studies with non-*Azotobacter* nitrogenases using allied techniques may help to define the latter areas. Next, we will summarize the information on the unusually diverse reactions catalyzed by these proteins.

Established characteristics, relevant models, and possible mode of action of nitrogenase turnover and substrate reduction

The nitrogenase reaction is catalyzed by a combination of the two nitrogenase proteins, in which ATP hydrolysis is coupled to electron transfer from ferredoxins or flavodoxins for reduction of

N_2 to $2NH_3$,
N_3^- to N_2+NH_3,
N_2O to N_2+H_2O,
RCCH to $RCHCH_2$,
RCN to RCH_3+NH_3,
RNC to RNH_2+alkanes and alkenes and/or $2H_3O^+$ to H_2+2H_2O.

H_2 is a specific competitive inhibitor of N_2 reduction and CO inhibits all reductions except that of H_3O as shown in Fig. 24.1 (Hardy & Burns, 1973). For discussion purposes Burns & Hardy (1975) have further divided the nitrogenase reaction into its two distinct functions. The first is called nitrogenase turnover and concerns the following sequence:

$$ATP + \text{reductant} \rightarrow ADP + P_i + e^{*-}$$

while the second is substrate reduction:

$$\text{reducible substrate} + e^{*-} \rightarrow \text{reduced product}$$

where e^{*-} is defined as the electron prepared for the reduction of nitrogenase. We will use the same division and present a summary of established characteristics, relevant models and possible modes of action for each function.

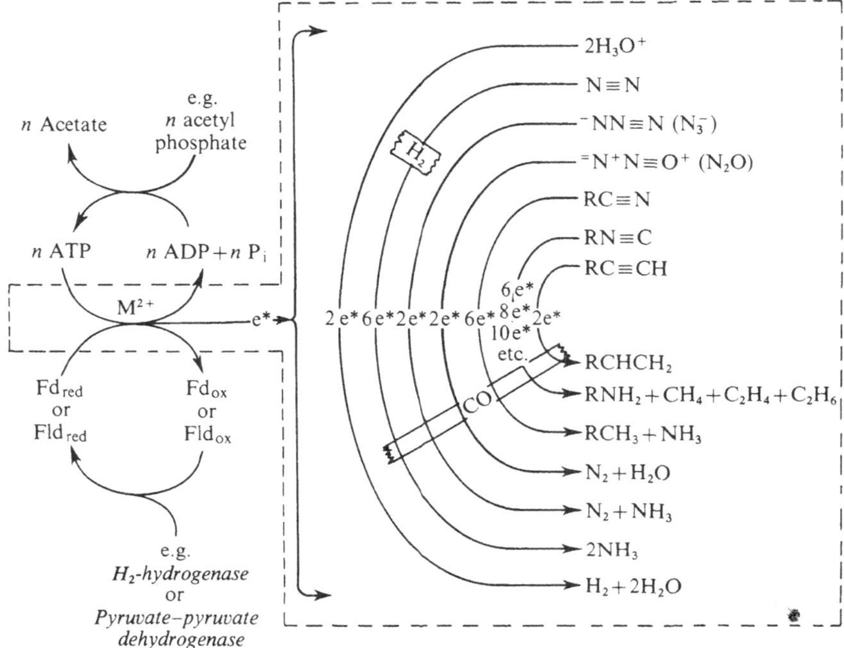

Fig. 24.1. The nitrogenase reaction (enclosed in dotted area) showing substrates, products, inhibitors, and natural electron donors with clostridial electron and phosphagen sources. Fd, ferredoxin; Fld, flavodoxin. (Hardy & Burns, 1973.)

Nitrogenase turnover

This function of the total nitrogenase reaction is much less understood than that of specific substrate reductions. However, all substrate reductions are dependent on nitrogenase turnover, thereby making the characteristics of the nitrogenase turnover function common to all substrate reduction functions regardless of the specific substrate. The following tabulation summarizes present knowledge.

1. *Established characteristics*

Reaction – $x\mathrm{H_2O} + x\mathrm{ATP} + 2\mathrm{e^-} \rightarrow x\mathrm{ADP} + x\mathrm{P_i} + x\mathrm{H^+} + 2\mathrm{e^{*-}}$ with the $2\mathrm{e^{*-}}$ measured by reduction of H-ions (H_3O^+) or exogenous reducible substrate.

Requirements – anaerobic conditions; nitrogenase (both proteins);

$$\mathrm{Mg^{2+} > Mn^{2+} > Co^{2+} > Fe^{2+} > Ni^{2+}};$$

reductant (as reduced ferredoxin, flavodoxin or viologen dyes or $Na_2S_2O_4$) for reductant-dependent reaction (a reductant-independent reaction occurs in the absence of reductant and possibly in its presence).

Stoichiometry – ATP/2e$^-$ variable; recorded values from 1 to 20 with an average of 4 to 5 *in vitro* but the ratio is increased at high Component I to Component II ratios, elevated temperature or decreased pH and may vary for certain reducible substrates such as RNC.

Rate – 1000–2000 molecules of ATP hydrolyzed/min/2 Mo atoms of nitrogenase.

pH maximum – broad from 6 to 8.

Reversal or exchange – not demonstrated.

Isotope effect – no effect of H_2O versus D_2O.

Activation energy – biphasic Arrhenius plot with a change of slope at about 20 °C; activation energy of 14 to 15 kcal/mole above 20 °C and 35 to 50 below 20 °C. (Nitrogenase turnover studies such as ATP binding and ESR spectral studies should attempt to examine the effect both above and below 20 °C.)

Iron interactions and ESR spectral changes – ATP·M^{2+} decreases the intensity of resonance at $g = 1.94$ of reduced Component II and at 4.3, 3.7 and 2.0 of Component I protein mixed with Component II protein, suggesting ATP-requiring electron transfer from iron of Component II to iron of Component I.

Mo interaction – none is established because Mo is spectrally inert and the major effect of V-nitrogenase versus Mo-nitrogenase is on the kinetics of reduction of exogenous reducible substrates and not on nitrogenase turnover where only rate is decreased.

ATP·M^{2+} binding – 2 moles of ATP·M^{2+} bind to each mole of Component II but not to Component I; ADP inhibits binding of one ATP.

Kinetics – complex; limiting reaction is usually bimolecular for ATP but mono- to penta-molecular reactions have been observed; apparent K_m of ATP varies with nitrogenase concentration; ATP concentration affects dilution effect (linearity of rate versus nitrogenase concentration) and the K_m of N_2 and C_2H_2 is reduced at decreased ATP concentrations.

Electron partitioning between reduction of exogenous reducible substrates and H_3O^+ – may be affected by ATP concentration.

Reducible substrates, their products or inhibitors – no effect of added N_2, NH_3, H_2, CO, etc.

Inhibitors – ADP but not AMP; thiol reagents, heavy metals and chelating agents; O_2 is uncompetitive with K_i of 0.025 atm; uncouplers of oxidative or photosynthetic phosphorylation and inhibitors of electron transport are generally ineffective; high salt concentrations produce mixed inhibitions; fluoride does not.

2. *Relevant models and possible modes of action*

Facilitation of reduction by good leaving group – rate of reduction by $S_2O_4^{2-}$ of Pt complex containing a tosylate leaving group is much faster than for a chloride leaving group. Tosylate may be considered to be similar to a phosphate group.

ATP involved in electron activation – e.g.

$$\text{M-OH}+\text{ATP} \xrightarrow{-\text{P}_i} \text{M·ADP} \xrightarrow{2e^-} \text{M: or MH} + \text{ADP}$$

$$\text{M-OH}+\text{ATP} \xrightarrow{-\text{ADP}} \text{M·P}_i \xrightarrow{2e^-} \text{M: or MH} + \text{P}_i$$

Formation of active nitrogenase by ATP in addition to its role in electron activation – e.g.

$$\text{Component II} + x\text{ATP} \rightarrow \text{Component II (ATP)}_x$$

$$y[\text{Component II (ATP)}_x] + \text{Component I} \rightarrow [\text{Component II (ATP)}_x]_y\ \text{Component I} = I$$

$$I + z\text{ATP} \rightarrow I\,(\text{ATP})_z$$

$$I\,(\text{ATP})_z + n\text{H}_3\text{O}^+ + ne^- \rightarrow I + z\text{P}_i + \frac{n}{2}\,\text{H}_2 + n\text{H}_2\text{O}$$

Other possible modes of action of ATP – dehydration by ATP hydrolysis at a site involved in a water sensitive reaction such as nitriding; formation of solvated electrons; site-specific proton source associated with reduction in a non-aqueous environment; conformational change of a dinuclear reduction site to accommodate the required elongation of the N—N bond during reduction.

3. *Selected references*

Bergersen & Turner (1973); Bui & Mortenson (1968); Bulen, LeComte, Burns & Hinkson (1965); Burns (1969); Burns & Hardy (1972; 1975; Chapter 28); Hadfield & Bulen (1969); Hardy, Burns & Parshall (1973); Hardy & D'Eustachio (1964); Hardy, Holsten, Jackson & Burns (1968); Hardy & Knight (1966; 1968); Mortenson (1964); Moustafa & Mortenson (1967); Orme-Johnson *et al.* (1972); Schrauzer, Kiefer, Doemeny & Hirsch (1973); Silverstein & Bulen (1970); Smith, Lowe & Bray (1972); Tso & Burris (1973); Winter & Burris (1968); Zumft *et al.* (1972).

Substrate reductions by nitrogenase

We present a detailed tabulation of the reduction of H_3O^+ and the exogenous reducible substrate N_2, and only briefly mention the alternate exogenous reducible substrates. The selection is based on space limitation and the biological emphasis of the volume while from a mechanistic viewpoint the alternate reducible substrates have provided more useful information than N_2 itself.

H_3O^+. This substrate is endogenous and in the absence of added reducible substrates utilizes all of the electrons transferred by nitrogenase turnover. Reduction of H_3O^+ is catalyzed by all nitrogenases *in vitro* while the presence of hydrogenase can obscure its in-vivo significance. No useful physiological role has been defined for the reaction although it has been suggested to provide a reductant to scavenge oxygen and thereby exclude it from the oxygen-sensitive nitrogenase. The reaction has been extensively used for manometric analyses of in-vitro nitrogenase activity but cannot be applied *in vivo* because of hydrogenases.

1. *Established characteristics*

Reaction – $2H_3O^+ + 2e^{*-} \rightarrow H_2 + 2H_2O$.

Requirements – as for nitrogenase turnover.

Rate – 200 to 400 molecules H_2/min/2 Mo atoms in nitrogenase.

Substrate – H_3O^+ is the ultimate source of H_2 and is not rate-limiting in the pH range of nitrogenase activity; there is no isotope effect of D_2O versus H_2O.

Reversibility – not reversible nor is H_2 at up to 1 atm an inhibitor.

Exchange – no definitive evidence for $D_2 + H_2O \rightarrow HD + HDO$ in absence of N_2 fixation.

Electron allocation – the ratio:

$$\frac{\text{electrons (exogenous substrate)}}{\text{electrons (exogenous substrate)} + \text{electrons } (H_3O^+)}$$

is affected by degree of saturation of exogenous substrate and in general the higher the K_m of exogenous substrate the lower the ratio; ratio varies from almost 1.0 for C_2H_2 to about 0.75 for N_2 to less than 0.01 for some of the larger nitriles; nature of nitrogenase affects the ratio with much higher values for Mo-nitrogenase than V-nitrogenase; concentration of ATP alters the ratio; ratio for nitriles is increased two- to five-fold in D_2O versus H_2O; high ionic strength and pH values removed from 7.0 decrease the ratio suggesting that the site for H_3O^+ reduction is less sensitive and presumably less complex than the site for reduction of exogenous reducible substrates.

Inhibitors – as for nitrogenase turnover; no agent has been found that specifically inhibits H_3O^+ reduction; neither CO nor H_2 inhibit H_3O^+ reduction; other reducible substrates may be described as inhibitors through consumption of electrons.

2. *Relevant models and possible mode of action*

H_2 evolution by acidic solutions of B_{12s} via hydride mechanism

Molybdo-thiol-reductant model – stimulation by nucleoside di- and triphosphates.

3. *Selected references*

Burns (1965); Burns & Bulen (1965); Das, Hill, Pratt & Williams (1967); Fuchsman & Hardy (1972); Hardy, Knight & D'Eustachio (1965); Jackson, Parshall & Hardy (1968); Schrauzer, Kiefer, Doemeny & Hirsch (1973); Strandberg & Wilson (1967).

N_2. Molecular nitrogen (N_2 or dinitrogen) is the physiological substrate of nitrogenase and this reaction is used extensively both *in vivo* and *in vitro* to measure nitrogenase activity and it is of course the ability of nitrogenase to facilitate the conversion of N_2 to ammonia that is responsible for both the scientific and practical interest.

1. *Established characteristics*

Reaction – $N_2 + 6e^{*-} + 6H^+ \rightarrow 2NH_3$.

Rate – 50 to 100 molecules of N_2 reduced/min/2 Mo atoms in nitrogenase.

Substrate – N_2 with no demonstrated isotope effect of $^{15}N_2$ versus $^{14}N_2$.

K_m – 0.04 to 0.20 atm or 0.03 to 0.12 mM N_2.

Intermediates – no enzyme-free intermediate such as diazene or hydrazine is found; added or *in situ* generated diazene and hydrazine are not reduced.

Product – ammonia, which shows no affinity for nitrogenase and is not an inhibitor.

Exchange – no $^{14}N_2 + ^{15}N_2 \rightleftarrows 2\ ^{14}N^{15}N$.

Reversibility – reaction has not been reversed.

Rate of electron addition – 0.75 relative to H_2 formation from H_3O^+ as 1.

Inhibitors – H_2 is competitive, with a K_I of 0.2 to 0.5 atm; CO (competitiveness in question) has a K_I of 6 to 40×10^{-4} atm or 5 to 35×10^{-6} M; NO has a K_I of 25×10^{-3} atm or 4.3×10^{-7} M; O_2 is uncompetitive with a K_I of 0.014 atm; other reducible substrates.

H_2 inhibition – H_2 is a specific competitive inhibitor and does not inhibit the reduction of any other substrates.

Hydrogen exchange – $D_2 + H_2O \rightleftarrows HD + HDO$; exchange requires conditions for nitrogenase turnover with N_2 reduction to NH_3; other reducible substrates are ineffective; CO inhibits exchange; HD/NH_3 ratio is linearly dependent on P_{D_2} with ratio of 4.5 at 0.6 atm D_2.

Metal interaction – K_m for N_2 and K_I for H_2 same for Mo- and V-nitrogenase but K_I for CO is 1 to 3×10^{-4} atm N_2 for Mo-nitrogenase and 4 to 6×10^{-4} atm N_2 for V-nitrogenase.

Selected references. Burns, Fuchsman & Hardy (1971); Burris, Winter, Munson & Garcia-Rivera (1965); Hwang & Burris (1972); Jackson, Parshall & Hardy (1968); Lockshin & Burris (1965); Hwang, Chen & Burris (1973); McKenna, Benemann & Traylor (1970); Parejko & Wilson (1971); Shilov & Lichtenshtein (1971); Stiefel (1973); Turner & Bergersen (1969); Wong & Burris (1972).

2. *Relevant models*

Four separate lines of chemical investigation are providing model information that may be relevant to the nitrogenase-catalyzed reduction of N_2 to NH_3. Much of this chemical investigation is pure chemistry while some is directed to the production of models that mimic the nitrogenase reaction.

Transition metal complexes and their reactions

(i) Linear or end-on orientation of metal and dinitrogen in all cases, e.g.

$M \leftarrow N{\equiv}N$; however, the first example of a side-on orientation has been found in matrix isolated $Co \cdot N_2$.

(ii) The number of metal atoms and the number of dinitrogen molecules (e.g. $M(N_2)$, $M(N_2)M$ and $M(N_2)_2$) have been studied.

(iii) The relative ligand affinity in $M \cdot N_2$ complexes is similar to that found with nitrogenase, e.g. $CO \gg N_2 \sim H_2 > H_2O$, NH_3, C_2H_4; H_2 and N_2 are competitive with each other as with nitrogenase.

(iv) Amine or phosphine ligand stabilized $M \cdot N_2$ complexes suggest amine, S^{2-}, or cysteine thiolate as similar types of biological ligands in nitrogenase.

(v) Adduct formation with reduction in N—N bond strength and length, $Re(N_2) + Mo^{4+} \rightarrow Mo = N \equiv N = Re$.

(vi) Partial reduction of N_2 to diazene and hydrazine in *trans*-$[M(N_2)_2]$ with M = Mo or W, thus – $M(N_2)_2 + 2HX \rightarrow MX_2(N_2H_2) \rightarrow MX_2(N_2H_4)$; $M(N_2)_2 + RCOCl \rightarrow [MCl_2({=}N{-}NH{-}COR)]$.

(vii) *Selected references*

Allen (1971); Chatt, Dilworth, Richards & Sanders (1969); Chatt, Heath & Richards (1972); Hardy & Burns (1973); Hardy, Burns & Parshall (1973); Ozin & Vander Voet (1973).

Homogeneous catalysis of N_2 to NH_3 with mild conditions in anhydrous media requiring strong reductants – these systems suffer as models since they require anhydrous conditions; example three utilizes Mo and Fe which are the metals of nitrogenase and a recent advance in this system functions in protonic media, supporting the site proposed by us earlier and based on the nitrogenase reaction biochemistry.

(i) Vol'pin and Shur

$$Cp_2{}^*TiCl_2 + RMgX + N_2 \rightarrow \text{nitride} \xrightarrow{H^+} NH_3$$

(ii) van Tamelen Cycle

$$\begin{array}{ccc} Ti(OR)_2 + N_2 & \rightarrow & [Ti(OR)_2N_2]_n \\ \uparrow 2e^- & & \downarrow 4e^- \\ 2NH_3 + Ti(OR)_4 & \xleftarrow{ROH} & \text{nitride} \end{array}$$

(iii) $Mo \cdot N_2$ and reduced ferredoxin-like moiety

$$Mo \cdot N_2 + \text{diphenyldithiolene Fe—S Cluster}^{4-} \rightarrow NH_3$$

(iv) *Selected references*

Marchon & Barbosa (1972); Vol'pin & Shur (1964; 1966); van Tamelen, Boche & Greeley (1968); van Tamelen, Gladysz & Miller (1973); van Tamelen, Rudler & Bjorklund (1971).

Homogeneous catalysis of N_2 to NH_3 with mild conditions in aqueous media – these models have an advantage with respect to comparison with nitrogenase, since they function in aqueous systems; most studied is the simple molybdo-thiol-reductant model whose similarities to nitrogenase are tabulated and described later in this paper.

* Cp_2 = biscyclopentadienyl.

(i) Schrauzer molybdo-thiol-reductant model

$$\text{Mo}\cdot\text{cysteine} + N_2 + BH_4^- \text{ or } S_2O_4^{2-} \xrightarrow{\text{XTP}} [N_2H_2 \text{ intermediate}] \downarrow 2NH_3$$

$$\text{Mo}\cdot\text{glutathione} + N_2H_4 + BH_4^- \xrightarrow{\text{XTP}} 2NH_3$$

$$N_2H_4 \xrightarrow{\text{binuclear } Mo^{3+}} 2NH_3$$

(ii) Mo or V complexes and strong reductants, e.g. Cr^{2+}, Ti^{3+}, V^{2+}

$$N_2 \rightarrow N_2H_4 \rightarrow 2NH_3$$

(iii) Various Fe complexes plus $NaBH_4$

$$N_2 \rightarrow 2NH_3$$

(iv) *Selected references*

Hill & Richards (1971); Mitchell & Searle (1972); Newton, Corbin, Schneider & Bulen (1971); Schrauzer, Doemeny, Frazier & Kiefer (1972); Schrauzer, Doemeny, Kiefer & Frazier (1972); Schrauzer, Kiefer, Doemeny & Hirsch (1973); Schrauzer, Schlesinger & Doemeny (1972); Shilov *et al.* (1971); Werner, Russell & Evans (1973).

Aryldiazonium model – one of the first designed to mimic substrate reduction of the nitrogenase reaction and its relationship to nitrogenase is still valid; diazene and hydrazine intermediates catalyze exchange between H_2 and H_2O; only bound intermediates between N_2 and NH_3 are found.

(i) Diazene and hydrazine intermediates

$$[\text{Ph-N}{\equiv}\text{N}]^+ + \text{HM} \rightarrow [\text{PhN}{=}\text{NHM}]^+$$

$$\downarrow H_2/Pt \text{ or } Na_2S_2O_4$$

$$[\text{PhNHNH}_2\text{M}]^+$$

$$\downarrow H_2/Pt \text{ or } Na_2S_2O_4$$

$$\text{PhNH}_2 + NH_3 \xleftarrow{H_2/Pt} \text{PhNHNH}_2 + \text{HM}$$

(ii) Substituted aryldiazonium salts and *trans*-$IrCl(CO)(PPh_3)_2$ yield a cyclic structure suggesting the possible involvement of three metals or that initial reduction may involve rearrangement with cleavage of a C—H bond in for example cysteine or an aromatic amino acid.

(iv) *Selected references*

Gilchrist, Rayner-Conham & Sutton (1972); Jackson *et al.* (1968); Parshall (1967).

3. *Possible modes of action*

Various proposals have been made to explain nitrogenase-catalyzed dinitrogen reduction. The earlier examples were simply theoretical chemistry while some of the more recent examples are based on either the biochemistry of N_2 fixation or the chemistry of abiological reactions of N_2. A few are based on both the chemistry and biochemistry. Earlier reviews (e.g. Hardy, Burns & Parshall, 1973; Hardy & Knight, 1968) can be consulted for detailed presentations of these proposals. For this discussion we will consider the possibilities that exist for three aspects of the mode of action: the nature of the site, the orientation of complexed N_2 and the reduction process and intermediates.

Site. All evidence indicates metal involvement but the only direct evidence is the comparison of the kinetics of Mo-nitrogenase versus V-nitrogenase for N_2, inhibitors and other exogenous reducible substrates; possibilities include a mono-nuclear site of Mo or Fe or a dinuclear homogeneous site of Mo or Fe or a dinuclear heterogeneous site of Mo and Fe; the differential inhibition pattern of H_2 and CO on substrate reduction and comparison of V-nitrogenase with Mo-nitrogenase supports a dinuclear heterogeneous site; possible but not confirmed complexity of interactions between different reducible substrates may suggest a more complex set of sites for different substrates.

Orientation. Possibilities include end on $M{-}N{\equiv}N$, $M{-}N{\equiv}N{-}M$ and side-on $M\overset{N}{\underset{N}{|||}}$, $M\overset{N}{\underset{N}{|||}}M$; many examples of either end-on type are found in abiological transition metal complexes of N_2 and one side-on example has now been found in a metal matrix: there is no biochemical information.

Reduction process including intermediates. The process appears to be completely reductive; possibilities include single-step six-electron reduction, e.g. $N_2+6e^-+6H^+ \rightarrow 2NH_3$ or multistep with nitrogenase-complexed intermediates, e.g. hydrazine intermediate

$$N_2+4e^-+4H^+ \rightarrow [N_2H_4]$$

and

$$[N_2H_4]+2e^-+2H^+ \rightarrow 2NH_3,$$

or diazene and hydrazine intermediates $N_2+2e^-+2H^+ \rightarrow N_2H_2$, or nitride intermediate, $M{-}N{\equiv}N+3e^-+3H^+ \rightarrow M{\equiv}N+NH_3$ and

$$M{\equiv}N+2H_2O \rightarrow \overset{O}{\overset{\|}{M}}OH+NH_3 \quad \text{and} \quad \overset{O}{\overset{\|}{M}}OH+3e^-+3H^+ \rightarrow M+2H_2O;$$

the involvement of stepwise two-electron reductions is supported by evidence for 2, 4, 6, 8, 10, 12 and 14 total electron additions for exogenous alternate substrates; a diazene intermediate is thermodynamically unfavorable as a free product but this may be overcome by co-ordination to metal and hydrogen bonding from enzyme; evidence for a diazene intermediate is seen in the molybdo-thiol-reductant model; hydrazine is a product of N_2 fixation in several abiological systems; a nitride intermediate would require almost complete exclusion of H_2O from the site; our best estimate, based on biochemical and abiological information, is a heterogeneous dinuclear site with initial side-on or end-on complexing of N_2 to a H_2-sensitive iron site, followed by interaction with a CO-sensitive molybdenum site by stepwise two-electron additions, to give successively, enzyme complexed diazene and hydrazine intermediates and finally ammonia which is readily released; the site is then regenerated for the next cycle.

Applications with emphasis on a novel abiological N_2 fixation system employing membranes

In general, biochemistry is a science that does not directly produce useful technologies and the biochemistry of the nitrogenase reaction is no exception. However, the rapid expansion of the knowledge of the biochemistry of N_2 fixation has provided the basis for a remarkable number of possible or demonstrated applications in related disciplines. This impressive list will be summarized below with one example, a novel abiological N_2-fixing system described in some detail.

Acetylene–ethylene assay

In N_2 fixation research the acetylene–ethylene assay is the most used application of the nitrogenase reaction (Hardy, Burns & Holsten, 1973). It is applied routinely to measure N_2-fixing activities of abiological, biochemical, biological, genetic, ecological and agronomic systems. Over 300 scientific papers have appeared utilizing this method and it is responsible for the increased research activity in biological nitrogen fixation. Several reviews of the method and applications to various systems are available (e.g. see Balandreau & Dommergues, 1973; Bergersen, 1970; Hardy, Burns & Holsten, 1973; Hardy & Holsten, 1975; Postgate, 1972).

Definition of opportunities for beneficial control of N_2 fixation

Definition of opportunities for beneficial control of biological N_2 fixation has been made feasible through the application of the acetylene–ethylene assay. Extensive measurements of symbiotic N_2 fixation may define approaches for its enhancement. For example, the experimental ability to increase symbiotic N_2 fixation in field-grown soybeans by increasing photosynthate through carbon dioxide enrichment (Hardy & Havelka, 1975) or to a lesser extent by light supplementation (Ham, Lawn & Brun, 1975) was described earlier at this meeting. Another

Table 24.1. In situ *N_2 fixation measurements with roots or non-legumes in Colombia**

Sample source	μg $N_2(C_2H_2)$ fixed/day/plant root
Mountain grasses (2/11)**	132–203
Tropical rain forest grasses (10/11)	9.2–59
Plateau grasses (3/16)	17–387

* In collaboration with P. Graham of CIAT. Freshly excavated roots were incubated in air containing 10 % C_2H_2 for 1 h.
** Number of roots showing $N_2(C_2H_2)$ fixation/total number of roots tested.

approach to increased nitrogen-input by N_2 fixation is to seek and domesticate heretofore unrecognized obligatory or non-obligatory symbioses. For example, a new symbiosis has been recognized between *Rhizobium* and the non-legume *Trema aspera* (Trinick, 1973). Research by Balandreau *et al.* (Chapter 4) on rice and Dobereiner and Day (Chapter 3) on tropical grasses indicates the large contributions which are possible from these non-obligatory symbioses. An exploratory investigation in Colombia by Graham & Hardy (unpublished) demonstrated $N_2(C_2H_2)$-fixing activity associated with the roots of several grasses (Table 24.1) while Raju, Evans & Seidler (1972) found N_2 fixation in the corn rhizosphere. Synthetic N_2-fixing systems are also being sought through extension of the legume symbioses by tissue culture techniques (Holsten, Burns, Hardy & Hebert, 1971; Child & LaRue, 1974; Phillips, 1974) and the transfer of *nif* genes (Dixon & Postgate, 1971). In other cases undesirable sites of N_2 fixation are being identified such as the N_2-fixing bacteria associated with decay in white fir trees (Seidler, Aho, Raju & Evans, 1972) and the low N_2-fixing activity in termites (Benemann, 1973; Breznak, Brill, Mertins & Coppel, 1973). In all of these examples, the acetylene- ethylene assay has made possible the definition of opportunities for beneficial control of N_2 fixation. However, absolute demonstration of N_2 fixation, especially in new systems, can only be made with ^{15}N-enrichment from $^{15}N_2$ as has been done for the tissue culture symbiosis (Hardy & Holsten, 1972) and the transfer of *nif* genes (Dixon & Postgate, 1971).

Molecular structures of flavodoxin and ferredoxin

Studies of the biochemistry of N_2 fixation led to the discovery of the new electron transfer proteins, ferredoxin and flavodoxin, described later by Evans and Phillips (Chapter 26). Application of these proteins to chemical structure studies has yielded the first molecular structure of a flavoprotein, flavodoxin (Watenpaugh *et al.*, 1972), and the structure of an iron–sulfur protein, ferredoxin (Adman, Sieker & Jensen, 1972) while model studies of the latter produced an Fe_4S_4 cluster compound (Herskovitz *et al.*, 1972). Presumably some of the iron of the nitrogenase proteins will be of the Fe_4S_4 cluster type but the unique ESR spectrum of the Mo–Fe protein (Burns & Hardy, 1972) shows the opportunity for additional chemical model work.

Novel abiological reactions

At the time of their discovery in 1965–67 several of the reductions of alternative substrates catalyzed by nitrogenase did not have a counterpart in abiological homogeneous catalysis. Application of the biochemical information has led to the development of novel abiological reactions. Such examples include homogeneous catalysis of the reduction of N_2O, RCN and RNC.

Novel abiological N_2-fixing systems

A number of novel abiological N_2-fixing systems that function under an ambient environment have been found (see e.g., Hardy, Burns & Parshall, 1973; Hardy & Burns, 1973). Some have their origin in pure chemistry, especially those that function in non-aqueous environments. Others have their origin in the nitrogenase reaction. The most relevant example of the latter case is the molybdo-thiol-reductant model developed by Schrauzer and his colleagues as a model of nitrogenase. The qualitative similarity between this abiological model and nitrogenase is most striking (Table 24.2). All of the reactions catalyzed by nitrogenase are performed by the model, and, in addition, the model catalyzes some other reactions including reduction of carbon monoxide to methane and ethylene oxide to ethylene. The quantitative similarity (Table 24.3) is much less dramatic with the reduction of N_2 to ammonia by the model, occurring at a rate of only 2×10^{-8} of the rate of enzymic reduction. However, one anticipates that further development of this or yet-to-be-discovered abiological catalysts will overcome the

Table 24.2. *Qualitative comparison of nitrogenase and molybdothiol-reductant model*

	Nitrogenase*	Model**
Substrates and products	$H_3O^+ \to H_2$ $N_2 \to 2NH_3$ $N_3^- \to N_2 + NH_3$ $N_2O \to N_2 + H_2O$ $RCN \to RCH_3 + NH_3$ $RNC \to CH_4, C_2H_6, C_2H_4, C_3H_8, C_3H_6 + RNH_2$	
		$C_2H_2 \to C_2H_4$ and C_2H_6
$\overset{O}{CH_2—CH_2}$ $\not\to$	$\overset{O}{CH_2—CH_2} \to C_2H_4$	
	$CO \not\to$	$CO \to CH_4$
$N_2H_4 \not\to$	$N_2H_4 \to 2NH_3$	
Electron additions	2, 4, 6, 8, 10, 12 and 14	
Requirements	Mo–Fe and Fe proteins ATP, M^{2+} Reductant ($S_2O_4^{2-}$)	Mo, Thiol Reductant ($S_2O_4^{2-}$ or BH_4^-)
Stimulators		XTP or XDP Fe
Metal specificity	Fe and Mo or V	Mo
Inhibitors	CO H_2 (for N_2 reduction)	CO is substrate No effect by H_2
Optimum pH	6–8	9+
Activation energy	14–15 (> 20 °C)	13–21

* Hardy, Burns & Parshall (1971).
** Schrauzer *et al.* (1973) and Werner *et al.* (1973).

current rate limitation. It can be concluded that application of information on the nitrogenase reaction has led to a most relevant qualitative model and could lead to a useful homogeneous catalyst for abiological N_2 fixation. Now we will describe a novel approach for utilizing such abiological systems as the one described above for localized production of fertilizer N which is more directly coupled to crop plants.

Novel abiological N_2 fixation system employing membranes

The expanding world acreage of new nitrogen-responsive cereals and the increasing environmental concern regarding soil/fertilizer nitrogen emphasize the need to seek new approaches for more effective produc-

Table 24.3. *Quantitative comparison of nitrogenase and molybdothiol-reductant model*

Substrate	K_m (mM)		Turnover no. (mole/min/mole Mo)	
	Nitrogenase*	Model**	Nitrogenase*	Model**
N_2	0.03–0.1	$\gg 1$	50	10^{-6} (10^{-2})***
N_3^-	0.2–1.0	~ 0.5	150	2.5×10^{-3}
N_2O	1.0	2.4	150	2.5×10^{-3}
HCN	0.4	12	25	5×10^{-5}
$CH_2{=}CHCN$	10–25	250	12	1×10^{-5}
CH_3CN	~ 500	~ 1000	1	5×10^{-6}
CH_3NC	0.2–1.0	8–100	40	4×10^{-5}
C_2H_2	0.1–0.4	0.33	200	0.05–6

* Hardy, Burns & Parshall (1971).
** Schrauzer *et al.* (1973) and references therein and Werner *et al.* (1973).
*** Turnover number for N_2 to N_2H_2 by model.

tion and utilization of fertilizer nitrogen in both developed and underdeveloped countries. Currently, the distribution costs for fertilizer nitrogen approach the cost of synthesis in a modern ammonia plant (Hardy, Burns, Hebert, Holsten & Jackson, 1971).

An innovative approach to synthesis and distribution of fixed nitrogen is provided by recent advances in homogeneous catalysis as described above and membrane technology. Molybdenum (Schrauzer *et al.*, 1971; Shilov *et al.*, 1971) and other transition metal catalysts recently developed as models of the N_2-fixing enzyme, nitrogenase, fix N_2 in aqueous media at ambient temperatures and pressures. Synthetic membranes have been developed for fractionation of gases in much the same way that solutions are concentrated by reverse osmosis (Gosser, Knoth & Parshall, 1973). We have now combined these new catalysts with membranes for N_2-enrichment and O_2-depletion of air and for fractionation of product ammonia from catalyst.

Our membrane-catalyst system, although limited to laboratory experimentation at this time, suggests an approach to nitrogen fertilization. For example, a unit containing a water-soluble catalyst contained within a selectively permeable membrane could be placed in an irrigation stream and operated to fix N_2 at the time of crop need, thereby improving crop use and eliminating distribution costs (Fig. 24.2). However, it is emphasized that major improvements in turnover rate of water-soluble, N_2-fixing catalysts and in the fractionation rates and

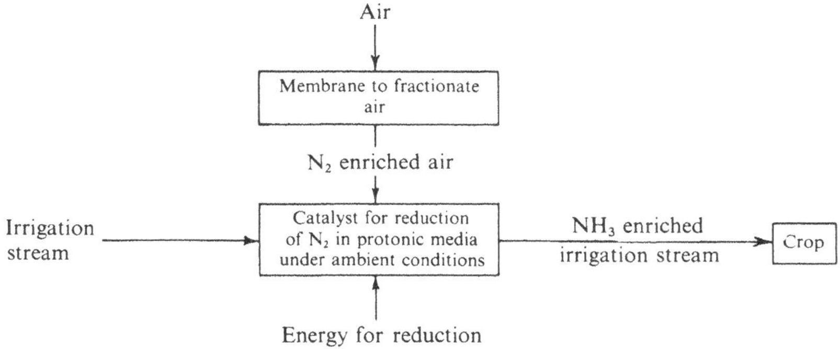

Fig. 24.2. Schematic of a possible system to better couple abiological N_2 fixation to crop nitrogen fixation. A membrane is used to produce N_2 enriched air, a catalyst contained in a membrane and inserted in an irrigation stream reduces the N_2 enriched air; and the product ammonia is carried by the irrigation stream to the crop.

effectiveness of the membranes as well as a practical supply of reductant are essential for this system to become a practical process.

The first step is dinitrogen enrichment of air to decrease competition between oxygen and N_2 for reductant. This is achieved in the laboratory with a standard high pressure permeation test cell (Richter & Hoehn, 1971) (*c.* 75 ml internal volume) fitted with a 50 mm disc of polychloral film with an O_2/N_2 gas transport selectivity of 4.6 (Vogl, Miller & Sharkey, 1972). The cell is pressured with air to 67 atm and gas permitted to diffuse through the membrane overnight. Mass spectrometric analysis of the retained gas showed 90.0 % N_2 (versus 78.4 % observed in ambient air) and 10.0 % oxygen.

In the second step, N_2 enriched gas is reduced to ammonia. A solution of 1.36 g 2-aminoethanethiol hydrochloride, 2.91 g sodium molybdate-(2 –) dihydrate and 0.060 g iron(2 +) sulfate heptahydrate in 300 ml of water was used to reduce N_2 enriched air at atmospheric pressure (Hill & Richards, 1971). Sodium tetrahydroborate (3 g) was added in small portions over 3.5 h while the N_2 enriched gas was bubbled into the mixture at 25 °C. The dark green reaction solution was stirred at room temperature for 16 h. Distillation of a 20 ml aliquot gave an alkaline distillate from which ammonium tetraphenylborate (infrared spectrum identical to an authentic sample) precipitated on addition of sodium tetraphenylborate solution. The yield of ammonium tetraphenylborate was 63 mg, after subtraction of ammonia background obtained from a matched control run under 90 % argon.

The third step is removal of the product ammonia and retention of the catalyst. The reaction mixture was filtered through a porous silver disc to remove a small amount of suspended solid and the filtrate transferred to a reverse osmosis apparatus and forced at 75 atm pressure against an aromatic polyamide membrane (a copolyamide of 3- and 4-aminobenzoyl hydrazides with isophthalic and terephthalic acids (Richter & Hoehn, 1971). The clear, colorless permeate (0.7 ml/min) and the residual solution were analyzed for molybdenum by X-ray fluorescence. The permeate gave 0.1 counts/sec versus 56.1 for the feed solution, indicating almost complete removal of the metal by the membrane, while the permeate contained ammonia.

To eliminate the need for subtraction of ammonia from nitrogen sources other than N_2, the model system described by Schrauzer, Schlesinger & Doemeny (1971) was employed for steps two and three. Using N_2 at 130 atm pressure, 1.7–5.0 μmoles of ammonia were formed in 50 ml of aqueous solution containing 5 nmoles of sodium molybdate, 2.5 mmoles of thioglycerol and 0.1 mmole of ferrous sulfate. The solution was diluted to 130 ml volume with water and was placed in a commercial Amicon model 420 reverse osmosis cell fitted with a permeable poly(*m*-phenyleneisophthalamide) membrane (Richter & Hoehn, 1971). Nitrogen pressure was applied and water was forced through the membrane. The ammonia concentration of the permeate equalled or exceeded that of the original 2.2×10^{-5} M solution throughout passage of 100 ml of solution while most of the molybdenum was retained.

In summary, the N_2 content of air can be enhanced by selective diffusion of oxygen through a synthetic membrane, the N_2 enriched air can be reduced to ammonia at room temperature and atmospheric pressure, and the aqueous ammonia can be separated from the catalyst by reverse osmosis. In addition to demonstrating a novel approach for production and distribution of fertilizer nitrogen, this in-vitro N_2-fixing system may simulate a membrane effect in biological N_2 fixation. The cellular localization of nitrogenase discussed earlier and the oxygen-insensitivity of particulate but not soluble preparations of nitrogenase (Oppenheim, Fisher, Wilson & Marcus, 1970) of the aerobe, *Azotobacter*, are consistent with natural membranes assisting in excluding oxygen but not nitrogen from the oxygen-sensitive nitrogenase. In other aerobic N_2-fixing systems, e.g. legume nodules, membranes such as those enclosing the bacteroids may serve a similar gas-partitioning role. In conclusion application of information from the nitrogenase reaction may provide the basis for increased nitrogen-input by abiological as well as biological N_2 fixation.

References

Adman, E. T., Sieker, L. C. & Jensen, L. H. (1972). Structure of a bacterial ferredoxin. *J. biol. Chem.*, **248**, 3987–96.

Allen, A. D. (1971). Developments in inorganic models of N_2 fixation. *Adv. Chem. Series*, **100**, 79–94.

Balandreau, J. & Dommergues, Y. (1973). Nitrogenase (C_2H_2) activity in the field. In *Symposium on Modern Methods in the Study of Microbial Ecology* (ed. Rosswall, T.), **17**, pp. 247–254. Stockholm: Swedish IBP Ecological Research Committee.

Benemann, J. R. (1973). Nitrogen fixation in termites. *Science*, **181**, 164–5.

Bergersen, F. J. (1970). The quantitative relationship between nitrogen fixation and the acetylene-reduction assay. *Aust. J. Biol. Sci.*, **23**, 1015–25.

Bergersen, F. J. & Turner, G. L. (1973). Kinetic studies of nitrogenase from soybean root-nodule bacteroids. *Biochem. J.*, **131**, 61–75.

Breznak, J. A., Brill, W. J., Mertins, J. W. & Coppel, H. C. (1973). Nitrogen fixation in termites. *Am. Soc. Microbiol. Abstr.*, p. 168.

Bui, P. T. & Mortenson, L. E. (1968). Mechanism of the enzymic reduction of N_2: the binding of adenosine 5′-triphosphate and cyanide to the N_2-reducing system. *Proc. natn. Acad. Sci. USA*, **61**, 1021–7.

Bulen, W. A., LeComte, J. R., Burns, R. C. & Hinkson, J. (1965). Nitrogen fixation studies with aerobic and photosynthetic bacteria. In *Non-Heme Iron Proteins: Role in Energy Conversion* (ed. San Pietro, A.), pp. 261–74. Antioch Press: Yellow Springs.

Burns, R. C. (1965). Adenosine triphosphate-dependent hydrogen evolution by cell-free preparations of *Clostridium pasteurianum*. In *Non-Heme Iron Proteins: Role in Energy Conversion* (ed. San Pietro, A.), pp. 289–97. Antioch Press: Yellow Springs.

Burns, R. C. (1969). The nitrogenase system from *Azotobacter*. Activation energy and divalent cation requirement. *Biochim. biophys. Acta*, **171**, 253–9.

Burns, R. C. & Bulen, W. A. (1965). ATP-dependent hydrogen evolution by cell-free preparations of *Azotobacter vinelandii*. *Biochim. biophys. Acta*, **105**, 437–45.

Burns, R. C., Fuchsman, W. H. & Hardy, R. W. F. (1971). Nitrogenase from vanadium-grown *Azotobacter*: Isolation, characteristics and mechanistic implications. *Biochem. biophys. Res. Comm.*, **42**, 353–8.

Burns, R. C. & Hardy, R. W. F. (1972). Purification of nitrogenase and crystallization of its Mo–Fe protein. *Methods in Enzymology*, **24***B*, 480–96.

Burns, R. C. & Hardy, R. W. F. (1975). *Nitrogen Fixation in Bacteria and Higher Plants*. Springer-Verlag: New York.

Burris, R. H., Winter, H. C., Munson, T. O. & Garcia-Rivera, J. (1965). Intermediates and cofactors in nitrogen fixation. In *Non-Heme Iron Proteins: Role in Energy Conversion* (ed. San Pietro, A.), pp. 315–21. Antioch Press: Yellow Springs.

Chatt, J., Dilworth, M. J., Richards, R. L. & Sanders, J. R. (1969). Chemical evidence concerning the function of molybdenum in nitrogenase. *Nature, Lond.*, **224**, 1201–2.

Chatt, J., Heath, G. A. & Richards, R. L. (1972). The reduction of ligating dinitrogen to yield a ligating N_2H_2 moiety. *J. chem. Soc., chem. Comm.*, pp. 1010–11.

Child, J. & LaRue, T. A. G. (1974). A simple technique for the establishment of nitrogenase in soybean callus culture. *Pl. Physiol.*, **53**, 88–90.

Das, P. K., Hill, H. A. O., Pratt, J. M. & Williams, R. J. P. (1967). The so-called hydridocobalamin. *Biochim. biophys. Acta*, **141**, 644–6.

Dixon, R. A. & Postgate, J. R. (1971). Transfer of nitrogen-fixation genes by conjugation in *Klebsiella pneumoniae. Nature, Lond.*, **234**, 47–8.

Fuchsman, W. H. & Hardy, R. W. F. (1972). Nitrogenase-catalyzed acrylonitrile reductions. *Bioinorgan. Chem.*, **1**, 195–213.

Gilchrist, A. B., Rayner-Conham, G. W. & Sutton D. (1972). Transition metal complexes of diazonium salts as models for nitrogenase. *Nature, Lond.*, **235**, 42–4.

Gosser, L. W., Knoth, W. H. & Parshall, G. W. (1973). Reverse osmosis in organometallic synthesis. *J. Am. Chem. Soc.*, **95**, 3436.

Hadfield, K. L. & Bulen, W. A. (1969). Adenosine triphosphate requirement of nitrogenase from *Azotobacter vinelandii. Biochemistry*, **8**, 5103–8.

Ham, G. E., Lawn, R. J. & Brun, W. A. (1975). Influence of inoculation, nitrogen fertilizers and photosynthetic source-sink manipulations on field-grown soybeans. In *Symbiotic Nitrogen Fixation in Plants* (ed. Nutman, P. S.), pp. 238–52. Cambridge University Press: London.

Hardy, R. W. F. & Burns, R. C. (1973). Comparative biochemistry of iron–sulfur proteins and dinitrogen fixation. In *Iron–Sulfur Proteins* (ed. Lovenberg, W.), pp. 65–110. Academic Press: New York.

Hardy, R. W. F., Burns, R. C., Hebert, R. R., Holsten, R. D. & Jackson, E. K. (1971). Biological nitrogen fixation: a key to world protein. *Plant and Soil, Special Volume*, 561–90.

Hardy, R. W. F., Burns, R. C. & Holsten, R. D. (1973). Applications of the acetylene–ethylene assay for measurement of nitrogen fixation. *Soil Biol. Biochem.*, **5**, 47–81.

Hardy, R. W. F., Burns, R. C. & Parshall, G. W. (1971). The biochemistry of N_2 fixation. *Adv. Chem. Series*, **100**, 219–47.

Hardy, R. W. F., Burns, R. C. & Parshall, G. W. (1973). Bioinorganic chemistry of dinitrogen fixation. In *Inorganic Biochemistry* (ed. Eichorn, G. L.), vol. **2**, pp. 745–93. Elsevier Publishing Company: Amsterdam.

Hardy, R. W. F. & D'Eustachio, A. J. (1964). The dual role of pyruvate and the energy requirement in nitrogen fixation. *Biochem. biophys. Res. Comm.*, **15**, 314–18.

Hardy, R. W. F. & Havelka, U. D. (1975). Photosynthate as a major factor limiting N_2 fixation by field-grown legumes with emphasis on soybeans. In *Symbiotic Nitrogen Fixation in Plants* (ed. Nutman, P. S.), pp. 421–39. Cambridge University Press: London.

Hardy, R. W. F. & Holsten, R. D. (1972). Symbiotic fixation of atmospheric nitrogen. U.S. Patent 3,704,546.

Hardy, R. W. F. & Holsten, R. D. (1975). Methods for measurement of N_2 fixation. In *Treatise on Dinitrogen* (N_2) *Fixation* (eds. Gibson, A., Hardy, R. W. F. & Silver, W.). Wiley–Interscience: New York (in press).

Hardy, R. W. F., Holsten, R. D., Jackson, E. K. & Burns, R. C. (1968). Acetylene–ethylene assay for nitrogen fixation: laboratory and field evaluation. *Pl. Physiol.*, **43**, 1185–207.

Hardy, R. W. F. & Knight, E., Jr (1966). Reductant-dependent adenosine triphosphatase of nitrogen-fixing extracts of *Azotobacter vinelandii. Biochim. biophys. Acta*, **122**, 520–31.

Hardy, R. W. F. & Knight, E., Jr (1968). Biochemistry and postulated mechanisms of nitrogen fixation. In *Progress in Phytochemistry* (eds. Reinhold, L. and Liwschitz, L.), vol. **1**, pp. 407–89. Wiley-Interscience: London.

Hardy, R. W. F., Knight E., Jr, & D'Eustachio, A. J. (1965). An energy-dependent hydrogen-evolution from dithionite in nitrogen-fixation of *Clostridium pasteurianum. Biochem. biophys. Res. Comm.*, **20**, 539–44.

Herskovitz, T., Averill, B. A., Holm, R. H., Ibers, J. A., Phillips, W. D. & Weiher, J. F. (1972). Structure and properties of a synthetic analog of bacterial iron–sulfur proteins. *Proc. natn. Acad. Sci. USA*, **69**, 2437–41.

Hill, S., Drozd, J. W. & Postgate, J. R. (1972). Environmental effects on the growth of nitrogen-fixing bacteria. *J. Appl. Chem. Biotechnol.*, **22**, 541–58.

Hill, R. E. E. & Richards, R. L. (1971). Reduction of dinitrogen in aqueous solution. *Nature, Lond.*, **233**, 114–15.

Holsten, R. D., Burns, R. C., Hardy, R. W. F. & Hebert, R. R. (1971). Establishment of symbiosis between *Rhizobium* and plant cells *in vitro. Nature, Lond.*, **232**, 173–6.

Hwang, J. C. & Burris, R. H. (1972). Nitrogenase-catalyzed reactions. *Biochim. biophys. Acta*, **283**, 339–50.

Hwang, J. C., Chen, C. H. & Burris, R. H. (1973). Inhibition of nitrogenase catalyzed reductions. *Biochim. biophys. Acta*, **292**, 256–70.

Jackson, E. K., Parshall, G. W. & Hardy, R. W. F. (1968). Hydrogen reactions of nitrogenase. *J. biol. Chem.*, **243**, 4952–8.

Lockshin, A. & Burris, R. H. (1965). Inhibitors of nitrogen fixation in extracts from *Clostridium pasteurianum. Biochim. biophys. Acta*, **111**, 1–10.

Marchon, J. C. & Barbosa, A. (1972). Reaction mechanisms of a nitrogenase model. Water and hydrogen as reducible substrates of the titanocene system. *Analyt. Lett.*, **5**, 897–904.

McKenna, C., Benemann, J. R. & Traylor, T. G. (1970). Vanadium-containing nitrogenase preparation: implications for the role of molybdenum in nitrogen fixation. *Biochem. biophys. Res. Comm.*, **41**, 1501–8.

Mitchell, P. C. H. & Scarle, R. D. (1972). Role of reactions of molybdenum compounds with hydrazine in nitrogen fixation. *Nature, Lond.*, **240**, 417–18.

Mortenson, L. E. (1964). Ferredoxin and ATP requirements for nitrogen fixation in cell-free extracts of *Clostridium pasteurianum*. *Proc. natn. Acad. Sci. USA*, **52**, 272–9.

Moustafa, E. & Mortenson, L. E. (1967). Acetylene reduction by nitrogen fixing extracts of *Clostridium pasteurianum*: ATP requirement and inhibition by ADP. *Nature, Lond.*, **216**, 1241–2.

Newton, W. E., Corbin, J. L., Schneider, P. W. & Bulen, W. A. (1971). On potential model systems for nitrogenase enzyme. *J. Am. Chem. Soc.*, **93**, 268–9.

Oppenheim, J., Fisher, R. J., Wilson, P. W. & Marcus, L. (1970). Properties of a soluble nitrogenase in *Azotobacter*. *J. Bacteriol.*, **101**, 292–6.

Oppenheim, J. & Marcus, L. (1970). Correlation of ultrastructure in *Azotobacter vinelandii* with nitrogen source for growth. *J. Bacteriol.*, **101**, 286–91.

Orme-Johnson, W. H., Hamilton, W. D., Jones, T. L., Tso, M.-Y. W., Burris, R. H., Shah, U. K. & Brill, W. J. (1972). Electron paramagnetic resonance of nitrogenase and nitrogenase components from *Clostridium pasteurianum* W5 and *Azotobacter vinelandii* OP. *Proc. natn. Acad. Sci. USA*, **69**, 3142–5.

Ozin, G. A. & Vander Voet, A. (1973). 'Sideways' bonded dinitrogen in matrix isolated cobalt dinitrogen, CoN_2. *Can. J. Chem.*, **51**, 637–40.

Parejko, R. A. & Wilson, P. W. (1971). Kinetic studies on *Klebsiella pneumoniae* nitrogenase. *Proc. natn. Acad. Sci. USA*, **68**, 2016–18.

Parshall, G. W. (1967). An inorganic analog of nitrogen reductase. *J. Am. chem. Soc.*, **89**, 1822–6.

Pate, J. L., Shah, V. K. & Brill, W. J. (1973). Internal membrane control in *Azotobacter vinelandii*. *J. Bacteriol.*, **114**, 1346–50.

Phillips, D. A. (1974). Factors affecting the reduction of acetylene by *Rhizobium*-soybean cell associations *in vitro*. *Pl. Physiol.*, **53**, 67–72.

Postgate, J. R. (1972). Acetylene reduction test for nitrogen fixation. *Methods in Microbiology* (ed. Norris, J. R.), **63**, 343–56.

Raju, P. N., Evans, H. J. & Seidler, R. J. (1972). An asymbiotic nitrogen-fixing bacterium from the root environment of corn. *Proc. natn. Acad. Sci. USA*, **69**, 3474–8.

Richter, H. J. & Hoehn, H. H. (1971). Permselective aromatic nitrogen-containing polymeric membranes. US Patent 3,567,632.

Schrauzer, G. N., Doemeny, P. A., Frazier, R. H. & Kiefer G. W. (1972). Chemical evolution of a nitrogenase model. v. Reduction of nitriles. *J. Am. chem. Soc.*, **94**, 7378–5.

Schrauzer, G. N., Doemeny, P. A., Kiefer, G. W. & Frazier, R. H. (1972). Chemical evolution of a nitrogenase model. iv. Reduction of isonitriles. *J. Am. chem. Soc.*, **94**, 3604–13.

Schrauzer, G. N., Kiefer, G. W., Doemeny, P. A. & Hirsch, H. (1973). Chemical evolution of a nitrogenase model. vi. The reduction of CN^-, N_3^-, N_2O, N_2, and other substrates by molybdocysteine catalysts in the presence of nucleoside phosphates. *J. Am. chem. Soc.*, **95**, 5582–92.

Schrauzer, G. N., Schlesinger, G. & Doemeny, P. A. (1971). Chemical evolution of a nitrogenase model. III. The reduction of nitrogen to ammonia. *J. Am. chem. Soc.*, **93**, 1803–4.

Seidler, R. J., Aho, P. E., Raju, P. N. & Evans, H. J. (1972). Nitrogen fixation by bacterial isolates from decay in living white fir trees (*Abies concolor*). *J. gen. Microbiol.*, **73**, 413–16.

Shilov, A. E., Denisov, N. T., Efimov, N. O., Shuvalov, N., Shuvalova, N. I. & Shilova, A. K. (1971). New nitrogenase model for reduction of molecular nitrogen in protonic media. *Nature, Lond.*, **231**, 460–1.

Shilov, A. E. & Lichtenshtein, G. I. (1971). Biological fixation of molecular nitrogen and its chemical models. *Izvest. Akademii Nauk, SSSR, Seriya Biologicheskaya*, **1**, 518.

Silverstein, R. & Bulen, W. A. (1970). Kinetic studies of the nitrogenase-catalyzed hydrogen evolution and nitrogen reduction reactions. *Biochemistry*, **9**, 3809–15.

Slepko, G. I., Uzenskaya, A. M., Linde, V. R. & Levchenko, L. A. (1971). Separation of nitrogenase components by gel-filtration on sephadex and their quaternary structure. *Izvest. Akademii Nauk, Seriya Biologicheskaya*, **1**, 86–91.

Smith, B. E., Lowe, D. J. & Bray, R. C. (1972). Nitrogenase of *Klebsiella pneumoniae*. Electron paramagnetic studies on the catalytic mechanism. *Biochem. J.*, **130**, 641–3.

Stasny, J. T., Burns, R. C. & Hardy, R. W. F. (1973). The immuno-electron microscopic localization of the Mo–Fe protein component of nitrogenase in cells of *Azotobacter vinelandii*. *Proc. Electron Microsc. Soc. America*, pp. 574–5.

Stasny, J. T., Burns, R. C., Korant, B. D. & Hardy, R. W. F. (1974). Electron microscopy of the Mo–Fe protein from *Azotobacter* nitrogenase. *J. Cell Biol.*, **60**, 311–16.

Stiefel, E. I. (1973). Proposed molecular mechanism for the action of molybdenum in enzymes: coupled proton and electron transfer. *Proc. natn. Acad. Sci. USA*, **70**, 988–92.

Strandberg, G. W. & Wilson, P. W. (1967). Molecular H_2 and the pN_2 function of *Azotobacter*. *Proc. natn. Acad. Sci. USA*, **58**, 1404–9.

Trinick, M. J. (1973). Symbiosis between *Rhizobium* and the non-legume, *Trema aspera*. *Nature, Lond.*, **244**, 459–60.

Tso, M.-Y. W. & Burris, R. H. (1973). Binding of ATP and ADP by nitrogenase components from *Clostridium pasteurianum*. *Biochim. biophys. Acta*, **309**, 263–70.

Turner, G. L. & Bergersen, F. J. (1969). Relation between nitrogen fixation and the production of HD from D_2 by cell-free extracts of soybean nodule bacteroids. *Biochem. J.*, **115**, 529–35.

van Tamelen, E. E., Boche, G. & Greeley, R. (1968). An organic–inorganic system for reaction with nitrogen of the air and operation of a facile nitrogen fixation–reduction cycle. *J. Am. chem. Soc.*, **90**, 1677–8.

van Tamelen, E. E., Gladysz, J. A. & Miller, J. S. (1973). Nonenzymic nitrogen fixation by an iron–molybdenum model for nitrogenase. *J. Am. chem. Soc.*, **95**, 1347–8.

van Tamelen, E. E., Rudler, H. & Bjorklund, C. (1971). Transition metal promoted organic reactions as models for nitrogenase behaviour. *J. Am. chem. Soc.*, **93**, 3526–7.

Vogl, O., Miller, H. C. & Sharkey, W. H. (1972). Monomer-cast chloral polymers. *Macromolecules*, **5**, 658–9.

Vol'pin, M. E. & Shur, V. B. (1964). Nitrogen fixation on complex catalysts. *Dokl. Akad. Nauk SSSR*, **156**, 1102–4.

Vol'pin, M. E. & Shur, V. B. (1966). Nitrogen fixation by transition metal complexes. *Nature, Lond.*, **209**, 1236.

Watenpaugh, K. D., Sieker, L. C., Jensen, L. H., LeGall, J. & Dubourdieu, M. (1972). Structure of the oxidized form of a flavodoxin at 2.5 Å resolution. *Proc. natn. Acad. Sci. USA*, **69**, 3185–8.

Werner, D., Russell, S. A. & Evans, H. J. (1973). Reduction of acetylene and hydrazine with a molybdenum–glutathione complex. *Proc. natn. Acad. Sci. USA*, **70**, 339–42.

Winter, H. & Burris, R. H. (1968). Stoichiometry of the adenosine triphosphate requirement for N_2 fixation and H_2 evolution by a partially purified preparation of *Clostridium pasteurianum*. *J. biol. Chem.*, **243**, 940–4.

Wong, P. P. & Burris, R. H. (1972). Nature of oxygen inhibition of nitrogenase from *Azotobacter vinelandii*. *Proc. natn. Acad. Sci. USA*, **69**, 672–5.

Zumft, W. G., Cretney, W. C., Huang, T. C., Mortenson, L. E. & Palmer, G. (1972). Structure and function of nitrogenase from *Clostridium pasteurianum*. *Biochem. biophys. Res. Comm.*, **48**, 1525–32.

25. Kinetics and mechanism of nitrogenase action

R. R. EADY, B. E. SMITH, R. N. F. THORNELEY, M. G. YATES & J. R. POSTGATE

Nitrogenases from both free-living bacteria and symbiotic systems require the co-operative action of a molybdenum–iron (Mo–Fe) protein and an iron (Fe) protein. For activity, ATP and Mg^{2+} are required as well as a reductant (Hardy & Burns, 1968; Burris, 1971; Dalton & Mortenson, 1972). A hydrogen-evolving reaction, resulting from interaction of hydrogen ions with the functioning enzyme, complicates all studies *in vitro* with substrates such as N_2 or acetylene. Because of its complexity, purely kinetic studies of the system have been few. Silverstein & Bulen (1970) fitted a computer model to steady-state kinetic studies of partly purified nitrogenase from *Azotobacter vinelandii* and Bergersen & Turner (1973) examined the kinetics of nitrogenase from *Rhizobium* bacteroids, studying, in particular, the variation of K_m with the concentrations of component proteins. Both groups reached plausible conclusions regarding the association of the component proteins and their interaction with ATP. With bacteroid nitrogenase, Bergersen & Turner (1973) assigned a substrate-binding function to *Rhizobium japonicum* bacteroid Component I.

Some evidence exists for the formation of an active complex of the nitrogenase components having a lifetime greater than the turnover time of the enzyme. The evidence is oblique and rests mainly on the following observations.

1. The dilution effect, whereby the enzyme activity falls off disproportionately at low protein concentrations (see Burns & Bulen, 1965; Hardy, Holsten, Jackson & Burns, 1968; Yates, 1970), suggests an equilibrium association of the two proteins, although involvement of equilibria between the subunits of either protein cannot be discounted.

2. The existence of an optimal stoichiometry between the two proteins supplemented by the fact that excess of Component I can be inhibitory (Vandecasteele & Burris, 1970; Eady, Smith, Cook & Postgate, 1972; Tso, Ljones & Burris, 1972; Mortenson, Zumft, Huang & Palmer, 1973) suggest that various associated species are formed, some

of which are less active. Ware (1972) showed that reduced benzyl viologen prevented this inhibition by excess *Klebsiella pneumoniae* Component I and he suggested that benzyl viologen displaced *K. pneumoniae* Component I from an inactive species.

3. Steady-state kinetic studies involving ATP utilization and substrate reduction by partly purified *Azotobacter vinelandii* nitrogenase, were consistent with equilibrium association of the two proteins followed by binding of two ATP molecules before enzyme activity is expressed (Silverstein & Bulen, 1970). With nitrogenase from *Rhizobium* bacteroids, the amount of ATP bound appeared closer to one molecule (Bergersen & Turner, 1973).

Kinetic data also suggest that the active nitrogenase has multiple substrate-binding sites because titration curves of *Azotobacter chroococcum* or *Klebsiella pneumoniae* Component I against homologous Component II show different optima with substrates such as cyanides from those with acetylene or N_2, and plateaux instead of optima if the ATPase function is measured (Kelly, 1969; Eady *et al.*, 1972; Eady, Smith, Thorneley, Ware & Postgate, 1973). In addition, acetylene can actually stimulate cyanide reduction by *K. pneumoniae* nitrogenase (Biggins & Kelly, 1970).

Their steady-state kinetic studies on *R. japonicum* bacteroid nitrogenase led Bergersen & Turner (1973) to propose that *R. japonicum* bacteroid Component II influences the reactivity of Component I, at which the substrate is bound. Co-chromatography led some workers to assign a substrate binding role to *Clostridium pasteurianum* Component I (see Dalton & Mortenson, 1972).

A possible reaction sequence

Recent studies of the interaction of nitrogenase proteins making use mainly of electron paramagnetic resonance (EPR) and γ-resonance (Mössbauer) spectroscopy have permitted more exact assignment of roles to the two proteins. Review of these data is not appropriate to the present contribution; they derive mainly from the work of groups at Wisconsin and Lafayette (USA) and Brighton (UK); earlier papers were cited by Smith, Lowe & Bray (1973) and Smith & Lang (1973). They suggest the following reaction sequence. The native Component II which can exist in two undamaged redox states, reacts with Mg–ATP. The reduced derivative of this complex then donates electrons to Component I, which itself can exist in at least three undamaged states, to give its most reduced state (with a distinctive Mössbauer spectrum but no EPR spectrum). This donates electrons to the substrate in what is probably the rate-determining step of the reaction. The detectable redox changes in this sequence involve the iron atoms of the two proteins; no data are available concerning the state of the molybdenum atoms though these are intuitively thought to be involved in substrate binding.

This communication correlates recent data bearing on the above scheme. The nomenclature for nitrogenase proteins will be as follows: Component I refers to the Mo–Fe protein and Component II to the Fe protein. Full experimental details will not be given unless they will not be published elsewhere.

Interaction of ATP with Component II proteins

Electron paramagnetic resonance (EPR) studies show that Mg–ATP specifically alters the symmetry of the iron chromophore in Component II from *A. vinelandii* (Orme-Johnson *et al.*, 1972). *Clostridium pasteurianum* (Zumft *et al.*, 1972) and *K. pneumoniae* (Smith, Lowe & Bray, 1973). Thorneley & Eady (1973) used stopped-flow kinetics to show that ATP increased the accessibility of sulphydryl groups in *K. pneumoniae* Component II to the group-specific reagent 5,5′-dithiobis-2-nitrobenzoate. ADP had a small effect and other nucleotide triphosphates had none. The oxygen sensitivity of this protein has now been shown to be increased by ATP (Fig. 25.1). Yates (1972) showed that ATP enhanced the oxygen sensitivity of *A. chroococcum* Component II; this effect was traced to non-specific chelation of metals such as Mg^{2+} or Ca^{2+}, since the specificity typical of the ATP requirement for nitrogenase action was absent and ethylenediaminetetraacetic acid (EDTA) treatment also increased oxygen sensitivity. Walker & Mortenson (1973) showed that ATP increased the accessibility of the iron atoms of *C. pasteurianum* Component II to α,α'-dipyridyl.

Thorneley & Eady (1973) showed that, in the absence of sodium dithionite Mg–ATP produced a change in the sedimentation behaviour of *K. pneumoniae* Component II: sedimentation velocity analysis showed that the single peak of 4.6 S characteristic of Component II protein became a double peak with a leading component of 6.6 S in the presence of Mg–ATP (Plate 25.1). They suggested tentatively that ATP might cause a dissociation of the normally dimeric *K. pneumoniae* Component II into its monomers, followed by immediate association

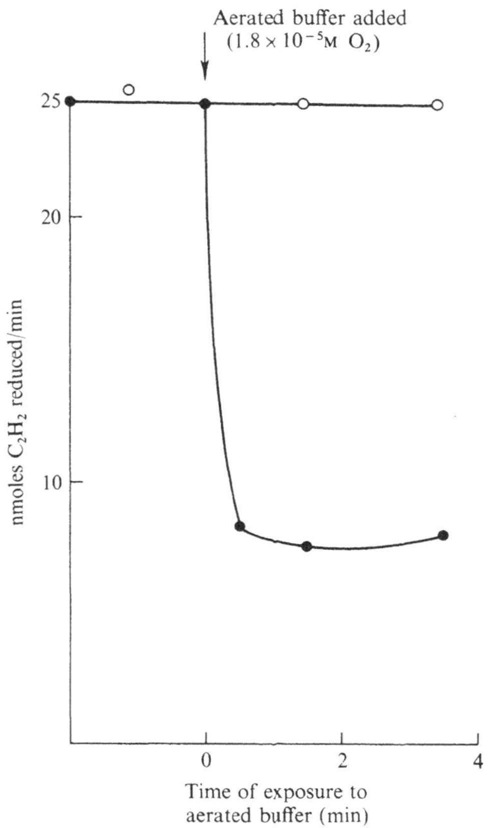

Fig. 25.1. ATP-induced oxygen-sensitivity of *K. pneumoniae* Component II. Component II (2.2 mg/ml) in 25 mM Tris HCl buffer pH 7.4 containing 10 mM $MgCl_2$ and 1 mM ATP was incubated under N_2. At the time indicated aerated buffer was added to the incubation mixture to give a final oxygen concentration of 1.8×10^{-5} M. Samples were withdrawn at intervals and assayed for acetylene reduction with Component I. The activity is compared with a control incubation not containing ATP. ○, Component II (3.3×10^{-5} M); ●, plus 1 mM ATP.

into trimers or possibly tetramers. This effect was reversed by sodium dithionite, a compound often present in nitrogenase preparations as a protective reagent. These experiments add to the cumulative evidence, from EPR and from binding of radioactive ATP to *C. pasteurianum* Component II (Bui & Mortenson, 1968) and to *A. chroococcum* Component II (Yates, 1972), that ATP reacts specifically with the Component II proteins of nitrogenases. They indicate that ATP may change

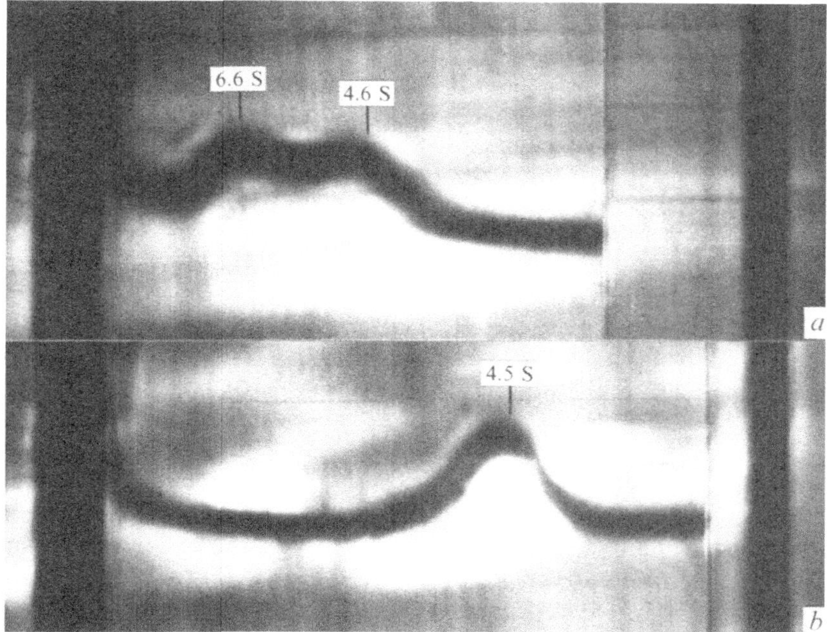

Plate 25.1. Effect of ATP on the sedimentation pattern of *K. pneumoniae* Component II. Component II (3.1 mg protein/ml in 25 mM Tris buffer, pH 7.4, containing 10 mM $MgCl_2$ and 0.5 mM ATP) was centrifuged at 20 °C under argon and 60000 revolutions/min. (*a*) Without sodium dithionite after 70 min; (*b*) with sodium dithionite (1 mM). $S^{\circ}_{20,w}$ Component II = 4.8 S.

their tertiary or quaternary structures, probably altering their reducing potential in a negative direction. W. G. Zumft & L. E. Mortenson (personal communication) have shown that Mg–ATP lowers the E_0' of *C. pasteurianum* Component II.

Interactions of substrates with Component I proteins

At present no technique is available for studying the molybdenum atoms in the nitrogenase Mo–Fe proteins which is comparable to the value of EPR and Mössbauer spectroscopy for the iron atoms. Mössbauer spectroscopy indicates that *K. pneumoniae* Component I contains three classes of Fe clusters corresponding to 8, 8 and 2 Fe atoms/mole (Smith & Lang, 1973), which is in satisfactory agreement with the analytical figure of 17.5 ± 1 Fe/mole (Eady *et al.*, 1972) and indicates a probable value of 18 Fe/mole. These iron clusters can undergo clear, reversible redox changes between the three undamaged states discussed by Smith

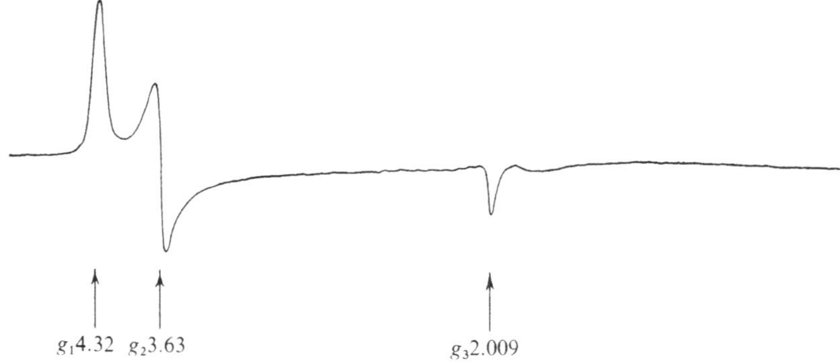

Fig. 25.2. Electronic paramagnetic resonance (EPR) spectrum of *K. pneumoniae* Component I; low pH form (after Smith *et al.*, 1973) 200 mW; 8.1 K.

& Lang (1973) but the substrates N_2 and acetylene and the inhibitor carbon monoxide did not affect the Mössbauer spectra. Thus, although small effects might have been missed in such complex spectra, a substrate-binding role for the Fe atoms of *K. pneumoniae* Component I protein is unlikely. The EPR spectrum of *K. pneumoniae* Component I is associated with the iron atoms since ^{95}Mo substitution did not affect the spectrum whereas ^{57}Fe substitution, in accordance with theory, broadened the sharpest line slightly (Smith, Lowe & Bray, 1973). *K. pneumoniae* Component I protein showed a pH-dependent change in EPR parameters with *g* values changing from 4.32, 3.63 and 2.009 at low pH (Fig. 25.2) to 4.27, 3.73 and 2.018 at high pH. The pK of 8.7 at 0 °C for this change was displaced in the presence of acetylene to about 8.2 (Fig. 25.3) which suggests, but does not prove, that acetylene competes with hydrogen ions for a binding site on *K. pneumoniae* Component I. Acetylene also largely suppresses the hydrogen-evolving function of nitrogenase, but the protons associated with the pK change are not necessarily the substrate protons. The acetylene is almost certainly not bound at the EPR active site since this would have resulted in much larger perturbations of the EPR spectrum. The above observations support the general opinion that substrate reduction occurs on the Mo–Fe protein of nitrogenase.

Interactions of Component I and Component II proteins

Eady (1973) showed complex formation between Component I and Component II proteins of *K. pneumoniae* in ultracentrifuge experiments.

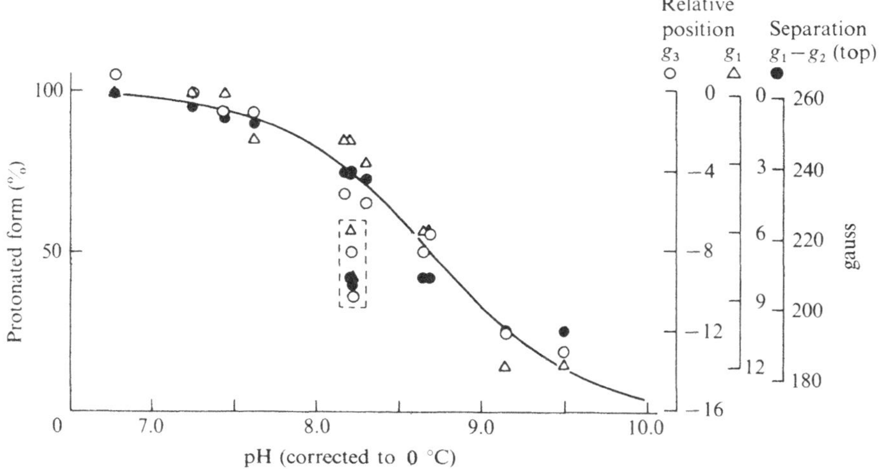

Fig. 25.3. Effect of pH on the parameters of the EPR spectrum of *K. pneumoniae* Component I compared with a theoretical curve for pK of 8.7. The points within the box refer to samples incubated under 87.5:12.5, C_2H_2:Ar (after Smith *et al.*, 1973).

Plate 25.2 illustrates the schlieren patterns of the two proteins sedimenting in the absence of both ATP and sodium dithionite. For comparison, a sedimentation profile in the presence of sodium dithionite is included: the two proteins sedimented independently with S values in good agreement with the $S_{20,w}$ of 11 and 4.8 already established for the *K. pneumoniae* Component I and Component II respectively. Without sodium dithionite the two proteins associated and formed a slightly asymmetrical peak at 12.4 S. The molar ratio for complex formation was 1:1; with excess Component II the 4.5 S peak, corresponding to the free protein, appeared and the sedimentation coefficient of the leading peak remained unchanged. Again, sodium dithionite prevented complex formation. The effect of ATP has yet to be studied in this context, but it is already evident that it can affect the quaternary structure of *K. pneumoniae* Component II, that the two components can form a univalent complex, and that both effects are prevented by sodium dithionite.

Titration of *K. pneumoniae* Component I against Component II monitored by acetylene-reducing activity, showed considerable inhibition by excess of Component I (Eady *et al.*, 1972) which could be reversed by benzyl viologen (Ware, 1972). Titrations of *A. chroococcum* Component I against Component II showed comparable inhibition, but though the shape of the titration curve was similar the slope of the

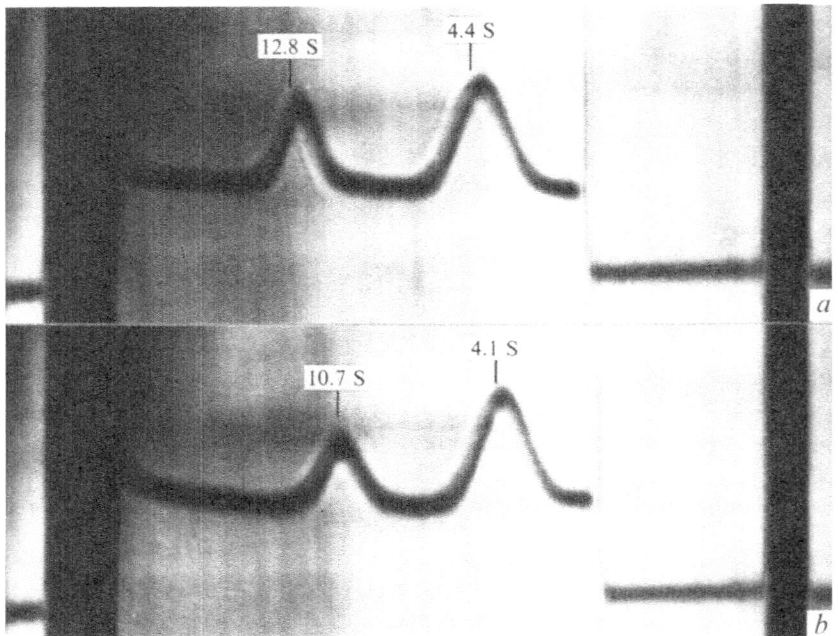

Plate 25.2. Complex formation by *K. pneumoniae* Components I and II. A 1:7 molar ratio of Component I : Component II (2.7 and 6 mg protein/ml in 25 mM Tris buffer, pH 7.4, containing 10 mM $MgCl_2$) was centrifuged under argon at 20 °C and 60000 revolutions/min. (*a*) Without sodium dithionite after 30 min; (*b*) with sodium dithionite (2 mM). $S^{\circ}_{20,w}$ Component I = 11 S; $S^{\circ}_{20,w}$ Component II = 4.8 S.

inhibition curve was shallower (Fig. 25.4). *K. pneumoniae* and *A. chroococcum* nitrogenase components cross-reacted completely when Components I were titrated with Components II but not in the reverse titrations. Figure 25.4 includes heterologous titrations of *K. pneumoniae* Component I and *A. chroococcum* Component I against *A. chroococcum* Component II and *K. pneumoniae* Component II and shows that the slope of the inhibition curve is determined by the nature of the Fe protein in each case.

Conclusion

Nitrogenase is a multi-component system which is a kineticist's nightmare. The known molecular species required for activity are two proteins, the one (Fe) a homomeric dimer and the other (Mo–Fe) a heteromeric tetramer, together with Mg–ATP and a reductant in an anaerobic environment. The reductant can be reduced ferredoxin, flavodoxin,

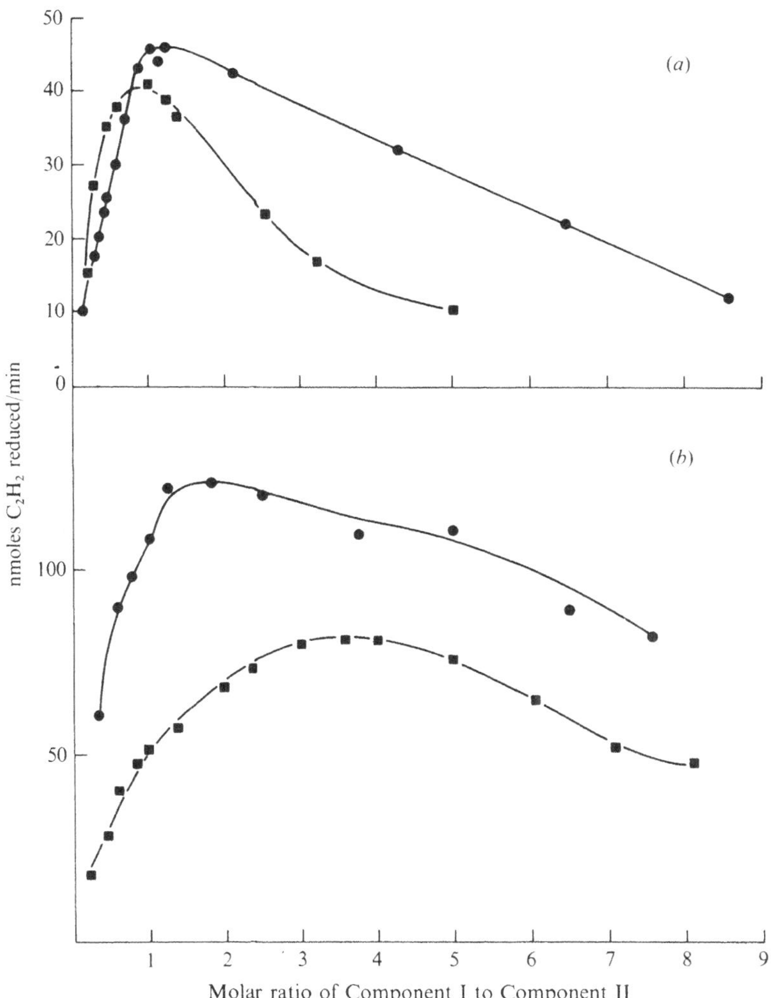

Fig. 25.4. Homologous and heterologous cross reactions of nitrogenase Components I and II from *K. pneumoniae* and *A. chroococcum*. In each assay Component II (*A. chroococcum* 48 μg; *K. pneumoniae* 52 μg) was constant and Component I varied. Molar ratios were determined using the following molecular weights in daltons: *A. chroococcum* Component I 200000; Component II 60000; *K. pneumoniae* Component I 218000; Component II 66800. ●, Homologous cross; ■, heterologous cross. (*a*) Constant *K. pneumoniae* Component II; (*b*) constant *A. chroococcum* Component II.

a reduced viologen or sodium dithionite. The solvent provides a reducible substrate, the hydrogen ion, which always competes *in vitro* with whichever of the numerous possible substrates (N_2, C_2H_2, HCN, CH_3NC, etc.) is available. Additional complications are (1) that the optimal stoichiometry of the enzyme proteins differs according to the substrate under test and the source of the enzyme; (2) that ATP utilization need not necessarily parallel substrate reduction (Ljones & Burris, 1972; Eady *et al.*, 1973); (3) that excess of the Mo–Fe protein is inhibitory and (4) that, in one instance (*C. pasteurianum* Component I), the native tetramer can dissociate on dilution (Mortenson *et al.*, 1973). Obtaining a complete kinetic account of functioning nitrogenase is a formidable undertaking. Silverstein & Bulen (1970) wrote of the nitrogenase: 'neither protein has yet been shown to possess any catalytic activity alone'. This statement is still true, but specific roles have now been assigned to the individual proteins which will enable the kinetics, and hence the mechanism, of nitrogenase action, to be approached analytically. An important practical consideration is that, of the reactivities reported here, the effect of ATP on the tertiary structure of *K. pneumoniae* Component II and the 1:1 association of the nitrogenase components from *K. pneumoniae* are both prevented by sodium dithionite, a reagent normally present in laboratory nitrogenase preparations. It is possible that some of the interactions crucial to nitrogenase action require the initial participation of oxidised forms of the component proteins, so a complete kinetic analysis must include this possibility, and an examination of the extent to which natural electron-donors mimic the effects of sodium dithionite.

We acknowledge the valuable co-operation of Drs R. C. Bray and D. J. Lowe in the *EPR* experiments and of Dr G. Lang in the Mössbauer experiments.

References

Bergersen, F. J. & Turner, G. C. (1973). Kinetic studies of nitrogenase from soya bean root nodule bacteroids. *Biochem. J.*, **131**, 61–75.

Biggins, D. R. & Kelly, M. (1970). Interaction of nitrogenase from *Klebsiella pneumoniae* with ATP or cyanide. *Biochim. biophys. Acta*, **205**, 288–99.

Bui, P. T. & Mortenson, L. E. (1968). Mechanism of the enzymic reduction of N_2: the binding of adenosine 5′-triphosphate to the N_2-reducing system. *Proc. natn. Acad. Sci. USA*, **61**, 1021–7.

Burns, R. C. & Bulen, W. A. (1965). ATP-dependent hydrogen evolution by cell-free preparations of *Azotobacter vinelandii*. *Biochim. biophys. Acta*, **105**, 437–45.

Burris, R. H. (1971). Fixation by free-living micro-organisms: Enzymology. In *The Chemistry and Biochemistry of Nitrogen Fixation* (ed. Postgate, J. R.), pp. 105–60. Plenum Press: London.

Dalton, H. & Mortenson, L. E. (1972) Dinitrogen (N_2) fixation (with a biochemical emphasis). *Bacteriol. Rev.*, **36**, 231–59.

Eady, R. R. (1973). Nitrogenase of *Klebsiella pneumoniae*. Interaction of the component proteins studied by ultracentrifugation. *Biochem. J.*, **135**, 531–35.

Eady, R. R., Smith, B. E., Cook, K. A. & Postgate, J. R. (1972). Nitrogenase of *Klebsiella pneumoniae*. Purification and properties of the component proteins. *Biochem. J.*, **128**, 655–75.

Eady, R. R., Smith, B. E., Thorneley, R. N. F., Ware, D. A. & Postgate, J. R. (1973). Nitrogenase of *Klebsiella pneumoniae*. *Biochem. Soc. Trans.*, **1**, 37–8.

Hardy, R. W. F. & Burns, R. C. (1968). Biological nitrogen fixation. *Ann. Rev. Biochem.*, **37**, 331–58.

Hardy, R. W. F., Holsten, R. D., Jackson, E. K. & Burns, R. C. (1968). The acetylene–ethylene assay for N_2 fixation: laboratory and field evaluation. *Pl. Physiol.*, **43**, 1185–207.

Kelly, M. (1969). Some properties of purified nitrogenase of *Azotobacter chroococcum*. *Biochim. biophys. Acta*, **171**, 9–22.

Ljones, T. & Burris, R. H. (1972). ATP hydrolysis and electron transfer in the nitrogenase reaction with different combinations of the iron protein and the molybdenum–iron protein. *Biochim. biophys. Acta*, **275**, 93–101.

Mortenson, L. E., Zumft, W. G., Huang, T. C. & Palmer, G. (1973). The structure and function of nitrogenase from *Clostridium pasteurianum*. Electron paramagnetic resonance studies. *Biochem. Soc. Trans.*, **1**, 35–7.

Orme-Johnson, W. H., Hamilton, W. D., Ljones, T., Burris, R. H., Shah, V. K. & Brill, W. J. (1972). Electron paramagnetic resonance of nitrogenase and nitrogenase components from *Clostridium pasteurianum* W5 and *Azotobacter vinelandii* O.P. *Proc. natn. Acad. Sci. USA*, **69**, 3142–5.

Silverstein, R. & Bulen, W. A. (1970). Kinetic studies of the nitrogenase-catalysed H_2 evolution and the N_2 reduction reaction. *Biochemistry*, **9**, 3809–15.

Smith, B. E. & Lang, G. (1973). Mössbauer spectroscopy of the nitrogenase proteins from *Klebsiella pneumoniae*. Structural assignments and mechanistic conclusions. *Biochem. J.*, **137**, 169–80.

Smith, B. E., Lowe, D. & Bray, R. C. (1973). Studies by electron paramagnetic resonance on the catalytic mechanism of nitrogenase of *K. pneumoniae*. *Biochem. J.*, **135**, 331–41.

Thorneley, R. N. F. & Eady, R. R. (1973). Nitrogenase of *Klebsiella pneumoniae*. Evidence for an ATP-induced association of the iron–sulfur protein. *Biochem. J.*, **133**, 405–8.

Tso, M.-Y. W., Ljones, T. & Burris, R. H. (1972). Purification of the nitrogenase proteins from *Clostridium pasteurianum*. *Biochim. biophys. Acta*, **267**, 600–4.

Vandecasteele, J. P. & Burris, R. H. (1970). Purification and properties of the constituents of the nitrogenase complex from *Clostridium pasteurianum*. *J. Bacteriol.*, **105**, 794–801.

Walker, G. A. & Mortenson, L. E. (1973). An effect of magnesium adenosine 5′-triphosphate on the structure of Azoferredoxin from *Clostridium pasteurianum*. *Biochem. biophys. Res. Comm.*, **53**, 904–9.

Ware, D. A. (1972). Nitrogenase of *Klebsiella pneumoniae*. Interaction with viologen dyes as measured by acetylene reduction. *Biochem. J.*, **130**, 301–2.

Yates, M. G. (1970). Effect of non-haem iron proteins and cytochrome *c* from *Azotobacter* upon the activity and oxygen sensitivity of *Azotobacter* nitrogenase. *FEBS Lett.*, **8**, 281–5.

Yates, M. G. (1972). The effect of ATP upon the oxygen sensitivity of nitrogenase from *Azotobacter chroococcum*. *Eur. J. Biochem.*, **29**, 386–92.

Zumft, W. G., Cretney, W. C., Huang, J. C., Mortenson, L. E. & Palmer, G. (1972). On the structure and function of nitrogenase from *Clostridium pasteurianum* W5. *Biochem. biophys. Res. Comm.*, **48**, 1525–31.

26. Reductants for nitrogenase and relationships to cellular electron transport

H. J. EVANS & D. A. PHILLIPS

Investigations into the pathways of electron transport to nitrogenase have been restricted to a relatively few organisms. Among the anaerobic and facultatively anaerobic N_2-fixing bacteria studied, the most intensive investigations have been conducted with *Clostridium pasteurianum*. Only fragmentary information exists on the pathways of electron transport to nitrogenase in facultative anaerobes such as *Bacillus polymyxa* and *Klebsiella pneumoniae*. Recently, investigations into electron transport pathways to nitrogenase in *Azotobacter* species and legume nodule bacteroids have been initiated and some success in reconstructing sequences of electron transport to nitrogenase has been discussed in general reviews of biological N_2 fixation by Burris (1966), Evans & Russell (1971), Bergersen (1971), Hardy & Burns (1968), and Dalton & Mortenson (1972). More specialized reviews concerning ferredoxins and flavodoxins in bacteria and discussions of sources of reductant for nitrogenases and related topics have been presented by Yoch & Valentine (1972), Benemann & Valentine (1971; 1972), and Wong, Evans, Klucas & Russell (1971). A consideration of the relationship of N_2 fixation to the biochemistry of iron-sulfur proteins was presented by Hardy & Burns (1972). Due to space restrictions in this volume our understanding of electron transport to nitrogenase in those organisms that have been studied in detail will be summarized briefly. Recent developments will be emphasized.

Electron transport to nitrogenase in non-photosynthetic anaerobic and facultatively anaerobic bacteria

Utilization of α-ketoacids and formate

The initial convincing evidence indicating the source of energy for N_2 fixation was supplied by Carnahan, Mortenson, Mower & Castle (1960*a*, *b*), who showed that the addition of high concentrations of pyruvate to reactions containing cell-free extracts of *C. pasteurianum* resulted in N_2 fixation. In these experiments vigorous production of gas was observed from the so-called phosphoroclastic reaction. Before cell-free N_2 fixation was accomplished Koepsell & Johnson (1942) demonstrated that *Clostridium* extracts catalyzed a reaction between pyruvate and phosphate yielding acetyl-phosphate, carbon dioxide and hydrogen. In a search for other metabolites, Carnahan *et al.* (1960*b*), found that α-ketobutyrate also functioned but rates of N_2 fixation with this substrate were less than half those obtained with pyruvate. Although there were conflicting reports on the capability of pyruvate to support N_2 fixation by whole cells and cell-free extracts of *B. polymyxa* (Grau & Wilson, 1962; 1963; Witz, Detroy, & Wilson, 1967) it now seems clear that low levels of N_2 fixation in reactions containing extracts of this bacterium are maintained by pyruvate (Fisher & Wilson, 1970). Reactions with cell-free extracts of *B. polymyxa* apparently utilize pyruvate effectively as a reductant but fail to derive sufficient ATP from pyruvate metabolism to support vigorous N_2 fixation. According to Fisher & Wilson (1970), substantial N_2 reduction rates were observed when either pyruvate or formate was added to reactions with *B. polymyxa* extracts and an ATP-generating system. Evidence supporting a role of pyruvate in N_2 fixation by whole cells of a strain of *Achromobacter* (later identified as *K. pneumoniae*) was provided by Hamilton, Burris & Wilson (1964) and by whole cells of *Aerobacter aerogenes* (also identified as *K. pneumoniae*) by Lindsey & Wilson (1963).

In 1966, Mortenson showed that formate functioned as a source of energy for N_2 fixation in reactions with extracts of *C. pasteurianum*. Since *C. pasteurianum* contains a formate:ferredoxin oxidoreductase, presumably electrons were transferred to nitrogenase via ferredoxin. Benemann & Valentine (1972) refer to unpublished work indicating that formate also serves as a substrate for N_2 reduction in reactions containing extracts of *K. pneumoniae* or *B. polymyxa*. In these experiments activity was greatest when freshly prepared extracts were utilized but deteriorated rapidly as the age of extracts increased. The details of the pathway of electron transport from formate to nitrogenase have not been elucidated.

Co-factors for the phosphoroclastic reaction and for N_2 fixation

Wolfe & O'Kane (1953) showed that the phosphoroclastic degradation of pyruvate in the presence of extracts of *C. butyricum* depended upon thiamine pyrophosphate, coenzyme A and divalent cations. A search for

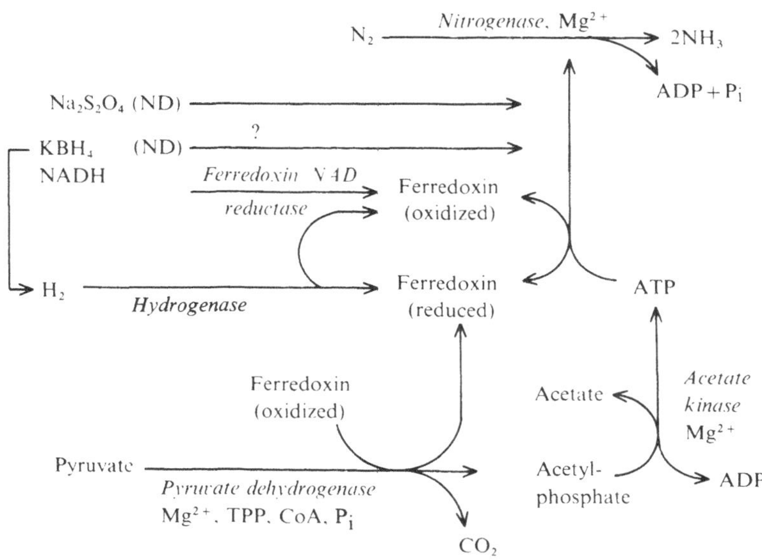

Fig. 26.1. A diagram illustrating the linkage of nitrogenase to the phosphoroclastic breakdown of pyruvate and to non-physiological electron-donors (designated ND).

potential donors capable of supplying electrons to nitrogenase via the phosphoroclastic reaction led to the isolation of ferredoxin from *C. pasteurianum* (Mortenson, Valentine & Carnahan, 1962). Mortenson (1963; 1964*a*, *b*) and Mortenson *et al.* (1963), using extracts of *C. pasteurianum*, obtained evidence for an essential role of ferredoxin as a carrier of electrons to nitrogenase. They also showed that ferredoxin was reduced during the phosphoroclastic reaction. Later, Munson, Dilworth & Burris (1965) demonstrated that the catalysis of N_2 fixation by extracts of *C. pasteurianum* required thiamine pyrophosphate, coenzyme A, ADP, Mg^{2+}, P_i and ferredoxin. These cofactors are essentially the same as those required for the phosphoroclastic reaction. A diagram illustrating the role of ferredoxin in the phosphoroclastic reaction and the relationship to the process of N_2 fixation is shown in Fig. 26.1. Recently Shethna, Stombaugh & Burris (1971) purified a ferredoxin from *B. polymyxa* that reversibly reduced hydrogen in presence of hydrogenase and enhanced nitrogenase-dependent acetylene reduction when pyruvate was utilized as the substrate. More recently, Yoch (1973) described two different ferredoxins from *B. polymyxa*. The molecular weight of each of these was about 8800 and the standard reduction potential of ferredoxin I, the only one measured, was reported

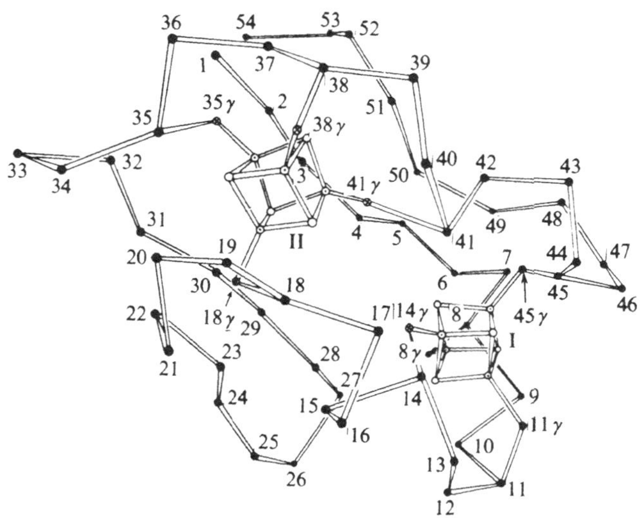

Fig. 26.2. Plot of the α-carbon, iron and sulfur positions in the three-dimensional structure of ferredoxin from *Peptococcus aerogenes*. Fe, ⊙; S_{inorg}, ○; S_{cys}, ⊗; C_{α}, ●. (After Adman *et al.*, 1973.)

to be -390 mV. The properties of ferredoxins in *Desulphovibrio gigas* and in other N_2-fixing bacteria have been discussed by Benemann & Valentine (1972) and Yoch & Valentine in 1972.

One of the most significant recent developments is the establishment at the 0.28 nm resolution of the three-dimensional structure of ferredoxin from *Peptococcus aerogenes* (Adman, Sieker & Jensen, 1973) (Fig. 26.2). Although this organism is not known to fix N_2, resolution of the structure of ferredoxin from this bacterium may provide considerable insight into the mechanism of ferredoxin-mediated electron transfer and also may serve as a model for structural investigations of other non-heme iron proteins including nitrogenase components.

An investigation of the N_2-fixing capability of *C. pasteurianum* cultured in an iron-deficient medium led Knight, D'Eustachio & Hardy (1966) to the discovery of flavodoxin, an electron carrier that substituted for ferredoxin in several reactions. These workers crystallized flavodoxin from iron-deficient *C. pasteurianum* cells and further demonstrated that the pure compound substituted for ferredoxin in the catalysis of N_2 fixation by extracts of *C. pasteurianum*. In the N_2 fixation reaction, flavodoxin, on an equimolar basis, was about 30 % as effective as ferredoxin. Flavodoxin also replaced ferredoxin in the evolution of hydrogen in the presence of hydrogenase, in the phosphoroclastic degradation of pyruvate, and in the reduction of NADP with a ferredoxin-NADP reductase (Mayhew, 1971; Knight & Hardy, 1966). Further research (Knight & Hardy, 1967) established a molecular weight of 14600 for *C. pasteurianum* flavodoxin and identified one mole of flavin mononucleotide (FMN) per flavodoxin molecule. FMN also is the flavin nucleotide prosthetic group of the other flavodoxins that have been examined (Yoch & Valentine, 1972). Investigations of the oxidation-reduction properties of flavodoxin from the non-N_2 fixer *Peptostreptococcus elsdensii* by Mayhew, Foust & Massey (1969), have established two steps in the reduction process. The first generates a blue-colored flavin semiquinone with a standard reduction potential for the couple oxidized flavodoxin–flavodoxin semiquinone of −115 mV. Further reduction produces the fully reduced flavodoxin hydroquinone with a standard reduction potential for this couple of −353 mV. Mayhew *et al.* (1969) and Mayhew (1971) also have reported that the standard reduction potentials of oxidized flavodoxin–flavodoxin semiquinone and the flavodoxin semiquinone–flavodoxin hydroquinone couples of flavodoxin from *Clostridium* MP are −92 mV and −399 mV respectively.

The corresponding two potentials for flavodoxin isolated from *C. pasteurianum* are −132 mV and −419 mV respectively. Although it is clear that flavodoxins are capable of functioning in the place of ferredoxins in reactions coupled to N_2 fixation it also is obvious (Yoch & Valentine, 1972) that flavodoxins are widely distributed in many fermentative and photosynthetic bacteria that apparently lack the capability of fixing N_2. Flavodoxin is present in *C. pasteurianum* cells cultured in an iron-deficient medium containing ammonia (Knight & Hardy, 1966). Under these conditions little if any N_2 fixation would be expected. Flavodoxin not only is capable of donating electrons to nitrogenase, but also participates in other metabolic pathways.

Other electron donors and carriers

In the early N_2 fixation experiments with cell-free extracts of anaerobic bacteria, pyruvate was added as a substrate for the supply of both reductant and high-energy phosphate. It was known that hydrogen was evolved during pyruvate-supported N_2 fixation reactions and also during N_2 fixation by intact organisms. The fact that hydrogen was known to inhibit N_2 fixation complicated the use of this gas as a reductant in nitrogenase assay systems. Since significant inhibition of nitrogenase is observed at relatively high partial pressures of hydrogen it has been possible to utilize hydrogen at a low partial pressure for the transfer of electrons, via hydrogenase and ferredoxin, to an ATP-dependent nitrogenase from *C. pasteurianum* (Mortenson, 1964*a*, *b*; D'Eustachio & Hardy, 1964). The relationship of hydrogenase to the phosphoroclastic reaction is illustrated in Fig. 26.1. In other experiments, Hardy & D'Eustachio (1964) replaced pyruvate in N_2 fixation reactions by adding an ATP-generating system and potassium borohydride (KBH_4). The possibility existed that KBH_4 reduced nitrogenase directly (Fig. 26.1) but Hardy & D'Eustachio concluded that the decomposition of KBH_4 produced H_2 which transferred electrons to nitrogenase in the presence of hydrogenase and ferredoxin. D'Eustachio & Hardy (1964) showed that methyl viologen functioned in place of ferredoxin as a carrier linking the phosphoroclastic reaction to nitrogenase. They also observed that NADH donated electrons to nitrogenase in reactions containing ferredoxin and extracts of *C. pasteurianum*. Relatively low activity was observed when NADPH was utilized as an electron donor.

As indicated in the following section, sodium dithionite is used in nitrogenase assays of extracts of anaerobic and facultatively anaerobic micro-organisms as well as aerobic micro-organisms (Witz *et al.*, 1967; Mahl & Wilson, 1968; Hardy, Knight & D'Eustachio, 1965).

Pathways in aerobic non-photosynthetic N_2-fixing bacteria and in bacteroids from legume nodules

In the early experiments the addition of substrates such as pyruvate and α-ketobutyrate that effectively supported N_2 fixation in cell-free systems from *C. pasteurianum*, failed to serve as substrates for N_2 fixation with extracts of *A. vinelandii* or nodule bacteroids (Bulen, Burns & LeComte, 1964; Koch, Evans & Russell, 1967). More recent experiments by Haaker, Bresters & Veeger (1972), however, have shown that pyruvate utilization under anaerobic conditions is capable of supplying the ATP for nitrogenase activity in extracts of *A. vinelandii*. In 1964, Bulen *et al.* devised a system in which a heated extract of *C. pasteurianum* containing ferredoxin and hydrogenase but lacking nitrogenase was used to transfer electrons to N_2 in an assay containing a source of ATP and an extract of *A. vinelandii*. A similar system was utilized by Wong *et al.* (1971) to transfer reducing power from H_2 to nitrogenase from soybean nodule bacteroids. The discovery by Bulen *et al.* (1965) that sodium dithionite was an excellent non-physiological donor for *Azotobacter* nitrogenase greatly simplified nitrogenase assays and now this is used widely in the study of properties of nitrogenase. Sodium dithionite apparently donates electrons directly to nitrogenase and therefore most and perhaps all of the natural electron transport components are bypassed when this donor is utilized. It has been essential, therefore, to search for physiological donors in order to elucidate the detailed steps in the transfer of electrons to nitrogenase systems.

Identification of co-factors

Efforts to utilize reduced pyridine nucleotides as electron donors for nitrogenase preparations from *Azotobacter* and soybean nodule bacteroids were complicated by the lack of an effective assay for the factors that transport electrons between reduced pyridine nucleotides and nitrogenase. In the reconstituted system in which electrons were transferred from NADH to nitrogenase, a viologen dye served as a carrier but most natural components were ineffective (Klucas & Evans, 1968). The development of a photochemical assay for factors capable of transporting electrons to nitrogenase resulted in the discovery of two previously unknown electron transport components (Yoch & Arnon, 1970; Benemann, Yoch, Valentine & Arnon, 1969). The assay system (Fig. 26.3) consisted of chloroplasts that had been heated or extracted by Tris-chloride (Koch *et al.*, 1970) in a way that Photosystem I activity was retained and Photosystem II inactivated. The other major component of the assay is a nitrogenase preparation that is free of electron transport factors. For this purpose Benemann *et al.* (1969) used a particulate nitrogenase that was prepared from a crude extract of *A. vinelandii* by differential centrifugation. Koch *et al.* (1970) employed a heat treatment and polypropylene glycol precipitation procedure for the preparation of a factor-free nitrogenase from soybean nodule bacteroids. In the assay system which contains nitrogenase, acetylene, an ATP-generating system, chloroplast fragments, 2,6-dichlorophenolindophenol (DCIP) ascorbate, light (for photoactivation) and an exogenous electron carrier, electrons are transferred from ascorbate to nitrogenase where acetylene is reduced to ethylene (see Fig. 26.3).

Flavodoxins from aerobes. The first component shown to function in the chloroplast nitrogenase assay was a flavoprotein named azotoflavin by Benemann *et al.* (1969) but referred to in this communication

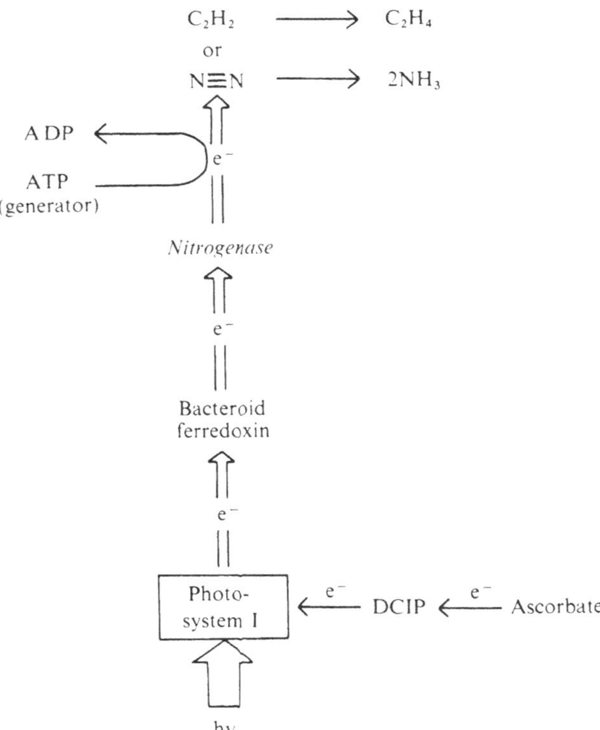

Fig. 26.3. An illustration of the sequence of electron flow from Photosystem I to nitrogenase in the chloroplast assay system. A ferrodoxin or flavodoxin is an essential co-factor.

as *Azotobacter* flavodoxin (Van Lin & Bothe, 1972). The *Azotobacter* flavodoxin of Benemann *et al.* (1969) apparently is the same protein that was discovered by Shethna, Wilson & Beinert (1966) and later crystallized and characterized by Hinkson & Bulen (1967). Neither Shethna *et al.* nor Hinkson & Bulen, however, identified a biological role for *Azotobacter* flavodoxin.

In a search for electron transport factors in extracts of *Rhizobium* bacteroids, Koch *et al.* (1970) identified a flavoprotein that exhibited a blue-colored semiquinone intermediate state, during reduction with sodium dithionite and functioned as an electron carrier in the chloroplast nitrogenase assay. No further characterization of this protein has been attempted.

Some of the major properties of flavodoxins from aerobic N_2-fixing organisms are summarized in Table 26.1. The spectra of oxidized, fully reduced, and semiquinone forms of *Azotobacter* flavodoxin are shown

Table 26.1. *Some properties of flavodoxins from aerobic N_2-fixing micro-organisms*

Organism	Name utilized	Semi-quinones detected	Molecular weight	Reduction potential	Biological activity*	References
Azotobacter vinelandii	Shethna flavoprotein	+	23000	—	None	Shethna *et al.* (1966) Edmonson & Tollin (1971*a*, *b*)
Azotobacter vinelandii	Free-radical flavoprotein	+	31200	—	None	Hinkson & Bulen (1967)
Azotobacter vinelandii	Azotoflavin	+	—	—	Chl-N_2	Benemann *et al.* (1969) Yoch & Valentine (1972)
				−270, −464		Yoch (1972)
Azotobacter vinelandii	Azotoflavin	+	—	—	NADPH-N_2 NADPH-N_2	Benemann *et al.* (1971) Wong *et al.* (1971)
Azotobacter vinelandii	Flavodoxin	+	—	—	Chl-N_2 PPNR (weak) H_2-NADP	Van Lin & Bothe (1972)
Azotobacter chroococcum	Flavodoxin	+	—	—	Flhq-N_2	Yates (1972)
Rhizobium japonicum bacteroids	Rhizobium flavoprotein (Rhizoflavin)	+	—	—	Chl-N_2	Koch *et al.* (1970) Yoch & Valentine (1972) (unpublished)
Anabaena cylindrica	Phytoflavin	+	16000	−180, −470	PPNR	Smillie (1965*b*) Citation by Van Lin & Bothe (1972)

* The following abbreviations for reactions are: PC, phosphoroclastic; Chl-N_2, chloroplast-nitrogenase; PPNR, photosynthetic pyridine nucleotide reductase; Flhq-N_2, flavodoxin hydroquinone to nitrogenase; H_2-NADP, hydrogen to nicotinamide adenine nucleotide phosphate; NADPH-N_2, nicotinamide adenine nucleotide phosphate (reduced) to nitrogenase.

† The two redox potentials cited are for the oxidized flavoprotein–flavoprotein semiquinone and the semiquinone flavoprotein–hydroquinone flavoprotein couples respectively.

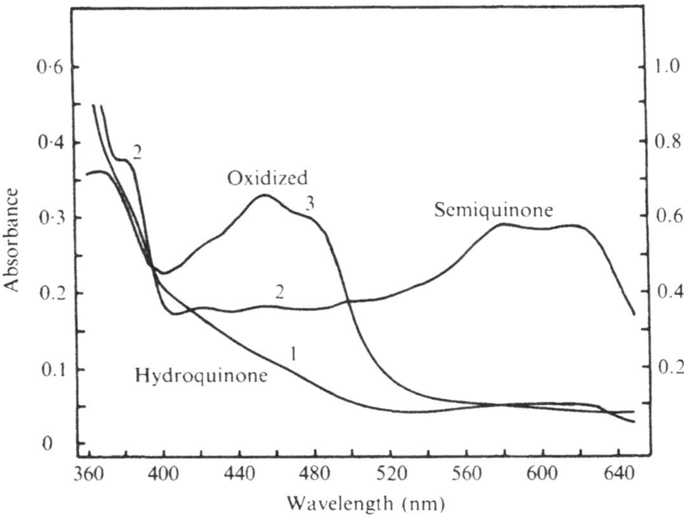

Fig. 26.4. The spectra of the oxidized (3), hydroquinone (1) and semiquinone (2) forms of *Azotobacter* flavodoxin (after Yates, 1972).

in Fig. 26.4. The molecular weight of *Azotobacter* flavodoxin, according to Edmonson & Tollin (1971*a*) is 23000. The standard reduction potentials of the oxidized flavodoxin–flavodoxin semiquinone and the flavodoxin semiquinone–flavodoxin hydroquinone couples are −270 mV and −464 mV respectively (Yoch, 1972). Yates (1972) has demonstrated that the fully reduced flavodoxin from *A. chroococcum* donates electrons directly to nitrogenase and Yoch (1972) has speculated that the potential of the oxidized flavodoxin–flavodoxin semiquinone couple of −270 mV is not sufficiently negative to serve as a nitrogenase donor. This opinion is supported by the experiments of Yates (1972). One of the interesting characteristics of the flavodoxin semiquinone is its resistance to oxidation in air. According to Edmonson & Tollin (1971*b*), the half-time for oxidation of *Azotobacter* flavodoxin semiquinone in air is 2000 minutes, a value that may be compared with 115 minutes for flavodoxin semiquinone from *C. pasteurianum*. Yoch (1972) also has observed that the rate of reduction of oxidized *A. vinelandii* flavodoxin to the semiquinone form and the rate of re-oxidation of the semiquinone was greatly stimulated by the addition of small amounts of methyl viologen. The activity of *Azotobacter* flavodoxin in the chloroplast nitrogenase reaction was stimulated substantially by the addition of catalytic amounts of either methyl viologen or *Azotobacter* ferredoxin.

The resistance of *Azotobacter* flavodoxin semiquinone to oxidation may be a unique adaption that is favorable to the N_2-fixing process in vigorous aerobes such as *Azotobacter*. Although a flavodoxin (referred to as phytoflavin) has been identified in *Anabaena cylindrica* (Smillie, 1965*a*, *b*), to our knowledge, no detailed investigations of the role of this flavoprotein in the process of N_2 fixation by blue-green algae have been conducted.

Many of the flavodoxins from aerobic N_2 fixers (Table 26.1) have been detected by their activities in the chloroplast-nitrogenase assay. *Azotobacter* flavodoxin functions weakly in the photosynthetic pyridine nucleotide reductase assay and in the reduction of NADP by H_2 in the presence of hydrogenase (Van Lin & Bothe, 1972). The role of flavodoxins as cofactors in the transfer of electrons from reduced pyridine nucleotides to nitrogenase will be discussed in a subsequent section of this paper.

Ferredoxins in aerobes. The chloroplast nitrogenase assay (Yoch & Arnon, 1970) also provided the first effective means of identifying ferredoxins in aerobic N_2-fixing micro-organisms. The use of this method enabled Yoch, Benemann, Valentine & Arnon (1969) to initially describe the purification and properties of ferredoxin in extracts of *A. vinelandii*. Prior to this, Shethna, Der Vartanian & Beinert (1968) had reported the occurrence of two proteins containing non-heme iron and acid-labile sulfur but molecular weights and other properties of these proteins distinguished them from the ferredoxin of Yoch *et al.* (1969). More recent work by Shethna (1970) has characterized a non-heme iron–sulfur protein III that appears to be identical with the *Azotobacter* ferredoxin. The occurrence of a ferredoxin in *A. chroococcum* was reported by Yates (1970).

In the initial investigation, Yoch *et al.* (1969) estimated a molecular weight of about 20000 for *Azotobacter* ferredoxin and reported that it contained six iron and six sulfur atoms per molecule. More recently Yoch & Arnon (1972) have resolved *Azotobacter* ferredoxin into two components, the major one (ferredoxin I) of which was reported to have a molecular weight of 14400 and contain eight atoms each of iron and sulfur per mole. The reduction potentials of ferredoxins I and II are −420 and −466 mV respectively (Yoch & Arnon, 1972). In addition to activity in the chloroplast nitrogenase assay, both *Azotobacter* ferredoxins function effectively in the photosynthetic pyridine nucleotide reductase assay and both are weakly active in the phosphoroclastic reaction. Ferredoxin from *A. vinelandii* also serves as a co-factor for the transfer of electrons from NADPH to N_2 (Benemann *et al.*, 1971) (see 'Sequences of electron transport to nitrogenase'). Reduced *Azotobacter* ferredoxin exhibits a resistance to oxidation, a property that seems analogous to that of *Azotobacter* flavodoxin semiquinone.

After the chloroplast nitrogenase assay became available Yoch *et al.* (1970) used a nodule bacteroid extract as a source of bacteroid nitrogenase and of electron transport factors in the chloroplast nitrogenase assay. In their experiments the addition of a relatively crude bacteroid factor preparation to an assay containing ascorbate, DCIP, chloroplast fragments, an ATP-generating system and crude bacteroid nitrogenase, approximately doubled the rate of acetylene reduction. The bacteroid factor also functioned in an assay in which *Azotobacter* nitrogenase was substituted for nodule bacteroid nitrogenase. Although Yoch *et al.* (1970) presented no evidence that the partially purified extract from soybean bacteroids contained non-heme iron or acid-labile sulfur this preparation was referred to by Benemann & Valentine (1972) as '*Rhizobium* ferredoxin'.

The purification and properties of an electron transport factor was described (Koch *et al.*, 1970) that functioned in the chloroplast nitrogenase and in the NADPH nitrogenase assay (Wong *et al.*, 1971). Separation of this non-heme iron protein (hereafter referred to as bacteroid ferredoxin) from the nitrogenase components of soybean bacteroids is illustrated in Fig. 26.5. The isolated bacteroid ferredoxin failed to function in either the phosphoroclastic reaction or the pyridine nucleotide reductase reaction. This oxygen-labile protein contained about three atoms each of non-heme iron and acid-labile sulfur per mole and exhibited an absorption spectrum similar to that of some other non-heme iron proteins. From sedimentation velocity measurements an $S_{20, w}$ value of 1.3 was reported and a molecular weight of 9400 was estimated. Benemann & Valentine (1972) have incorrectly referred to this investigation (Koch *et al.*, 1970) as being conducted by 'Burton *et al.* 1970' and have concluded that 'no data on its physical properties are yet available'.

Aerobically cultured *Rhizobium* species fail to synthesize factors capable of transferring electrons to nitrogenase in the chloroplast nitrogenase assay (Table 26.2). Phillips, Daniel, Appleby & Evans (1973) have shown, however, that *Rhizobium japonicum* cultured essentially anaerobically in a medium containing nitrate (Daniel & Appleby, 1972) produced factors capable of coupling the reducing power of

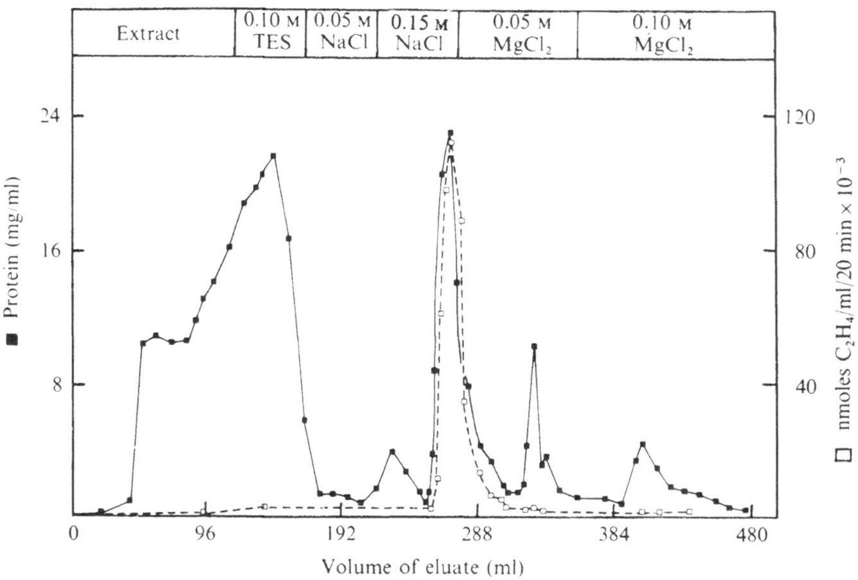

Fig. 26.5. An illustration of the separation of nodule bacteroid extract from one kilogram of soybean nodules into Mo–Fe protein, Fe protein and bacteroid ferredoxin. Peaks of elution of these proteins are at 273, 327, and 402 ml of eluate respectively. (Unpublished experiments by D. Israel and H. J. Evans, 1973.)

Table 26.2. *Capacities of* Rhizobium *extracts to function as factors in the chloroplast-nitrogenase assay.** (Unpublished experiments by David Biggins, Sterling Russell and H. J. Evans)

Source of factor	Preparation	Activity (nmoles C_2H_4/ 20 min/ mg protein)
R. meliloti, with NO_3^-	Crude extract	< 1.5
R. meliloti, with NO_3^-	30 % acetone extract	< 1.5
R. japonicum, with glutamate	30 % acetone extract	3.2
R. japonicum, with NO_3^-	Crude extract	< 1.5
R. japonicum, anaerobic with NO_3^-	DEAE eluate	203.0
R. japonicum bacteroids	Partially purified	—
	ferredoxin	1246.0
Factor omitted	—	2.0

* The assay procedure was described by Koch *et al.* 1970. Experiments conducted in the dark or without an ATP-generating system produced less than 2 nmoles C_2H_4/min/mg protein.

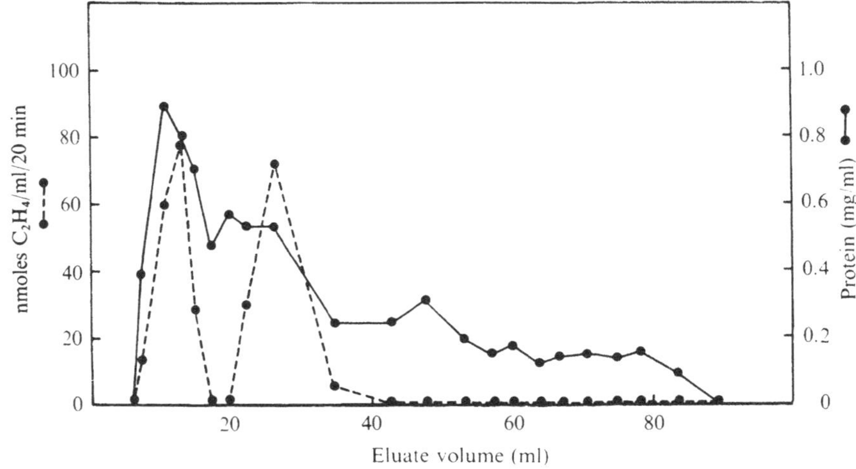

Fig. 26.6. Separation of electron transport factors from an extract of nitrate-grown *R. japonicum* by chromatography on DEAE-cellulose. Proteins were eluted with a linear gradient of $MgCl_2$ ranging from 50 mM to 0.15 M (from Phillips *et al.*, 1973).

Photosystem I to soybean bacteroid nitrogenase. The chromatography of these factors on DEAE-cellulose (Fig. 26.6) provided convincing evidence for the existence of two different active components in the nitrate grown *R. japonicum* cells. The yield of *R. japonicum* cells under near anaerobic conditions is limited and, therefore, insufficient protein has been available to permit a definitive characterization of these factors. The evidence obtained so far suggests that at least one of the components is similar to bacteroid ferredoxin (Phillips *et al.*, 1973). The major significance of this research is the demonstration that *R. japonicum* in pure culture possesses the genetic capability of synthesizing proteins that function in the transfer of electrons to nitrogenase. These electron transport factors, however, undoubtedly participate in other reactions that have not been identified. We expect these cofactors to participate in the transfer of electrons to the nitrate reductase of *R. japonicum*. In the experiments conducted so far we have obtained no positive evidence of the capability of free-living *R. japonicum* cultured with NO_3^- under near anaerobic conditions to synthesize either component of nitrogenase.

Properties of ferredoxins from aerobic N_2-fixing micro-organisms are summarized in Table 26.3.

Table 26.3. *Some properties of ferredoxins from aerobic N_2-fixing micro-organisms*

Organism	Molecular weight	Reduction potential (mV)	Fe and S^- per mole	Biological activity	References
A. vinelandii	20000	—	6	Chl-N_2;* PPNR PC (weak)	Yoch *et al.* (1969)
A. vinelandii Fd I	14400	−420	8	PPNR: Chl-N_2 PC (weak) NADPH-N_2	Yoch & Arnon (1972) Benemann *et al.* (1971)
A. vinelandii Fd II	—	−460	—	PPNR: Chl-N_2 PC (weak)	Yoch & Arnon (1972)
A. chroococcum	—	—	—	$Na_2S_2O_4$-N_2	Yates (1970)
A. vinelandii	13000	—	6 to 7	None	Shethna (1970)
R. japonicum bacteroids	—	—	—	Chl-N_2	Yoch *et al.* (1970)
R. japonicum bacteroids	9400	—	3	Chl-N_2 NADPH-N_2	Koch *et al.* (1970) Wong *et al.* (1971)
R. japonicum (with NO_3^-)	—	—	—	Chl-N_2	Phillips *et al.* (1973)
Anacystis nidulans	10000	—	2 (1)	PPNR	Yamanaka, Takenami, Nada & Okunuki (1969)

* The following abbreviations for reactions are used: PC, phosphoroclastic; Chl-N_2, chloroplast nitrogenase; PPNR, photosynthetic pyridine nucleotide reductase; NADPH-N_2, nicotinamide adenine dinucleotide phosphate (reduced) nitrogenase; $Na_2S_2O_4$-N_2, $Na_2S_2O_4$ nitrogenase; Fd, ferredoxin.

Sequences of electron transport to nitrogenase

Progress toward an elucidation of sequences of steps in the transfer of electrons from cellular transport chains to nitrogenase demands detailed information on electron transport factors and requirements for them in reconstituted transport chains. The possibility of using NADH as a donor for nitrogenase in aerobes was suggested by D'Eustachio & Hardy's (1964) success with this coenzyme in reactions with extracts of *C. pasteurianum*. This report and the consistent detection of a powerful NAD-specific β-hydroxybutyrate dehydrogenase in bacteroid extracts led to tests in which bacteroid nitrogenase was provided with NADH, generated by the β-hydroxybutyrate dehydrogenase system (Klucas & Evans, 1968). Nitrogenase-dependent acetylene reduction proceeded provided that an ATP-generating system and a dye such as benzyl viologen was added. Convincing evidence was supplied that the reduction potential of the NAD–NADH couple of −320 mV was sufficient for maintenance of nitrogenase activity of crude extracts of either soybean bacteroids or *A. vinelandii*. In further investigations (Wong *et al.*, 1971), the only natural components found to function in place of the dyes were FAD, or FMN and an unidentified factor(s) in nodule extracts. In these experiments, which were conducted with impure enzyme preparations, NADH dehydrogenase markedly stimulated the transfer of electrons to nitrogenase. The enzyme was apparently needed in the link between NADH and either dye or flavin nucleotide. More definitive experiments were conducted by Yates (1971), who prepared a homogeneous NADH dehydrogenase from *A. chroococcum* and clearly established its essentiality in the benzyl viologen-dependent transfer of electrons from NADH to nitrogenase.

In the flow of electrons from NADH to nitrogenase one must consider the reduction potential of NAD–NADH and of oxidized benzyl viologen-reduced benzyl viologen which are −320 mV (Sober, 1968) and −315 mV (Mann Laboratories, New York) respectively. Under conditions where the concentration of NADH relative to NAD is maintained at a high level the actual reduction potential of NAD–NADH couple would be considerably more negative than the standard reduction potential of −320 mV. In the presence of NADH-generating system and NADH dehydrogenase the flow of electrons from NADH to benzyl viologen would be expected. In unpublished experiments from our laboratory Dr Robert Klucas determined the ratio of oxidized to reduced benzyl viologen during the steady-state flow of electrons from NADH (generated by the *β*-hydroxybutyrate dehydrogenase system) to acetylene in presence of an ATP-generating system and bacteroid nitrogenase. Electrons were transferred to nitrogenase at an observed reduction potential of −315 mV at pH 7.5 and 25 °C. In other experiments Wong *et al.* (1971) have shown that flavin mononucleotides (FMN) or flavin–adenine dinucleotide (FAD) with a standard reduction potential of −219 mV (Sober, 1968) functioned in place of benzyl viologen as a carrier between NADH and bacteroid nitrogenase. Partially purified extracts of bacteroids were utilized in these experiments and therefore FAD or FMN may have combined with an unidentified apoenzyme resulting in a flavoprotein with a reduction potential considerably more negative than that of free FMN or FAD. In this regard it is known that FMN (standard reduction potential of −219 mV) is the prosthetic group of *Azotobacter* flavodoxin and that the standard reduction potential of flavodoxin semiquinone–flavodoxin hydroquinone couple is −464 mV (Yoch, 1972).

Reconstituted electron transfer sequences have been described (Table 26.4) in which NADPH (provided by a generating system) transferred electrons to nitrogenase in systems containing ferredoxin-NADP reductase, and flavodoxin and ferredoxin from *A. vinelandii* or *R. japonicum* bacteroids (Table 26.4). The sequence of transfer in this system is outlined in Fig. 26.7 but the order of flavodoxin and ferredoxin in the scheme has not been established and the labile factors in the system have not been identified and located precisely (Benemann & Valentine, 1972). Whether or not the heat stable factor that Yates (1970) has shown to stimulate nitrogenase activity in the presence of sodium dithionite is related to factors referred to by Benemann *et al.* (1971) remains to be established. In the reconstituted systems involving NADPH (Table 26.4) ferredoxin–NADP reductase from spinach has been utilized by Benemann *et al.* (1971) and by Wong *et al.* (1971). Although the existence of weak activities of ferredoxin–NADP reductase in extracts of *Azotobacter* and nodule bacteroids have been detected, the enzyme has not been characterized from these sources and there is no convincing evidence for its presence in sufficient quantity to be of physiological importance. Active ferredoxin–NADP reductases, however, have been observed in *B. polymyxa* (Yoch, 1973) and *A. cylindrica* (Smith, Noy & Evans, 1971). In the schemes of electron flow from NADPH to nitrogenase (Table 26.4) in aerobes, ferredoxins I and II with reduction potentials of −420 and −460 mV respectively and *Azotobacter* flavodoxin with potentials of −270 and −464 mV for E_1

Table 26.4. *Summary of some reconstituted electron transport sequences to nitrogenase*

Source of nitrogenase	Electron donor	Reductase	Electron carriers	Other factors	Acceptor from nitrogenase*	References
R. japonicum bacteroids	NADH†	Endogenous	Benzyl or methyl viologen	—	C_2H_2	Klucas & Evans (1968)
A. chroococcum (particles)	NADH	Endogenous	Endogenous	—	C_2H_2	Yates & Daniel (1970)
A. chroococcum (soluble extract)	NADH	NADH dehydrogenase	Benzyl viologen	—	C_2H_2	Yates (1971)
A. vinelandii	NADPH‡	Fd–NADP (spinach)	*Azotobacter* flavodoxin and ferredoxin	Labile *A. vinelandii* component	C_2H_2	Benemann *et al.* (1971)
M. flavum (soluble extract)	NADH	Endogenous	Benzyl viologen	—	C_2H_2	Biggins & Postgate (1971)
R. japonicum bacteroids	NADPH‡	Fd–NADP (spinach)	*Azotobacter* flavodoxin *R. japonicum* ferredoxin	Endogenous	C_2H_2	Wong *et al.* (1971)
R. japonicum bacteroids	NADPH‡	Fd–NADP (spinach)	FMN or FAD	Endogenous	C_2H_2	Wong *et al.* (1971)
R. japonicum	NADH†	Endogenous	FAD	Labile bacteroid component	C_2H_2	Wong *et al.* (1971)
A. chroococcum	—	—	*Azotobacter* flavodoxin hydroquinone	—	C_2H_2	Yates (1972)
A. cylindrica	NADPH	Fd–NADP (spinach)	Ferredoxin (*A. cylindrica*)	—	C_2H_2	Bothe (1970)
A. cylindrica	NADPH	Fd–NADP (*A. cylindrica*)	Ferredoxin (*A. cylindrica*)	—	C_2H_2	Smith *et al.* (1971)

* Nitrogenase-dependent C_2H_2 reduction in all cases required an ATP-generating system with the exception of the particulate nitrogenase preparation from *A. chroococcum*.

† Generated by β-hydroxybutyrate dehydrogenase system.

‡ Generated by glucose-6-phosphate dehydrogenase or *iso*citric dehydrogenase systems.

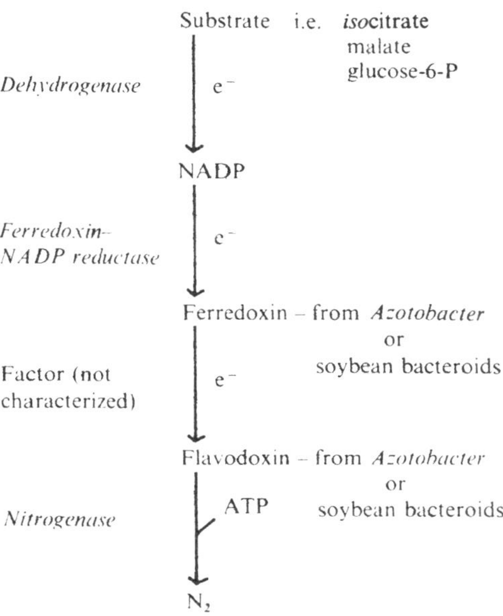

Fig. 26.7. An outline of a sequence of electron transport from NADP-linked dehydrogenases to nitrogenase. In the investigations of electron transfer in aerobic N_2 fixers, ferredoxin–NADP reductase from spinach has been added. The order of ferredoxin and flavodoxin in the sequence has not yet been established.

and E_2 respectively (Table 26.2) are proposed to participate in some poorly defined way as intermediate electron carrier proteins. According to Benemann & Valentine (1971), the highly negative standard reduction potentials of intermediate carriers apparently pose no thermodynamic barrier to electron flow in coupled electron transport systems where the ratio of NADPH to NADP in the donor system is maintained at a high level and the final nitrogenase acceptor system which requires ATP and evolves hydrogen is virtually irreversible. Of course, the ratio of oxidized to reduced forms of intermediate carriers during the steady-state transfer of electrons to nitrogenase may be maintained at a proportion whereby the observed reduction potential of a particular carrier is considerably more positive than standard reduction potentials which are based upon ratios of oxidant to reductant of 1:1. Yates (1972) has shown that *Azotobacter* flavodoxin hydroquinone donates electrons directly to nitrogenase but the role, if any, of the oxidized flavodoxin–flavodoxin semiquinone couple and the roles of the two different ferredoxins remains to be clarified.

Physiological significance of pathways

There is considerable indirect evidence that is pertinent to an assessment of the physiological importance of pathways whereby reductant is provided for N_2 fixation. In 1959, Mumford, Carnahan & Castle reported that the addition of *iso*citrate and other citric acid cycle acids restored N_2-fixing capability to a mutant of *A. vinelandii* that failed to grow and fix N_2 in media containing glucose. According to Benemann *et al.* (1971) the addition of NADP and either glucose-6-phosphate or *iso*citrate greatly stimulated nitrogenase activity of a dialyzed extract of *A. vinelandii.* The observation that NADP-linked *iso*citrate dehydrogenase may constitute up to 1 % of the total soluble protein in *Azotobacter* cells is consistent with the opinion that this enzyme is an important electron donor system in N_2-fixing *Azotobacter* species (Chung & Franzen, 1969). According to Bergersen & Turner (1968) both N_2 fixation and hydrogen evolution by washed nodule bacteroids was stimulated by the addition of succinate, fumarate or pyruvate. Dixon (1968; 1972) observed hydrogen uptake by washed bacteroids from nodules of *Pisum sativum* and has discussed a role of hydrogenase in maintaining N_2-fixing efficiency. The addition of succinate or β-hydroxybutyrate, according to Koch *et al.* (1967), stimulated nitrogenase activity of washed soybean nodule bacteroids. Rokosh, Kurz & Wetter (1973) described experiments in which activities of citrate synthetase, malic dehydrogenase and *iso*citric dehydrogenase in nodule bacteroids from *Pisum sativum* were monitored during the growth on N_2 and have observed that peak activities of nitrogenase in bacteroids coincided with peak activities of *iso*citric dehydrogenase. These results are consistent with the argument for a role of NADP-linked *iso*citric dehydrogenase as a source of reductant for N_2 fixation in this legume. In an attempt to assess the importance of a NADH donor system, Wong *et al.* (1971) investigated the role of poly-β-hydroxybutyrate and β-hydroxybutyrate dehydrogenase as a possible source of reductant for nitrogenase in soybean nodule bacteroids and concluded that neither the utilization of poly-β-hydroxybutyrate by nodules nor the activity of β-hydroxybutyrate dehydrogenase in bacteroids during different experimental conditions, was correlated with nitrogenase activity.

The physiological significance of the reconstituted electron transport sequences (Table 26.4) remains an open question until all components of a reconstituted system are demonstrated from a single N_2-fixing organism and acetylene reduction by a particular reconstituted system is shown to proceed at a rate that would be expected to support N_2 fixation *in vivo*. In those experiments where the NADPH donor system was coupled to *Azotobacter* nitrogenase, rates of acetylene reduction were no more than 40 % of the rates in comparable reactions where electrons were provided by ascorbate, DCIP, and illuminated chloroplasts (Benemann *et al.*, 1971). Maximum rates of acetylene reduction in reactions where NADPH was coupled to soybean bacteroid nitrogenase (Wong *et al.*, 1971) usually were less than 10 % of those observed in reactions supplied with sodium dithionite or with the chloroplast ascorbate donor system. The relatively sluggish activities and the requirement for exogenous ferredoxin–NADP reductase of the reconstituted NADPH donor systems raise unanswered questions about their physiological importance. Additional evidence is needed to determine whether or not electron carriers, and enzymes that make up reconstituted systems are present in N_2-fixing organisms in sufficient concentrations to account for N_2 fixation rates during normal growth conditions.

Electron transport in photosynthetic N_2 fixers

The occurrence of photosynthetic organisms which reduce N_2 has promoted interest in the possible relationship between these two processes. Presumably the production of reductant and ATP during photosynthesis could be adapted for use in N_2 fixation by the organism. This mechanism recently was shown to operate in reconstituted systems of the blue-green alga, *Anabaena cylindrica* (Smith *et al.*, 1971) and the photoreduction of ferredoxin has been demonstrated during photosynthesis in *Chlorobium thiosulfatophilum* (Buchanan & Evans, 1965). The reduction of N_2, however, in nitrogen-starved, blue-green algae is not closely dependent upon light (Cox, 1966), and both *Chlorogloea fritschii* and *Anabaenopsis circularis* can grow heterotrophically in nitrogen-free medium in the dark (Fay, 1965; Watanabe & Yamamoto, 1967). It seems probable that several pathways link nitrogenase to a reductant in these organisms. In-vitro studies have used many of the same electron donors which function with nitrogenase in other organisms.

Photosynthetic bacteria

Arnon, Losada, Nozaki & Tagawa (1961) demonstrated marginal fixation of N_2 in extracts of the purple bacterium *Chromatium* with either NADH or H_2 serving as a reductant. Whole cells showed increased N_2 fixation in the light but not in the dark when thiosulfate, succinate or oxalacetate was supplied. Pyruvate also stimulated light-dependent N_2 reduction in whole cells of this organism (Bennett, Rigopoulos & Fuller, 1964). This compound is degraded in *Chromatium* by a clostridial type of phosphoroclastic cleavage (Bennett *et al.*, 1964). More recently Winter & Arnon (1970) have shown more active and more reproducible N_2 fixation in extracts of *Chromatium* by adding sources of ATP and reductant. Sodium dithionite functioned alone as a reductant, and H_2 was effective when coupled with either a viologen dye or ferredoxin from *Chromatium* or *Clostridium*. The *Chromatium* ferredoxin which was crystallized by Bachofen & Arnon (1966) transferred electrons from H_2 to NAD or NADP in the presence of both soluble and particulate fractions of the cell extract (Buchanan & Bachofen, 1968). An artificial N_2 fixation system has been constructed with *Chromatium* nitrogenase using ATP generated by cyclic photophosphorylation in *Chromatium* or *R. rubrum* chromatophores and bacterial ferredoxin reduced by heat-treated spinach chloroplasts. Ferredoxin from *Chromatium* or *C. pasteurianum* was much more active than that isolated from spinach chloroplasts (Yoch & Arnon, 1970).

Schneider *et al.* (1960) prepared extracts from *R. rubrum* which reduced N_2 in both the light and the dark. Addition of pyruvate, α-ketoglutarate, oxalacetate, NADPH and NADH resulted in no clear indication of the source of reducing power normally used by these extracts. Bulen *et al.* (1965) obtained active cell-free preparations of *R. rubrum* that functioned with sodium dithionite as the electron donor. Wide variations in the N_2-fixing activity of different preparations were controlled by Burns & Bulen (1966) by maintenance of pH and removal of interfering substances on an anaerobic Sephadex column (Munson *et al.*, 1965). Munson & Burris (1969), however, reported erratic results using these methods and attributed the variation to the condition of the cells. Schick (1971) encountered similar difficulties and conducted his studies with intact cells.

A role for reduced ferredoxin in N_2 fixation is implicated by recent work with the green sulfur bacterium *Chloropseudomonas ethylicum*. Evans & Smith (1971) coupled illuminated chromatophores and ferredoxin from *C. ethylicum* with nitrogenase from this organism reducing acetylene to ethylene. Addition of pyruvate and coenzyme A to crude extracts promoted acetylene reduction in the dark. There is some evidence that ferredoxin from *C. ethylicum* has a close linkage to nitrogenase in this organism (Evans & Smith, 1971). When reactions containing DEAE-cellulose treated extracts used sodium dithionite as a reductant, the addition of ferredoxin from *C. ethylicum* promoted ethylene production by a factor of 3–4. Under these conditions *C. pasteurianum* and *Chromatium* ferredoxins were slightly stimulatory, spinach ferredoxin failed to function and methyl viologen (0.67 μM) was inhibitory. Evans & Smith (1971) suggested that *C. ethylicum* grown autotrophically may use photoreduced ferredoxin for N_2 fixation. Under certain heterotrophic conditions it appears that pyruvate oxidation may supply electrons for N_2 fixation (Evans & Smith, 1971). It must be concluded that although ferredoxin has been directly implicated in the carbon dioxide-reducing mechanism of photosynthetic bacteria (Buchanan & Evans, 1965; Evans & Buchanan, 1965), its role as a reductant of nitrogenase in N_2-fixing bacteria of this type is largely circumstantial. There is no doubt that ferredoxin is capable of transferring electrons to nitrogenase *in vitro* (Buchanan & Bachofen, 1968; Yoch & Arnon, 1970), but only indirect evidence exists for a role of ferredoxin as the reductant for nitrogenase *in vivo* in the light.

Blue-green algae

The first report of substantial cell-free N_2 fixation in blue-green algae was that of Schneider *et al.* (1960). Active extracts were obtained from *Mastigocladus laminosus*, *Nostoc muscorum*, *Anabaena cylindrica*, *Gloeotrichia echinulata* and *Calothrix parietina*. In all cases the extracts exhibited nitrogenase activity in the light without added factors.

An observation that pyruvate stimulated N_2 fixation in intact *A. cylindrica* cells (Cox, 1966) was the first evidence suggesting that pyruvate may supply energy and reductant for N_2 fixation in this organism. Subsequent work by Cox & Fay (1967; 1969) supported this hypothesis. Exogenous pyruvate stimulated N_2 fixation in reactions with cell-free extracts of *A. cylindrica.* Decarboxylation of pyruvate was stimulated under N_2-fixing conditions and the decarboxylation activity of cell-free extracts was positively correlated with N_2 fixation activity (Cox & Fay, 1967). If the nitrogen content of the algal cells was depleted by growing the organisms in a nitrogen-free medium under carbon dioxide and argon, then a 90 % inhibition of carbon dioxide fixation by *p*-chlorophenyl-1,1-dimethylurea (CMU) produced no decrease in N_2 fixation (Cox & Fay, 1969). Similar treatment of normal cells inhibited N_2 fixation by 50 %. Presumably carbon skeletons which were stored during nitrogen starvation were used to accept the products of N_2 reduction when photosynthesis was inhibited. Light greatly stimulated acetylene reduction under these same conditions which suggested that the cyclic photophosphorylation which occurs in the presence of CMU was important. From these results, Cox & Fay (1969) suggested that N_2 fixation in *A. cylindrica* was independent of the reducing potential generated by photosynthesis but was dependent upon photosynthesis for ATP.

Although progress had been made in understanding aspects of electron transport to nitrogenase in blue-green algae, it was not possible to use the non-physiological donor, sodium dithionite, until Smith & Evans (1970) demonstrated that the optimal concentrations of sodium dithionite in assays was 0.0016 M. The higher concentration normally used in bacterial nitrogenase assays was inhibitory. Haystead, Robinson & Stewart (1970) confirmed that sodium dithionite was effective as a donor for nitrogenase in *Plectonema boryanum* provided that the concentration was maintained at a low level.

Work by several laboratories suggests that ferredoxin isolated from *A. cylindrica* plays a role in N_2 fixation (Bothe, 1970; Smith *et al.*, 1971). Bothe (1970) coupled nitrogenase actively in *A. cylindrica* extracts to spinach of *A. cylindrica* ferredoxin reduced by photoactivated spinach chloroplast fragments and DCIP and ascorbate. In addition, acetylene reduction was observed with extracts supplemented with glucose-6-phosphate, NADP, ferredoxin, and ferredoxin–NADP reductase from spinach. Attempts to substitute pyruvate, *iso*citrate, or malate for glucose-6-phosphate resulted in no nitrogenase activity. In similar experiments with this organism Smith *et al.* (1971) have reconstituted a system which reduces acetylene in the light. Under these conditions native ferredoxin, chromatophores, DCIP, ascorbate, an ATP-generating system and a crude algal extract were essential. An effective acetylene-reducing system without light was constructed with a Sephadex G75-treated extract, NADP, *iso*citrate, $MgCl_2$, ferredoxin and an ATP-generating system. In this reaction, pyruvate substituted for *iso*citrate in the dark. Separate assays of the Sephadex G75-treated extract revealed the presence of ferredoxin–NADP reductase, and *iso*citrate dehydrogenase. Undoubtedly, pyruvate oxidase also was present.

With the exception of the capacity to carry out light-dependent production of both reductant and ATP, there are no reasons to believe that the essential processes whereby electrons are transported to nitrogenase in blue-green algae are fundamentally different from those in non-photosynthetic N_2-fixing organisms.

Summary

A review is presented of our understanding of mechanisms whereby electrons are transported from metabolic processes to nitrogenase. In the early research with cell-free extracts of the anaerobe, *Clostridium pasteurianum*, the phosphoroclastic degradation of pyruvate was shown to provide the reductant and ATP for N_2 fixation. Pyruvate also supports N_2 fixation in *Bacillus polymyxa*, *Chloropseudomonas ethylicum* and certain *Klebsiella* species. In the catalysis of the phosphoroclastic reaction by enzyme systems from anaerobic bacteria, ferredoxin is reduced and apparently serves directly as the electron donor to nitrogenase. Nitrogenase activity in *Clostridium* extracts also may be supported by hydrogen, NADH, or by systems that maintain low potential dyes in the reduced state. In the presence of hydrogenase, electrons from H_2 are transferred to nitrogenase via ferredoxin. The effectiveness of low potential dyes as electron donors to nitrogenase is due to a capacity to replace ferredoxin in electron transport chains. When the synthesis of ferredoxin in *C. pasteurianum* is limited by iron deficiency the bacterium produces flavodoxin, a flavoprotein with a redox potential sufficiently low, to substitute for ferredoxin as a couple between the phosphoroclastic reaction and N_2 fixation. Although flavodoxins are known to occur in several anaerobic, facultatively anaerobic and photosynthetic bacteria the general role of proteins of this type in electron transport to nitrogenase requires further clarification.

In the early experiments the activities of cell-free nitrogenase preparations from aerobes such as *Azotobacter* and *R. japonicum* bacteroids were not supported by pyruvate. More recent work, however, indicates that pyruvate utilization by extracts of *A. vinelandii* under anaerobic conditions is capable of providing ATP for N_2 fixation. The non-physiological reductant, sodium dithionite, has been used extensively as a source of electrons for the enzyme from these and other sources. Sodium dithionite apparently donates electrons directly to the nitrogenase without involvement of intermediate carriers. Recent investigations have revealed that *A. vinelandii* and *R. japonicum* bacteroids contain ferredoxins and flavoproteins that are capable of transferring electrons from Photosystem I from spinach chloroplasts to nitrogenase. Also NADH has the capability of providing electrons for the activity of nitrogenase from either *A. vinelandii* or *R. japonicum* bacteroids provided that benzyl viologen and a crude extract containing NADH dehydrogenase are added. Electron transport chains from NADPH to nitrogenase from either *A. vinelandii* or nodule bacteroids have been reconstructed. The constitutents in these chains include flavodoxin (also known as azotoflavin) and ferredoxin from *A. vinelandii* or from nodule bacteroids, NADPH–ferredoxin reductase from spinach and an extract containing an unknown heat-labile factor(s). NADPH generated by dehydrogenase systems also will function as an electron donor for N_2 fixation by extracts of blue-green algae. So far it has not been possible to reconstitute highly active electron transport chains from NADPH to nitrogenase by use of components isolated from a single N_2-fixing species. The rates of nitrogenase-dependent acetylene reduction in all reconstituted systems involving NADPH are relatively low and thus the question of whether or not these represent physiologically important electron donor systems remains to be established. Some results of physiological experiments support the role of pathways to nitrogenase that include NADPH but there is less evidence for physiologically important electron pathways to nitrogenase that utilize NADH.

The preparation of this review is part of a research program supported by a grant (GB 29600X) from the National Science Foundation and by the Oregon Agricultural Experiment Station (Technical Paper No. 3667). The authors express their appreciation to Mrs Mariana Frick for editorial assistance and drafting, and to Mrs Donna Boydston for typing the manuscript.

References

Adman, T., Sieker, L. C. & Jensen, L. H. (1973). The structure of a bacterial ferredoxin. *J. biol. Chem.*, **248**, 3987–96.

Arnon, D. I., Losada, M., Nozaki, M. & Tagawa, K. (1961). Photoproduction of hydrogen, photofixation of nitrogen and a unified concept of photosynthesis. *Nature, Lond.*, **190**, 601–6.

Bachofen, R. & Arnon, D. I. (1966). Crystalline ferredoxin from the photosynthetic bacterium *Chromatium. Biochim. biophys. Acta*, **120**(2), 259–65.

Benemann, J. R. & Valentine, R. C. (1971). High-energy electrons in bacteria. *Adv. microbiol. Physiol.*, **5**, 135–72.

Benemann, J. R. & Valentine, R. C. (1972). The pathways of nitrogen fixation. *Adv. microbiol. Physiol.*, **8**, 59–104.

Benemann, J. R., Yoch, D. C., Valentine, R. C. & Arnon, D. I. (1969). The electron transport system in nitrogen fixation by *Azotobacter*. I. Azotoflavin as an electron carrier. *Proc. natn. Acad. Sci. USA*, **64**, 1079–86.

Benemann, J. R., Yoch, D. C., Valentine, R. C. & Arnon, D. I. (1971). The electron transport system in nitrogen fixation by *Azotobacter*, III. Requirements for NADPH-supported nitrogenase activity. *Biochim. biophys. Acta*, **226**, 205–12.

Bennett, R., Rigopoulos, N. & Fuller, R. C. (1964). The pyruvate phosphoroclastic reaction and light-dependent nitrogen fixation in bacterial photosynthesis. *Proc. natn. Acad. Sci. USA*, **52**, 762–8.

Bergersen, F. J. (1971). Biochemistry of symbiotic nitrogen fixation in legumes. *Ann. Rev. Pl. Physiol.*, **22**, 121–39.

Bergersen, F. J. & Turner, H. L. (1968). Comparative studies of nitrogen fixation by soybean root nodules, bacteroid suspensions and cell-free extracts. *J. gen. Microbiol.*, **53**, 205–20.

Biggins, D. R. & Postgate, J. R. (1971). Nitrogen fixation by extracts of *Mycobacterium flavum* 301. *Eur. J. Biochem.*, **19**, 408–15.

Bothe, H. (1970). Photosynthetische Stickstoffixierung mit einem zellfreien Extrakt aus der Blaualge *Anabaena cylindrica*. *Ber. dt. bot. Ges.*, **83**, 421–32.

Buchanan, B. B. & Bachofen, R. (1968). Ferredoxin-dependent reduction of nicotinamide-adenine dinucleotides with hydrogen gas by subcellular preparations from the photosynthetic bacterium, *Chromatium*. *Biochim. biophys. Acta*, **162**, 607–10.

Buchanan, B. B. & Evans, M. C. W. (1965). The synthesis of α-ketoglutarate from succinate and carbon dioxide by a subcellular preparation of a photosynthetic bacterium. *Proc. natn. Acad. Sci. USA*, **54**, 1212–18.

Bulen, W. A., Burns, R. C. & LeComte, J. R. (1964). Nitrogen fixation: cell-free system with extracts of *Azotobacter*. *Biochem. biophys. Res. Comm.*, **17**, 265–71.

Bulen, W. A., Burns, R. C. & LeComte, J. R. (1965). Nitrogen fixation: hydrosulfite as electron donor with cell-free preparations of *Azotobacter vinelandii* and *Rhodospirillum rubrum*. *Proc. natn. Acad. Sci. USA*, **53**, 532–9.

Burns, R. C. & Bulen, W. A. (1966). A procedure for the preparation of extracts from *Rhodospirillum rubrum* catalyzing N_2 reduction and ATP-dependent H_2 evolution. *Arch. Biochem. Biophys.*, **113**, 461–3.

Burris, R. H. (1966). Biological nitrogen fixation. *Ann. Rev. Pl. Physiol.*, **17**, 155–79.

Carnahan, J. E., Mortenson, L. E., Mower, H. F. & Castle, J. E. (1960*a*). Nitrogen fixation in cell-free extracts of *Clostridium pasteurianum*. *Biochim. biophys. Acta*, **38**, 188–9.

Carnahan, J. E., Mortenson, L. E., Mower, H. F. & Castle, J. E. (1960*b*). Nitrogen fixation in cell-free extracts of *Clostridium pasteurianum*. *Biochim. biophys. Acta*, **44**, 520–35.

Chung, A. E. & Franzen, J. S. (1969). Oxidized triphosphopyridine nucleotide specific isocitrate dehydrogenase from *Azotobacter vinelandii*. Isolation and characterization. *Biochemistry*, **8**, 3175–84.

Cox, R. M. (1966). Physiological studies on nitrogen fixation in the blue-green alga *Anabaena cylindrica*. *Arch. Mikrobiol.*, **53**, 263–76.
Cox, R. M. & Fay, P. (1967). Nitrogen fixation and pyruvate metabolism in cell-free preparations of *Anabaena cylindrica*. *Arch. Mikrobiol.*, **58**, 357–65.
Cox, R. M. & Fay, P. (1969). Special aspects of nitrogen-fixation by blue-green algae. *Proc. R. Soc. Lond. B*, **172**, 357–66.
Dalton, H. & Mortenson, L. E. (1972). Dinitrogen (N_2) fixation (with a biochemical emphasis). *Bacteriol. Rev.*, **36**, 231–60.
Daniel, R. M. & Appleby, C. A. (1972). Anaerobic-nitrate, symbiotic and aerobic growth of *Rhizobium japonicum*: effects on cytochrome P_{450}, other haemoproteins, nitrate and nitrite reductases. *Biochim. biophys. Acta*, **275**, 347–54.
D'Eustachio, A. J. & Hardy, R. W. F. (1964). Reductants and electron transport in nitrogen fixation. *Biochem. biophys. Res. Comm.*, **15**, 319–23.
Dixon, R. O. D. (1968). Hydrogenase in pea root nodule bacteroids. *Arch. Mikrobiol.*, **62**, 272–83.
Dixon, R. O. D. (1972). Hydrogenase in root nodule bacteroids: occurrence and properties. *Arch. Mikrobiol.*, **85**, 193–201.
Edmonson, D. E. & Tollin, G. (1971*a*). Chemical and physical characterization of the Shethna flavoprotein and apoprotein and kinetics and thermodynamics of flavin analog binding to the apoprotein. *Biochemistry*, **10**, 124–32.
Edmonson, D. E. & Tollin, G. (1971*b*). Flavoprotein interactions and the redox properties of the Shethna flavoprotein. *Biochemistry*, **10**, 133–45.
Evans, H. J. & Russell, S. A. (1971). Physiological chemistry of symbiotic nitrogen fixation by legumes. In *Chemistry and Biochemistry of Nitrogen Fixation* (ed. Postgate, J. R.), pp. 191–244. Plenum Press: London, New York.
Evans, M. C. W. & Buchanan, B. B. (1965). Photoreduction of ferredoxin and its use in carbon dioxide fixation by a subcellular system from a photosynthetic bacterium. *Proc. natn. Acad. Sci. USA*, **53**, 1420–5.
Evans, M. C. W. & Smith, R. V. (1971). Nitrogen fixation by the green photosynthetic bacterium *Chloropseudomonas ethylicum*. *J. gen. Microbiol.*, **65**, 95–8.
Fay, R. (1965). Heterotrophy and nitrogen fixation in *Chlorogloea fritschii*. *J. gen. Microbiol.*, **39**, 11–20.
Fisher, R. J. & Wilson, P. W. (1970). Pyruvate-supported nitrogen fixation by cell-free extracts of *Bacillus polymyxa*. *Biochem. J.*, **117**, 1023–4.
Grau, F. H. & Wilson, P. W. (1962). Physiology of nitrogen fixation by *Bacillus polymyxa*. *J. Bacteriol.*, **83**, 490–6.
Grau, F. H. & Wilson, P. W. (1963). Hydrogenase and nitrogenase in cell-free extracts of *Bacillus polymyxa*. *J. Bacteriol.*, **85**, 446–50.
Haaker, H., Bresters, T. W. & Veeger, C. (1972). Relation between anaerobic ATP synthesis from pyruvate and nitrogen fixation in *Azotobacter vinelandii*. *FEBS Lett.*, **23**, 160–2.
Hamilton, I. R., Burris, R. H. & Wilson, P. W. (1964). Hydrogenase and

nitrogenase in a nitrogen-fixing bacterium. *Proc. natn. Acad. Sci. USA*, **52**, 637–41.

Hardy, R. W. F. & Burns, R. C. (1968). Biological nitrogen fixation. *Ann. Rev. Biochem.*, **37**, 331–58.

Hardy, R. W. F. & Burns, R. C. (1972). Comparative biochemistry of iron–sulfur proteins and dinitrogen fixation. In *Iron–sulfur Proteins* (ed. Lovenberg, W.), pp. 65–110. Academic Press: New York.

Hardy, R. W. F. & D'Eustachio, A. J. (1964). The dual role of pyruvate and the energy requirement in nitrogen fixation. *Biochem. biophys. Res. Comm.*, **15**, 314–18.

Hardy, R. W. F., Knight, E. & D'Eustachio, A. J. (1965). An energy-dependent hydrogen-evolution from dithionite in nitrogen fixing extracts of *Clostridium pasteurianum*. *Biochem. biophys. Res. Comm.*, **20**, 539–44.

Haystead, A., Robinson, R. & Stewart, W. D. P. (1970). Nitrogenase activity in extracts of heterocystous and non-heterocystous blue-green algae. *Arch. Mikrobiol.*, **74**, 235–43.

Hinkson, J. W. & Bulen, W. A. (1967). A free radical flavoprotein from *Azotobacter*. *J. biol. Chem.*, **242**, 3345–51.

Klucas, R. V. & Evans, H. J. (1968). An electron donor system for nitrogenase-dependent acetylene reduction by extracts of soybean nodules. *Pl. Physiol.*, **43**, 1458–60.

Knight, E., D'Eustachio, A. J. & Hardy, R. W. F. (1966). Flavodoxin: a flavoprotein with ferredoxin activity from *Clostridium pasteurianum*. *Biochim. biophys. Acta*, **113**, 626–8.

Knight, E. & Hardy, R. W. F. (1966). Isolation and characteristics of flavodoxin from nitrogen-fixing *Clostridium pasteurianum*. *J. biol. Chem.*, **241**, 2752–6.

Knight, E. & Hardy, R. W. F. (1967). Isolation and characteristics of flavodoxin from nitrogen-fixing *Clostridium pasteurianum*. *J. biol. Chem.*, **242**, 1370–4.

Koch, B., Evans, H. J. & Russell, S. (1967). Reduction of acetylene and nitrogen gas by breis and cell-free extracts of soybean root nodules. *Pl. Physiol.*, **42**, 466–8.

Koch, B., Wong, P., Russell, S. A., Howard, R. & Evans, H. J. (1970). Purification and some properties of a non-haem iron protein from the bacteroids of soy-bean (*Glycine max* Merr.) nodules. *Biochem. J.*, **118**, 773–81.

Koepsell, H. J. & Johnson, M. J. (1942). Dissimilation of pyruvic acid by cell-free preparations of *Clostridium butylicum*. *J. biol. Chem.*, **145**, 379–86.

Lindsey, H. L. & Wilson, P. W. (1963). Optimum conditions and inhibitors of nitrogen fixation by *Aerobacter aerogenes*. *Bacteriol. Proc.*, p. 122.

Mahl, M. C. & Wilson, P. W. (1968). Nitrogen fixation by cell-free extracts of *Klebsiella pneumoniae*. *Can. J. Microbiol.*, **14**, 33–8.

Mayhew, S. G. (1971). Properties of two Clostridial flavodoxins. *Biochim. biophys. Acta*, **235**, 276–88.

Mayhew, S. G., Foust, G. P. & Massey, V. (1969). Oxidation-reduction properties of flavodoxin from *Peptostreptococcus elsdenii*. *J. biol. Chem.*, **244**, 803–10.

Mortenson, L. E. (1963). Nitrogen-fixation: role of ferredoxin in anaerobic metabolism. *Ann. Rev. Microbiol.*, **17**, 115–38.

(1964*a*). Ferredoxin and ATP, requirements for nitrogen fixation in cell-free extracts of *Clostridium pasteurianum*. *Proc. natn. Acad. Sci. USA*, **52**, 272–9.

(1964*b*). Ferredoxin requirement for nitrogen fixation by extracts of *Clostridium pasteurianum*. *Biochim. biophys. Acta*, **81**, 473–8.

(1966). Components of cell-free extracts of *Clostridium pasteurianum* requirements for ATP-dependent H_2 evolution from dithionite and for N_2 fixation. *Biochim. biophys. Acta*, **127**, 18–25.

Mortenson, L. E., Valentine, R. C. & Carnahan, J. E. (1962). An electron transport factor from *Clostridium pasteurianum*. *Biochem. biophys. Res. Comm.*, **7**, 448–52.

Mortenson, L. E., Valentine, R. C. & Carnahan, J. E. (1963). Ferredoxin in the phosphoroclastic reaction of pyruvate acid and its relation to nitrogen fixation in *Clostridium pasteurianum*. *J. biol. Chem.*, **238**, 794–800.

Mumford, F. E., Carnahan, J. E. & Castle, J. E. (1959). Nitrogen fixation in a mutant of *Azotobacter vinelandii*. *J. Bacteriol.*, **77**, 86–90.

Munson, T. O. & Burris, R. H. (1968). Nitrogen fixation by *Rhodospirillum rubrum* grown in nitrogen-limited continuous culture. *J. bacteriol.*, **97**, 1093–8.

Munson, T. O., Dilworth, M. J. & Burris, R. H. (1964). Method for demonstrating cofactor requirements for nitrogen fixation. *Biochem. biophys. Acta*, **104**, 278–81.

Phillips, D. A., Daniel, R. M., Appleby, C. A. & Evans, H. J. (1973). Isolation from *Rhizobium* of factors which transfer electrons to soybean nitrogenase. *Pl. Physiol.*, **51**, 136–8.

Rokosh, D. A., Kurz, W. G. W. & Wetter, L. R. (1973). Reductant for symbiotic nitrogen fixation. *Pl. Physiol.*, *Suppl.*, **51**, 185.

Schick, H. J. (1971). Substrate and light dependent fixation of molecular nitrogen in *Rhodospirillum rubrum*. *Arch. Mikrobiol.*, **75**, 89–101.

Schneider, K. C., Bradbeer, C., Singh, R. N., Wang, L. C., Wilson, P. W. & Burris, R. H. (1960). Nitrogen fixation by cell-free preparations from microorganisms. *Proc. natn. Acad. Sci. USA*, **46**, 726–33.

Shethna, Y. I. (1970). Non-heme (iron–sulfur) proteins of *Azotobacter vinelandii*. *Biochim. biophys. Acta*, **205**, 58–62.

Shethna, Y. I., DerVartanian, D. V. & Beinert, H. (1968). Non-heme (iron–sulfur) proteins of *Azotobacter vinelandii*. *Biochem. biophys. Res. Comm.*, **31**, 862–8.

Shethna, Y. I., Stombaugh, N. A. & Burris, R. H. (1971). Ferredoxin from *Bacillus polymyxa*. *Biochem. biophys. Res. Comm.*, **42**, 1108–16.

Shethna, Y. I., Wilson, P. W. & Beinert, H. (1966). Purification of a non-heme iron protein and other electron transport components from *Azotobacter* extracts. *Biochim. biophys. Acta*, **113**, 225–34.

Smillie, R. M. (1965*a*). Isolation of two proteins with chloroplasts ferredoxin activity from a blue-green alga. *Biochem. biophys. Res. Comm.*, **20**, 621–9.

Smillie, R. M. (1965*b*). Isolation of phytoflavin, a flavoprotein with chloroplast ferredoxin activity. *Pl. Physiol.*, **40**, 1124–8.

Smith, R.V. & Evans, M. C.W. (1970). Soluble nitrogenase from vegetative cells of the blue-green alga *Anabaena cylindrica. Nature, Lond.*, **225**, 1253–4.

Smith, R., Noy, R. J. & Evans, M. C. W. (1971). Physiological electron donor systems to the nitrogenase of the blue-green alga *Anabaena cylindrica. Biochim. biophys. Acta*, **253**, 104–9.

Sober, H. (1968). *Handbook of Biochemistry Selected Data Molecular Biology*, pp. J-27. The Chemical Rubber Co.: Cleveland, Ohio.

Van Lin, B. & Bothe, H. (1972). Flavodoxin from *Azotobacter vinelandii. Arch. Mikrobiol.*, **82**, 155–72.

Watanabe, A. & Yamamoto, Y. (1967). Heterotrophic nitrogen fixation by the blue-green alga *Anabaenopsis circularis. Nature, Lond.*, **214**, 738.

Winter, H. C. & Arnon, D. I. (1970). The nitrogen fixation system of photosynthetic bacteria (I. Preparation and properties of a cell-free extract from *Chromatium*). *Biochim. biophys. Acta*, **197**, 170–9.

Witz, H. F., Detroy, R. W. & Wilson, P. W. (1967). Nitrogen fixation by growing cells and cell-free extracts of the Bacillaceae. *Arch. Mikrobiol.*, **55**, 369–81.

Wolfe, R. S. & O'Kane, D. J. (1953). Cofactors of the phosphoroclastic reaction of *Clostridium butyricum. J. biol. Chem.*, **205**, 755–65.

Wong, P., Evans, H. J., Klucas, R. & Russell, S. (1971). Investigations into the pathway of electron transport to the nitrogenase from nodule bacteroids. *Plant and Soil, Special Volume*, 525–43.

Yamanaka, T., Takenami, S., Nada, K. & Okunuki, K. (1969). Purification and some properties of ferredoxin derived from the blue-green alga, *Anacystis nidulans. Biochim. biophys. Acta*, **180**, 196–8.

Yates, M. G. (1970). Effect of non-haem iron proteins and cytochrome c from *Azotobacter* upon the activity and oxygen sensitivity of *Azotobacter* nitrogenase. *FEBS Lett.*, **8**, 281–5.

(1971). Electron transport to nitrogenase in *Azotobacter chroococcum.* Purification and some properties of NADH dehydrogenase. *Eur. J. Biochem.*, **24**, 347–57.

(1972). Electron transport to nitrogenase in *Azotobacter chroococcum*: *Azotobacter* flavodoxin hydroquinone as an electron donor. *FEBS Lett.*, **27**, 63–7.

Yates, M. G. & Daniel, R. M. (1970). Acetylene reduction with physiological electron donors by extracts and particulate fractions from nitrogen-fixing *Azotobacter chroococcum. Biochim. biophys. Acta*, **197**, 161–9.

Yoch, D. C. (1972). The electron transport system in nitrogen fixation by *Azotobacter*. IV. Some oxidation–reduction properties of azotoflavin. *Biochem. biophys. Res. Comm.*, **49**, 335–42.

Yoch, D. C. (1973). Purification and properties of two ferredoxins from the nitrogen-fixing bacterium *Bacillus polymyxa. Archs. Biochem. Biophys.*, **158**, 633–40.

Yoch, D. C. & Arnon, D. I. (1970). The nitrogen fixation system of photosynthetic bacteria. II. *Chromatium* nitrogenase activity linked to photochemically generated assimilatory power. *Biochim. biophys. Acta*, **197**, 180–4.

Yoch, D. C. & Arnon, D. I. (1972). Two biologically active ferredoxins from the aerobic nitrogen-fixing bacterium, *Azotobacter vinelandii*. *J. biol. Chem.*, **247**, 4514–20.

Yoch, D. C., Benemann, J. R., Arnon, D. I., Valentine, R. C. & Russell, S. A. (1970). An endogenous electron carrier for the nitrogenase system of *Rhizobium* bacteroids. *Biochem. biophys. Res. Comm.*, **38**, 838–42.

Yoch, D. C., Benemann, J. R., Valentine, R. C. & Arnon, D. I. (1969). The electron transport system in nitrogen fixation by *Azotobacter*. II. Isolation and function of a new type of ferredoxin. *Proc. natn. Acad. Sci. USA*, **64**, 1404–10.

Yoch, D. C. & Valentine, R. C. (1972). Ferredoxins and flavodoxins of bacteria. *Ann. Rev. Microbiol.*, **26**, 139–62.

27. Relationship between nitrogenase systems and ATP-yielding processes

R. O. D. DIXON

The energy requirements for the fixation of nitrogen have been considered in detail by Wilson & Burris (1947), Bayliss (1956) and at the last IBP meeting by Bergersen (1971*a*). The purpose of these discussions has been to try to arrive at a figure for the amount of substrate utilised for each mole of nitrogen fixed.

The fixation of nitrogen by the equation given below is exergonic when standard conditions are used.

$$N_2 + 3H_2 = 2NH_3aq \quad \Delta G = -53 \text{ kjoules} \tag{1}$$

This shows that, if electrons are provided at the potential of the hydrogen electrode, the reduction of nitrogen will occur spontaneously provided a suitable catalyst is supplied. Thus, in micro-organisms, it would seem that energy from the oxidation of substrate should only be required to produce electrons at this low potential. However, in order for nitrogen to be fixed by the nitrogenase complex, energy, from ATP, must be supplied in addition to the energy of the low potential electrons.

Nitrogen may not be reduced in one step. It has been postulated (Burris, 1971) that the reduction of nitrogen occurs in two-electron steps giving di-imide, hydrazine and ammonia complexes on the enzyme active site. Although the complete reaction of the reduction of nitrogen to ammonia is exergonic, it has been pointed out (Pratt, 1962; Leigh, 1971) that the first steps themselves may be endergonic and it could be that energy is required at these points. However nitrogen has been reduced non-enzymically under mild conditions, one atmosphere and, presumably, room temperature, without provision of energy other than that in the low potential electrons ($NaBH_4$) (Hill & Richards, 1971).

If we knew that there was an energy requirement at any stage, we would have no better basis for calculating the amount of ATP required as this would depend upon the efficiency with which the free energy of hydrolysis of the terminal phosphate bond was used. Very similar reactions may require different amounts of ATP. An instance, of interest to those concerned with nitrogen metabolism, is the formation of carbamyl phosphate:

$$NH_4^+ + HCO_3^- + 2ATP = NH_2CO.OPO(OH)_2 + 2ADP + H_3PO_4 \quad (2)$$

$$NH_4^+ + HCO_3^- + ATP = NH_2CO.OPO(OH)_2 + ADP + H_2O \quad (3)$$

Equation 2 is catalysed by a synthetase which occurs in liver and which requires N-acetyl glutamate as a cofactor. Equation 3 is that of a reaction catalysed by carbamyl kinase present in some bacteria and which has no cofactor requirement. These equations are obtained from a stimulating review about ATP utilisation by Banks & Vernon (1970).

It is plain from the above discussion that we cannot calculate a theoretical requirement for ATP. However, if we know the number of ATP molecules hydrolysed per mole of nitrogen fixed, the number of electrons required in the reduction process and if the metabolic pathways producing the electrons and ATP are known, we should be able to estimate the amount of substrate that must be utilised in order to fix one mole of nitrogen. Unfortunately the answers to these questions are not known in most systems. Therefore this paper must be concerned with discussing our knowledge so far, and with attempting to assess the reasons for the difficulties which are hindering our gain of knowledge in this field.

The ATP requirement

Partitioning of electrons

The utilisation of ATP is connected to the transfer of electrons from a donor molecule through the nitrogenase system on to an acceptor molecule. The uncertainty with regard to the number of electrons required, and hence ATP consumed, is due to the fact that both nitrogen and protons act as electron acceptors in the natural system. The total amount of ATP expended for each mole of nitrogen fixed will depend upon the partitioning of electrons between nitrogen and protons.

The ratio of electrons donated to nitrogen and protons is governed by a number of factors: ATP concentration, pH, substrate and substrate concentration.

Silverstein & Bulen (1970) assayed hydrogen evolution and ammonia production from a nitrogenase preparation of *Azotobacter* using concentrations of ATP ranging from 0.025 M to 0.125 M. They found that the curves, obtained by plotting the rates of hydrogen evolution and ammonia production, differed over this ATP concentration range. The curve obtained for hydrogen evolution was hyperbolic while that for ammonia production was sigmoidal. This indicates that the partitioning of the electrons between the two acceptors differs according to the ATP concentration. About 80 % of the electrons were donated to protons at the lowest ATP concentration and just over 50 % at the highest.

Hydrogen evolution, as a manifestation of nitrogenase activity is affected by pH (Burns & Bulen, 1965). Using acrylonitrile instead of nitrogen as the alternative substrate to protons, Fuchsman & Hardy (1972) have shown that the partitioning of electrons between substrate and protons is also affected by pH.

If no alternative substrate is supplied, electrons are all donated to protons. As the concentration of nitrogen increases the proportion of electrons donated to protons decreases (Hadfield & Bulen, 1969). The proportion of electrons donated to protons when nitrogenase is under an atmosphere of 100 % nitrogen is between 25 % and 33 % (Mortensen, 1966; Bulen & LeComte, 1966; Hadfield & Bulen, 1969). The proportion of hydrogen evolved to nitrogen reduced does not appear to differ significantly from this when the nitrogen tension is reduced to 0.75 atmosphere (Hadfield & Bulen, 1969). Thus in-vitro experiments show that a quarter to a third of the electrons are donated to protons and a similar proportion of the ATP expended on nitrogen fixation must be used for this process.

I have been unable to find figures in the literature for the partitioning of electrons in in-vivo experiments where 80 % or 100 % nitrogen was used. This is because in most cases hydrogenases preclude the estimation of hydrogen evolved from the nitrogenase system either by also evolving hydrogen, as in *Clostridium*, or by taking up the hydrogen evolved by nitrogenase, as in *Azotobacter* and some *Rhizobium* bacteroids. Where it is possible to examine this, in systems such as in the soya bean nodule, which does not have significant hydrogenase activity, it has not been done presumably because of the cost of the ^{15}N necessary for short-term nitrogen fixation experiments. Where figures are available they are for nitrogen tensions of about 10 %. Thus it is not possible to say what the partitioning of the electrons is likely to be in in-vivo systems. This is a pity because differences in partitioning may well be part of the explanation of the differences in ATP demand by in-vivo systems which are discussed below.

ATP per electron pair

Since the discovery of the requirement of ATP for the fixation process (Hardy & D'Eustachio, 1964), there have been many attempts to determine the stoichiometric ratio of ATP utilised per electron or electron pair transferred through the nitrogenase system. As discussed above there is no way of predicting the requirement for ATP. Furthermore the mechanism of action of ATP is not understood. This is discussed elsewhere (Chapters 17 and 25). Recent experimental evidence indicates that it is required for the association of subunits of Component II of *Klebsiella aerogenes* and it is postulated that the complexes so formed may be important for the functioning of nitrogenase (Thorneley & Eady, 1973). Whether this is the sole requirement for ATP remains to be discovered.

In-vitro studies give varying estimates of the amount of ATP hydrolysed for each electron pair transferred. They vary from two (Jeng, Morris & Mortensen, 1970) to about four or five (Bulen & LeComte, 1966; Mortensen, 1966; Winter & Burris, 1968; Hadfield & Bulen, 1969). This ratio, however, varies not only between laboratories but also with temperature (Hadfield & Bulen, 1969), pH (Winter & Burris, 1968) and with different ratios of nitrogenase components (Kelly, 1969; Ljones & Burris, 1972). This has led to the suggestion that there is not in fact any stoichiometric relationship between ATP hydrolysis and electron transfer (Ljones & Burris, 1972) in spite of the fact that they both depend upon each other.

It is of interest to find a plausible reason for the discrepancies of the estimations of the ATP/2e ratios between various investigators. Care has been taken in these investigations to correct for reductant-independent ATPase activity. However Kelly (1969) has found that nitrogenase Component I alone has a dithionite-dependent ATPase activity not connected with electron transfer. This complication cannot be taken into account when preparations are used in which the fractions of nitrogenase are not separated and neither is it possible to determine the effect of the addition of Component II upon this activity.

Ljones & Burris (1972) have found that with a constant Component II concentration and increasing Component I concentration the ATP/2e ratio increased from around 4 to over 11. The hydrolysis of ATP remained at a constant level once a 1:1 ratio of proteins was achieved. The increasing ATP/2e ratio is caused by inhibition of electron transfer at the higher levels of Component I. In the reverse case, when Component I was held constant and the amounts of Component II were varied, the ATP/2e ratio declined slightly from 5 to 4 at the highest Component II concentration and electron transfer was not inhibited by excess Component II. When excess Component I is added to the reaction mixture the hydrolysis of ATP may be said to be uncoupled from electron transfer. Similarly the activity shown by reduced Component I alone is uncoupled.

The differences between various laboratories in estimates of ATP/2e could possibly be explained by such effects. However in intact cells, where ATP concentration, pH and nitrogenase component concentrations are controlled there may be a stoichiometric ATP/2e ratio. Even if there is more than one reaction in which ATP participates each should have its own stoichiometry.

The figure of practical interest is the amount of ATP expended on nitrogen fixation *in vivo* as this will determine, in large part, the amount of the organism's resources that will have to be expended on nitrogen fixation. A few attempts have been made to ascertain this but due to various difficulties few estimates have been made.

Comparing chemostat cultures of *Clostridium pasteurianum* growing on ammonium or nitrogen, Daesch & Mortensen (1968) found that the rate at which sucrose was utilised by the cultures was the same regardless of nitrogen source. The amount of growth, Y_{sub} (g dry wt cells/mole of substrate consumed) was much lower in the case of the nitrogen fixing culture ($Y_{sub} = 36$) than in the culture grown with ammonia in excess ($Y_{sub} = 63$). The fermentation products of the two cultures were the same. Knowing the reactions which led to these products a figure of 7.64 mole of ATP/mole of sucrose was calculated. This gives a Y_{ATP} of 10.3 for the cells grown with ammonia, a figure in excellent agreement with the average value of 10.5 for cells (Bauchop & Elsden, 1960). Using a similar calculation for the nitrogen-fixing cells the Y_{ATP} is reduced to 6.3. Taking the figure of 10.3 as that being the ATP actually used for the growth of the nitrogen fixing cells it was possible to calculate the amount of ATP used for nitrogen fixation as 20 moles ATP consumed/mole N_2 fixed. Comparable experiments, with *Klebsiella pneumoniae*, by Hill (Postgate, 1971) using similar calculations gave a value approaching 30 moles ATP consumed/mole N_2 fixed.

Senez (1962) working with *Desulphovibrio desulfuricans* found a similar lower value of Y_{sub} for nitrogen-fixing batch cultures than for cultures grown with combined nitrogen. However, unaware of the ATP requirement for nitrogen fixation, he attributed the lower value to uncoupled substrate (lactate) metabolism due to nitrogen shortage in these organisms. Little or no uncoupled substrate catabolism occurred in the experiments of Daesch & Mortensen.

The large difference between the Y_{sub} values of organisms grown on combined nitrogen and those fixing nitrogen indicates that a large proportion of the organisms resources are devoted to nitrogen fixation. In the case of *C. pasteurianum* (Daesch & Mortensen, 1968) it was estimated that 39 % of the ATP produced was expended on this one aspect of the cell's metabolism.

Dalton & Postgate (1969) worked with *Azotobacter chroococcum* in continuous culture. Here again there was a lower yield of cells (Y_{sub}) with nitrogen-fixing cells than with ammonia grown cells. In *Azotobacter* respiratory protection maintains the nitrogenase in an aerobic environment (Dalton & Postgate, 1968). Thus differences in Y_{sub} will be due to substrate consumption for this as well as the ATP consumption of the nitrogen-fixing system. With increasing growth rate the need for respiratory protection declines. By extrapolating their results to an infinite growth rate they obtained a yield difference between the nitrogen and ammonia cultures which thus represented the carbon consumed in metabolism leading to ATP production for the nitrogenase. Using the figure for Y_{ATP} of 10.5, this yield difference gave a value of 4 to 5 moles ATP/mole N_2 fixed. This low value can be partly explained by the fact that *Azotobacter* contains a hydrogenase which takes up the hydrogen which is evolved from the nitrogenase and oxidises it with the production of ATP (Hyndman, Burris & Wilson, 1953), thus saving some or all of the ATP used for the production of this hydrogen. In this sense the fixation of nitrogen by *Azotobacter* is more efficient than that of the anaerobes, but it ignores the substantial consumption of carbon in respiratory protection.

In considering the ATP requirement for nitrogen fixation in whole organisms, any changes in the assimilatory pathway for the ammonia produced must be taken into account. Nitrogen-fixing organisms appear to assimilate most of the ammonia by way of glutamine synthetase and glutamate synthase (Nagatini, Shimzu & Valentine, 1971) a process which requires ATP where the more normal route by glutamic dehydrogenase does not. Where this route of nitrogen assimilation occurs, two extra moles ATP will be consumed per mole of nitrogen. However Dilworth (personal communication) has estimated the activities of enzymes concerned with ammonia assimilation in lupin nodules and came to the conclusion that the route by glutamine synthetase and glutamate synthase could not account for all the ammonia uptake.

In contrast to the results with free-living bacteria, Gibson (1966) was unable to find any statistical difference between the carbohydrate requirements of nodulated *Trifolium subterraneum* and plants assimilating the same amount of nitrogen from ammonium nitrate. However, as pointed out by Gibson and subsequently more fully calculated by Bergersen (1971*a*), the reduction of nitrate requires a high expenditure of energy which is more or less equal to that necessary for nitrogen fixation. Thus while Gibson's experiments do not enable us to arrive at conclusions with regard to the energy, or carbohydrate, expenditure of the plant on nitrogen fixation, they show that these demands are not inordinate.

In more direct experiments bearing on the amount of substrate required for the fixation of nitrogen Minchin & Pate (1973) have examined the carbon economy of *Pisum sativum* with particular attention to the economy of the root nodules. They found that there was no significant difference in the respiration of nodulated roots and of roots taking up nitrate with the same nitrogen flux. Most of this nitrate would be reduced in the roots (Wallace & Pate, 1965). This is in accord with Gibson's results and agrees with a value of about 4ATP/2e transferred. They also found that during their experiment, in which 21.8 ml of nitrogen were fixed, that 65.1 ml of oxygen were consumed and 75.4 ml of carbon dioxide were evolved. If one assumes a P/O ratio of 2.0 these figures give a value of 12 moles of ATP/N_2 fixed using the figure for oxygen uptake. Oxygen uptake will also have to contribute to respiration concerned with nodule maintenance and growth so that the actual value will be less than this. Similar calculations using the quoted figure for carbon dioxide evolution and based on the metabolism of glucose by the Embden–Meyerhof pathway and the Krebs' cycle give a value of 9.5 moles ATP/N_2 fixed. The evolved carbon dioxide however is a low estimate as carbon dioxide will be fixed in carboxylation reactions concerned with the generation of C_4 nitrogen acceptors. Taking this into account one can calculate a value of 11.6 moles ATP/N_2 fixed. This again ignores that respiration used for growth and maintenance.

There is reason to believe that the rates of respiration as measured may be low compared to the values for nitrogen fixation. The latter were obtained from undisturbed plants while respiration measurements were done on roots disturbed by the attachment of microrespirometers. Experiments (Dixon, unpublished) have shown that disturbance of the nodule root system can decrease acetylene reduction to one third of its original rate. While respiration may not be as sensitive, it is reasonable to suppose that it also will have decreased. Such a decrease even if small could be equivalent to the ATP which would be expected to be needed for growth and maintenance.

Thus neither in-vitro nor in-vivo studies give us an unequivocal answer to the question of how much ATP is necessary for nitrogen fixation. If we take the minimum value 2ATP/2e and 25 % electrons donated to protons, then 8 moles ATP/N_2 fixed will be the requirement for actual fixation with possibly two extra moles ATP required, for glutamine synthetase, and for subsequent ammonia assimilation.

The production of ATP

The electrons required by nitrogenase must be at a low oxidation reduction potential and thus can only be provided by a limited number of the redox reactions available in respiratory metabolism. The free energy of hydrolysis of ATP however only varies between narrow limits *in vivo* as the ADP/ATP ratio, and thus ATP concentration, and its ionic environment are usually closely controlled. Any reaction which will generate ATP will serve the purpose provided that it is not spacially separated from the nitrogenase site. There is no reason to suppose that organisms will have developed special systems for ATP production for nitrogen fixation. The reactions which contribute ATP will depend upon the different metabolic pathways followed by the organisms concerned.

The photosynthetic bacteria and the blue-green algae may obtain their ATP as a result of photophosphorylation. Without wishing to go into the question of whether nitrogenase is solely sited in the heterocysts, in the algae which possess them, which is discussed elsewhere (Chapter 9), it has been shown that a large proportion, at least of the nitrogenase, resides in these cells (Wolk & Wojcuich, 1971). The heterocysts only contain the apparatus for the catalysis of Photosystem I (Donze, Haveman & Schiereck, 1972) a finding predicted by Cox & Fay (1969). They contain very little or no phycocyanin, the accessory pigment necessary for Photosystem II (Cox & Fay, 1969), and they do not fix carbon dioxide (Wolk, 1968). These facts together with the lack of sensitivity to DCMU (Cox & Fay, 1969; Haystead & Stewart, 1971) an inhibitor of Photosystem II, show that the negative finding of Donze *et al.* is real and not due to technique. There is strong presumptive evidence then that, in the light, ATP for nitrogen fixation is produced by cyclic photophosphorylation by Photosystem I.

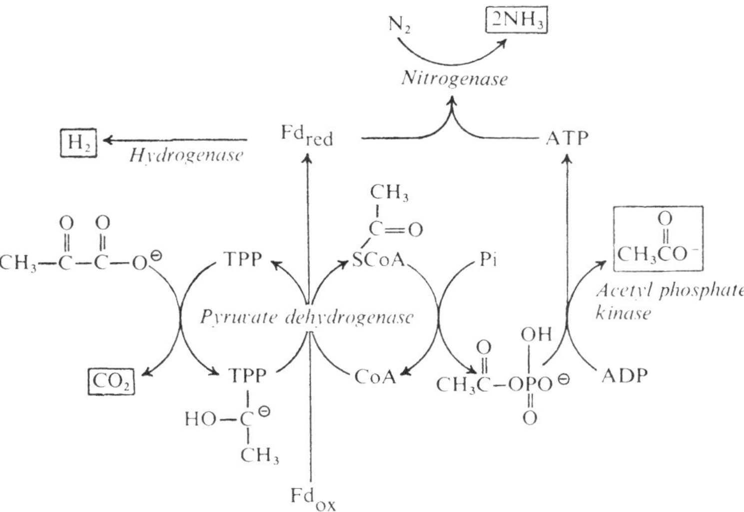

Fig. 27.1. The phosphoroclastic reaction in nitrogen fixation in clostridia. Fd indicates ferredoxin. Taken from 'The pathways of nitrogen fixation', Benemann, J. R. & Valentine, R. C. *Advances in Microbial Physiology*, **8**, 71 (1972).

Bennet, Rigopoulos & Fuller (1964) showed that nitrogen fixation in *Chromatium* was strongly dependent upon light and the provision of pyruvate. Hydrogen evolution was depressed in the light compared with the rate in the dark. Using the argument following it can be inferred from this that in *Chromatium*, in the light, the reducing power is provided by the pyruvate and that photophosphorylation is providing the ATP.

The phosphoroclastic reaction of pyruvate (Fig. 27.1) produces one ATP for every two electrons. If the requirement for ATP is higher than this ratio then there is an imbalance with surplus electrons. These electrons are removed by hydrogenase with the evolution of hydrogen in some organisms such as *Chromatium*. Thus, in the light, extra ATP is supplied by photophosphorylation, there is no surplus of electrons and hydrogen evolution will be mainly due to the nitrogenase itself. When pyruvate is the source of both ATP and electrons, in the dark, hydrogen evolution will be increased because of the imbalance of demand and supply of ATP and electrons. Extracts of *Chromatium* can support nitrogen fixation by photogenerated ATP (Yoch & Arnon, 1970).

In aerobic organisms, ATP-yielding processes and the production of electrons for reductive processes need not be so carefully balanced because excess electrons are passed to oxygen with concomitant oxidative phosphorylation. As far as *Azotobacter* is concerned, the supply of ATP would be in excess if electron transfer was tightly coupled to phosphorylation even when nitrogen was being fixed. This is because a large component of the respiration is concerned with the respiratory protection of the nitrogenase (Dalton & Postgate, 1968). Ackrell, Erickson & Jones (1972) have investigated the oxidative phosphorylation of electron transfer particles isolated from *Azotobacter* grown under conditions of high and low oxygen partial pressures. In the particles from bacteria grown under high oxygen partial pressure, when respiratory protection is most needed, electron transfer was much less tightly coupled to phosphorylation than in similar particles from bacteria grown under low oxygen partial pressures. Presumably even when largely uncoupled, the high rate of respiration produces sufficient ATP for normal cell requirements. At high oxygen partial pressure nitrogenase is also switched off (Drozd & Postgate, 1970), due to a conformational change making it less sensitive to oxygen inactivation. There is thus no need to postulate any other source of ATP for the nitrogenase of *Azotobacter* other than that derived from respiration. The oxidation of the hydrogen derived from nitrogen, mentioned earlier, will have a sparing effect on the carbon metabolised. However the proportion of this is likely to be small. Some *Rhizobium* bacteroids similarly oxidise the hydrogen from nitrogenase with the formation of ATP. The activity of the nitrogenase in these organisms may be limited by the supply of substrate (see below) and in this case the sparing effect of the ATP produced by hydrogen oxidation may be significant (Dixon, 1972).

The carbon compounds which are metabolised to provide ATP and electrons for nitrogenase in legume nodules are not certainly known. There is evidence that the rate of nitrogenase activity, as measured by acetylene reduction, is closely controlled by the rate of photosynthesis of the host (e.g. Hardy, Holsten, Jackson & Burns, 1968). Soya bean nodule bacteroids contain a high concentration of poly-β-hydroxybutyrate (PBH) which may be up to 50 % of their dry weight (Klucas & Evans, 1968) but there is evidence that this is not used to support nitrogen fixation to any substantial degree (Wong & Evans, 1971). Although the concentration of PBH remained high in the bacteroids, nitrogenase activity declined rapidly when nodules were excised or the intact plants placed in the dark. *Iso*citrate lyase, needed for the functioning of the glyoxylate cycle which is necessary for the catabolism of β-hydroxybutyrate was not active until the nitrogenase activity had declined to almost zero in ageing nodules. Even when PBH could be shown to be metabolised, nitrogenase activity did not increase.

Failure to utilise storage PBH is in accord with the finding that host photosynthesis rates directly affect the nitrogenase activity. What is more surprising is the finding that glucose does not appear to support nitrogenase activity in soya bean bacteroids. Rigaud, Bergersen, Turner & Daniel (1973) found that glucose did not support nitrogenase activity in soya bean bacteroids and in fact acted as an osmoticum which suggests that the bacteroids are either impermeable or very poorly permeable to glucose. Succinate and pyruvate however supported nitrogenase activity in these bacteroids. Pea root nodule bacteroids similarly do not utilise glucose (R. O. D. Dixon, unpublished) and can use pyruvate and succinate (Dixon, 1968). The results, obtained by using whole bacteroids *in vitro*, suggest that the bacteroids function best for nitrogen fixation when supplied with organic acids rather than with more primary substrates such as glucose or β-hydroxybutyrate. The suggestion has been made that, as the bacteroids contain enzymes capable of metabolising glucose and can utilise it after an incubation period of 7–10 hours, the glucose transport mechanism had been lost during isolation of the bacteroids (cited in review by Bergersen 1971*b*).

A number of studies have indicated that the nitrogenase activity of legume nodules is related to the availability of oxygen (Bergersen, 1971*b*) and thus respiration and oxidative phosphorylation can be presumed to supply the ATP necessary for fixation. In this context, the role of leghaemoglobin has been shown to be important. Experiments by Bergersen, Turner & Appleby (1973) were done using soya bean bacteroids and leghaemoglobin. Oxygen uptake and nitrogenase activity were followed with different degrees of aeration, controlled by the degree of shaking, in the presence and absence of leghaemoglobin. The results showed that when leghaemoglobin was supplied, the rate of oxygen consumption increased by a factor of two while acetylene reduction was increased by as much as a factor of 21. Subsequent experiments by Wittenberg, Bergersen, Appleby & Turner (1974) have probed this system further. In these experiments a number of oxygen-binding proteins were tried in place of leghaemoglobin. Twelve of these augmented both oxygen uptake and acetylene reduction by bacteroid suspensions. The role of leghaemoglobin is not then specific and the ability of the bacteroid suspensions to utilise other oxygen-binding proteins lends further support to the idea that leghaemoglobin acts as an oxygen carrier, facilitating the diffusion of oxygen through the solution to the bacteroid surface.

The bacteroids have a high rate of oxygen uptake even in the absence of leghaemoglobin but this is relatively ineffective in supporting nitrogenase activity. The addition of leghaemoglobin, and other oxygen-binding proteins, brings about an increment of oxygen uptake which increases nitrogenase activity by a large amount. In these experiments which differed from the earlier ones, the ratio of the increment of oxygen uptake, due to leghaemoglobin, to the increment of acetylene reduction was about 2:1. Taking a figure of 4 ATP/2e transferred it was calculated that the incremental oxygen was utilised with a P/O ratio of between 1 and 2. As a result of these experiments it was postulated that the oxygen uptake without leghaemoglobin maintains a low oxygen partial pressure at the bacteroid surface and that oxygen is taken up by a terminal oxidase of high oxygen affinity but which is ineffective for nitrogenase activity. When leghaemoglobin is present, the oxygen tension at the bacteroid surface is raised and extra oxygen can be taken up by a terminal oxidase of lower oxygen affinity but which is effective for nitrogenase activity.

There are at least two reasons for a difference in effectiveness for nitrogenase activity by two different terminal oxidase systems.

1. The oxidase ineffective for nitrogen may be spacially separated from it.

2. The ineffective oxidase system may have a much lower P/O ratio. It would be interesting to know whether the steady-state levels of ATP differ in the bacteroids provided with leghaemoglobin from those without.

The provision of ATP by anaerobic micro-organisms which catalyse the phosphoroclastic reaction and which evolve hydrogen or reduce organic acceptors in order to rid themselves of excess electrons is fairly clear and is summarised in Fig. 27.1.

This review has served to show our ignorance rather than our degree of enlightenment on this topic. It may be that the demand for ATP in the nitrogen-fixing system will only be known when the reasons for its involvement are more precisely understood.

References

Ackrell, B. A. C., Erickson, S. K. & Jones, C. W. (1972). The respiratory-chain NADPH dehydrogenase of *Azotobacter vinelandii*. *Eur. J. Biochem.*, **26**, 387–92.

Banks, B. E. C. & Vernon, C. A. (1970). Reassessment of the role of ATP *in vivo*. *J. theoret. Biol.*, **29**, 301–26.

Bauchop, T. & Elsden, S. R. (1960). The growth of micro-organisms in relation to their energy supply. *J. gen. Microbiol.*, **23**, 457–69.

Bayliss, N. S. (1956). The thermochemistry of biological nitrogen fixation. *Aust. J. biol. Sci.*, **9**, 364–70.

Bennett, R., Rigopoulos, N. & Fuller, R. C. (1964). The pyruvate phosphoroclastic reaction and light dependent nitrogen fixation in bacterial photosynthesis. *Proc. natn. Acad. Sci. USA*, **52**, 762–8.

Bergersen, F. J. (1971*a*). The central reactions of nitrogen fixation. *Plant and Soil, Special Volume*, 511–24.

Bergersen, F. J. (1971*b*). Biochemistry of symbiotic nitrogen fixation in legumes. *Ann. Rev. Pl. Physiol.*, **22**, 121–40.

Bergersen, F. J., Turner, G. L. & Appleby, C. A. (1973). Studies of the physiological role of leghaemoglobin in soybean root nodules. *Biochim. biophys. Acta*, **292**, 271–82.

Bulen, W. A. & LeComte, H. R. (1966). The nitrogenase system from *Azotobacter*: Two enzyme requirement for N_2 reduction, ATP-dependent H_2 evolution and ATP hydrolysis. *Proc. natn. Acad. Sci. USA*, **56**, 979–86.

Burns, R. C. & Bulen, S. A. (1965). ATP-dependent hydrogen evolution by cell-free preparations of *Azotobacter vinelandii*. *Biochim. biophys. Acta*, **150**, 437–45.

Burris, R. H. (1971). Fixation by free-living micro-organisms: Enzymology. In *The Chemistry and Biochemistry of Nitrogen Fixation* (ed. Postgate, J. R.), pp. 105–60. Plenum Press: London, New York.

Cox, R. M. & Fay, P. (1969). Special aspects of nitrogen fixation by blue-green algae. *Proc. R. Soc. Lond. B*, **172**, 357–66.

Daesch, G. & Mortensen, L. E. (1968). Sucrose catabolism in *Clostridium pasteurianum* and its relation to N_2 fixation. *J. Bacteriol.*, **96**, 346–51.

Dalton, H. & Postgate, J. R. (1968). Effect of oxygen on growth of *Azotobacter chroococcum* in batch and continuous culture. *J. gen. Microbiol.*, **54**, 463–73.

Dalton, H. & Postgate, J. R. (1969). Growth and physiology of *Azotobacter chroococcum* in continuous culture. *J. gen. Microbiol.*, **56**, 307–19.

Dixon, R. O. D. (1968). Hydrogenase in pea root nodule bacteroids. *Arch. Mikrobiol.*, **62**, 272–83.

Dixon, R. O. D. (1972). Hydrogenase in legume root nodule bacteroids: occurrence and properties. *Arch. Mikrobiol.*, **85**, 193–201.

Donze, M., Haveman, J. & Shiereck, P. (1972). Absence of photosystem 2 in heterocysts of the blue green algae *Anabaena*. *Biochim. biophys. Acta*, **256**, 157–61.

Drozd, J. & Postgate, J. R. (1970). Interference by oxygen in the acetylene-reduction test for aerobic nitrogen-fixing bacteria. *J. gen. Microbiol.*, **66**, 427–9.

Fuchsman, W. H. & Hardy, R. W. F. (1972). Nitrogenase-catalysed acrylonitrile reductions. *Bioinorgan. Chem.*, **1**, 195–213.

Gibson, A. H. (1966). The carbohydrate requirements for symbiotic nitrogen fixation: A 'whole-plant' growth analysis approach. *Aust. J. biol. Sci.*, **19**, 499–515.

Hadfield, K. L. & Bulen, W. A. (1969). Adenosine triphosphate requirement of nitrogenase from *Azotobacter vinelandii*. *Biochemistry*, **8**, 5103–8.

Hardy, R. W. F. & D'Eustachio, A. H. (1974). The dual role of pyruvate and the energy requirement in nitrogen fixation. *Biochem. biophys. Res. Comm.*, **15**, 314–18.

Hardy, R. W. F., Holsten, R. D., Jackson, E. K. & Burns, R. C. (1968). The acetylene–ethylene assay for N_2 fixation: laboratory and field evaluation. *Pl. Physiol.*, **43**, 1185–1207.

Haystead, A. & Stewart, W. D. P. (1972). Characteristics of the nitrogenase system of the blue-green alga *Anabaena cylindrica*. *Arch. Mikrobiol.*, **82**, 325–36.

Hill, R. E. E. & Richards, R. L. (1971). Reduction of dinitrogen in aqueous solution. *Nature, Lond.*, **233**, 114–15.

Hyndman, L. A., Burris, R. H. & Wilson, P. W. (1953). Properties of hydrogenase from *Azotobacter vinelandii*. *J. Bacteriol.*, **65**, 522–31.

Jeng, D. Y., Morris, J. A. & Mortensen, L. E. (1970). The effect of reductant in inorganic phosphate release from adenosine 5′-triphosphate by purified nitrogenase of *Clostridium pasteurianum*. *J. biol. Chem.*, **245**, 2809–13.

Kelly, M. (1969). Some properties of purified nitrogenase of *Azotobacter chroococcum*. *Biochim. biophys. Acta*, **171**, 9–22.

Klucas, R. V. & Evans, H. J. (1968). An electron donor system for nitrogenase-dependent acetylene reduction by extracts of soybean nodules. *Pl. Physiol.*, **43**, 1906–12.

Leigh, G. J. (1971). Abiological fixation of molecular nitrogen. In *The Chemistry and Biochemistry of Nitrogen Fixation* (ed. Postgate, J. R.), pp. 19–56. Plenum Press: London, New York.

Ljones, T. & Burris, R. H. (1972). ATP hydrolysis and electron transfer in the nitrogenase reaction with different combinations of iron protein and the Mo-iron protein. *Biochim. biophys. Acta*, **275**, 93–101.

Minchin, F. R. & Pate, J. S. (1973). The carbon balance of a legume and the functional economy of its root nodules. *J. exp. Bot.*, **24**, 259–71.

Mortensen, L. E. (1966). Components of cell free extract of *Clostridium pasteurianum* required for ATP-dependent H_2 evolution from dithionite and for N_2 fixation. *Biochim. biophys. Acta*, **127**, 18–25.

Nagatini, H., Shimizu, M. & Valentine, R. C. (1971). The mechanism of ammonia assimilation in nitrogen fixing bacteria. *Arch. Mikrobiol.*, **79**, 164–75.

Postgate, J. R. (1971). Fixation by free living microbes: Physiology. In *The Chemistry and Biochemistry of Nitrogen Fixation* (ed. Postgate, J. R.), pp. 161–90. Plenum Press: London, New York.

Pratt, J. M. (1962). The fixation of nitrogen. *J. theoret. Biol.*, **2**, 251–8.

Rigaud, J., Bergersen, F. J., Turner, G. L. & Daniel, R. M. (1973). Nitrate dependent anaerobic acetylene-reduction and nitrogen-fixation by soybean bacteroids. *J. gen. Microbiol.*, **77**, 137–44.

Senez, J. E. (1962). Some considerations on the energetics of bacterial growth. *Bacteriol. Rev.*, **26**, 95–107.

Silverstein, R. & Bulen, W. A. (1970). Kinetic studies of the nitrogenase-catalysed hydrogen evolution and nitrogen reduction reactions. *Biochemistry*, **9**, 3809–15.

Thorneley, R. B. F. & Eady, R. R. (1973). Nitrogenase of *Klebsiella pneumoniae*: evidence for an adenosine triphosphate-induced association of the iron sulphur protein. *Biochem. J.*, **133**, 405–8.

Wallace, W. & Pate, J. S. (1965). Nitrate reductase in the field pea (*Pisum arvense* L.). *Annls Bot.* **29**, 655–71.

Wilson, P. W. & Burris, R. H. (1947). The mechanism of biological nitrogen fixation. *Bacteriol. Rev.*, **11**, 41–73.

Winter, H. C. & Burris, R. H. (1968). Stoichiometry of the adenosine triphosphate requirement for N_2 fixation and H_2 evolution by a partially purified preparation of *Clostridium pasteurianum*. *J. biol. Chem.*, **243**, 940–4.

Wittenberg, J. B., Bergersen, F. J., Appleby, C. A. & Turner, G. L. (1974). The role of leghemoglobin in nitrogen fixation by bacteroids isolated from soybean root nodules. *J. biol. Chem.* **249**, 4057–66.

Wolk, C. P. (1968). Movement of carbon from vegetative cells to heterocysts in *Anabaena cylindrica*. *J. Bacteriol.*, **96**, 2138–43.

Wolk, C. P. & Wojcuich, E. (1971). Photoreduction of acetylene by heterocysts. *Planta, Berlin*, **97**, 126–33.

Wong, P. P. & Evans, H. J. (1971). Poly-β-hydroxybutyrate utilisation by soybean (*Glycine max* Merr.) nodules and assessment of its role in maintenance of nitrogenase activity. *Pl. Physiol.*, **47**, 750–5.

Yoch, D. C. & Arnon, D. I. (1970). The nitrogen fixation of photosynthetic bacteria. II. *Chromatium* nitrogenase activity linked to photochemically generated assimilatory power. *Biochim. biophys. Acta*, **197**, 180–4.

28. *Azotobacter* nitrogenase: ATP kinetics and inhibition by high ionic strength*

R. C. BURNS & R. W. F. HARDY

In early work with cell-free nitrogenase systems it was observed that the relationship between nitrogenase activity and nitrogenase concentration was non-linear, resulting in disproportionately low activity at low enzyme levels (Mortenson, 1965; Bulen, Burns & LeComte, 1965). This phenomenon has been referred to as the 'dilution effect' of nitrogenase, and has been attributed to a deficiency in one or the other of the nitrogenase proteins, since it can be overcome by supplementation with iron (Fe) protein or molybdenum–iron (Mo–Fe) protein (Sorger & Trofimenkoff, 1970; Shah, Davis & Brill, 1972). The dilution effect and its correction by supplemental Fe protein is illustrated in Fig. 28.1. Preparations of nitrogenase that do not exhibit this effect may be presumed to contain balanced levels of Mo–Fe protein and Fe protein. Our results show that the dilution effect can be produced in these balanced preparations either by decreasing the ATP concentration or by increasing ionic strength. The kinetics of ATP utilization suggest involvement of two molecules of ATP in nitrogenase turnover. The implications of these observations for the nitrogenase reaction are discussed.

Methods

The purification of *Azotobacter vinelandii* nitrogenase through the protamine sulfate step and methods for the determination of hydrogen evolution, acetylene reduction and N_2 reduction have been described (Burns & Hardy, 1972*b*). Except where indicated otherwise, reaction mixtures contained 5 μmoles ATP, 10 μmoles $MgCl_2$, 60 μmoles creatine phosphate, 20 μmoles dithionite, 25 μmoles TES, pH 7.0, 10 units of creatine phosphokinase, and nitrogenase preparation to 1.0 ml; the temperature was 30 °C. For acetylene reduction 0.05 atm C_2H_2 was used, and for N_2 reduction 1.0 atm N_2 was used. When low concen-

* A preliminary report was presented at the 164th Annual Meeting of the American Chemical Society, New York, 1972; *Biol. Abstr.*, **201** (Burns & Hardy, 1972*a*).

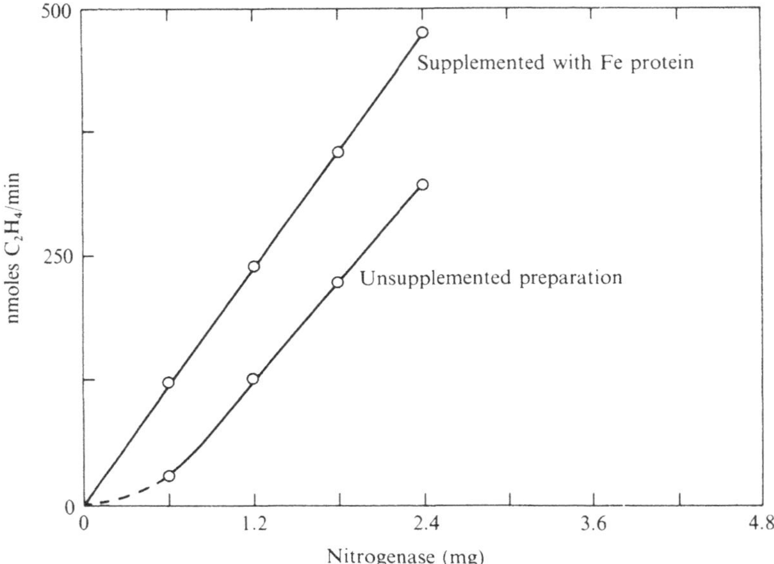

Fig. 28.1. Dilution effect in nitrogenase activity and its correction by supplemental Fe protein.

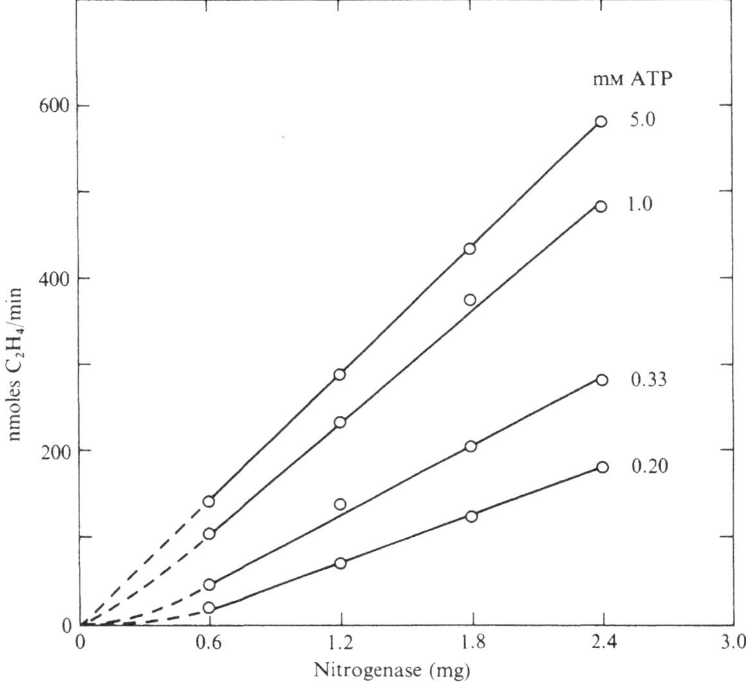

Fig. 28.2. Influence of ATP concentration on the dilution effect.

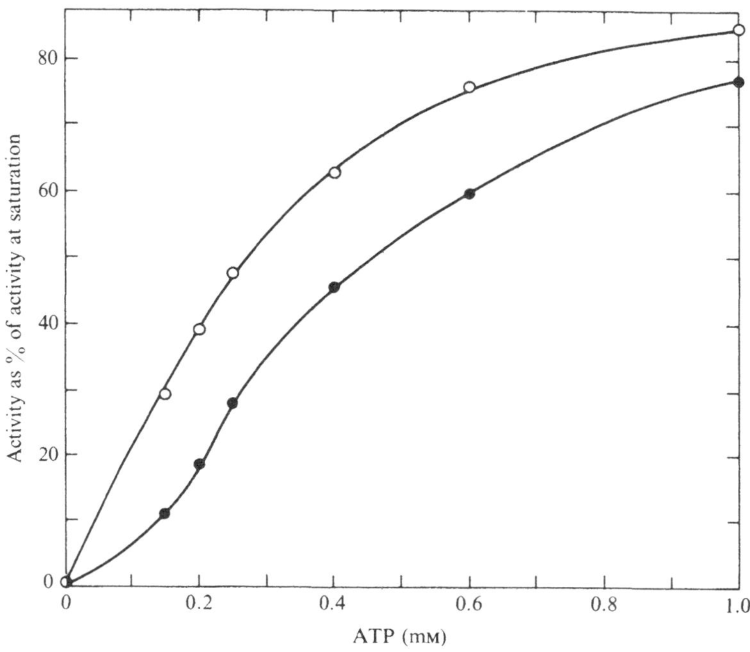

Fig. 28.3. Effect of nitrogenase concentration on ATP saturation kinetics. Nitrogenase: ●, 0.6 mg; ○, 2.4 mg.

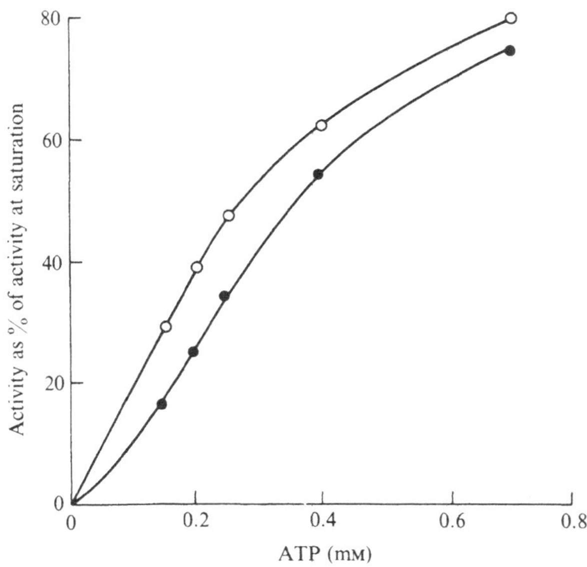

Fig. 28.4. Effect of pH on ATP saturation kinetics; ○, pH 8.4; ●, pH 5.7.

trations of ATP were employed, the concentration of kinase was optimized to assure that it was not limiting. Similarly the effect of lowered $MgCl_2$ concentration was tested in selected low-ATP reactions; no effects were observed.

Results

ATP: concentration effects and kinetics

The dilution effect of nitrogenase was strongly influenced by the concentration of ATP (Fig. 28.2). The nitrogenase preparation which was in balance at the saturation level of 5 mM ATP showed an increasingly intensive dilution effect as the ATP concentration was lowered. The data are recast in Fig. 28.3 to show the variation in saturation kinetics of ATP at two different nitrogenase concentrations. At the high enzyme level the saturation curve was almost hyperbolic, whereas sigmoidicity is clearly evident for the low concentration of enzyme.

Sigmoidicity in ATP saturation curves was somewhat more pronounced at low pH than at high (Fig. 28.4). Additionally, nitrogenase sensitivity to excess ATP was observed at low but not high pH. At pH 5.7 activity was maximum at 2 mM ATP and decreased strikingly at higher ATP levels (Fig. 28.5), while at pH 8.4, maximum activity occurred at about 5 mM ATP and higher levels up to 10 mM did not inhibit.

The sigmoidal ATP saturation curves produce non-linear Lineweaver-Burke plots of $1/v$ versus 1/[ATP] (Fig. 28.6*a*), and such curved plots have been reported (Kennedy, 1970; Bergersen & Turner, 1973). If only relatively high ATP concentrations are used for the reciprocal plot, the curvature is not very apparent and K_m values for ATP based on the extrapolation of such plots have been reported (Burns, 1969; Biggins & Kelly, 1970; Gallon, LaRue & Kurz, 1972). However, a linear relationship is observed between $1/v$ and $1/[ATP]^2$ over a wide range of ATP concentrations (Fig. 28.6*b*). Dixon & Webb (1964) have considered two explanations for this kind of kinetic response. In one case the reaction is bimolecular in the varied substrate, and in the other the varied substrate serves as an activator as well as a substrate. The kinetics do not distinguish between these alternatives.

Salt effects

High ionic strength has long been known to decrease nitrogenase activity (Burns & Bulen, 1965; Hardy, Holsten, Jackson & Burns,

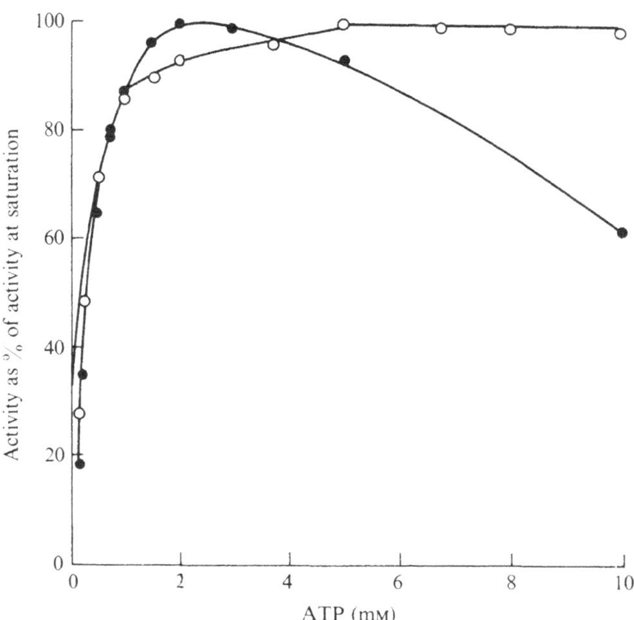

Fig. 28.5. Effect of pH on inhibition of nitrogenase by ATP: ○, pH 8.4; ●, pH 5.7.

1968; Sorger, 1969), but the nature of this inhibition has not received much study. The data in Fig. 28.7 explain the inhibition in terms of a dilution effect produced by high salt levels. The effect increases in intensity as salt concentration is increased (Fig. 28.7), and is produced by a variety of mono and divalent salts (Fig. 28.8).

The similarity in the effects produced by high salt concentration and by low ATP concentration suggested that the salt might be interfering with an ATP function. Fig. 28.9 shows that salt strongly alters the ATP saturation curve; the effect is similar to that demonstrated for ADP (Moustafa & Mortenson, 1967). The reciprocal plot in Fig. 28.10 suggests that salt inhibits ATP non-competitively; however, since plots of $1/v$ versus salt concentration at varying ATP concentrations did not intercept at a common point on the x-axis, as would be expected for non-competitive inhibition, it is likely that the inhibition is mixed. The nature of the curves shown in Fig. 28.10 suggests that the inhibition pattern contains a strong competitive component, and it is noted that the inhibitor does not influence the linearity of the relationship between $1/v$ and $1/[ATP]^2$.

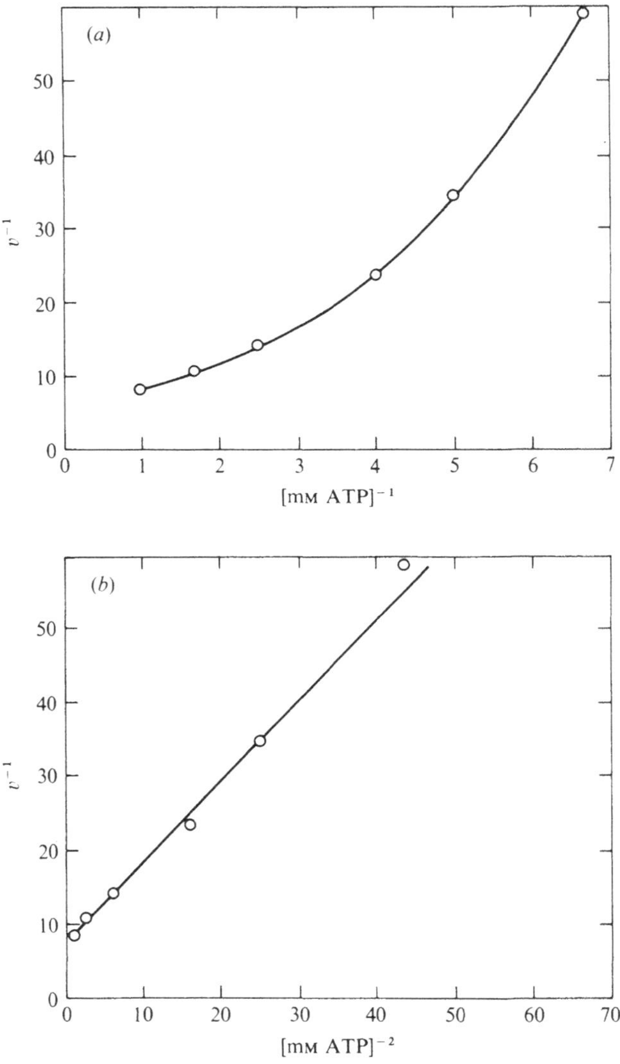

Fig. 28.6. Lineweaver-Burke plots of (*a*) $1/v$ versus $[ATP]^{-1}$ and (*b*) $1/v^{-1}$ versus $[ATP]^{-2}$: v is μmoles C_2H_2 reduced/min × 0.6 mg protein.

Discussion

The effects described here can be accommodated by a nitrogenase reaction sequence such as the following:

$$\text{Inactive complexes} \rightleftharpoons \text{f+m} \rightleftharpoons \text{N} \underset{\text{Product}}{\overset{\text{ATP}}{\rightleftharpoons}} \text{N}^*$$

$$\text{N, N}^* \rightleftharpoons \text{N–N, N}^*\text{–N, N}^*\text{–N}^*$$

$$\text{N–N} \underset{\text{Product}}{\overset{\text{ATP}}{\rightleftharpoons}} \text{N}^*\text{–N} \underset{\text{Product}}{\overset{\text{ATP}}{\rightleftharpoons}} \text{N}^*\text{–N}^*$$

in which f is Fe protein (nitrogenase Component II), m is Mo–Fe protein (nitrogenase Component I), N is a catalytically active species, N* is an activated intermediate enzyme species, and N–N, N*–N and N*–N* are various multimeric species. At low ATP concentration and particularly at low enzyme concentration the major species would be

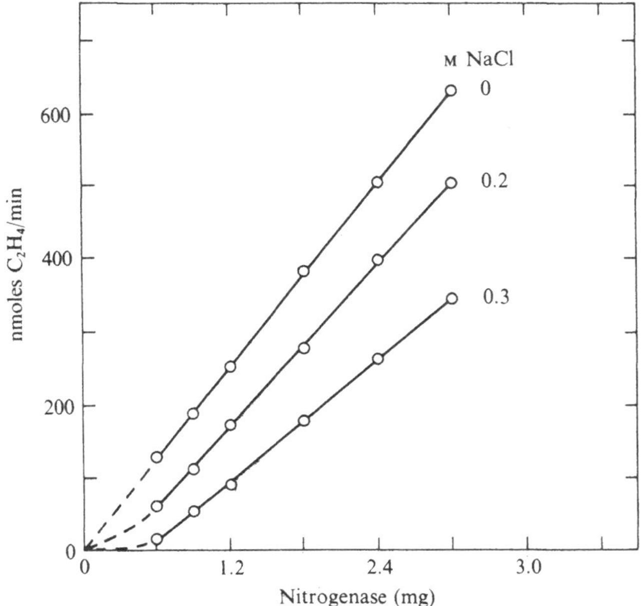

Fig. 28.7. Influence of NaCl on dilution effect.

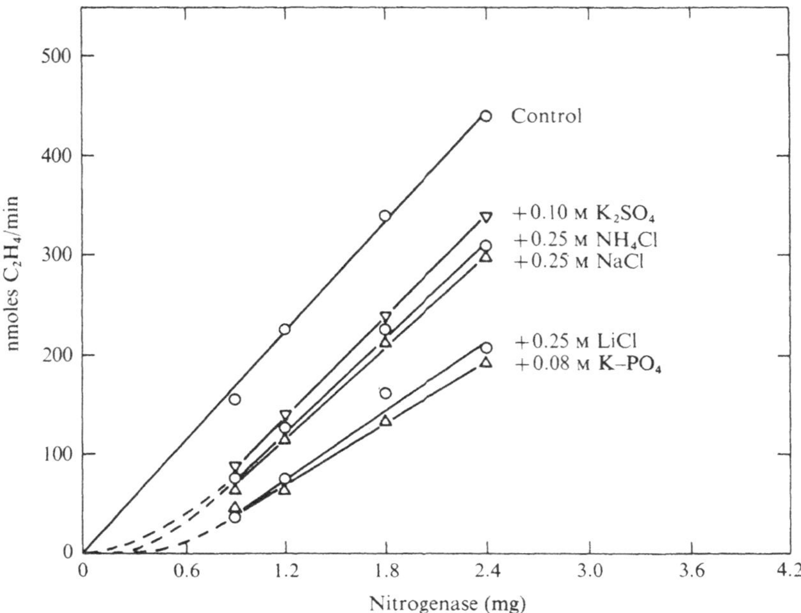

Fig. 28. 8. Influence of various salts on dilution effect.

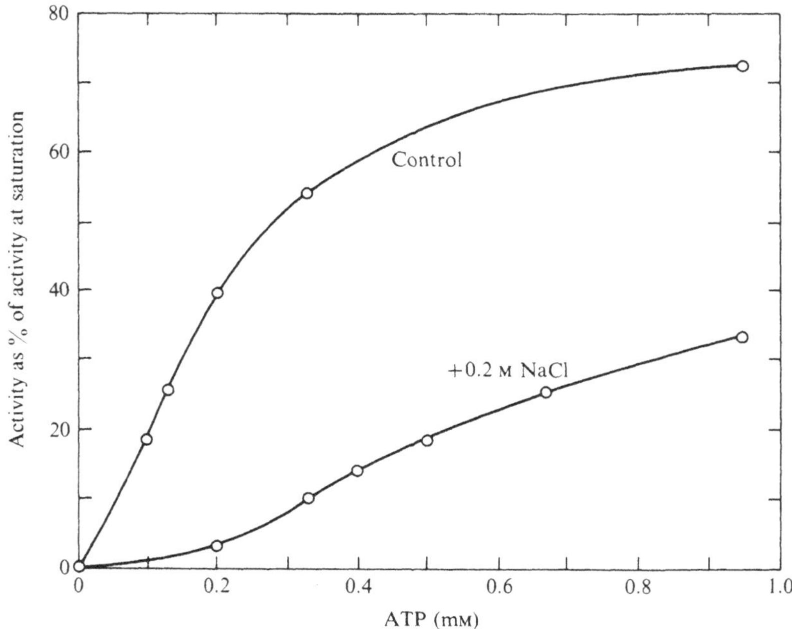

Fig. 28.9. Effect of NaCl on ATP saturation curve.

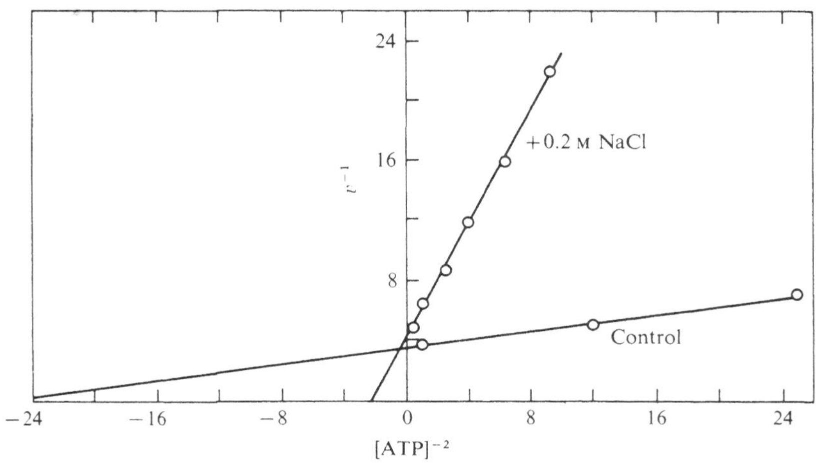

Fig. 28.10. Inhibitory effect of NaCl at variable ATP concentration.

the unassociated, catalytically inactive nitrogenase proteins and the reactive species N. Higher concentrations of either ATP or enzyme would result in an increasing fraction of enzyme proteins in the reactive dimeric species. In the case where Component I protein and Component II protein were in balance, essentially all enzyme protein would be maintained in the dimeric form at optimal ATP concentration, and no dilution effect would be evident.

The effect of nitrogenase concentration on ATP saturation kinetics

Figure 28.3 indicates more effective utilization of ATP by the presumed dimeric species. Homotropic co-operative binding of ATP to the dimeric species is a reasonable explanation for this observation and thus the present work supports the concept of nitrogenase as an allosteric enzyme. Heterotropic ATP effects have also been observed, in which the apparent K_m values of reducible substrates were influenced by ATP concentration (Burns & Hardy, 1972*a*; Bergersen & Turner, 1973).

The locus of salt inhibition can only be speculated on, but it is not unreasonable to consider that salt would inhibit the formation of the multimeric species by interfering with ATP binding and/or with protein–protein interactions.

The authors gratefully acknowledge C. J. Madden for excellent technical assistance.

References

Bergersen, F. J. & Turner, G. L. (1973). Kinetic studies of nitrogenase from soybean root-nodule bacteroids. *Biochem. J.*, **13**, 61–75.

Biggins, D. R. & Kelly, M. (1970). Interaction of nitrogenase from *Klebsiella pneumoniae* with ATP or cyanide. *Biochim. biophys. Acta*, **205**, 288–99.

Bulen, W. A., Burns, R. C. & LeComte, J. R. (1965). Nitrogen fixation: hydrosulfite as electron donor with cell-free preparation of *Azotobacter vinelandii* and *Rhodospirillum rubrum*. *Proc. natn. Acad. Sci. USA*, **53**, 532–9.

Burns, R. C. (1969). The nitrogenase system from *Azotobacter*: activation energy and divalent cation requirement. *Biochim. biophys. Acta*, **171**, 253–9.

Burns, R. C. & Bulen, W. A. (1965). Nitrogen fixation studies with aerobic and photosynthetic bacteria. *Biochim. biophys. Acta*, **105**, 437–55.

Burns, R. C. & Hardy, R. W. F. (1972*a*). Allosteric nature of *Azotobacter* nitrogenase: ATP function and variation in K_m of ATP, N_2 and C_2H_2. 164th Annual Meeting, American Chemical Society, *Biol. Abstr.* p. 201.

Burns, R. C. & Hardy, R. W. F. (1972*b*). Purification of nitrogenase and crystallization of its Mo–Fe protein. In *Photosynthesis and nitrogen fixation*, Part B (ed. San Pietro, A.). *Meth. Enzym.*, **24**, 480–96.

Dixon, M. & Webb, E. C. (1964). *Enzymes*. Academic Press: New York.

Gallon, J. R., LaRue, T. A. & Kurz, W. G. W. (1972). Characteristics of nitrogenase activity in broken cell preparation of the blue-green alga *Gloeocapsa* sp. LB 795. *Can. J. Microbiol.*, **18**, 327–32.

Hardy, R. W. F., Holsten, R. D., Jackson, E. K. & Burns, R. C. (1968). The acetylene–ethylene assay for N_2 fixation: laboratory and field evaluation. *Pl. Physiol.*, **43**, 1185–1207.

Kennedy, I. R. (1970). Kinetics of acetylene and cyanide reduction by the N_2 fixing system of *Rhizobium lupini*. *Biochim. biophys. Acta*, **222**, 135–44.

Mortenson, L. E. (1965). Nitrogen fixation in extracts of *Clostridium pasteurianum*. In *Non-Heme Iron Proteins: Role in Energy Conversion* (ed. San Pietro, A.), pp. 243–59. Antioch Press: Yellow Springs, Ohio.

Moustafa, E. & Mortenson, L. E. (1967). Acetylene reduction by nitrogen fixing extracts of *Clostridium pasteurianum*: ATP requirement and ADP inhibition. *Nature, Lond.*, **216**, 1241–2.

Shah, V. K., Davis, L. C. & Brill, W. J. (1972). Nitrogenase. I. Repression and derepression of the Fe–Mo and Fe-proteins of nitrogenase in *Azotobacter vinelandii*. *Biochim. biophys. Acta*, **256**, 498–511.

Sorger, G. J. (1969). Regulation of nitrogen fixation in *Azotobacter vinelandii* OP: the role of nitrate reductase. *J. Bacteriol.*, **98**, 56–61.

Sorger, G. J. & Trofimenkoff, D. (1970). Nitrogenaseless mutants of *Azotobacter vinelandii*. *Proc. natn. Acad. Sci. USA*, **65**, 74–80.

29. *Azotobacter* nitrogenase: mechanism and kinetics of allene reduction

R. C. BURNS, R. W. F. HARDY & W. D. PHILLIPS

The reduction of allene to propylene was shown to be catalyzed by the nitrogenase of *Clostridium pasteurianum* in a reaction which required ATP and reductant (Hardy & Jackson, 1967). Although allene lacks a triple bond and thus differs from other nitrogenase substrates, its reactivity with nitrogenase has not stimulated much interest. This is probably because it was considered likely that allene underwent isomerization to propyne, a known and effective nitrogenase substrate, prior to reduction. Evidence is now presented that favors a mechanism of direct reduction of allene with no prior isomerization. Additionally, some of the kinetic characteristics of allene reduction are reported.

Methods

The purification of *Azotobacter vinelandii* nitrogenase through the PS-2 step and methods for the analysis of hydrogen evolution, acetylene reduction and N_2 reduction have been described (Burns & Hardy, 1972). Unless indicated otherwise, reaction mixtures contained 5 μmoles ATP, 10 μmoles $MgCl_2$, 60 μmoles creatine phosphate, 20 μmoles sodium dithionite, 25 μmoles TES buffer, pH 7.0, 10 units creatine phosphokinase, and enzyme solution to 1.0 ml; temperature was 30 °C. For acetylene reduction, 0.05 atm acetylene was used and for N_2 reduction 1.0 atm N_2 was used. Argon served as diluent gas where needed to give 1 atm total pressure. Allene was purified from commercial allene by preparative gas chromatography using a Perkin-Elmer 900 gas chromatograph (Atkinson, Russell & Stuart, 1967). Allene and propylene were assayed with a Perkin-Elmer F-11 gas chromatograph as described earlier for use in acetylene-reduction assays (Hardy, Holsten, Jackson & Burns, 1968).

Results and discussion

To investigate whether allene was converted to propyne prior to reduction, allene reduction reactions were carried out in D_2O and the labelling pattern of product propylene was determined. Presumably, direct reduction of allene in D_2O would yield 2,3-dideuteropropylene,

$$CH_2{=}C{=}CH_2 \xrightarrow{2D} CH_2{=}CD{-}CH_2D,$$

whereas prior isomerization to propyne would lead to deuteration in the 1 and 2 positions of propylene.

$$CH_2{=}C{=}CH_2 \longrightarrow CH{\equiv}C{-}CH_3 \xrightarrow{2D} CHD{=}CD{-}CH_3.$$

Involvement of exchangeable hydrogen(s) could lead to multiply-deuterated species, such as the following:

$$CH_2{=}C{=}CH_2 \xrightarrow{-H,\,+D} CH{\equiv}C{-}CH_2D \xrightarrow{2D} CHD{=}CD{-}CH_2D.$$

Reaction mixtures of 5.0 ml total volume were prepared with D_2O to give a theoretical H_2O content of $< 0.1\,\%$. Incubations were conducted in 20 ml vessels under 1.0 atm allene at 30 °C for 16 hours. At termination the atmosphere of each reaction flask was displaced by D_2O into cold $CDCl_3$ or CS_2 which was contained in an NMR sample tube. The spectrum of the dissolved gas was then determined with a Varian 220 MHz NMR spectrometer using the pulse Fourier transform method. Control reactions, lacking nitrogenase, were also run; it had previously been established that in similar incubations, but using H_2O and omitting either ATP or dithionite, no detectable allene conversion to other hydrocarbons, including propyne and propylenes occurred.

Representative NMR spectra are shown in Fig. 29.1. Comparison of the spectra obtained from the allene reduction flasks with the computed spectra for propylene and 2,3-dideuteropropylene (Bothner-By & Noor-Colin, 1961) showed that nitrogenase produced the 2,3-dideutero species as would be expected from direct reduction of allene with no prior isomerization. This was established by (1) the appearance of a doublet in the spectral position for methylene hydrogens of propylene and by (2) a value of 1.1 for the ratio of the peak areas of the methyl hydrogens versus the methylene hydrogens; the theoretical values are 1.0 for 2,3-dideuteropropylene and 3.0 for the 1,2-dideuteropropylene. Multiply-deuterated species are not indicated.

To test whether any significant deuterium isotope effects might

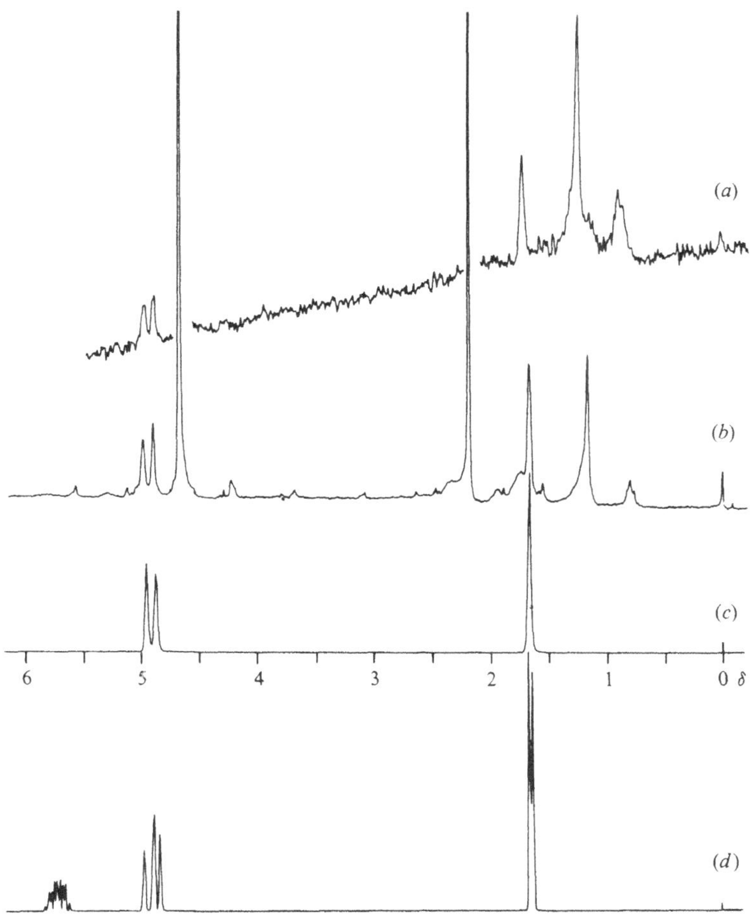

Fig. 29.1. NMR spectra for identification of labelling pattern in deuterated propylene. (*a*) Reaction gas mixture collected in CS_2. (*b*) Reaction gas mixture collected in $CDCl_3$. (*c*) Calculated spectrum of $CH_2 = CD-CH_2D$. (*d*) Calculated spectrum of $CH_2 = CH-CH_3$. Spectral positions are as follows: Methyl-H of propylene at 1.7δ; methylene-H of propylene at 4.9δ; methylene-H of allene at 4.7δ. Other resonances in (*a*) and (*b*) are due to solvent impurities.

occur in these reactions, the rate of propylene formation from allene was compared for reaction mixtures that were 100 % H_2O versus 67 % D_2O–33 % H_2O. In both cases the rate of propylene formation was found to be 38 nmoles per minute per mg protein, and thus no significant D_2O effect was indicated.

The K_m of allene was evaluated over the range 0.02 to 0.08 atm allene. The data presented in Fig. 29.2 indicate a K_m value of about 0.4 atm.

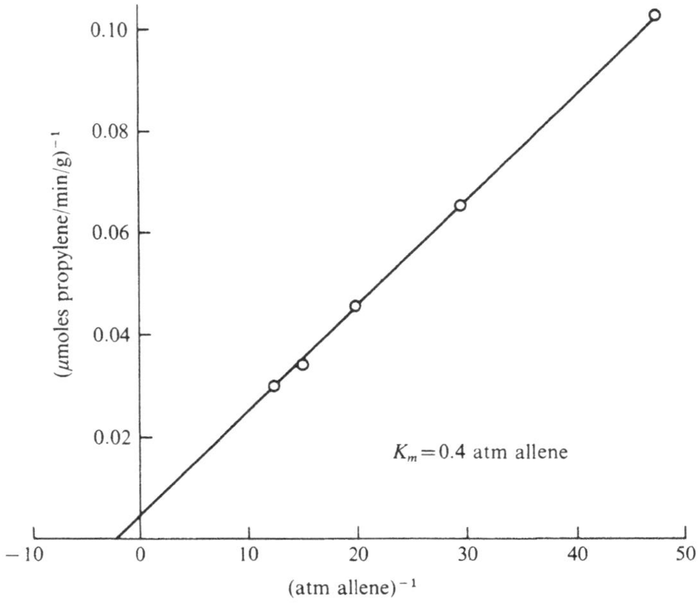

Fig. 29.2. Lineweaver–Burke plot for evaluation of K_m for allene.

This K_m value is in the range known to be common for numerous other nitrogenase substrates (Hardy, Burns & Parshall, 1971).

After demonstrating by gas chromatography (Fuchsman & Hardy, 1972) that hydrogen was evolved by nitrogenase during allene reduction, quantitative estimates of hydrogen evolution and of propylene formation were made at various partial pressures of allene. Hydrogen evolution was measured by respirometry and propylene was measured on samples of the respirometer flask atmospheres withdrawn at termination. The results (Table 29.1) show that the rate of total electron transfer is inhibited slightly at 0.09 atm allene and is inhibited severely at 1.0 atm. At the lower partial pressure of allene one-fifth of the electrons activated by nitrogenase are used for allene reduction. At the higher partial pressure it is interesting to note that almost three-fourths of the activated electrons are used for allene reduction, a consequence of more selective inhibition of hydrogen evolution versus allene reduction. Table 29.2 shows that allene reduction is inhibited by carbon monoxide but not by H_2, thus exhibiting the inhibition pattern characterstic of all nitrogenase substrates other than N_2 and H_3O+ (Hardy *et al.*, 1971).

Table 29.1. *Electron allocation and substrate inhibition in allene reduction*

Reaction atmosphere		Products (nmoles/min/ mg protein)		Electron transfer		Electron allocation coefficient
P_{allene}	P_{argon}	Propylene	H_2	Total	(%)	
—	1.00	—	200	200	100	—
0.09	0.91	37	151	188	94	0.20
1.00	—	34	13	47	24	0.72

Table 29.2. *Allene reduction in presence of H_2 and CO*

P_{allene}	P_{H_2}	P_{CO}	Propylene (nmoles/min/ mg protein)
0.11	—	—	40
0.11	1.0	—	41
0.11	—	0.02	1

In summary, the observation that allene reduction in D_2O yields 2,3-dideuteropropylene supports a reduction mechanism in which allene is reduced directly without prior isomerization to propyne. With regard to K_m, inhibition pattern and concomitant hydrogen evolution, the reduction of allene is quite similar to the reduction of other non-N_2 substrates of nitrogenase. The relatively high electron allocation to allene, coupled with the fact that both allene and product propylene are gases, make allene an attractive nitrogenase substrate for analytical purposes. It is suggested that an allene $\longrightarrow$ propylene assay, which could be set up with minor modifications of the acetylene $\longrightarrow$ ethylene assay apparatus, would be particularly useful for qualitative verification of acetylene $\longrightarrow$ ethylene data and for application with test systems that consume ethylene or produce it independently of acetylene reduction.

The authors thank Dr R. C. Ferguson and F. Ferrari for assistance in the NMR analyses and C. J. Madden for competent technical assistance.

References

Atkinson, J. G., Russell, A. A. & Stuart, R. S. (1967). Gas chromatographic studies of isotopically labelled ethylenes. *Can. J. Chem.*, **45**, 1963–9.

Bothner-By, A. A. & Noor-Colin, C. (1961). The proton magnetic spectra of olefins. I. Propene, butene-1 and hexene-1. *J. Am. chem. Soc.*, **83**, 231–6.

Burns, R. C. & Hardy, R. W. F. (1972). Purification of nitrogenase and crystallization of its Mo–Fe protein. In *Photosynthesis and nitrogen fixation*, Part B (ed. San Pietro, A.). *Meth. Enzym.*, **24**, 480–96. Academic Press: New York.

Fuchsman, W. H. & Hardy, R. W. F. (1972). Nitrogenase-catalyzed acrylonitrile reductions. *Bioinorgan. Chem.*, **1**, 197–215.

Hardy, R. W. F., Burns, R. C. & Parshall, G. W. (1971). The biochemistry of N_2 fixation. *Adv. Chem. Series*, **100**, 219–47.

Hardy, R. W. F. & Jackson, E. K. (1967). Reduction of model substrates – nitriles and acetylenes – by nitrogenase (N_2ase). *Fedn Proc.*, **28**, 725.

Hardy, R. W. F., Holsten, R. D., Jackson, E. K. & Burns, R. C. (1968). The acetylene–ethylene assay for N_2 fixation: laboratory and field evaluation. *Pl. Physiol.*, **43**, 1185–207.

Index